ALIEN PLANTS

OF THE BRITISH ISLES

ALIEN PLANTS

of the

BRITISH ISLES

A provisional catalogue of vascular plants
(excluding grasses)

E. J. CLEMENT

M. C. FOSTER

with guidance on nomenclature by

D. H. KENT

Botanical Society of the British Isles
London
1994

Published by the Botanical Society of the British Isles
London

© BSBI 1994

ISBN 0 901158 23 2

Camera-ready copy produced by M. C. Foster

Index produced by R. Gwynn Ellis

Printed in Great Britain by The Devonshire Press,
Torquay

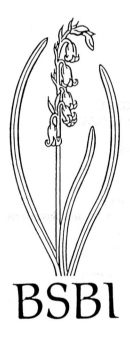

The cover illustration shows Thorn-apple (*Datura stramonium* L.)
drawn by G. M. S. Easy

CONTENTS

PREFACE

In the recently published monumental *New Flora of the British Isles,* Professor C.A.Stace includes alien plants to be found 'in the wild' at the present time, either naturalised or, if casual, frequently recurrent. Our book may be regarded as complementary, as it not only gives additional details for every alien plant included in that work, but also attempts to cover all other alien species (except grasses), however short-lived or infrequent, that have ever been reliably recorded.

We began compiling a catalogue of aliens nearly twenty years ago when, as amateur botanists particularly interested in the subject, we were frustrated by the scattered and fragmentary nature of the data available. At first, we gathered all manner of material relating to aliens. It soon became clear, however, that in order to produce a book of moderate length and within a reasonable number of years, the categories of information given would need to be restricted; supplying, for each species, the date of the first correct record and a list of vice-counties proved too time-consuming and too prone to the incorporation of errors, particularly from older sources; providing keys or descriptions was found to be equally impossible.

Nevertheless, the resulting volume is much more than just a checklist and a source of localities for the less common established species. It can be used as an aid to the identification of specimens. If the genus is known or suspected, clues to the species most likely to be found are given by the number of bullet symbols used and by the means of introduction (which may often be deduced from associated plants). The references to illustrations and descriptions should be useful to readers having access to even a small botanical library; matching a plant with a clear illustration is one of the shortest routes to a correct identification, provided it is followed up by checking against a detailed description. Comparing with correctly named, pressed specimens is also most instructive; here the list of locations of voucher specimens should be helpful.

All errors are the responsibility of the authors. The conflicting views of experts on the subjects of correct identification and correct nomenclature of taxa have, at times, obliged us to make our own decisions, especially where the impossibility of separating old records of an aggregate species into its modern segregates has forced us to use the broader sense. Our affirmative approach, sparing in the use of qualifying terms such as 'perhaps' and 'probably', does not imply that we consider our statements indisputable. We have often been compelled to rely on incomplete records and somewhat dubious identifications, and our investigations have been too limited to be conclusive. Neither do both of us always agree in the conclusions we present.

The regrettable omission of Poaceae (the family of grasses) from this volume is due to several factors, not the least of which are their large number and the lengthy and difficult problem of equating older names with their modern equivalents.   To include the grasses in this volume would have caused undue delay. We hope that a supplementary volume devoted to them will be published at a later date.

Being well aware of the defects of our work, we offer this book primarily as a basis for further study: we hope it will encourage others to add to the store of knowledge of the ever-changing alien flora. S.T.Dunn's *Alien flora of Britain* of 1905 was the last comprehensive account of our plant invaders: let us not wait another ninety years for the next!

ERIC CLEMENT   SALLY FOSTER
Gosport, Hampshire   Bexleyheath, Kent

March 1994

ACKNOWLEDGEMENTS

Since we began studying alien plants some thirty-five years ago we have received generous help from numerous botanists, both professional and amateur, mostly members of the Botanical Society of the British Isles, the Wild Flower Society, or the London Natural History Society: we are greatly indebted to them all. Sincere thanks are due to those correspondents, listed under Personal Communications (see reference numbers 400-485), who have given us a great deal of useful information; we especially appreciate the time and trouble taken by several B.S.B.I. vice-county recorders in supplying detailed records and answering our many enquiries.

In the course of preparing this book we have also had more specific assistance: D.H.Kent's boundless knowledge of the botanical literature has been invaluable; he has also carefully checked our first typescript, guiding us in all matters of nomenclature; Prof. C.A.Stace has generously passed on to us much new material that he acquired while writing his *New Flora of the British Isles*; Dr A.C.Leslie has read and commented on various parts of our manuscript, saving us from making many errors; a few specialists on particular families or genera have advised us, most notably Dr L.A.S.Johnson on Juncaceae, Dr T.C.G.Rich on Brassicaceae, M.J.Southam on Apiaceae and Ms J.Fryer on *Cotoneaster*; D.McClintock has given advice, over many years, on garden escapes and other matters; C.G.Hanson and Mrs R.Cracknell have assisted in finding illustrations; and Mrs E.M.Hyde has searched several local journals for alien records.

The major British herbaria are largely-untapped sources of unpublished records. The tedious work of abstracting some of these records and locating voucher specimens was undertaken by Miss R.Sedergreen (BM), G.M.S.Easy (CGE), R.G.Ellis (NMW), Ms S.K.Marner and the late J.Milton (both OXF). We also wish to thank the keepers and staff of the few herbaria that we have been able to visit, namely BM, K, OXF, RNG and SLBI; in particular, A.O.Chater, Ms M.Dowlen, Dr S.L.Jury and A.R.Vickery have taken an interest in our work and have solved a variety of problems for us. Keepers of other herbaria have very kindly lent us pressed specimens to inspect, or sent details of specimens in their care; especially helpful have been Dr J.R.Edmondson (LIV) and Dr G.Halliday (LANC).

We are greatly indebted to Ms G.Douglas, the librarian at the Linnean Society of London, for the loan of many rare books and journals. We have also been priviledged to work, on occasions, in the great libraries at the Natural History Museum, South Kensington, and at the Royal Botanic Gardens, Kew.

We also wish to thank R.G.Ellis for compiling the index and for his expert help in editing this work and guiding it through publication.

Finally, we remember with gratitude the late Dr J.G.Dony and J.E.Lousley who aroused our interest in alien plants and gave us much encouragement in the early years; nor do we forget those who helped us with formative field-work in the years that followed, our most frequent companions being C.G.Hanson, Dr A.C.Leslie, Dr J.L.Mason, J.R.Palmer, T.B.Ryves, B.Wurzell, and the late A.L.Grenfell and Mrs J.McLean.

INTRODUCTION

The flora of the British Isles is ever-changing. While alterations in land use have, over many years, contributed to the decline, or even loss, of native species, trade with other countries has encouraged the slow and insidious infiltration of the indigenous flora by species of foreign origin.

Some of these alien plants have, in various ways, already had a considerable impact. Many have become permanent members of the flora, some to the extent of becoming threats to native vegetation; the vigorously dominant *Rhododendron ponticum* is a well-known example, now requiring drastic control measures. Hybridisation with native species has given rise to new taxa, such as the fertile, spreading *Senecio cambrensis*. Back-crossing has produced hybrid swarms, e.g., *Calystegia sepium* × *sylvatica*; and in the case of *Centaurea jacea* × *nigra*, an alteration in the genetic make-up of the native species, even though the alien species has usually died out. In all these instances the alien species concerned would at first have been considered casual introductions of no importance: their effects on the native flora would not have been foreseen. Likewise, some of the immigrants which are today found only as casual plants may in the future bring about dramatic changes in the flora.

The search for these new arrivals, and the attempt to identify them, to trace their means of entry and their countries of origin, has proved a rewarding hobby for many generations of amateur botanists, introducing them to a wide selection of the world's flora without the necessity of world travel.

From the 16th century, when recording of plants in the British Isles may be said to have begun, observations of unusual plants growing in fields, at ports, and by mills and factories, led to a slow accumulation of knowledge until, by the end of the 19th century, about a thousand species of aliens had been identified. Prominent among these were species inadvertently brought in with ships' ballast.

In the present century, three inventories have been published. The first, *Alien flora of Britain* (1), written by S.T.Dunn in 1905, gives not only the names of species found, but also their countries of origin and their means of introduction. This was followed in 1908 by G.C.Druce's *List of British plants* (2), of which a second and much enlarged edition under the title *British plant list* (7) was published in 1928, by which time the number of aliens recorded had risen to about 2000. Later authors have restricted their lists to established species, or to particular categories of aliens.

With the decline in the use of ships' ballast, the most important vector by which aliens arrived in the early years of this century was adulterated foreign grain. In particular, wheat and oats from Russia and barley from eastern Europe, Turkey and the Middle East were heavily contaminated with seeds of foreign cornfield weeds. The study of alien plants germinating from these seeds around the breweries at Burton-on-Trent led to the publication of lists of grain aliens by R.Curtis (8) and R.C.L.Burges (11).

Also early in this century large numbers of aliens were found by rivers near the woollen mills at Galashiels in southern Scotland. These, mainly species with hooked or spined fruits or adhesive seeds, were found to have arrived with imported wool and to be germinating where wool waste had been dumped or washed down by rivers. *The adventive flora of Tweedside* (3) by I.M.Hayward and G.C.Druce, published in 1919, listed some 350 species from this source. Another group of botanists had been studying wool aliens near Bradford but, in both places, recording came to an end about 1920. Few were then recorded - a lack of observers rather than plants - until the discovery by Dr J.G.Dony in the late 1940s of a large number of wool aliens in Bedfordshire. The origin of these was traced to the use of wool waste as a soil conditioner on market gardens and farms, and it was then realised that this use might be widespread. Enthusiasm for searching for and recording wool aliens spread rapidly as more species and more localities were found so that, by 1961, J.E.Lousley (14) was able to list about 500 species.

Although commercial seed mixtures for caged birds have been used for a long time, it was not until the 1960s that much attention was given to aliens brought in by this means. The occurrence of foreign plants in areas where grain had been scattered for chickens and pheasants had long been known, but plants germinating from bird-seed in gardens and on tips had largely been ignored. Several short accounts and illustrations then appeared in journals giving details of the most frequent or most bizarre species. Later, N.S. and G.D.Watts (273) analysed the composition of bird-seed mixtures, comparing the results with the weed flora of a Norwich rubbish tip. The cultivation by C.G.Hanson and Dr J.L.Mason of seeds of a wide variety of commercial mixtures, and from seed-cleaning waste, revealed that among them could be found some 30 species deliberately imported as bird-food, together with many impurities. That work, combined with increased recording of plants occurring as bird-seed aliens, culminated in the publication in 1985 of *Bird seed aliens in Britain* (19), which lists over 400 species and gives an account of their origins and the factors influencing their frequency.

In the 1940s alien plants were discovered in the vicinity of Soya Foods Ltd at Springwell in Middlesex, their seeds having been brought in as impurities with soya beans from America. These were recorded in *A hand list of the plants of the London area* by D.H.Kent and J.E.Lousley (12). Other North American plants, usually in association with soya bean plants, were found by J.R.Palmer to be plentiful on rubbish tips in the Dartford area of Kent in 1973. Further investigation traced the source to an oil-milling factory nearby, and visits by other botanists helped to increase the number of species seen. When a checklist (18) was published in 1977, more than 50 species had been attributed to the oil-milling industry.

In addition to the authors mentioned above and others who studied particular groups of aliens, there were enthusiasts who produced lists for localised areas such as ports and municipal tips - sources of a wide range of species - and writers of county floras who included many aliens in their plant lists. Furthermore, a vast number of individual records were published in national and local journals: in particular, records for over 1000 species of 'exotics' have been published in the *Wild Flower Magazine* (371) since 1960. Records of many very rare species, some new to the British Isles have, since 1975, been published in *B.S.B.I. News* (303), usually in the series entitled *Adventive news*.

The need to collate the records from all these sources, and to supply an up-to-date comprehensive list of aliens, has been apparent for a long time. In undertaking this work, we have become aware that it does not give an entirely reliable picture of the distribution and quantity of aliens. Under-recording is a major defect. One reason for this is the uneven distribution of observers: aliens in Ireland, for instance, have been grossly under-recorded. Also, many botanists have considered casual aliens to be of no importance and not worth recording in print, while others have been deterred by the problems of naming them. Correctly identifying alien plants, especially those from the more distant and less well botanised parts of the world, or those selected and hybridised in gardens, is difficult. The re-examination of herbarium specimens by experts has brought numerous errors to light, often many years later. Where no herbarium specimen exists, there is no means of checking the identity of a suspect record: too few specimens of difficult groups have been preserved.

Despite these drawbacks the number of records is enormous - evidence of the enthusiasm of field botanists, almost all of them amateurs, and the patience and industry of referees. In this catalogue it has therefore been necessary to condense considerably the bulk of records abstracted. On the other hand, indications of countries of origin, the locations of voucher specimens, and a selection of references to published records have been added. It was found impracticable to give keys or descriptions of the species but, instead, references are given to works in which descriptions and illustrations are to be found. Because the last attempt at a complete list of aliens was published in the late 1920s, our main concern has been to provide a summary of the information which has accrued in the succeeding years. Inevitably, in the search through the formidable number of books, journals and herbaria, some records will have been missed. We hope not many.

In his *British plant list* Druce named 1999 alien species, 293 as established and 1706 as adventive: this catalogue, which excludes grasses, lists 3586 species (or groups of closely related species), of which 885 are established. In addition, 146 taxa of lower rank are treated, and a further 355 taxa are excluded for reasons that are specified. Furthermore, some 60 alien taxa are reported to have hybridised in the wild with native species. Yet another 59 taxa, most of which have occurred as aliens in NW Europe, are mentioned as being likely to occur. Thus, in total, the catalogue deliberates on some 4200 taxa.

Besides the increase in the number of species - which perhaps reflects both a greater interest in aliens and an increased diversity of commerce - changes in the relative quantities of the different categories are apparent. The reasons for the major changes are clear. Improved cleaning of agricultural seed and the use of herbicides has reduced the number of foreign weeds on arable land: such cornfield weeds as *Agrostemma githago* and *Adonis annna*, once well known, are seldom seen nowadays. In horticulture, the infertility of many highly-bred modern plants and the trend towards labour-saving gardening, have reduced the quantities of annual and short-lived species seen as garden escapes but increased the number of shrub species; for instance, only four species of *Cotoneaster* were listed in 1928 whereas about 70 are listed here. The locally higher water temperatures in some of our rivers, due to heated discharges from industry, has led to the establishment of additional exotic water plants. The selection and quantity of aliens due to commerce has varied through the years in response to fluctuating markets and more produce being cleaned before shipment. Containerised cargoes bring in few seeds, and ballast is no longer dumped in large quantities at ports.

The introduction of additional aliens and the gradual establishment of some of those already occurring as casuals can be expected to continue. It might be thought that species classified as casual and recorded repeatedly over many years were now unlikely to become established, but different races or varieties, perhaps from other countries, might have the necessary characteristics for survival and spread; in agriculture and horticulture, breeding programmes and genetic engineering may in the future produce hardy strains of species at present too tender to survive our winters; nor should we forget the predicted long-term warming of our climate. The potential aggressiveness of the very rare species has yet to be put to the test. Some of the invaders which are today found only as casual plants may well become, as has happened so often in the past, established members of the flora.

PLAN OF THE CATALOGUE

In the following catalogue we attempt to give an annotated systematic list of all alien species of vascular plants, excluding grasses, which have been reliably recorded as growing wild in the British Isles (including the Channel Islands). The word 'alien' is used in a broad sense: it denotes all plants, whether established or not, that are thought to have arrived as a result of human activities, and includes plants referred to by other authors as 'adventives', 'casuals', 'ephemerals', 'exotics', 'introductions' and 'volunteers'.

In the tradition of British floras, species probably introduced in ancient times with the advent of cultivation are treated as though native. Also, many potential records, lacking a reliable name, have had to be excluded, as have a few of the names listed in our *Alphabetical tally list* (1061), a hasty compilation of unchecked information intended solely as a working document.

As mentioned in the Introduction, the last comprehensive list of aliens was that of Druce in the late 1920s, so emphasis is given in this catalogue to the period from 1931 to 1990 - a period long enough to put into perspective the 'fashions' in recording, such as the enthusiasm for wool aliens from the 1950s to 1970s and the recent concentration on bird-seed aliens and garden escapes.

For species recorded since 1930, full accounts are given in the following sequence:

> **Scientific name; English name; frequency; means of introduction; status; origin; location of voucher specimens; references; and synonyms.**

These terms are explained in detail below. A few 1991-1993 records have also been incorporated; for these the date of the record has been added. For species known before 1930 but not recorded since, shorter accounts are given.

Owing to the many changes in taxonomic rank resulting from revisions of genera, some published records cannot now be allocated to a particular taxon: where two or more taxa were at one time considered conspecific but are now treated as separate species, they are 'lumped' as one entry. Similarly, allied species which have been confused sometimes appear under a group heading, for example, *Sidalcea malviflora* group. This may not be sound taxonomy, but it indicates where more research is needed; the species remain inseparable without the redetermination of voucher specimens by a specialist.

For similar reasons the treatment of taxa below the rank of species is not entirely consistent. In general, alien subspecies of native species are given a full account. The varieties and cultivars of native species recorded as of alien origin are listed as 'alien variants', although some of these may be only extremes of the variation present in our native populations. Minor variants such as colour forms are not usually mentioned.

Hybrids are variously treated. Notable hybrids are given separate entries, but those of less importance are mentioned briefly in notes under the parent species. For most of them more detailed information is readily available in *Hybridization and the flora of the British Isles* (1311), where descriptions and references are given.

Taxa recorded in error, or represented only by dubious records, are enclosed in square brackets.

SEQUENCE AND NOMENCLATURE

The circumscription and sequence of families and genera follow those of the *New Flora of the British Isles* (20) whenever they appear in that work. Within each genus the species are listed alphabetically.

Scientific name   Species names in bold type are those considered to be currently accepted and in accordance with the latest version of the *International code of botanical nomenclature* (1373), the authors' names being abbreviated in accordance with *Draft index of author abbreviations* (1223).

English name   The vernacular names given are, with very few exceptions, those adopted in the *New Flora of the British Isles* (20). Additional names are taken from a variety of sources, but have occasionally been amended to follow the guidelines suggested in *English names of wild flowers* (1086). For some species, no English name has been found; exceptionally we have coined one of our own, based on some obvious characteristic of the plant.

SUMMARY OF RECORDS, 1931-1990

There follows, for species recorded since 1930, a summary of the frequency, means of introduction and status for the 60-year period, and where appropriate a note on distribution, habitat, localities or recent changes of frequency.

Frequency   Terms such as 'rare' and 'scarce', normally used to indicate the frequency of alien plants, are not comparable with the same terms as used for native species, as most aliens are sporadic in appearance and impermanent; nor is the number of records appropriate, as a species may be recorded from a particular locality many times. The system used here is based on the number of separate localities from which the species has been recorded, in print, on herbarium labels, or privately communicated, for the period January 1931 to December 1990, represented by bullet symbols as follows:

●	1-4 localities	
●●	5-14 localities	
●●●	15-49 localities	Above 15, the
●●●●	50-499 localities	number of localities
●●●●●	over 500 localities	is an estimate

Bullet symbols have also been allotted to post-1990 records if the species was almost certainly present in 1990.

Means of introduction   Seeds and propagules coming into the British Isles may be divided into three groups:

CULTIVATION; FOODSTUFFS; and OTHER COMMODITIES

The terms used in the catalogue, and the categories of aliens included in them, are listed below.

CULTIVATION

Agricultural seed: crop plants growing from spilt or discarded seed, or self-sown, away from cultivated land; plants germinated from seed impurities, especially in grain and, during and after the 1939-45 war, American carrot seed.

Grass seed: alien grasses and impurities introduced with grass seed mixtures and now found wild.

Garden escape: plants discarded, or growing in the wild from spilt or discarded seed, or wind-blown or bird-sown from gardens. Where escapes are only from botanic gardens or water-gardens this is stated. For foreign weed species accidentally dispersed with nursery plants, details are given.

Greenhouse escape: plants usually grown under glass, house plants and pot plants discarded or growing from seed or spores.

Relic: crop plants persisting in fields; garden plants persisting on sites of former gardens or nursery gardens.

Introduced: trees or other plants deliberately planted or sown in wild habitats; water plants introduced into ponds or waterways; grasses and legumes sown on roadsides and reclaimed land, especially after mining operations.

FOODSTUFFS

Grain: cereals found near flour mills and granaries; cereals from corn fed to domestic fowl and game birds; and weed species from impurities in these.

Bird-seed: plants germinating, mainly in gardens and on tips, from cage-bird and wild-bird seed mixtures - mostly cereal grasses and oily-seeded species, but also a wide variety of impurities.

Animal feed: species found on farms, at racecourses and at army camp sites, probably brought in with fodder; plants found in gardens and on tips from food for hamsters and other small mammals - mostly cereals and oily-seeded species.

Oil-seed: soya-bean, sunflower, flax, etc. growing on tips and waste ground from seed imported for extraction of oil and manufacture of oil-cake, together with weeds from the countries where these were cultivated.

Food refuse: marrow, tomato, citrus fruits, date, mung bean, etc. growing on tips; seeds from kitchen waste germinating at sewage works.

Spice: plants on tips and waste ground from fruits or seeds used as condiments, including a few impurities. A few of these may have been deliberately sown.

OTHER COMMODITIES

Plants found on waste ground and tips near factories and mills have been attributed to impurities in the following commodities:

Wool: mainly fleeces from Australia, New Zealand and S Africa, but also vicuna and alpaca wools from S America, wool for carpets from central Asia and angora goat hair. Plants from these sources have also been found in fields and orchards where wool cleanings have been used as a soil conditioner.

Esparto: various grasses used in the manufacture of high-quality paper.

Coir: recently used in horticulture as a substitute for peat.

Cotton, rag shoddy and jute.

Hides and furs.

Tanning and dyeing materials, especially Valonia oak cupules.

Ballast, granite and iron ore.

Timber and wood products.

It will be realised that these categories overlap to some extent. Also, a species may have arrived by several means: sunflowers found on tips may have germinated from garden throw-outs, or have come from bird-cage refuse or oil-milling waste. In the catalogue entries, the most frequently recorded vector is given first; and where the vector is unknown, this is stated.

Within the British Isles alien species may be further dispersed by human activities, such as the long-distance transport of topsoils, manures and hardcore containing seeds or propagules, and the distribution of nursery plants with their accompanying weeds. Local dispersal may be by accidental carriage on vehicles or on footwear, as well as by natural means such as wind, water currents and birds.

Status   The degree of permanence of alien species ranges from a single short-lived plant occurring untended in an artificial habitat to many large, long-established colonies overwhelming the native vegetation. During the years since 1930 several systems have been used to divide this range into categories, with the result that recorders have differed widely in their interpretation of the terms used.

In particular, the word 'naturalised' has been used to describe a wide range of conditions, some records refering to only a single plant amongst native vegetation. Moreover, the extent to which trees and shrubs reproduce is seldom recorded; nor is it always clear from the records whether an annual species is persisting from self-sown seed or by repeated re-introduction.

However, we have usually been able to allocate species to one of the categories of permanence used in this catalogue, which are as follows:

Casual	Not persisting in a locality for more than two years without re-introduction.
Persistent	Remaining longer than two years but unlikely to be permanent, not reproducing by seed or vigorously spreading vegetatively.
Established	Likely to remain permanently; at least one colony either reproducing by seed or vigorously spreading vegetatively.

The status may vary in different localities, but the category given is the highest achieved during the period 1931-1990: it does not necessarily represent the status at the present time. Other terms used are:

Introduced	Deliberately planted or sown in the wild; if any degree of reproduction has been recorded, this is mentioned.
Naturalised	Established extensively amongst native vegetation so as to appear native.
Recorded	A published record, but no status mentioned.
Reported	No published record; a personal communication only.
Collected	No published record; a specimen apparently claimed to be from the wild has been found in a herbarium.

Distribution in general terms is given for commonly established species; for other established and persistent species, examples of localities of special interest may be mentioned, and for the rarest (•) every locality is named. When known, the locality is followed, in parentheses, by the vice-county name. The vice-county boundaries are shown in the end-papers of *New Flora of the British Isles* (20); larger-scale versions of this map can be found in *Watsonian vice-counties of Great Britain* (1076) and *Census catalogue of the flora of Ireland* (252). The advantage of using these old boundaries is that, unlike county boundaries, they remain constant.

Habitats are given if they differ from those in which casual aliens are commonly found - waste ground, tips, dockyards - or cannot be deduced from the vector, but neither habitats nor localities are given for casual species, except in the few cases where the means of introduction is unknown.

ADDITIONAL INFORMATION

Following the summary of records for 1931-1990, a brief note giving other information from the records is sometimes added. 'Formerly' here means before 1931; 'modern', since 1960; and 'recent' implies within the last 10 years.

Origin   The countries, regions or continents mentioned indicate the native area(s) of the species. Although many aliens have come directly from their native areas, it is certain that some have arrived from places where they have become naturalised; for example, some species, native in the Mediterranean area but growing from impurities in wool, will have come directly from Australia or elsewhere in the southern hemisphere. Similarly, an alien fodder plant may have been brought in, not from its native area, but from a country where it is cultivated on a large scale. Such areas from which the species may have come directly are added in parentheses. If the species is now widely naturalised, perhaps in several continents, the term 'widespread' is used.

Location of voucher specimens   Institutional herbaria are abbreviated according to the recommendations in *British and Irish herbaria* (1178). Where possible, six locations are given. The most frequently cited are those listed below, in which important collections of aliens are housed.

BM	British Museum (Natural History), London (now re-named Natural History Museum)
CGE	University of Cambridge, Botany School
E	Royal Botanic Garden, Edinburgh
NMW	National Museum of Wales, Cardiff
OXF	University of Oxford, Botany School
RNG	University of Reading, Plant Sciences Laboratories

None of these herbaria can be assumed to have been exhaustively covered, as new determinations are continually being made and new material added. Other herbaria cited have been quoted from the literature and from personal communications. If specimens have been located in less than two institutional herbaria, privately owned herbaria are quoted, in abbreviated form, as follows:

HbAB	Lady Anne Brewis, Blackmoor, Hants
HbACL	Dr A.C.Leslie, Guildford, Surrey
HbAMB	Mrs A.M.Boucher, Hoddesdon, Herts
HbARGM	A.R.G.Mundell, Fleet, Hants
HbCGH	C.G.Hanson, Ware, Herts
HbDJH	Dr D.J.Hambler, Bradford, W.Yorks
HbEC	Dr E.Chicken, Driffield, North Humberside
HbEJC	E.J.Clement, Gosport, Hants
HbFEC	Dr F.E.Crackles, Hull, North Humberside
HbGMSE	G.M.S.Easy, Milton, Cambridge
HbJDF	Ms J.D.Fryer, Petersfield, Hants
HbKWP	K.W.Page, Leatherhead, Surrey
HbLJM	L.J.Margetts, Honiton, Devon
HbMARK	M.A.R.Kitchen, Bevington, Gloucs
HbPM	Dr P.Macpherson, Newlands, Glasgow
HbRPB	R.P.Bowman, Southampton, Hants
HbTGE	T.G.Evans, Chepstow, Gwent

The following private herbaria are now dispersed and we have not had the opportunity to locate individual specimens:

HbALG	†A.L.Grenfell
HbCAS	Prof. C.A.Stace, Ullesthorpe, Leics
HbJDG	†J.D.Groves
HbJPMB	†J.P.M.Brenan
HbMMcCW	†Miss M.McCallum Webster
HbTBR	T.B.Ryves, Kingston Hill, Surrey
HbUKD	†Miss U.K.Duncan

No herbaria are cited for the most frequent species (●●●●● and ●●●●), as these are presumed to be represented in all major collections.

References   All reference numbers, whether to published records, private communications, descriptions or illustrations, are in one sequence, the lowest numbers (1-99) normally being for those most frequently used.

From the list at the end of the book it will be seen that reference numbers 1-19 refer to major lists of aliens published this century, arranged in chronological order. For each species the earliest of these is always quoted, giving an indication of the period in which it was first recorded. Other important references to published records are numbered from 100-399. For the rarest (●) established or long-persistent species a reference to each post-1930 locality is given; for others a selection is made.

Reference numbers 20-99 refer to readily available floras and other books aiding identification and are among the most generally useful for botanists who do not have access to large botanical libraries. These are often supplemented by other references to good descriptions and illustrations. We make no apology for the occasional lengthy strings of reference numbers: being well aware of the difficulties of gaining access to foreign publications, we have attempted to give a wide choice. Number 20, the current standard flora for the British Isles, is always cited.

All books in the list of references are presented in the form used in *Watsonia*, the journal of the Botanical Society of the British Isles; the page number is quoted only if the index is absent or faulty. The somewhat cryptic abbreviations of periodicals are those of the traditional *World list of scientific periodicals* (1044) and the increasingly used, and recommended, *B-P-H* (1192); for these references the volume and page number(s) are given, e.g. 370(11:245-247) refers to *Watsonia*, volume 11, pages 245-247.

Illustrations are indicated by an asterisk (*) after the reference number, or after the volume number if more than one volume is quoted. Where illustrations are traditionally referred to by their plate numbers, or are on un-numbered pages, the plate (or tableau) number is given, e.g. 312*(t.415) refers to *Curtis's Botanical Magazine*, plate number 415. A selection of both analytical line drawings and coloured photographs or paintings has, where possible, been chosen to demonstrate

all attributes of the plant.   Preference has been given to works likely to be found
in natural history society libraries; in the absence of these, obscure references
from such works as *Index Londinensis* (1312, 1365) are quoted.   The excellent
series of portrayals of alien plants in *B.S.B.I. News* (303) are nearly always cited
(up to December 1993).

No attempt has been made to list the innumerable monographs and generic
revisions that exist; many of these may be found in *Index to botanical monographs*
(1176), *The plant-book* (1327), *Illustrierte Flora von Mitteleuropa* (595a), *The new
Royal Horticultural Society dictionary of gardening* (51b), volume 4, pp.844-852,
The European garden flora (50), and *The families and genera of vascular plants*
(1185).

Synonyms   A generous but selected synonymy follows, quoting all
binomials used in British works and those under which descriptions or illustrations
are referenced.   Additional synonyms are used to clarify problems of
nomenclature.   Misapplied names are followed by 'auct.', where several authors
are concerned, or by 'sensu' where only one author has erred.

We have also drawn attention to some homonyms, that is, differing plants
given the same specific name by different authors; these can cause much
confusion, especially in early or popular works where author citations are often
omitted.   By indicating such potential pitfalls, we hope that readers will
circumvent them.

ABBREVIATIONS AND SYMBOLS

aff.	*affinis*: akin to; bordering
agg.	aggregate species
append.	appendix(es)
auct.	*auctorum*: of authors, but not of the original author
C	central
c.	*circa*: about
cf.	*confer*: comparable with
comp.	compiler; compiled by
cv.	*cultivarietas*: cultivar
E	East; eastern
ed.	editor; edition
em.	*emendatus:* emended; *emendavit*: he emended
et al.	*et alii*: and others
f.	*filius*: son
fasc.	fascicle(s); parts
hort.	*hortulanorum*: of gardeners
incl.	includes; including
ined.	*ineditus*: unpublished; not yet published in accordance with the rules of the *International code of botanical nomenclature*
Is	Isle(s); Island(s)
l.c.	*loco citato*: at the place cited
Medit.	region centred on the Mediterranean Sea, and typically includes parts of southern Europe, N Africa and SW Asia
N	North; northern
NE	northeast
No.	number
nom. illeg.	*nomen illegitimum*: illegitimate name
nom. inval.	*nomen invalidum*: invalid name
nom. nud.	*nomen nudum*: name unaccompanied by a description or reference to a published description
n.s.	new series (of a periodical)
NW	northwest
p.	page
pp.	pages
q.v.	*quod vide*: which see
rev.	revision; revised by
S	South; southern
SE	southeast

ser.	series
sp.	species (singular)
sphalm.	*sphalmate*: by mistake
spp.	species (plural)
subsp.	subspecies (singular)
subspp.	subspecies (plural)
suppl.	supplement
suppls	supplements
SW	southwest
syn.	synonym; synonymy
t.	*tabula*: plate; illustration
trans.	translation; translated by
var.	*varietas*: variety
vol.	volume
W	West; western

*	indicates an illustration (in a referenced work)
†	indicates deceased (before a personal name)
' '	indicates a cultivar name
×	indicates a hybrid name
[ ]	indicates a taxon recorded in error

Numbers of post-1930 localities:
• 1-4;   •• 5-14;   ••• 15-49;   •••• 50-499;   ••••• over 500.

The most commonly cited herbaria are listed on page xiv; others may be found in *British and Irish herbaria* (1178). JDFC, which is not listed in that work, we have used for the herbarium at the John Dony Field Centre, Luton, Bedfordshire.

PTERIDOPHYTA

SELAGINELLACEAE

SELAGINELLA P.Beauv.

[*denticulata* (L.) Spring   In error for *S. kraussiana*.   7, 266, 354(6:586,640).]

[*helvetica* Link   Dubious records.   352(51:51), 354(5:798-9; 12:68).]

kraussiana (Kunze) A.Braun   **Krauss's Clubmoss**

••• An established greenhouse escape; naturalised in woodland and churchyards, by rivers and ponds, in scattered localities throughout most of the British Isles, especially in the west. Reported to have persisted for at least 50 years in Kilberry Castle lawn (Kintyre). Azores, tropical and S Africa; (widespread). BM, CGE, E, NMW, OXF, RNG. 7, 20, 50*, 73*, 137, 546a*, 1168*. *S. denticulata* auct., non (L.) Spring; *Didiclis kraussiana* (Kunze) Rothm.; *Lycopodium kraussianum* Kunze.

EQUISETACEAE

EQUISETUM L.

ramosissimum Desf.   **Branched Horsetail**

• An established ballast alien in riverside grass near Boston (S Lincs) since 1947; also in grass near the sea at Weston-Super-Mare (N Somerset) since c.1963. Eurasia, Africa. RNG. 16, 20, 160, 370*(1:149-153), 546*, 737*, 1168*. N American plants are referable to *E. hyemale* L. (*E. ramosissimum* auct. amer., non Desf.).

[*OPHIOGLOSSACEAE*

BOTRYCHIUM Sw.

lanceolatum (S.Gmelin) Angström   In error for native *B. lunaria*.   187, 320(1898:291-297), 354(5:797), 1167.

matricariifolium A.Braun ex Koch   In error for native *B. lunaria*. 320(1898:291), 354(5:797), 1167.

multifidum (S.Gmelin) Rupr. In error for native *B. lunaria*. 203, 354(5:797), 1167. *B. matricariae* (Schrank) Sprengel; *B. rutaceum* Sw.]

ADIANTACEAE

ADIANTUM L.

pedatum L.   **American Maidenhair-fern**

• A garden escape established on a bridge near Virginia Water (Berks), probably now gone; also on a wall of a fern nursery at Kingston Lacy (Dorset). N America, Asia. BM, RNG. 38*, 50*, 51*, 54*, 349(7:632), 403. Some records of native *A. capillus-veneris* may be referable to the S American

A. raddianum C.Presl (*A. cuneatum* Langsd. & Fischer) or to the Himalayan *A. venustum* G.Don. 21a, 303(35:12).

[*reniforme* L. In error for native *Asplenium ruta-muraria*. 358(1870:43,139).]

PTERIDACEAE

PTERIS L.

cretica L. Ribbon Fern

●●● An established greenhouse escape; on sheltered damp walls and rocks in widely scattered localities. Tropics, subtropics. BM, LTN, OXF, RNG. 20*, 50*, 66*, 73*, 126, 349(5:121), 1168*.

incompleta Cav. Spider Brake

● A greenhouse escape which persisted for a few years in a basement enclosure in Bristol (W Gloucs). Medit. 20, 50*, 371(374:9), 737*. *P. palustris* Poiret; *P. serrulata* auct., non Forsskål.

[*longifolia* L. In error for *P. vittata*. 7, 151, 349(7:387).]

tremula R.Br. Tender Brake

● A casual greenhouse escape. Australasia. OXF. 50*, 303(14:15), 371(394:36), 558*, 569*, 625*.

vittata L. Ladder Brake

● A greenhouse escape established since 1924 on an external wall of the tropical greenhouses, Oxford Botanic Garden, and long persistent on a colliery tip at Bream (W Gloucs), now gone; also on brickwork by a greenhouse at Chelsea Physic Garden (Middlesex). Tropics, subtropics. BM, K, NMW, OXF, RNG. 7, 20, 44*, 50*, 182, 303*(35:1,12), 349(5:336-338; 7:387), 737*, 1168*. *P. longifolia* auct., non L.

HYMENOPHYLLACEAE

TRICHOMANES L.

venosum R.Br. Veined Bristle-fern

● Established in a garden near Truro (W Cornwall) for over a century, perhaps brought in with *Dicksonia antarctica*, its characteristic host in New South Wales. Australasia. 51b, 314(14:155-160), 539*, 590*, 1001*, 1045*. *Polyphlebium venosum* (R.Br.) Copel.

POLYPODIACEAE

PHYMATODES C.Presl

diversifolia (Willd.) Pichi Serm. Kangaroo Fern

●● An established garden escape; naturalised in woodland on Tresco (Scilly), on a wall near the Caledonia Nursery and elsewhere in Guernsey (Channel Is); also a garden relic spreading by vegetative means at Rossdohan (S Kerry). Australasia. BM, RNG. 20*, 50*, 220, 303(29:12), 334(2:39-42), 370(8:183; 9:185), 558*, 569*. *Microsorum diversifolium* (Willd.) Copel.; *Phymatosorus diversifolia* (Willd.) Pichi Serm.; *Polypodium diversifolium* Willd.

CYATHEACEAE
CYATHEA Smith
dealbata (Forster f.) Sw.   Silver Tree-fern
- A garden relic reproducing at Rossdohan (S Kerry). New Zealand. 20, 50*, 51*, 325*(114:292), 334(2:39-42), 1045*. *Alsophila tricolor* (Colenso) Tryon.

DICKSONIACEAE
DICKSONIA L'Hér.
antarctica Labill.   Australian Tree-fern
- A garden escape established on roadbanks near Falmouth (W Cornwall), in Scilly and on Valentia Is (S Kerry); also a garden relic reproducing at Rossdohan (S Kerry), and young plants self-sown near St Ives (W Cornwall) and St Just-in-Roseland (E Cornwall). Australia, Tasmania. BM, OXF, RNG. 20*, 50*, 212, 213, 303(28:15; 35*:1,12), 334(2:39-42), 558*, 569*, 389(2:102-103). Some records may be referable to *D. fibrosa* Colenso. 389(2:102-103), 1168.

DENNSTAEDTIACEAE
LEPTOLEPIA Prantl
novae-zelandiae (Colenso) Mett. ex Diels
Pre-1930 only. New Zealand. 7, 36, 51, 320(15:78), 625*, 1045*, 1073*, 1363*. *Davallia novae-zelandiae* Colenso.

ASPLENIACEAE
ASPLENIUM L.
[*cuneifolium* Viv. In error for native *A. adiantum-nigrum*. 20, 370(11:278).]
fontanum (L.) Bernh.   Smooth Rock-spleenwort
Pre-1930 only. Europe. BM, LANC, OXF. 1, 16, 21, 50*, 73*, 126, 150, 188, 201, 218(p.105), 825(16:101-106), 1349. *A. halleri* Willd.; *Aspidium fontanum* (L.) Sw.; *Athyrium fontanum* (L.) Roth; *Polypodium fontanum* L.

WOODSIACEAE
Athyriaceae
MATTEUCCIA Tod.
struthiopteris (L.) Tod.   Ostrich Fern
- An established garden escape; naturalised in damp woods, by streams and lakes, mainly in the north and west. Known since 1928 as two large colonies by the shore of Lough Neagh (Co Antrim), and persistent by the burn in the old garden at Blairadam, Kinross (Fife) since 1834. Eurasia. BM, RNG. 13, 20*, 21, 50*, 73*, 114, 319(14:250), 1168*, 1242*. *Osmunda struthiopteris* L.

ONOCLEA L.
sensibilis L.   Sensitive Fern
- An established garden escape; naturalised in damp woodland, by ponds, lakes and streams, in widely scattered localities; in abundance along the lower

reaches of the river Brathay above Lake Windermere (Westmorland). E Asia, N America. BM, CGE, E, LANC, OXF, RNG. 1, 20*, 50*, 330(124:2471), 1168*.

DAVALLIACEAE

DAVALLIA Smith

bullata group   Squirrel's-foot Ferns
- A greenhouse escape on a wall at Claremont, Esher (Surrey), now gone. E Asia. K, RNG. 43, 50*, 199, 625*, 1171*. Incl. *D. bullata* Wallich; *D. mariesii* T.Moore; *D. trichomanoides* Blume.

canariensis (L.) Smith   Hare's-foot Fern
- A garden or greenhouse escape established since about 1970 on a wall at Coin Colin, St Martin, Guernsey (Channel Is). SW Europe, Madeira, Canary Is. 20, 21, 50*, 66*, 357(21:614-618), 370(16:233), 625*, 737*, 1218.

DRYOPTERIDACEAE

Aspidiaceae pro parte

POLYSTICHUM Roth

[*braunii* (Spenner) Fée   In error for native *P. setiferum.* 7, 354(3:398), 1075.]

munitum (Kaulfuss) C.Presl   Western Sword-fern
- A garden escape self-sown on the bank of a sunken lane at Hascombe (Surrey), now gone. Western N America. HbACL. 20, 41*, 50*, 199, 303(28:24), 625*.

CYRTOMIUM C.Presl

falcatum (L.f.) C.Presl   House Holly-fern
- ●●● An established greenhouse escape; on walls, in basements of houses and in churchyards, mainly in the south and west; naturalised on the shore near Porth Minick, St Mary's (Scilly) since 1956. S Africa, E Asia. BM, DBN, NMW, RNG. 13, 20*, 50*, 54*, 186, 207*, 877*(3:261-264), 1168*. *Phanerophlebia falcata* (L.f.) Copel.; *Polystichum falcatum* (L.f.) Diels. The similar *C. fortunei* J.Smith occurs in Europe and could be overlooked. 877(3:261-264).

BLECHNACEAE

BLECHNUM L.

cordatum (Desv.) Hieron.   Chilean Hard-fern
- ●●● A garden or greenhouse escape established in a few places in SW England, W Scotland and SW Ireland. Known since 1953 in a ditch near Salakee, St Mary's (Scilly), and a garden relic long established by vegetative spread at Rossdohan (S Kerry). Temperate southern hemisphere. BM, K, OXF, RNG. 7, 20*, 50*, 207, 212, 303*(35:1,12), 334(2:39-42), 374(1953:35), 1168*(fig. as *B. penna-marina*). *B. capense* auct., non (L.) Schldl.; *B. chilense* (Kaulf.) Mett.; *B. magellanicum* auct., non Mett.; *B. tabulare* auct., non (Thunb.) Kuhn; *Lomaria cordata* Desv.; *L. magellense* Desv.

penna-marina (Poiret) Kuhn   Antarctic Hard-fern
- Recorded as naturalised in SW England, without locality.   Temperate
southern hemisphere, Antarctic.   20, 36, 50*, 302(14:12), 354(5:797; 8:37),
558*, 1168*(fig. as *B. cordatum*).   *B. alpinum* Mett.; *Lomaria alpina* (Poiret)
Sprengel.

WOODWARDIA Smith
radicans (L.) Smith   Chain Fern
- A garden relic reproducing both vegetatively and by spores at Rossdohan
(S Kerry) and also self-sown in Guernsey (Channel Is).   SW Europe, Canary Is.
20, 21, 50*, 66*, 334(2:39-42), 1218.

AZOLLACEAE

AZOLLA Lam.
filiculoides Lam.   Water Fern
••••• An established aquatic plant.   Formerly introduced into ponds and
streams, now widely naturalised, especially in southern England; often
abundant, but varying in quantity from year to year.   Tropical and subtropical
America; (widespread).   1, 20, 21, 33*, 73*, 703*, 683*, 1168*.
A. caroliniana Willd.; *A. pinnata* auct., non R.Br.   *A. mexicana* A.Br.
(*A. caroliniana* auct., non Willd.) occurs in NW Europe and could be
overlooked.   683*, 826*(25:14-19), 1168.

SPERMATOPHYTA

GYMNOSPERMAE

GINKGOACEAE
GINKGO L.
biloba L.   Maidenhair-tree
- Introduced. China. 50*, 58*, 59*, 60*, 119, 166.

PINACEAE
ABIES Miller
alba Miller   European Silver-fir
●●●● Introduced; reproducing in woodland in Scotland, Man and Ireland.
C & S Europe. RNG. 3, 21, 50, 58*, 59*, 60*, 102, 319(19:369).
A. pectinata DC., nom. illeg.; *Picea pectinata* Don.

cephalonica Loudon   Greek Fir
- Introduced; seedlings have been reported from the New Forest (S Hants) and
from an old flint wall near Wentworth Hall, Headley (Surrey). SE Europe.
20, 21, 50, 58*, 59*, 119, 199, 458.

cilicica (Antoine & Kotschy) Carrière   Cicilian Fir
- Introduced. Turkey, Syria, Lebanon. 50, 58*, 60*, 119.

concolor (Gordon) Lindley ex Hildebr.   Colorado White-fir
- Introduced. Western N America. 50, 58*, 59*, 60*, 119. Incl. *A. lowiana*
(Gordon) A.Murray.

delavayi group   Delavay's Silver-firs
- Introduced. China. 50, 58*, 59*, 60*, 119, 204. Incl. *A. delavayi*
Franchet; *A. forrestii* Craib ex Coltm.Rog.

firma Siebold & Zucc.   Momi Fir
- Introduced. Japan. 50, 58*, 59*, 60*, 119.

grandis (Douglas ex D.Don) Lindley   Giant fir
●●● Introduced; reproducing in W Kent, Surrey, Berks, W Norfolk,
Co Durham and Perthshire. Western N America. 16, 20*, 58*, 59*, 60*,
119, 166, 199, 235, 458. *Pinus grandis* Douglas ex D.Don.

homolepis Siebold & Zucc.   Nikko Fir
- Introduced. Japan. 50, 58*, 59*, 60*, 119.

lasiocarpa (Hook.) Nutt.   Subalpine Fir
- Introduced. Western N America. 21, 50, 58*, 59*, 60*, 119.

nordmanniana (Steven) Spach   Caucasian Fir
- Introduced. Turkey, Caucasus. RNG. 20, 21, 58*, 59*, 60*, 119, 166.

pinsapo Boiss.   Spanish Fir
- Introduced.   SW Spain, N Africa.   LANC.   21, 50, 58*, 60*, 119, 312*(n.s. t.242 as var. or hybrid), 737*.

procera Rehder   Noble Fir
- ●● Introduced; reproducing at Hamsterley (Co Durham), Polecat Hill and Craig's Wood, Hindhead (both Surrey) and in Perthshire. Western N America. LANC. 16, 20*, 58*, 59*, 60*, 199. *A. nobilis* (Douglas ex D.Don) Lindley, non A.Dietr.

veitchii Lindley   Veitch's Silver-fir
- Introduced.   Japan.   50, 58*, 59*, 60*, 119.

PSEUDOTSUGA Carrière
menziesii (Mirbel) Franco   Douglas Fir
- ●●●● Introduced; reproducing in many places, especially in the south and east, sometimes abundantly, as in Breckland (W Norfolk) and Fonthill Abbey Wood (S Wilts). Western N America. 16, 20*, 58*, 59*, 60*, 235, 261. *P. douglasii* (Lindley) Carrière; *P. taxifolia* Britton, nom. illeg.; *Abies douglasii* Lindley; *A. menziesii* Mirbel.

TSUGA (Antoine) Carrière
canadensis (L.) Carrière   Eastern Hemlock-spruce
- ●● Introduced.   Eastern N America. 20, 50, 58*, 59*, 60*, 119, 156, 166, 826*(13:90-92).

heterophylla (Raf.) Sarg.   Western Hemlock-spruce
- ●●● Introduced; reproducing in many widely scattered localities. Western N America. 13, 20*, 58*, 59*, 60*, 303(32:20), 349(4:40). *T. albertiana* (A.Murray) Sénécl.; *T. mertensiana* auct., non (Bong.) Carrière; *Abies heterophylla* Raf.

PICEA A.Dietr.
abies (L.) Karsten   Norway Spruce
- ●●●● Introduced; reproducing in a few widely scattered localities. N & C Europe, W Asia. 3, 20*, 58*, 59*, 60*, 102, 156, 235, 1242. *P. excelsa* Link; *Abies abies* (L.) Rusby; *A. excelsa* Poiret, nom. illeg.; *A. picea* Miller; *Pinus abies* L.

asperata Masters   Dragon Spruce
- Introduced.   W China. 21, 50, 58*, 59*, 60*, 119. Incl. *P. retroflexa* Masters.

breweriana S.Watson   Brewer Spruce
- Introduced.   California, Oregon. 59*, 60*, 61*, 312*(t.9543), 1058.

engelmannii Parry ex Engelm.   Engelmann Spruce
- Introduced.   Western N America. 20, 50, 58*, 59*, 60*, 156, 166.

glauca (Moench) Voss   White Spruce
- ●● Introduced.   Canada. 20, 21, 50, 58*, 59*, 60*, 119, 156. *P. alba* (Aiton) Link.

glehnii (F.Schmidt) Masters   Sakhalin Spruce
• Introduced. Japan, Sakhalin. 50, 60*, 61, 119.
montigena Masters
• Introduced.   China.   61, 119, 632*, 1053*.   Possibly *P. asperata* ×
P. likiangensis (Franchet) Pritzel.
omorika (Pančić) Purkyně   Serbian Spruce
• Introduced. Yugoslavia, Bulgaria. 20, 50, 58*, 59*, 60*, 156, 166.
orientalis (L.) Link   Oriental Spruce
• Introduced. Turkey, Caucasus. 21, 50, 58*, 59*, 60*, 119, 276.
polita (Siebold & Zucc.) Carrière   Tiger-tail Spruce
• Introduced.   Japan.   50, 58*, 59*, 60*, 61, 119.   *P. torano* (Siebold)
Koehne.
pungens Engelm.   Colorado Spruce
• Introduced. SW USA. 21, 50, 58*, 59*, 60*, 119.
sitchensis (Bong.) Carrière   Sitka Spruce
•••• Introduced; reproducing in widely scattered localities. Regenerating after
felling over extensive areas of Forestry Commission plantations in
SW Scotland.   Western N America.   13, 20, 50, 58*, 59*, 60*, 126,
303(49:36), 360(30:134-142).   *P. menziesii* (Douglas ex D.Don) Carrière;
Pinus sitchensis Bong.
smithiana (Wallich) Boiss.   Morinda Spruce
• Introduced. Himalayas. 50, 58*, 59*, 60*, 119, 204. *P. morinda* Link.

LARIX Miller
decidua Miller   European Larch
••••• Introduced; reproducing and well naturalised in many places throughout
the British Is. C Europe. 2, 20*, 50, 58*, 59*, 60*.   *L. communis* Lawson
ex Gordon; *L. europaea* DC., nom. illeg.; *L. larix* (L.) Karsten; *Abies larix*
Lam.
gmelinii (Rupr.) Rupr. ex Kusen.-Proch.   Dahurian Larch
• Introduced. Siberia, Manchuria, Korea. 21, 22*, 50, 59*, 119. *L. dahurica*
Trautv.
kaempferi (Lindley) Carrière   Japanese Larch
•••• Introduced; reproducing in a few widely scattered localities. Japan. 13,
20*, 50, 58*, 59*, 60*, 156, 199, 370(17:184).   *L. leptolepis* (Siebold &
Zucc.) Endl.; *Abies kaempferi* Lindley.
laricina (Duroi) K.Koch   American Larch
• Introduced. N America. 50, 58*, 59*, 60*, 119. *L. americana* Michaux.
× **marschlinsii** Coaz   Hybrid Larch
•• (but perhaps overlooked). Introduced; reproducing near Wick Pond at
Virginia Water and Deerleap Wood at Wotton (both Surrey) and in Perthshire.
Spontaneous with the parents in plantations. Originated in cultivation. BM,
NMW. 20*, 50, 58*, 59*, 60*, 156, 199. A fertile hybrid which backcrosses
with both parents. *L. decidua* × *L. kaempferi*; *L.* × *eurolepis* A.Henry, nom.
illeg.; *L.* × *henryana* Rehder.

CEDRUS Trew
atlantica (Endl.) Carrière   Atlas Cedar
  •• Introduced; self-sown at West Clandon and Weston Green (both Surrey) and
  in north London (Herts & Middlesex). Algeria, Morocco. 20*, 50, 58*, 59*,
  60*, 119, 184, 199, 265, 303(64:41). *C. libani* A.Rich. subsp. *atlantica*
  (Endl.) Battand. & Trabut; *Pinus atlantica* Endl.
deodara (Roxb. ex D.Don) Don   Deodar
  •• Introduced; seedlings have been reported from the New Forest (S Hants).
  Himalayas. 20*, 50, 58*, 59*, 60*, 119, 166, 458. *C. libani* A.Rich.
  subsp. *deodara* (Roxb. ex D.Don) Sell; *Pinus deodara* Roxb. ex D.Don.
libani A.Rich.   Cedar-of-Lebanon
  ••• Introduced. Turkey, Syria, Lebanon. 16, 20*, 50, 58*, 59*, 60*, 119,
  135, 156. *C. libanensis* Mirbel.

PINUS L.
ayacahuite Ehrend.   Mexican White-pine
  • Introduced. Mexico, Guatemala. 50, 58*, 59*, 60*, 122.
banksiana Lambert   Jack Pine
  • Introduced. N America. 21, 50, 58*, 59*, 60*, 119.
cembra L.   Arolla Pine
  • Introduced. Alps, Carpathians, Urals. 21, 50, 58*, 59*, 60*, 119, 166.
contorta Douglas ex Loudon   Lodgepole Pine
  ••• Introduced; reproducing in Breckland (W Norfolk), Cards and Scilly;
  seedlings reported from Co Carlow and from Cromford Moss (Derbys) and
  elsewhere. Western N America. CGE, NMW, RNG. 13, 20*, 50, 51b, 58*,
  59*, 60*, 235, 402, 458. Incl. subsp. *latifolia* (Engelm. ex S.Watson) Critchf.
  (var. *latifolia* Engelm. ex S.Watson).
densiflora Siebold & Zucc.   Japanese Red-pine
  • Introduced. Japan, Korea, China. 50, 58*, 60*, 61, 119.
mugo Turra   Dwarf Mountain-pine
  •• Introduced in shelter-belts; reproducing on Towyn Burrows (Carms).
  Europe. BM. 16, 20, 22*, 26*, 27*, 50, 156, 219, 276. *P. montana* Miller.
muricata D.Don   Bishop Pine
  • Introduced; seedlings reported from Herm (Channel Is). California, Baja
  California. 59*, 60*, 61*, 500*, 1143*, 1218. *P. edgariana* Hartw.
nigra Arnold   Black Pines, incl. Austrian & Corsican Pines
  •••• Introduced; reproducing in widely scattered localities, colonising heaths
  and sand-dunes, especially in East Anglia. Europe, Turkey. 7, 20, 50, 58*,
  59*, 60*, 235, 258. *P. austriaca* Höss; *P. laricio* Poiret; *P. poiretiana* hort.
  ex Gordon.
peuce Griseb.   Macedonian Pine
  • Introduced. Balkans. LSR. 20, 50, 58*, 59*, 60*, 119, 156.
pinaster Aiton   Maritime Pine
  ••• Introduced; sometimes reproducing in the south; long naturalised at
  Blackheath (Surrey) and on the heathlands around Poole (Dorset) and

Bournemouth (S Hants). SW Europe, N Africa. CGE, LANC, OXF, RNG.
2, 20, 50, 58*, 59*, 60*, 165, 199, 737*. *P. maritima* Duroi, non Miller.

pinea L.   Stone Pine
• Reported from Ashburnham (E Sussex). Medit. 50, 58*, 59*, 402, 737*.
The record for Caerns was in error for *P. radiata.* 156, 473.

ponderosa Douglas ex Lawson & P.Lawson   Western Yellow-pine
• Introduced. Western N America. RNG. 20, 50, 58*, 59*, 60*, 119, 156.

radiata D.Don   Monterey Pine
••• Introduced; reproducing in a few places in the south; abundant saplings on
an old railway track at Egloskerry (E Cornwall); seedlings on Appletree Banks,
Tresco (Scilly), near Marazion (W Cornwall) and in the forest at Bedgebury
(W Kent) and elsewhere. California. RNG. 13, 20, 50, 58*, 59*, 60*, 212,
232, 458. *P. insignis* Douglas ex Loudon.

rigida Miller   Northern Pitch-pine
• Introduced. Eastern N America. 58*, 59*, 60*, 61, 119.

strobus L.   Weymouth Pine
••• Introduced; reproducing at Bunkers Hill (W Norfolk), on a golf course
edge at St George's Hill, Weybridge and plentifully at Longcross, Chobham
Common (both Surrey). Eastern N America. HbEJC, HbKWP. 13, 20, 50,
58*, 59*, 60*, 199, 235.

sylvestris L. **alien variants**
Our native plant, probably restricted to Scotland, has been separated by some
authors as var. *scotica* Beissn. (*P. scotica* Willd. ex Endl., nom. inval.), those
elsewhere being the alien var. *sylvestris*, or intermediate between the two taxa.

thunbergii Parl.   Japanese Black-pine
• Introduced. Japan, Korea. 50, 58*, 59*, 60*, 119, 220, 312*(n.s. t.558).
The existence of the homonym *P. thunbergii* Lambert does not necessitate a
change of name to *P. thunbergiana* Franco as advocated by 511.

wallichiana A.B.Jackson   Bhutan Pine
•• Introduced; apparently self-sown on a steep bank above Leatherhead by-pass
(Surrey). Himalayas. HbKWP. 16, 20, 50, 58*, 59*, 60*, 119, 156, 199.
P. chylla Lodd., nom. nud.; *P. excelsa* Wallich ex D.Don, non Lambert;
P. griffithii McClell.

TAXODIACEAE

SCIADOPITYS Siebold & Zucc.

verticillata (Thunb.) Siebold & Zucc.   Umbrella-pine
• Introduced. Japan. 50, 58*, 60*, 61*, 119, 166.

SEQUOIA Endl.

sempervirens (D.Don ex Lambert) Endl.   Coastal Redwood
••• Introduced; self-sown seedlings recorded recently from Hatchford and
Ockham (both Surrey) and from Worthy Combe (S Somerset). California,
Oregon. 16, 20, 50, 58*, 59*, 60*, 119, 156, 224, 342(Feb93), 384(1993:19).

SEQUOIADENDRON Buchholz
giganteum (Lindley) Buchholz   Wellingtonia
●●● Introduced. California. 16, 20*, 50, 58*, 59*, 60*, 119, 156. *Sequoia gigantea* (Lindley) Decne., non Endl.; *S. wellingtonia* Seemann.

METASEQUOIA Miki ex Hu & W.C.Cheng
glyptostroboides Hu & W.C.Cheng   Dawn Redwood
● Introduced. China. 20, 50, 58*, 59*, 60*, 119, 166.

TAXODIUM Rich.
distichum (L.) Rich.   Swamp Cypress
●● Introduced. SE USA. 20, 21, 50, 58*, 59*, 60*, 119, 166.

CRYPTOMERIA D.Don
japonica (L.f.) D.Don   Japanese Red-cedar
●● Introduced; self-sown seedlings recorded from three localities in the Weald of Kent. China, Japan. NMW, RNG. 20*, 50, 58*, 59*, 60*, 119, 224, 236.

CUPRESSACEAE

CUPRESSUS L.
duclouxiana Hickel   Chinese Cypress
● Introduced. China, Tibet. 51, 61, 61a, 119, 312*(t.9049).
goveniana Gordon   Californian Cypress
● Introduced. California. 50, 51*, 58*, 60*, 224.
macrocarpa Hartweg ex Gordon   Monterey Cypress
●●● Introduced; reproducing in a few places in the south. California. 16, 20*, 50, 58*, 59*, 60*, 156, 220, 236.
sempervirens L.   Italian Cypress
● Introduced. E Medit. 21, 50, 58*, 59*, 60*, 119.

× **CUPRESSOCYPARIS** Dallimore
leylandii (A.B.Jackson & Dallimore) Dallimore   Leyland Cypress
●● Introduced; also spontaneous in parks where both parents are grown. Originated in cultivation. 20*, 50, 58*, 59*, 60*, 135, 156, 166. *Cupressus macrocarpa* × *Chamaecyparis nootkatensis*.

CHAMAECYPARIS Spach
lawsoniana (A.Murray) Parl.   Lawson's Cypress
●●●● Introduced; often reproducing on heathland and self-sown on old walls in the south. Western N America. 16, 20*, 58*, 59*, 60*, 192, 349(4:40; 6:236). *Cupressus lawsoniana* A.Murray.
nootkatensis (Lambert) Spach   Nootka Cypress
●● Introduced. Western N America. 20, 50, 58*, 59*, 60*, 119, 156, 166. *Cupressus nootkatensis* Lambert.
obtusa (Siebold & Zucc.) Endl.   Hinoki Cypress
● Introduced. Japan. 21, 50, 58*, 59*, 60*, 119, 166. *Cupressus obtusa* (Siebold & Zucc.) K.Koch; *Retinospora obtusa* Siebold & Zucc.

pisifera (Siebold & Zucc.) Siebold & Zucc.   Sawara Cypress
* Introduced; reproducing in Mosses Wood, Leith Hill (Surrey). Japan. 20, 50, 58*, 59*, 60*, 119, 166, 199. *Cupressus pisifera* (Siebold & Zucc.) K.Koch; *Retinospora pisifera* Siebold & Zucc.

PLATYCLADUS Spach
orientalis (L.) Franco   Chinese Thuja
** Introduced; self-sown on old walls at Kew and Haslemere (both Surrey). China, Manchuria, Korea. HbACL. 21, 50*, 59*, 60*, 119, 199, 331(40:42). *Thuja orientalis* L.

THUJA L.
occidentalis L.   Northern White-cedar
* Introduced. Eastern N America. 21, 50*, 58*, 59*, 60*, 119.
plicata Donn ex D.Don   Western Red-cedar
*** Introduced; reproducing in woodlands in widely scattered localities, as at New Abbey (Kirkcudbrights) and Box Hill (Surrey). Western N America. NMW, OXF, RNG. 13, 20*, 58*, 59*, 60*, 303(32:20).

THUJOPSIS (L.f.) Siebold & Zucc. ex Endl.
dolabrata (L.f.) Siebold & Zucc.   Hiba
* Introduced. Japan. 50, 58*, 59*, 60*, 119, 166. *Thuja dolabrata* L.f.

JUNIPERUS L.
chinensis L.   Chinese Juniper
* Introduced. Mongolia, China, Japan. 20, 50, 58*, 59*, 119, 649*.
recurva Buch.-Ham. ex D.Don   Himalayan Juniper
* Introduced. Himalayas, Burma, China. 50, 58*, 59*, 60*, 119.
[*sabina* L. In error for native *J. communis* subsp. *alpina* Čelak. 218, 239, 255, 354(5:796).]
virginiana L.   Virginian Juniper
* Introduced. Eastern N America. 21, 50, 58*, 59*, 60*, 119.

LIBOCEDRUS Endl.
chilensis (D.Don) Endl.   Chilean Incense-cedar
* Introduced. Chile, Argentina. 50, 59*, 60*, 119. *Austrocedrus chilensis* (D.Don) Florin & Boutelje; *Thuja chilensis* D.Don.

CALOCEDRUS Kurz
decurrens (Torrey) Florin   Californian Incense-cedar
** Introduced; self-sown on a garden wall at Oxford Botanic Garden (Oxon) and in Moor Park, Farnham (Surrey). Western N America. 50, 58*, 59*, 60*, 119, 342(Feb 89), 481. *Libocedrus decurrens* Torrey.

ARAUCARIACEAE
ARAUCARIA A.L.Juss.
araucana (Molina) K.Koch   Monkey-puzzle
*** Introduced. Chile, Argentina. BM. 16, 20*, 50, 58*, 59*, 60*, 119, 224. *A. imbricata* Pavón; *Pinus araucana* Molina.

CEPHALOTAXACEAE

CEPHALOTAXUS Siebold & Zucc.

fortunei Hook.   **Chinese Plum-yew**
- Introduced; also a garden relic on the site of Warley Place gardens (S Essex).
China.   50, 58*, 59*, 60*, 409.

harringtonia (Knight ex Forbes) K.Koch   **Japanese Plum-yew**
- Introduced.   China, Japan.   50, 58*, 59*, 60*, 119.   *Taxus harringtonia*
Knight ex Forbes; incl. *C. drupacea* Siebold & Zucc.

ANGIOSPERMAE

DICOTYLEDONAE

MAGNOLIACEAE

MAGNOLIA L.
acuminata L.   Cucumber-tree
   ● Introduced. Eastern N America. 38*, 50, 58*, 59*, 60*, 119.

LIRIODENDRON L.
tulipifera L.   Tulip-tree
   ●● Introduced or a garden relic. Eastern N America. LANC, LIV. 21, 50*,
   58*, 59*, 60*, 119, 166.

LAURACEAE

LAURUS L.
nobilis L.   Bay
   ●●● A persistent or established garden escape; in woods and scrub, on cliffs
   and roadsides in the south and west; bird-sown in various places, as at Headon
   Warren (Wight). Medit. BM, LANC, NMW, OXF. 20, 32*, 44*, 59*, 60*,
   156, 256, 1156*.

UMBELLULARIA (Nees) Nutt.
californica (Hook. & Arn.) Nutt.   California Bay
   ● Long-persistent as a garden relic on the site of Warley Place gardens
   (S Essex). Western N America. 50, 51*, 52*, 59*, 60*, 370(9:411).

PERSEA Miller
americana Miller   Avocado
   ●● A food refuse casual. C America; (widespread as a crop). 50, 68*, 69*,
   199, 236, 331(62:109). *P. gratissima* C.F.Gaertner.

ARISTOLOCHIACEAE

[*ASARUM* L.
europaeum L.   Accepted, with reservations, as native.]

ARISTOLOCHIA L.
bodamae Dingler   Thracian Birthwort
   ● Established since 1969 on the site of former deciduous woodland at Elveden
   (W Suffolk). Greece, Turkey. 21, 39, 258, 269, 344*(25:t.46), 371(425:8).
clematitis L.   Birthwort
   ●●● An established garden escape and relic; long-naturalised in the grounds of
   abbeys, in churchyards and woods, on grassy banks and in derelict gardens, in

widely scattered localities; decreasing. Formerly cultivated as a medicinal plant. SE Europe, Turkey, Caucasus. BM, E, LANC, NMW, OXF, RNG. 1, 20*, 22*, 26*, 33*, 235, 277, 595*. *A. longa* auct., non L.

rotunda L.   Smearwort
- Established; possibly originally a garden escape, naturalised for more than 70 years on a chalk hillside near Woldingham (Surrey). Formerly also on chalk downs near Shoreham (W Kent). Medit. BM, E, RNG, SLBI. 7, 20*, 24*, 29*, 32*, 199, 209, 354(4:428), 595*.

NYMPHAEACEAE

NYMPHAEA L.

garden hybrids   Hybrid Waterlilies
- • Introduced or garden escapes, persistent in ponds and lakes in a few widely scattered localities. Originated in cultivation. RNG, HbEJC. 20, 34*, 50, 52*, 54*, 55*, 243, 432, 458. Incl. *N.* × *marliacea* Latour-Marl. Modern hardy cultivars, which may be introduced in lakes, are derived from nine or more species.

odorata Aiton   Fragrant Waterlily
- Introduced and established in Boldermere Lake, Wisley (Surrey) and in mature pits at Trilakes, Sandhurst (Berks). USA, Mexico, Guyana. BM. 50, 55*, 353(40:28), 312*(t.819; t.1652; t.6708), 594*, 1326*.

NUPHAR Smith

advena (Aiton) Aiton f.   Spatter-dock
- Introduced; well established in Cow Pond, Windsor Great Park, and in lakes at Pain's Hill Park, Cobham and at Godstone (all Surrey); also recorded from Loch Ard (W Perth) and, in 1992, from Carlingwark Loch, Castle Douglas (Kirkcudbrights). N America. 20, 38*, 50, 303(24:16), 370(14:188), 475, 1307. *N. lutea* (L.) Smith subsp. *advena* (Aiton) Kartesz & Gandhi; *Nymphaea advena* Aiton.

CABOMBACEAE

CABOMBA Aublet

caroliniana A.Gray   Carolina Water-shield
- Introduced or an escape from aquaria; persistent for a few years in water heated by factory effluent in the Forth and Clyde Canal at Dalmuir (Dunbarton); also found in 1990-1991 in several places in the Basingstoke Canal between Colt Hill Bridge and Broad Oak (N Hants). SE USA. BM, CGE, MNE, RNG. 38*, 40, 50, 87*, 316(18:565-8), 371(367:29), 426, 436.

RANUNCULACEAE

CALTHA L.

palustris L. subsp. **polypetala** (Hochst. ex Lorent) Velen.   Giant Kingcup
- A garden escape established near Chypons, Lizard (W Cornwall); introduced and persistent for some years on Marazion Marsh (W Cornwall). Bulgaria, SW Asia. HbEJC. 21, 39, 212, 213, 347(11:253), 730*, 808(21:119-150),

1172. *C. polypetala* Hochst. ex Lorent.   Confused in horticultural literature
with other variants of *C. palustris*, and probably better considered as var.
polypetala (Hochst. ex Lorent) Huth.

HELLEBORUS L.
argutifolius Viv.
  • A garden escape established on a rocky outcrop at Stow-on-the-Wold, on the
side of Fosse Way near Broadwell (both E Gloucs), at Crayford (W Kent) and
on a wall in Jersey (Channel Is); also reported from W Lancs.   Corsica,
Sardinia.   21, 24*, 50, 52*, 85*, 201, 204, 220, 323(38:403), 347*(3:5),
371(422:7), 458. *H. corsicus* Willd., nom. nud.; *H. lividus* subsp. *corsicus*
(Briquet) P.Fourn., nom. nud.   The closely related *H. lividus* Aiton, from
Majorca, is unlikely to escape.

dumetorum Waldst. & Kit.
  • An established garden escape; self-seeding in a copse near Wendover
(Bucks).   C Europe.   21, 31*, 50, 85*, 312*(n.s. t.545), 325*(113:110),
371(385:34; 391:33). Incl. *H. atrorubens* Waldst. & Kit., non hort.

× **hybridus** Voss   Hybrid Lenten-roses
  • Seedlings on a wall in Guernsey; also recorded from Sark (both Channel Is).
Originated in cultivation.   85*, 220, 1218.   A complex of hybrids involving
H. orientalis and other species.

orientalis Lam.   Lenten-rose
  •• A relic of cultivation or garden escape persistent or established in a few
woods and copses in the south; long persistent on the site of Warley Place
gardens (S Essex).   SE Europe, Turkey, Caucasus.   OXF.   7, 20, 31*, 50, 85*,
199, 303(33:12), 313(31:370), 347*(3:1), 354(8:20).   Incl. *H. caucasicus*
A.Braun; *H. olympicus* Lindley.

ISOPYRUM L.
thalictroides L.   Isopyrum
  • A garden relic long persistent on the site of Warley Place gardens (S Essex).
Europe.   21, 22*, 29*, 371(419:14), 546*, 683*.

ERANTHIS Salisb.
hyemalis (L.) Salisb.   Winter Aconite
  ••••• An established garden escape; naturalised in woods, plantations,
churchyards and derelict gardens, especially in S and E England.   S Europe.
1, 20*, 22*, 26*, 27*.   *Cammarum hyemale* (L.) E.Greene; *Helleborus
hyemalis* L.   Some records may be referable to *E. cilicica* Schott & Kotschy or
E. × tubergenii Bowles (*E. cilicica × E. hyemalis*).   312*(n.s. t.196).

NIGELLA L.
arvensis L.   Nigella
  • A grain and bird-seed casual.   Eurasia, N Africa.   OXF.   1, 21, 22*, 32*,
44*, 347*(4:229-235), 370(2:36), 371(361:26), 546*.

damascena L.   Love-in-a-mist
  •••• A persistent garden escape, mainly in the south.   Medit.   1, 20*, 22*,
25*, 53*, 347*(4:229-235), 546*.

gallica Jordan   Pale Fennel-flower
- (but confused with *N. hispanica*). A bird-seed casual. SW Europe. 7, 16, 19, 21, 191, 354(7:763), 371(364:29), 536*, 546*, 1028*. *N. hispanica* auct., non L.

hispanica L.   Fennel-flower
- (but confused with *N. gallica*). A bird-seed casual. SW Europe, NW Africa. HbACL. 21, 30*, 45*, 46*, 50, 201, 303(11:10), 354(7:860), 648*.

sativa L.   Black-cumin
- A bird-seed and grain casual. SW Asia, N Africa. HbGMSE. 1, 21, 29*, 50, 303(54:24-5), 336(33:51), 347(4:229-235), 354(9:682), 371(357:24), 648*, 1368*.

ACONITUM L.
angustifolium Bernh.
- Recorded from a dry ditch at North Somercotes Warren (N Lincs). Yugoslavia. 21, 51, 371(331:20), 692*, 693*, 695*.

× **cammarum** L.   Hybrid Monk's-hood
- ••• A persistent garden escape; in woods and by roadsides in the north, mainly in central Scotland; known since 1970 on a roadside at Forest Moor (MW Yorks). Originated in cultivation. BM, CGE, E, NMW, RNG. 2, 20*, 277, 335(113:78), 371(352:26; 355:28), 1255*. *A. napellus* × *A. variegatum*; *A.* × *intermedium* DC.; *A.* × *stoerckianum* Reichb.; *A. variegatum* auct., non L. Probably all records are referable to *A.* 'Bicolor', non *A.* × *bicolor* Schultes.

carmichaelii Debeaux
- A casual garden escape. E Asia. 50*, 51, 51a, 55*, 57*, 371(346:29). *A.* 'Arendsii'; *A. fischeri* hort., non Reichb.; incl. *A. wilsonii* Stapf.

napellus L. subsp. **firmum** (Reichb.) Gáyer
Pre-1930 only. C Europe. BM. 7, 21a, 354(4:402), 623a*, 822(80:1-76). *A. amoenum* Reichb.; *A. firmum* Reichb.

napellus L. subsp. **vulgare** (DC.) Rouy & Fouc.
- •• (but mostly in error for *A.* × *cammarum*). A persistent or established garden escape. Pyrenees, Alps. LANC, OXF. 1, 21, 51b, 103, 156, 354(5:273-4), 693*. *A. compactum* (Reichb.) Gáyer; *A. vulgare* DC.

[*variegatum* L. (*A. cammarum* Jacq., non L.) Unconfirmed: all records appear to be in error for *A.* × *cammarum*. RNG. 20.]

vulparia Reichb.   Wolf's-bane
- •• A garden escape established in woods, by roadsides and streams, in a few places in the north. Europe. BM, E, OXF. 2, 20*, 22*, 27*, 28*, 303(28:14), 370(9:375; 14:420). *A. lycoctonum* L. subsp. *vulparia* (Reichb.) Nyman. Early records of *A. lycoctonum* are probably referable to this species.

DELPHINIUM L.
× **cultorum** Voss   Garden Delphinium
- A garden escape, shortly persistent, possibly self-sown. Originated in cultivation. 51b, 53*, 54*, 511, 594*, 1118. *D.* × *hybridum* hort., non

Stephan ex Willd.   A collective name for garden hybrids between several
species.

elatum group   Alpine Delphinium
  • A garden escape on waste ground at Biggin Hill airfield (W Kent) and at the
  foot of a wall in Guildford (Surrey). Eurasia; hybrids originating in cultivation.
  HbACL. 23*, 29*, 51, 458, 546*. *D. elatum* L. and hybrids with several
  species. A polymorphic group, not clearly separable in cultivation from
  D. × *cultorum*. *D. elatum*, as a pure species, is now seldom cultivated.
exaltatum Aiton
  Pre-1930 only. N America. OXF. 7, 38*, 40, 51, 354(8:300), 1319.
fissum Waldst. & Kit.
  Pre-1930 only. S Europe. SLBI. 2, 21, 31, 50*, 300(1907:37), 546*.
  D. hybridum Stephan ex Willd.
nudicaule Torrey & A.Gray
  Pre-1930 only. Western N America. 2, 50*, 52*, 53*, 55*, 184(p.22),
  300(1904:175).
peregrinum L.   Violet Larkspur
  Pre-1930 only. Medit. 2, 21, 24*, 29*, 39*, 44*, 149, 546*.
verdunense Balbis
  • A casual on a tip at Oxford (Oxon), vector unknown. No modern records.
  SW Europe. OXF. 21, 30*, 403, 525*, 1225. *D. cardiopetalum* DC.

CONSOLIDA (DC.) Gray
ajacis (L.) Schur   Larkspur
  ••••• A persistent garden escape and grain alien. Medit.; (N America).
  1, 20*, 22*, 35*, 303(10:17-18; 45*:22-23).   *C. ambigua* auct., non (L.)
  P.Ball & Heyw.; *C. gayanum* (Wilm.) Laínz; *Delphinium addendum* Macnab;
  D. ajacis L.; *D. ambiguum* auct., non L.; *D. gayanum* Wilm. A putative
  hybrid with a perennial *Delphinium* species has been recorded. HbEJC. 126.
orientalis (Gay) Schroedinger   Eastern Larkspur
  ••• (but mostly in error for *C. ajacis*). Some pre-1930 specimens have been
  confirmed as this species. Medit., SW Asia. BM, CGE, LIV, NMW, OXF,
  RNG. 1, 20, 22*, 30*, 39*, 51b, 303(10:17-18), 312*(n.s. t.186).
  Delphinium ajacis auct., non L.; *D. orientale* Gay.
pubescens (DC.) Soó
  Pre-1930 only. W & C Medit. NMW. 2, 21, 46*, 300(1907:37),
  354(10:927), 546*, 683*. *Delphinium pubescens* DC.
regalis Gray   Forking Larkspur
  •• (but over-recorded for *C. ajacis*). A grain casual. Europe, SW Asia. BM,
  CLE, LIV, NMW, OXF. 1, 20, 22*, 25*, 26*, 39*, 51b, 235, 303(10:17-18),
  1242*. Incl. *Delphinium consolida* L.; *D. divaricatum* Ledeb.

ACTAEA L.
erythrocarpa Fischer
  • A casual garden escape. NE Europe, Asia. 43, 50, 51, 589*, 861*(16:t.4),
  1041, 1217*. *A. spicata* L. var. *erythrocarpa* (Fischer) Ledeb. Possibly in
  error for *A. rubra* (Aiton) Willd.

ANEMONE L.

apennina L.   **Blue Anemone**

    •••• An established garden escape and relic; naturalised in woods, parks and hedgerows in widely scattered localities. S Europe. 1, 20, 22*, 23*, 25*, 50, 209. *Anemonoides apennina* (L.) Holub.

blanda Schott & Kotsky

    • (but perhaps overlooked as *A apennina*). A garden escape by the car park at Tewkesbury Abbey (E Gloucs) and on a roadside between Kentmere and Staveley (Westmorland); a garden relic long established on the site of Warley Place gardens (S Essex); also apparently self-sown in pavement at Kingswood (Surrey). SE Europe, Turkey. LANC, RNG. 20, 29*, 53*, 126, 303(30:14), 370(18:420), 584*, 700*.

coronaria L.   **Poppy Anemone**

    • A grain casual and garden escape. Medit. BM, E, NMW, OXF. 8, 21, 24*, 44*, 50, 54*, 135, 143, 156, 253, 457.

× **fulgens** Gay

    Pre-1930 only. Originated in cultivation. OXF. 7, 21, 50, 53*, 54*, 104, 354(8:102). *A. hortensis* L. × *A. pavonina* Lam.

× **hybrida** Paxton   **Japanese Anemone**

    ••• A persistent garden escape and relic; on waste ground, roadsides and spoil heaps in widely scattered localities; known for many years on the site of Warley Place gardens (S Essex). Originated in cultivation. BM, NMW, OXF. 2, 20, 53*, 55*, 126, 220. *A. hupehensis* (Lemoine) Lemoine × *A. vitifolia* Buch.-Ham. ex DC.; *A.* × *elegans* Decne.; *A. hupehensis* hort. var. *elegans*; *A.* × *japonica* hort., non (Thunb.) Siebold & Zucc.

ranunculoides L.   **Yellow Anemone**

    ••• An established garden escape; naturalised in woods and on river banks in scattered localities; known since 1916 in a wood at Ovington (W Norfolk). Europe, W Asia. BM, K, LANC, LSR, OXF, RNG. 1, 20, 22*, 23*, 26*, 264. A hybrid with native *A. nemorosa*, (*A.* × *lipsiensis* Beck; *A.* × *intermedia* Winkler, non *A. intermedia* G.Don) may have been overlooked. 51b, 826(18:98-100).

sylvestris L.   **Snowdrop Anemone**

    Pre-1930 only. Europe, W Asia. OXF. 7, 16, 21, 22*, 23*, 26*, 50, 57*, 354(8:299).

HEPATICA Miller

nobilis Schreber   **Liverleaf**

    • A persistent garden relic on the site of Warley Place gardens (S Essex), in woodland at Edge (E Gloucs), in an old garden at Frilford (Berks) and at Sunderlandwick (SE Yorks). Europe, W Asia. OXF. 1, 20, 22*, 25*, 27*, 30*, 119, 126, 182, 1242*. *H. triloba* Chaix; *Anemone hepatica* L.

[*PULSATILLA* Miller
pratensis (L.) Miller   In error for native *P. vulgaris*. 148.]

CLEMATIS L.
armandii Franchet
● A relic of cultivation persistent for a time on an old wall at a bombed site at the Barbican, London (Middlesex), now gone. China. 50, 63*, 65*, 74*, 126, 371(367:30).

cirrhosa L.   Early Virgin's-bower
● A garden escape self-sown on outer walls of the Caledonia Nursery, Guernsey (Channel Is). Medit. 20, 50, 63*, 65*, 74*, 220. Incl. *C. balearica* Rich.

flammula L.   Virgin's-bower
●● An established garden escape; locally naturalised on dunes and sea cliffs on the south coast of England and at Abersoch (Caerns); known at Sandwich Bay (E Kent) since 1927. Medit. BM, K, LANC, NMW, OXF, RNG. 7, 20*, 22*, 24*, 74*, 236, 354(8:384), 370(13:132), 1270.

montana Buch.-Ham. ex DC.   Himalayan Clematis
●● A long-persistent or established garden escape; modern records are from the site of Warley Place gardens (S Essex), from waste ground at Weetwood, Headingly (MW Yorks), self-sown on a wall top at Haslemere (Surrey), at Brighton (E Sussex) and in the village of Eastbridge, Leiston (E Suffolk). Himalayas, China. 7, 20*, 50, 63*, 65*, 74*, 122, 199, 313(31:370), 371(419:14), 432.

[*orientalis* L. In error for *C. vernayi* in the *C. tangutica* group. 347(7:193-204).]

tangutica group   Golden Clematis
● Persistent garden escapes; in a disused gravel pit near Hoddesdon (Herts) and on dunes at Findhorn (Moray); seedlings have been reported about gardens in E Suffolk and E Norfolk; now gone from both Marazion (W Cornwall) and a quarry at Cothill (Berks). Asia. OXF, RNG. 7, 20*, 50, 63*, 65*, 74*, 212, 303(28:15), 347*(7:193-204), 371(346:28; 374:9), 442. Incl. *C. orientalis* hort., non L.; *C. tangutica* (Maxim.) Korsh.; *C. tibetana* Kuntze; *C. vernayi* Fischer and putative hybrids. The cultivar *C.*'Orange-peel' is now considered to be a selection from *C. vernayi* (*C. tibetana* subsp. *vernayi* (Fischer) Grey-Wilson), not *C. orientalis* L. 347(7:193-204).

viticella L.   Purple Clematis
●● A garden escape persistent in a few places, mainly in southern England. SE Europe, SW Asia. BM, CGE, LANC, OXF, RNG, SLBI. 7, 20*, 24*, 50, 63*, 65*, 74*, 192, 349(3:53).

RANUNCULUS L.
aconitifolius L.   Aconite-leaved Buttercup
● A garden escape established on the banks of a stream at New Earswick, York, and in open woods at Kirkcaldy (Fife); casual elsewhere. C Europe. OXF. 7, 20*, 22*, 23*, 26*, 113, 253, 371(397:35). Most records are referable to *R. aconitifolius* 'Flore Pleno'.

[*alpestris* L.   A dubious record. 187, 354(5:739), 361(7:127).]

cordiger Viv.

Pre-1930 only. Corsica, Sardinia. 7, 21, 354(6:272), 683*. Treated by some authors as conspecific with native *R. sardous.*

[*gramineus* L. Dubious records. 141, 188, 193(p.27), 354(5:740), 370(2:304).]

marginatus Urv.   St Martin's Buttercup

•• A grain and bird-seed alien; established for many years as an abundant weed in a few arable fields on St Martin's and St Mary's (both Scilly); still on St Martin's in 1988, though much reduced; casual elsewhere. E Medit. BM, OXF, RNG. 13, 19, 20*, 39*, 44*, 160, 207*, 212, 370(2:104), 371(416:7). Incl. *R. trachycarpus* Fischer & C.Meyer.

monspeliacus L.

Pre-1930 only. W Medit. 21, 216, 300(1908:101), 546*, 728*.

muricatus L.   Rough-fruited Buttercup

•• An established bird-seed, grain and wool alien; long naturalised as an abundant weed in bulbfields and gardens in Scilly, now spreading to the mainland of W Cornwall; casual elsewhere. Medit.; (Australasia). ABD, BM, CGE, E, OXF, RNG. 1, 19, 20*, 24*, 44*, 160, 207*, 212.

pensylvanicus L.f.   Bristly Buttercup

Pre-1930 only. N America. 1, 38*, 40*, 41*, 205(p.6).

rupestris Guss.

Pre-1930 only. SW Europe. 2, 21, 300(1908:101), 683*. *R. spicatus* Desf.

sessiliflorus R.Br. ex DC.   Australian Small-flowered-buttercup

• A wool casual. Australia; (New Zealand). E. 36c, 328*(1956), 514, 533, 549, 743. Collected as var. *pilulifer* (Hook.) Melville.

[*sphaerospermus* Boiss. & Blanche In error for native *R. penicillatus* subsp. *pseudofluitans.* 174.]

trilobus Desf.

• A grain casual. Formerly a wool casual. Medit. OXF, SLBI. 1, 3, 21, 182, 249, 300(1907:37), 525*, 556*, 737*. *R. sardous* Crantz subsp. *trilobus* (Desf.) Rouy.

CERATOCEPHALA Moench

falcata (L.) Pers.

• A grain casual at docks. No modern records. Medit., SW Asia. LCN. 1, 21, 30*, 44*, 134, 160, 354(11:464), 546*. *Ranunculus falcatus* L.

testiculata (Crantz) Roth

Pre-1930 only. Eurasia. 21, 31, 39, 136, 695*. *C. orthoceras* DC.

ADONIS L.

aestivalis L.   Summer Pheasant's-eye

• A wool and grain casual. Eurasia, N Africa; (Australia). E, K, LANC, NMW, OXF, SLBI. 2, 8, 9, 12, 16*, 21, 22*, 25*, 26*, 44*, 50, 354(10:927). Some old records, especially of early-flowering plants, are in error for *A. annua.* K.

annua L.   Pheasant's-eye

•••• A grain alien, formerly well naturalised as a cornfield weed in southern England, now much reduced but persisting in a few places; also a casual garden

escape. Medit., SW Asia. 1, 20*, 22*, 23*, 35*, 50, 172, 236.
A. autumnalis L.

flammea Jacq. **Large Pheasant's-eye**
 • A casual. Eurasia, N Africa. 1, 21, 22*, 25*, 39*, 50, 121, 348(20:29).
[*vernalis* L. In error for *A. aestivalis*. 384*.]

AQUILEGIA L.
[*alpina* L. In error for *A. pyrenaica* and native *A. vulgaris*. 354(5:740).]
garden hybrids
 •• (but probably under-recorded). Casual garden escapes. Originated in
cultivation. 51a, 54*, 253, 331(29:86; 37:203-4), 416. A complex of hybrids
involving many species; incl. *A.* 'Haylodgensis'; *A.* × *hybrida* hort.; *A.* Long-
spurred Hybrids.

olympica Boiss.
 • A garden escape established since 1984 on the chalky banks of Wrotham Hill
(W Kent). S Russia, Turkey, Caucasia, Iran. 39, 50, 303(62:45), 630*, 674*,
730*, 842*(1896:108), 882*(7:1-150), 888*(9:120). *A. caucasica* Ledeb. ex
Rupr.

pubescens Cov.
Pre-1930 only. California. OXF. 7, 42, 354(5:547), 500*, 1143*.

pyrenaica DC. **Pyrenean Columbine**
 • Introduced and established since 1895 on rock ledges at the head of
Caenlochan Glen (Angus). Pyrenees. BM, OXF. 7, 20*, 23*, 27*, 28*, 50,
187, 312*(n.s. t.435), 349(3:89,182). *A. alpina* auct., non L.

vulgaris L. **alien variants**
Several cultivars have been recorded, as well as unnamed variants.

THALICTRUM L.
aquilegiifolium L. **French Meadow-rue**
 •• An established garden escape; on roadsides and railway banks in a few
widely scattered localities in England and Scotland; known since 1949 on a
railway bank near Tunbridge Wells West Stn. (W Kent). Eurasia. ABD, BM,
E, RNG. 1, 20, 22*, 23*, 26*, 29*, 50, 51b, 236, 349(1:161).

delavayi Franchet **Chinese Meadow-rue**
 • A garden escape at Fulbourn (Cambs). China. HbGMSE. 20, 50, 51b*,
54*, 55*, 303(32:20), 634*, 657*. *T. dipterocarpum* hort., non Franchet.

lucidum L. **Shining Meadow-rue**
 • Probably a garden escape, on a bank of the river Whiteadder, Edrington
Levels (Berwicks). Eurasia. E, HbEJC. 20, 22*, 26*, 50, 303(32:20).
T. angustifolium auct., non L.

simplex L. **Small Meadow-rue**
Pre-1930 only. Eurasia. 7, 21, 22*, 27*, 50, 354(2:208; 5:547). Incl.
T. bauhinii Crantz; *T. gallicum* Rouy & Fouc.

speciosissimum L.
 • Reported as a relic of cultivation by a pond near Whetstead and at Stone
(both W Kent). W Medit. LTR. 21, 30*, 50, 54*, 458, 473. *T. flavum* L.
subsp. *glaucum* (Desf.) Battand.; *T. glaucum* Desf.

BERBERIDACEAE
Incl. *Podophyllaceae*

PODOPHYLLUM L.
hexandrum Royle   Himalayan May-apple
- A persistent garden escape or relic. No modern records. Himalayas. 50, 52*, 57*, 325*(113:233-238), 351(79:25), 1300*. *P. emodi* Wallich ex Hook.f. & Thomson.

CAULOPHYLLUM Michaux
thalictroides (L.) Michaux   Blue Cohosh
- A garden relic long persistent on the site of Warley Place gardens (S Essex). Eastern N America. 38*, 50, 347*(4:1-15), 370(9:411).

EPIMEDIUM L.
alpinum L.   Barrenwort
- •• An established garden escape and relic; in woods, parklands and derelict gardens, especially in the north; known since 1926 at Aberfeldy (Mid Perth). Italy, Yugoslavia, Albania. BM, E, K, NMW, OXF, RNG. 1, 20, 22*, 23*, 27*, 29*, 50, 354(8:104), 370(2:408), 432.

perralderianum Cosson   Algerian Barrenwort
- A persistent relic of cultivation in a derelict garden in Bishop's Walk, Fulham (Middlesex). Algeria. 46*, 50, 303(33*:12-13 as *E. pubigerum*; 34:24).

pinnatum Fischer   Caucasian Barrenwort
- A persistent relic of cultivation in a few derelict gardens.   Turkey, Caucasus, Iran. 20, 50, 55*, 323(33:170), 371(343:29; 367:30), 650*. Incl. subsp. *colchicum* (Boiss.) Busch. Some records may be in error for a hybrid with *E. grandiflorum* Morren (*E.* × *versicolor* Morren). 20.

[*pubigerum* (DC.) Morren & Decne. In error for *E. perralderianum*. 303(33:12; 34:24).]

VANCOUVERIA Morren & Decne.
cf. **planipetala** Calloni   Redwood-ivy
- Introduced and established at Carbury Towers (E Lothian).   Western N America. RNG. 50, 51, 327*(51:409-535), 500*, 699*, 1143*, 1257*. *V. parviflora* E.Greene. Originally mis-named as *Epimedium pinnatum* subsp. *colchicum*.

BERBERIS L.
aggregata C.Schneider   Clustered Barberry
- A persistent garden escape high on a wall at Ardlussa in the Isle of Jura (S Ebudes), in a rampant hedge at Otley (MW Yorks) and in a hanger at Steep (N Hants). W China. HbJDF. 20*, 50*, 54*, 370(18:215), 371(334:22; 409:40), 613*. Superseded in gardens by its hybrids with *B. wilsoniae* and other species, (*B.* × *carminea* Ahrendt and *B.* × *rubrostilla* Chitt.). *B.* × *rubrostilla* has been recorded as self-sown in Jersey (Channel Is).

[*aristata* DC. In error for *B. glaucocarpa*. 248, 354(12:479).]

buxifolia Lam.   Box-leaved Barberry

•• A long-persistent garden escape; on roadsides and commons in a few widely
scattered localities; known from East Walton Common near King's Lynn
(W Norfolk) since the 1950s. Southern S America. BM, DUE, OXF, RNG.
7, 20*, 50, 63*, 312*(t.6505), 327*(57:1-410), 370(12:348; 14:231). Incl.
B. buxifolia 'Nana'; *B. dulcis* Sweet.

darwinii Hook.   Darwin's Barberry

••• An established garden escape; naturalised in woodland, scrub and
hedgebanks in widely scattered localities, as at Brandon (W Suffolk) and near
Ballinaboy (W Galway). Chile, Argentina. BM, E, NMW, OXF, RNG. 7,
20*, 50, 54*, 55*, 63*, 258, 319(18:82).

gagnepainii C.Schneider   Gagnepain's Barberry

• A persistent garden escape; on a wall top in Guildford (Surrey), and in
hedges near Cwmamman (Carms) and at Otley (MW Yorks). China. 20*, 50,
51*, 55*, 199, 305(41,suppl:11), 312*(n.s. t.504), 370(409:40).

glaucocarpa Stapf   Great Barberry

•• An established garden escape; planted for hedging and sometimes bird-
sown; long naturalised in hedges in the Porlock to Minehead area (S Somerset).
C Asia. BM, CGE, K, NMW, OXF, RNG. 7, 20*, 50, 209, 248, 327*(57:1-
410), 349(5:127), 354(12:479). *B. aristata* auct., non DC.

hookeri Lemaire   Hooker's Barberry

• Recorded from woodland at Merthyr Mawr (Glam). Himalayas. NMW. 50,
61, 63*, 156. *B. wallichiana* Hook., non DC. Confused in horticulture with
B. manipurana.

× **hybrido-gagnepainii** Sur.

• Recorded from a hedgerow near Halstead (W Kent). Originated in
cultivation. 50, 61, 61a, 303(50:27), 327(57:1-410), 631*. *B. candidula*
C.Schneider × *B. gagnepainii*; *B.* × *wokingensis* Ahrendt, nom. nud.; incl.
B. × *chenaultii* Ahrendt (*B. gagnepainii* × *B. verruculosa* Hemsley &
E.Wilson).

julianae C.Schneider   Chinese Barberry

• A garden escape on waste ground at Cambuslang (Lanarks) and reported as
seedlings on old walls at Orpington (W Kent); casual elsewhere. China. 20,
50, 61, 62*, 445, 458.

manipurana Ahrendt   Manipur Barberry

• Introduced and persistent for many years on Tunbridge Wells Common
(W Kent); also at the edge of a wood at Buckler's Hard (S Hants).
Indo-Burmese border.   BM, RNG.   50, 61, 303(48:35), 349(4:41),
371(331:18-19).   *B. hookeri* var. *latifolia* hort.; *B. knightii* hort., non
(Lindley) K.Koch; *B. wallichiana* auct., non Hook.

× **ottawensis** C.Schneider

• A garden escape reported in 1987 as probably bird-sown in W Kent.
Originated in cultivation. 50, 61, 63*, 371(413:7), 631*, 1116*. *B. thunbergii*
× *B. vulgaris*; incl. *B.* 'Superba'.

pruinosa Franchet
- A garden escape reported in 1987 as probably bird-sown in Jersey (Channel Is). SW China. 50, 61, 371(413:7), 631*, 902*(9:39).

× **stenophylla** Lindley Hedge Barberry
- •• Introduced and established by suckering on roadsides in Kent and Suffolk; known for over 50 years on the banks of the A12 road at Martlesham and Woodbridge (E Suffolk). A very fertile hybrid; it appeared spontaneously in 1965 on an eagle's eyrie cliff on the Mull of Kintyre (Kintyre). Originated in cultivation. BM, DHM, DUH, OXF, RNG. 20*, 50, 56*, 63*, 258, 371(343:29; 391:33). *B. darwinii* × *B. empetrifolia* Lam.

thunbergii DC. Thunberg's Barberry
- •• An established garden escape; in abundance on rocky slopes near Llandudno (Caerns) and bird-sown in a few places in Kent and Surrey. Japan. 20*, 62*, 63*, 199, 303(39:9; 46:29), 1128*.

verruculosa Hemsley & E.Wilson
- A garden escape reported in 1987 as probably bird-sown in Jersey (Channel Is). W China. 50, 61, 63*, 312*(t.8454), 327*(57:1-410, figs 7,23), 371(413:7), 631*, 824(81:226).

[*vulgaris* L.   Accepted, with reservations, as native.]

wilsoniae Hemsley Mrs Wilson's Barberry
- •• An established garden escape; bird-sown in scrub and hedges, and on sand dunes, mainly in the south and west; known on Great Ormes Head (Caerns) since 1964. China. CGE, E, NMW, OXF. 20*, 50, 62*, 156, 199, 303(24:14), 371(374:8-10). Incl. *B. stapfiana* C.Schneider.

[× *MAHOBERBERIS* C.Schneider
neubertii (Baum.) C.Schneider (*Berberis vulgaris* × *Mahonia aquifolium*)   In error for *B. glaucocarpa*. 1311.]

MAHONIA Nutt.
aquifolium (Pursh) Nutt. Oregon-grape
- ••••• (but confused with hybrids). An established garden escape widely and sometimes abundantly naturalised, especially in SE England; also planted as game cover. Western N America. 1, 20, 22*, 50, 52*, 63*. *Berberis aquifolium* Pursh. A hybrid with *M. pinnata* (Lagasca) Fedde (*M.* × *wagneri* (Jouin) Rehder) has been recorded. HbACL.

× **decumbens** Stace Newmarket Oregon-grape
- •• Introduced, now extensively established on roadsides and tracksides in the Newmarket to Thetford area (W Suffolk); also recorded for Wilts, Cambs and Man. Originated in cultivation. BM(holotype). 20, 370(18:320), 371(346:27). *M. aquifolium* × *M. repens*; *M. repens* auct., non (Lindley) G.Don.

japonica (Thunb.) DC.
A garden escape reported in 1992 as self-sown on a path in Eaglesfield Park, Greenwich (W Kent). China. 50, 331(72:114), 347*(1:12-20), 617*. *Berberis japonica* (Thunb.) R.Br.; *Ilex japonica* Thunb. Much confused in horticultural literature with *M. bealei* (Fortune) Carrière.

[*repens* (Lindley) G.Don (*Berberis repens* Lindley)   In error for *M.* ×
decumbens. 370(18:320).]

LARDIZABALACEAE
AKEBIA Decne.
quinata (Houtt.) Decne.   Five-leaf Akebia
  • A persistent garden escape or relic at Sandling (E Kent). E Asia. 50, 61*,
  63*, 65*, 371(352:28; 382:23).

PAPAVERACEAE
PAPAVER L.
aculeatum Thunb.   Bristle Poppy
  • A wool casual. S Africa; (Australia). 17, 37, 51, 666*, 1124.
cf. **arenarium** M.Bieb.
  • Recorded in 1975 as an abundant weed in a bulbfield at Middle Town,
  St Martin's (Scilly). S Russia, SW Asia. RNG. 21, 39*, 51, 212, 630*,
  1124*.
atlanticum (Ball) Cosson   Atlas Poppy
  •••• An established garden escape; on roadsides, railway banks, walls and
  waste ground in widely scattered localities, especially in S and E England;
  increasing. Known since before 1930 about Godstone and Limpsfield (both
  Surrey). N Africa. 7, 20*, 51*, 192, 236, 258, 303*(41:1), 370(1:117),
  1124*. *P. lateritium* auct., non K.Koch; *P. rupifragum* var. *atlanticum* Ball.
commutatum Fischer & C.Meyer
  • Introduced or a garden escape; casual. No modern records. Crete,
  SW Asia. BM. 21, 39*, 288(p.lvi), 354(9:552), 1124*. *P. umbrosum* hort.
  Perhaps not specifically distinct from native *P. rhoeas.*
glaucum Boiss. & Hausskn.   Tulip Poppy
  • A wool casual. SW Asia. 17, 39*, 51, 55, 1124*.
laevigatum M.Bieb.
  Pre-1930 only. SE Europe, Turkey, C Asia.. 7, 21, 51, 354(1:199),694*,
  695*, 1172*. Perhaps not specifically distinct from native *P. dubium.*
[*lateritium* K.Koch   Unconfirmed. Probably always in error for *P. atlanticum.*
  20.]
nudicaule L.   Iceland Poppy
  • A casual garden escape. Northern subarctic regions. K, LIV, OXF. 2, 21,
  236, 300(1906:101), 319(10:264-6), 506*, 650*, 685*.
orientale L. sensu lato   Oriental Poppy
  ••• A persistent garden escape; on dunes, railway banks, roadsides and waste
  ground in widely scattered localities. SW Asia. BM, CGE, E, LANC, OXF,
  RNG. 7, 20*, 22*, 39*, 56*. Incl. *P. bracteatum* Lindley; *P. pseudo-
  orientale* (Fedde) Medw.
pavoninum Meyer   Peacock Poppy
  Pre-1930 only. C Asia. 2, 51, 267, 300(1905:225), 824*(t.1095), 1124*.

[*pinnatifidum* Moris   Unconfirmed. Probably always in error for *P. atlanticum.*
   BM. 354(10:21).]
rhoeas L. **alien variants**
   Var. *wilkesii* Druce, Shirley Poppy, and the double-flowered Simpson's strain
   occur as garden escapes. 674a*, 1124*.
[*rupifragum* Boiss. & Reuter   Unconfirmed.   Probably always in error for
   P. atlanticum. LANC, OXF. 354(8:301).]
somniferum L. sensu lato   Opium Poppy
   ●●●●● Long-established on arable land in the south, perhaps as a relic of
   former cultivation, now much reduced; also a spice casual, garden escape and
   bird-seed alien in widely scattered localities. Turkey; (widespread). 1, 19,
   20*, 22*, 29*, 39*, 209.   Incl. *P. hortense* Hussenot; *P. nigrum* DC.;
   P. officinale Gmelin; *P. setigerum* DC. A putative hybrid with *P. rhoeas* has
   been recorded. BM. 192, 1311.

ARGEMONE L.
mexicana L.   Mexican Poppy
   ●●● A wool and bird-seed casual and garden escape. Tropical & subtropical
   America; (widespread). BM, CGE, E, NMW, OXF, RNG. 1, 14, 19, 20*,
   29*, 53*, 55*. Incl. *A. ochroleuca* Sweet (*A. mexicana* var. *ochroleuca*
   (Sweet) Lindley.

ROEMERIA Medikus
hybrida (L.) DC.   Violet Horned-poppy
   ●● A grain casual, decreasing. Formerly persistent in arable fields; also a wool
   casual. Medit., SW Asia. BM, E, LCN, NMW, OXF, RNG. 1, 3, 20, 21,
   29*, 44*, 45*, 46*, 258. *R. violacea* Medikus; *Chelidonium hybridum* L.;
   Glaucium violaceum Juss.

GLAUCIUM Miller
corniculatum (L.) Rudolph   Red Horned-poppy
   ●●● A wool, grain and esparto casual, and a garden escape, sometimes
   persisting for a few years; possibly also a bird-seed alien; decreasing. Medit.,
   SW Asia. BM, E, LANC, NMW, OXF, RNG. 1, 14, 15, 19, 20, 22*, 25*,
   34*, 44*, 235.   *G. grandiflorum* sensu Hayek, non Boiss. & A.Huet;
   G. phoenicium Crantz, nom. illeg.; *Chelidonium corniculatum* L.
grandiflorum Boiss. & A.Huet
   ● A casual at Bath (N Somerset), vector unknown. No modern records.
   SW Asia. 39, 44*, 354(11:464), 1124*.

[*CHELIDONIUM* L.
majus L.   Accepted, with reservations, as native.]

ESCHSCHOLZIA Cham.
caespitosa Benth.   Tufted Californian-poppy
   Pre-1930 only.   California.   42, 51, 53*, 54, 197(p.5), 500*, 1124*.
   E. tenuifolia Benth.

californica Cham.   Californian-poppy
●●●● A persistent or established garden escape and grain alien; on dunes, cliffs, roadsides, by railways and in quarries, mainly in south-east England. Western N America. 1, 11, 20*, 22*, 29*, 54*, 126, 236, 500*. Incl. *E. crocea* Benth.; *E. douglasii* (Hook. & Arn.) Walp.

MACLEAYA R.Br.
cordata (Willd.) R.Br.   Five-seeded Plume-poppy
●● (but confused with *M.* × *kewensis*). A persistent garden escape. E Asia. BM, OXF. 7, 43, 51, 126, 312*(t.1905), 331(54:65), 564*, 594*, 650*. *Bocconia cordata* Willd.
× **kewensis** Turrill   Hybrid Plume-poppy
● (but probably under-recorded). A garden escape long-persistent by suckering; known for many years near Inverary (Main Argyll). China, or originated in cultivation. RNG. 20, 312*(n.s. t.321), 371(407:42), 450. *M. cordata* × *M. microcarpa* (Maxim.) Fedde; *M.* × *cordata* auct., non (Willd.) R.Br.; *Bocconia* × *cordata* sensu G.C.Brown ex Druce et auct., non Willd.

PLATYSTEMON Benth.
californicus Benth.   Cream-cups
Pre-1930 only. Western USA. 1, 42, 51*, 53*, 55*, 254, 500*. Incl. *P. leiocarpus* Fischer & C.Meyer.

HYPECOUM L.
imberbe Sibth. & Smith
Pre-1930 only. Medit., SW Asia. BM, CGE, CLE. 2, 21, 29*, 44*, 300(1904:237), 354(3:228-9; 4:470), 1172*. *H. aequilobum* Viv.; *H. grandiflorum* Benth.
pendulum L.
● An esparto casual. Medit., SW Asia. BM, LCN, NMW, OXF, RNG. 2, 9, 10, 21, 44*, 134, 354(5:637; 6:716), 546*, 666*, 1172*.
procumbens L.
Pre-1930 only. Medit. 1, 21, 24*, 32*, 44*, 121, 152, 546*, 1172*.

FUMARIACEAE
DICENTRA Bernh.
canadensis (Goldie) Walp.   Squirrel-corn
Pre-1930 only. Eastern N America. 7, 38*, 40, 51, 354(6:716; 7:555), 1319*. *Capnorchis canadensis* (Goldie) Druce.
eximia (Ker Gawler) Torrey   Turkey-corn
●● (but confused with *D. formosa*). A persistent or established garden escape. Eastern N America. DEE, RNG, TOR. 7, 20, 28*, 38*, 40, 55*, 57*, 199, 217, 370(16:440), 864(13:1-57). *Bicuculla eximia* (Ker Gawler) Millsp.; *Bikukulla eximia* (DC.) Druce; *Capnorchis eximia* (DC.) Druce; *Dielytra eximia* DC. Hybrids with *D. formosa* and *D. spectabilis* have been recorded. E, OXF. 436, 1315.

formosa (Andrews) Walp.   Western Bleeding-heart
●●● (but confused with *D. eximia*). An established garden escape; naturalised
in woods and by streams, mainly in Wales and Scotland. Western N America.
BM, CGE, E, LANC, NMW, RNG. 1, 20, 41*, 42, 51a, 55*, 339(16:215),
371(370:29), 864(13:1-57). *Bikukulla formosa* (Andrews) Cov.; *Capnorchis
formosa* (Andrews) Planchon; *Fumaria formosa* Andrews.   A hybrid with
D. eximia has been collected and may be overlooked. E, OXF.

spectabilis (L.) Lemaire   Asian Bleeding-heart
●● (but over-recorded for other species). A casual garden escape. E Asia. E.
7, 20, 21, 22*, 54*, 55*, 57*, 354(6:601), 356(1979:14).   *Capnorchis
spectabilis* (L.) Borkh. A hybrid with *D. eximia* has been reported. 436.

CORYDALIS DC.

capnoides (L.) Pers.   Beaked Corydalis
Pre-1930 only. E Europe. 21, 22*, 180, 683*, 715*. *Fumaria capnoides* L.
pro parte.

cava (L.) Schweigger & Koerte   Tuberous Corydalis
●● (but confused with *C. solida*). A persistent garden escape; known for many
years as a relic on the site of Warley Place gardens (S Essex). Eurasia. BM,
CGE, LANC, NMW, RNG, SLBI.   1, 20, 22*, 23*, 26*, 27*, 191.
C. bulbosa auct., non (L.) DC.; *C. marschalliana* (Pallas) Pers.; *C. tuberosa*
DC.; *Capnoides bulbosa* (L.) Druce; *C. cava* (L.) Moench, nom. illeg.;
Fumaria bulbosa L. var. *cava* L.; *Neckeria bulbosa* (L.) N.E.Br.

chelianthifolia Hemsley   Chinese Corydalis
● A garden escape established on walls at Winchester (S Hants) and Guildford
(Surrey); also on waste ground at Wisley (Surrey). China. HbEJC. 51, 52*,
53*, 199, 303(30:12), 325*(246), 347(2:129-131), 1181*.

ophiocarpa Hook.f. & Thomson
● A casual garden escape. Himalayas. 51, 342(Feb 89), 617*.

sempervirens (L.) Pers.   Harlequin Corydalis
● A casual garden escape. N America. HbACL. 21, 22*, 38*, 41*, 199,
371(364:29), 1319*. *C. glauca* Pursh; *Capnoides sempervirens* (L.) Borkh.

solida (L.) Clairv.   Bird-in-a-bush
●●●● An established garden escape; naturalised on roadsides and river banks,
in woods and churchyards, sometimes in abundance, as at Ovington
(W Norfolk) and Daglingworth (E Gloucs). Eurasia. 2, 20, 22*, 23*, 27*,
30*, 182, 264.   *C. bulbosa* (L.) DC., nom. illeg.; *Capnoides solida* (L.)
Moench, nom. illeg.; *Fumaria bulbosa* var. *solida* L.; *F. halleri* Willd.;
F. solida (L.) Miller.

PSEUDOFUMARIA Medikus
alba (Miller) Lidén   Pale Corydalis
●● A garden escape sometimes established on walls in the south, as at Godstone
(Surrey) and Ashmore (Dorset). SE Europe. BM, RNG. 20, 22*, 25*, 29*,
192, 199, 303(30:12), 347*(2:129-131), fig. as *Corydalis solida*).
P. ochroleuca Holub; *Corydalis ochroleuca* Koch, nom. illeg.; *Fumaria alba*
Miller.

lutea (L.) Borkh.   Yellow Corydalis
●●●●● An established garden escape; on old walls, cliffs and roadsides
throughout the British Is; known at Godalming (Surrey) since 1849. Europe,
perhaps native only in N Italy. 1, 20, 22*, 23*, 25*, 33*, 209. *Capnoides
lutea* (L.) Gaertner; *Corydalis capnoides* var. *lutea* (L.) DC.; *Corydalis lutea*
(L.) DC.; *Fumaria capnoides* Miller, non L.; *F. lutea* L.

FUMARIA L.
agraria Lagasca
● A grain casual. No modern records. Medit. LANC, LSR, OXF, RNG,
SLBI. 1, 21, 45*, 354(11:238; 12:26).
[*bella* P.D.Sell (*F. major* Badarò). In error for native *F. reuteri* subsp. *martinii*
(*F. paradoxa* Pugsley). 354(2:535; 3:68).]

PLATYCAPNOS (DC.) Bernh.
spicata (L.) Bernh.
Pre-1930 only. Medit. 1, 21, 29, 546*, 737*. *Fumaria spicata* L.

PLATANACEAE
PLATANUS L.
× **hispanica** Miller ex Muenchh.   London Plane
●●●●● Introduced and freely self-seeding in the London area, now well
established at intervals by the river Thames (Surrey); also reproducing in Black
Spinney, near Brandon (Warks). Origin obscure. 12, 20, 34*, 59*, 60*, 100,
126, 192, 546a*, 826*(18:4-17). Probably *P. occidentalis* × *P. orientalis*;
P. × *acerifolia* (Aiton) Willd.; *P.* × *hybrida* Brot.
occidentalis L.   American Plane
Pre-1930 only. Eastern N America. LIV. 38*, 61, 148, 505*, 606*.
According to 546a* the plant commonly in cultivation is *P. densicoma* Dode
(*P.* × *acerifolia* var. *pyramidalis* A.Henry & Flood; *P. occidentalis* hort.,
non L.; *P. pyramidalis* A.Henry).
orientalis L.   Oriental Plane
Pre-1930 only. SE Europe, SW Asia. LANC. 21, 59*, 148, 505*, 546a*,
606*, 662*. The Cripplegate (Middlesex) record was probably in error for
P. × *hispanica*. 331(29:87).

ULMACEAE
ULMUS L.
laevis Pallas   European White-elm
● Introduced. Europe, Turkey, Caucasus. LIV. 12, 21, 22*, 50, 58*, 59*,
60*, 151, 202, 221. *U. effusa* Willd.; *U. pedunculata* Foug.; *U. racemosa*
Borkh.
thomasii Sarg.   Rock Elm
● Recently recorded, as apparently self-sown, from a disused gravel quarry at
Marford, near Wrexham (Denbs). Eastern N America. 38*, 40*, 59*, 61,
370(18:428).

ZELKOVA Spach
carpinifolia (Pallas) K.Koch   Caucasian Elm
  • Introduced; reported as spreading vegetatively in Northumberland. SW Asia.
  NMW. 50*, 58*, 59*, 60*, 156, 347*(11:80-86), 403, 476. *Z. crenata*
  (Michaux f.) Spach.
serrata (Thunb.) Makino   Japanese Zelkova
  • Introduced; reported as self-seeding at Penzance (W Cornwall). Japan,
  Korea, Taiwan. RNG. 50*, 58*, 59*, 60*, 347*(11:80-86), 458, 482.

CANNABACEAE
CANNABIS L.
sativa L.   Hemp
  ●●●● A bird-seed and agricultural seed casual. Formerly cultivated for fibre.
  Asia; (widespread as a crop). 1, 19, 20, 22*, 25*, 29*, 34*, 50, 161,
  303(51:8), 1242*.

HUMULUS L.
japonicus Siebold & Zucc.   Japanese Hop
  • A casual garden escape. SE Asia. NMW, OXF, RNG. 4, 43, 50, 66*,
  126, 156, 331(38:20). *H. scandens* (Lour.) Merr.

MORACEAE
MORUS L.
alba L.   White Mulberry
  • Introduced. China. CGE, LANC. 21, 50, 58*, 59*, 60*, 125, 248.
nigra L.   Black Mulberry
  ●● A garden escape, food-refuse alien and relic of cultivation; plants on Nore
  Hill tip (Surrey) and on the river wall near Chiswick Bridge (Middlesex)
  originated from seed. C Asia. BM. 20, 58*, 59*, 60*, 126, 331(41:12;
  60:92).

FICUS L.
carica L.   Fig
  ●●●● A persistent food refuse alien and garden relic; bird-sown on cliffs and
  walls in widely scattered localities, especially in the south; also on banks of
  rivers heated by industrial discharges, the trees by the river Don in Sheffield
  (SW Yorks) being at least 60 years old. SW Asia. 2, 20, 50*, 58*, 59*, 60*,
  209, 370(18:85), 1083.
pumila L.   Creeping Fig
  • A persistent greenhouse escape on old garden walls in the south-west, at
  Launceston (E Cornwall) and at two unspecified localities; also on a wall at the
  Caledonian Nursery in Guernsey (Channel Is). E Asia. RNG. 50, 56*, 67*,
  371(352:27), 436, 594*, 719*. *F. repens* hort., non Missb. ex Miq.

URTICACEAE
URTICA L.
flabellata Kunth
Pre-1930 only. S America. 7, 354(6:622), 608, 820*(17:266,270).

incisa Poiret   Scrub Nettle
•• A wool casual. Australasia. BM, CGE, E, LTN, OXF, RNG. 7, 14, 20, 36c, 133, 145, 354(6:622), 569*, 590*, 725*.

cf. **massaica** Mildbr.
• A wool casual. Tropical Africa. E, K. 17, 48*, 618*.

pilulifera L.   Roman Nettle
• A persistent weed for some years in a garden at Wandsworth (Surrey); also recorded from Iken (E Suffolk); decreasing or perhaps now gone. Formerly a ballast and grain alien. Medit. ABD, BM, CGE, NMW, OXF, RNG. 1, 16, 21, 22*, 24*, 32*, 50, 235, 258. *U. dodartii* L.

GIRARDINIA Gaudich.
diversifolia (Link) Friis   Nilghiri Nettle
• A wool casual. Africa, India, SE Asia. K. 14, 48*, 736*. *G. condensata* (Hochst. ex Steudel) Wedd.; *G. heterophylla* Decne.

PILEA Lindley
microphylla (L.) Liebm.   Artillery-plant
Pre-1930 only. Eastern tropical America; (Australia). E, RNG. 3, 21, 50, 54*, 303(33:10), 590*, 593*, 600*, 725*. *P. muscosa* Lindley.

PARIETARIA L.
officinalis L.   Erect Pellitory-of-the-wall
• Recorded from the site of Warley Place gardens (S Essex) and the wild grounds of Friern Hospital, Friern Barnet (Middlesex), probably an ancient herbal introduction. Europe, SW Asia. RNG. 21, 22*, 29*, 303(63:27), 370*(6:365-370), 371(361:27), 825*(3:96-102,109-118). *P. erecta* Mert. & Koch; *P. officinalis* var. *erecta* (Mert. & Koch) Wedd. Possibly overlooked elsewhere as native *P. judaica*.

SOLEIROLIA Gaudich.
soleirolii (Req.) Dandy   Mind-your-own-business
••••• An established garden or greenhouse escape; widespread on damp walls, roadsides, in churchyards, and a persistent weed in gardens, mainly in SW England and Ireland. W Medit. 7, 20, 22*, 50, 590*. *Helxine soleirolii* Req.

JUGLANDACEAE
JUGLANS L.
nigra L.   Black Walnut
• Introduced. Eastern N America. 21, 22*, 50, 59*, 60*, 119, 199, 338(1979).

regia L.   Walnut
●●●● An established garden escape; naturalised in quarries, hedges and on
railway banks, mainly in the south. SE Europe, C Asia. 1, 20, 22*, 59*, 60*,
160, 258. Incl. *J. sinensis* (DC.) Dode.

CARYA Nutt.
cordiformis (Wangenh.) K.Koch   Bitternut Hickory
● Introduced. Eastern N America. 21, 50, 59*, 60*, 119, 347*(9:133-153).
ovata (Miller) K.Koch   Shagbark Hickory
● Introduced. Eastern N America. 21, 50, 59*, 60*, 119, 347*(9:133-153).
C. alba Nutt., non K.Koch.

PTEROCARYA Kunth
fraxinifolia (Poiret) Spach   Caucasian Wingnut
●● Introduced and established by suckering in woodland on Lowfield Heath
(Surrey); near Culgruff House at Crossmichael (Kirkcudbrights); on the edge
of a field at Warley (S Essex) and in Melbury Park (Dorset). Caucasus, Iran.
RNG. 20, 59*, 60*, 61*, 126, 199, 331(69:140), 371(391:33), 403. *Juglans
fraxinifolia* Poiret. Some records may be referable to the hybrid with
P. stenocarpa C.DC. (*P.* × *rehderiana* C.Schneider). 450.

MYRICACEAE
MYRICA L.
cerifera L. sensu lato   Bayberry
● An established garden escape; long-naturalised on heathland near Holmsley
in the New Forest (S Hants) and in a marshy wood near Fleet (N Hants);
apparently spreading vegetatively only. Eastern N America. BM, OXF, RNG,
SLBI. 13, 20, 22*, 38*, 40, 50, 354(11:278 as *M. cerasifera*, 503).
M. caroliniensis sensu Wangenh., non Miller; *M. pensylvanica* Lois. ex
Duhamel, nom. illeg. Recorded in error at Leonardslee (W Sussex). 122.
Often divided into two species, but intermediates occur (*M. heterophylla* Raf.).

FAGACEAE
NOTHOFAGUS Blume
antarctica (Forster f.) Oersted   Antarctic Beech
● Introduced. Chile, Argentina. 50*, 58*, 59*, 60*, 458.
dombeyi (Mirbel) Blume   Dombey's Beech
●● Introduced. Chile, Argentina. 20, 50*, 58*, 59*, 60*, 119, 370(12:344).
nervosa (Philippi) Krasser   Rauli
●●● Introduced; reproducing on Forestry Commission land. Chile, Argentina.
HbLJM. 20*, 58*, 59*, 60*, 119, 156, 370(12:344-5). *N. alpina* (Poeppig &
Endl.) Oersted; *N. procera* (Poeppig & Endl.) Oersted; *Fagus nervosa* Philippi;
F. procera Poeppig & Endl., non Salisb. A hybrid with *N. obliqua* has been
recorded. 370(12:344-5).

obliqua (Mirbel) Blume   **Roblé**
    ●●● Introduced; reproducing on Forestry Commission land and at the edge of
    Wisley Common (Surrey). Chile, Argentina. RNG. 20*, 50*, 58*, 59*, 60*,
    119, 199, 370(12:344-5). *Fagus obliqua* Mirbel.   Hybrids with *N. menziesii*
    (Hook.f.) Oersted and *N. nervosa* have been recorded. 370(12:344).

CASTANEA Miller
crenata Siebold & Zucc.   **Japanese Chestnut**
    ● Introduced. Japan. 21, 50, 61, 119, 563*, 606*, 649*. *C. japonica* Blume.
sativa Miller   **Sweet Chestnut**
    ●●●●● Introduced; reproducing and long-naturalised, mainly on light soils in
    southern England. Medit. 2, 20*, 22*, 34*, 58*, 59*.   *C. castanea* (L.)
    Karsten; *C. vesca* Gaertner, nom. illeg.; *C. vulgaris* Lam., nom. illeg.; *Fagus
    castanea* L.

QUERCUS L.
alba L.   **White Oak**
    ● Introduced. Eastern N America. LTN. 50*, 58*, 59*, 60*, 143.
canariensis Willd.   **Algerian Oak**
    ● Self-sown saplings recorded in 1990 from Holmbury St Mary (Surrey).
    Spain, N Africa. 20, 58*, 60*, 342(Feb 90). *Q. mirbeckii* Durieu.
castaneifolia C.Meyer   **Chestnut-leaved Oak**
    ● A few saplings in scrub at the top of the Black Rock Gully, Clifton Down
    (W Gloucs), probably bird-sown. Caucasus, Iran. 20, 58*, 59*, 60*, 61*,
    303(50:30). According to 478 this is an error for *Q. ilex*.
cerris L.   **Turkey Oak**
    ●●●●● Introduced; reproducing and well-naturalised in many places, especially
    on light soils in the south. S Europe, Turkey. 1, 20*, 58*, 59*, 60*.
    Cut-leaved variants, now known as forma *laciniata* (Loudon) C.Schneider, and
    a hybrid with native *Q. robur* have been recorded. RNG.
coccinea Muenchh.   **Scarlet Oak**
    ●● Introduced; self-sown seedlings in a churchyard at Ashtead (Surrey).
    Eastern N America. LNHS. 12, 20, 58*, 59*, 60*, 118, 119, 191.
ilex L.   **Evergreen Oak**
    ●●●●● Introduced; reproducing and well-naturalised in many places in the
    south, sometimes extensively so near the coast. Medit. 1, 20*, 58*, 59*, 60*,
    189, 370(13:271-286), 1097*. This species is locally becoming a threat to
    native vegetation. A putative hybrid with *Q. cerris* has been recorded from
    Moreton (Dorset). RNG.
imbricaria Michaux   **Shingle Oak**
    ● Introduced. Eastern N America. 50*, 58*, 59*, 60*, 119.
mas Thore
    ● Introduced and possibly established; seedlings reported in 1992 from along
    naturally wooded roadsides near Stone and Bean (both W Kent); putative
    hybrids with native *Q. petraea* were also present.   N Spain. 21, 61,
    303(62:44), 331(72:115), 546a, 822*(Suppl.D. 1(5):t.7), 851(44:555-558).
    According to 536*, the correct name for this Spanish endemic is *Q. petraea*

(Mattuschka) Liebl. subsp. *huguetiana* Franco & G.López (*Q. mas* auct.), with
Q. mas Thore being a synonym of our native subsp. *petraea*.

palustris Muenchh.   Pin Oak
  • Introduced. Eastern N America. RNG, SLBI. 21, 50*, 58*, 59*, 60*.

phellos L.   Willow Oak
  • Introduced. Eastern N America. 50*, 58*, 59*, 60*, 353(23:34).

phillyreoides A.Gray
  • A garden escape, collected from a limestone outcrop in Marl Hall Wood,
  near Llandudno (?Caerns). China, Japan. RNG. 51*, 61, 631*(p.100; the
  entry on p.101 refers to *Q. phellos*), 649*. *Q. ilex* var. *phillyreoides* (A.Gray)
  Franchet.

× **pseudosuber** group   Lucombe Oak group
  •• (but confused with *Q. cerris*). Introduced; also self-sown in SE England.
  Originated in cultivation. LANC, OXF, RNG. 16, 20, 58*, 59*, 60*, 199,
  303(60:35). *Q. cerris* × *Q. suber* and backcrosses; incl. *Q.* × *hispanica* auct.,
  non Lam.; *Q.* × *lucombeana* Sweet, nom. nud.; *Q.* × *pseudosuber* Santi.

pubescens Willd.   Downy Oak
  • Introduced.   Europe, SW Asia.   21, 50*, 58*, 59*, 60*, 197, 438.
  Q. lanuginosa (Lam.) Thuill.

pyrenaica Willd.   Pyrenean Oak
  • Introduced. W Medit. 21, 50*, 59*, 60*, 119. *Q. toza* Touss.Bast.

rubra L. sensu lato   Red Oak, incl. Northern Red Oak
  •••• Introduced; reproducing and naturalised in several places in southern
  England and Wales. Eastern N America. 16, 20*, 58*, 59*, 60*, 126, 192,
  209. Incl. *Q. ambigua* Michaux f.; *Q. borealis* Michaux f.; *Q. maxima*
  (Marshall) Ashe.

suber L.   Cork Oak
  • Introduced. Medit. LANC. 21, 50*, 58*, 59*, 60*, 128, 191, 201, 403.
  Incl. *Q. occidentalis* Gay.

velutina Lam.   Black Oak
  • Recently recorded from Petworth (W Sussex). Eastern N America. 50*,
  58*, 59*, 60*, 122.

BETULACEAE
Corylaceae

BETULA L.

[*forrestii* (W.Smith) Hand.-Mazz. A dubious record; the true species may not be
  in cultivation. 119. *B. delavayi* Franchet var. *forrestii* W.Smith.]

papyrifera Marshall   Paper Birch
  • Introduced. Northern N America. 20, 21, 50*, 59*, 60*, 119.

pendula Roth **alien variants**   Swedish Birch
  Two cut-leaved variants have occurred as introductions: *B.*'Dalecarlica'
  (var. *dalecarlica* (L.f.) C.Schneider; *B. alba* L. var. *dalecarlica* L.f.) and
  B.'Laciniata' (var. *laciniata* (Wahl.) ined.; *B. alba* var. *laciniata* Wahl.).
  58*, 60*, 61, 61a, 458, 613*, 915*(51:417-436).

ALNUS Miller
cordata (Lois.) Duby   Italian Alder
•• Introduced; reproducing in W Kent, Surrey, Middlesex and E Norfolk. Corsica, Italy. BM. 20*, 58*, 59*, 60*, 212, 303(46:29), 331(65:198), 458. *Betula cordata* Lois.

incana (L.) Moench   Grey Alder
•••• Introduced; reproducing and naturalised on river banks, boggy ground and railway banks, especially in Ireland. Europe, W Asia. 2, 20*, 58*, 59*, 60*, 252, 352(45: 247-8), 1242*. *Betula alnus* L. var. *incana* L. A hybrid with native *A. glutinosa* (*A.* × *pubescens* Tausch) has been recorded. NMW. 259, 371(425:9), 1311.

maritima (Marshall) Nutt.   Seaside Alder
• Introduced. No modern records. Eastern USA. 38*, 40, 197.

orientalis Decne.   Oriental Alder
• Introduced. E Medit. 39*, 51, 61, 174.

rubra Bong.   Oregon Alder
• Introduced. Western N America. 41*, 50*, 59*, 60*, 119. *A. oregana* Nutt.

viridis (Chaix) DC.   Green Alder
• Introduced. Eurasia, N America. 13, 20, 21, 58*, 60*, 166, 174, 354(8:638). Incl. *A. crispa* (Aiton) Pursh.

CORYLUS L.
colurna L.   Turkish Hazel
• Introduced. SE Europe, SW Asia. 21, 60*, 61*, 166.

maxima Miller   Filbert
•• Introduced or a relic of cultivation; sometimes self-sown near gardens. SE Europe, SW Asia. LANC, RNG. 16, 20, 80*, 126, 199, 258, 312*(n.s. t.268), 606*, 684*. A hybrid with native *C. avellana* has been recorded. 258.

PHYTOLACCACEAE
PHYTOLACCA L.
acinosa group   Asian Pokeweeds
•••• Established garden escapes in a few places in the south; also introduced for pheasant food. Recorded in the Ipswich area (E Suffolk) at intervals since 1927, and for many years on Kew Green (Surrey). E Asia. 20, 50, 52*, 54*, 303(31:28; 32:22-23), 369(5:154; 26*:t.6), 371(370:29). *P. americana* auct., non L.; *P. decandra* auct., non L. A complex of intergrading species and possible hybrids, incl. *P. acinosa* Roxb.; *P. esculenta* Van Houtte; *P. latbenia* (Buch.-Ham.) H.Walter.

[*americana* L. (*P. decandra* L.)   Unconfirmed. Probably always in error for other species. 303(32:22-23).]

icosandra L.   Button Pokeweed
Pre-1930 only. C and S America. 348(20:21), 582*.

octandra L.   Inkweed
- A wool casual. Tropical America and Africa; (Australasia). LTN, RNG. 36c, 145, 519*, 520*, 666*, 725*.

polyandra Batalin
- (but probably overlooked as *P. acinosa* group). An established garden escape near Ipswich and at Felixstowe (both E Suffolk) and on Jethou (Channel Is). China. 51b, 312*(t.8978), 315*(1922:39), 325*(118:459-460), 369(29:41), 370(16:233), 600*, 732, 1152*, 1218. *P. clavigera* W.Smith.

NYCTAGINACEAE
MIRABILIS L.
jalapa L.   Marvel-of-Peru
- • A garden escape persisting for a few years on tips in the south. Mexico, tropical C & S America. RNG. 20, 36c*, 53*, 54*, 69*, 126, 201, 212, 220.

AIZOACEAE
Incl. *Tetragoniaceae*

GALENIA L.
africana L.
- A wool casual. S Africa. ABD, BM, E, LTN, OXF, RNG. 145, 349(4:469), 371(382:25), 592, 830*(64:53), 831*(34:269).

MESEMBRYANTHEMUM L.
crystallinum L.   Common Ice-plant
- • A wool and esparto casual; possibly also a garden escape. S Africa; (Medit., Australia). BM, LTN, RNG. 7, 14, 15, 21, 24*, 50, 44*, 286, 615*. *Cryophytum crystallinum* (L.) N.E.Br.

nodiflorum L.
- A wool casual. Medit., N & S Africa, SW Asia. BM, RNG. 17, 21, 44*, 546*, 683*, 737*. *Cryophytum nodiflorum* (L.) L.Bolus.

DELOSPERMA N.E.Br.
burtoniae L.Bolus
- A wool casual. S Africa. CGE, E, RNG. 17, 619.

[*davyi* N.E.Br.   In error for *D. burtoniae*. RNG.]

CLERETUM N.E.Br.
bellidiforme (Burman f.) G.D.Rowley   Livingstone-daisy
- A casual garden escape. S Africa. RNG. 50, 51*, 52*, 56*, 336(22:51), 371(382:24).   *Dorotheanthus bellidiformis* (Burman f.) N.E.Br.; *Mesembryanthemum bellidiforme* Burman f.; *M. criniflorum* L.f.

APTENIA N.E.Br.
cordifolia (L.f.) Schwantes   Heart-leaf Ice-plant
- • A garden escape established on banks and walls in Scilly and the Channel Is, apparently naturalised under coniferous trees near Le Château des Roches at St Brelade's Bay, Jersey (Channel Is); a casual elsewhere. S Africa;

(widespread). BM, CGE, E, K, RNG. 20*, 36c*, 37c*, 50, 117*, 201, 207, 220, 370(9:189; 17*:217-245), 590*. *Mesembryanthemum cordifolium* L.f.

RUSCHIA Schwantes
caroli (L.Bolus) Schwantes   **Shrubby Dew-plant**
  • A garden escape introduced and established on walls on St Mary's and Tresco (Scilly). S Africa. CGE, K, RNG. 20*, 207, 370(17:217-245). *Mesembryanthemum caroli* L.Bolus.

LAMPRANTHUS N.E.Br.
conspicuus (Haw.) N.E.Br.
  • Introduced on walls in Scilly. S Africa. 20, 50, 207, 208, 619*.
falciformis (Haw.) N.E.Br.   **Sickle-leaved Dew-plant**
  •• An established garden escape; naturalised on coastal rocks, cliffs and walls in several places in Scilly; introduced and long-naturalised in a disused quarry at The Gann, St Ishmael and on nearby sea cliffs at Dale Point (both Pembs). Plants at Three Castle Head (W Cork) and on sandy cliffs at Roches Point (E Cork) are probably also referable to this species. S Africa. BM, CGE, E, K, NMW, RNG. 20*, 50, 156, 207, 309(19:17), 370(17:217-245). *Mesembryanthemum falciforme* Haw.
roseus (Willd.) Schwantes   **Rosy Dew-plant**
  •• An established garden escape; naturalised by seed and by fragmentation on cliffs in the Channel Is and W Cornwall; also recorded for E and W Cork. S Africa. BM, CGE. 16, 20*, 32*, 50, 117*, 212, 213, 220, 370(13:169; 17:217-245), 590*, 704*. *L. multiradiatus* (Jacq.) N.E.Br.; *Mesembryanthemum multiradiatum* Jacq.; *M. roseum* Willd.
scaber (L.) N.E.Br.
  • Reported from woodland in Jersey (Channel Is), probably a garden escape. Formerly a ballast alien. S Africa. 7, 20, 50, 51, 286, 354(4:198), 458, 645*(t.1043), 822*(63:59), 1022*, 1218, 1287*. *L. falcatus* (L.) N.E.Br.; *L. glomeratus* (L.) N.E.Br.; *Mesembryanthemum falcatum* L.; *M. glomeratum* L.
spectabilis (Haw.) N.E.Br.
  • Introduced on walls in Scilly. S Africa. 36c*, 50, 55*, 207, 208, 680*. *Mesembryanthemum spectabile* Haw.

OSCULARIA Schwantes
[*caulescens* (Miller) Schwantes Unconfirmed. Recorded from Scilly, probably in error for *O. deltoides*. 370(17:228), 371(355:25). *Mesembryanthemum caulescens* Miller.]
deltoides (L.) Schwantes   **Deltoid-leaved Dew-plant**
  •• A garden escape established on rocks and walls in Scilly; also recorded for Guernsey (Channel Is). S Africa. CGE, RNG. 20*, 207*, 370*(17:217-245), 600*. *O. caulescens* auct., non (Miller) Schwantes; *Lampranthus deltoides* (L.) Glen; *Mesembryanthemum deltoides* L. According to 863(16:55), this genus is best considered to be monotypic.

DISPHYMA N.E.Br.

[*australe* (Sol. ex G.Forster) N.E.Br.   In error for *D. crassifolium.*
370(17:230).]

crassifolium (L.) L.Bolus   **Purple Dew-plant**
••• An established garden escape; naturalised and locally abundant on sea
cliffs, dunes and walls in Anglesey, E Sussex, W Cornwall, Scilly and the
Channel Is; also a wool casual. S Africa, Australia. BM, CGE, RNG. 20*,
37c*, 207*, 212, 220, 303(30:12), 370(9:45; 17:217-245). *D. australe* auct.,
non (Sol. ex G.Forster) N.E.Br.; *Mesembryanthemum crassifolium* L. Incl.
D. clavellatum (Haw.) Chinn.

DROSANTHEMUM Schwantes
attenuatum (Haw.) Schwantes
Pre-1930 only. Africa. OXF. 354(9:270), 619, 1029*. *Mesembryanthemum
attenuatum* Haw.; *D. striatum* (Haw.) Schwantes var. *attenuatum* Salm.

floribundum (Haw.) Schwantes   **Pale Dew-plant**
••• An established garden escape; on coastal rocks, banks and walls in Scilly,
the Lizard area (W Cornwall), and the Channel Is; also a wool casual.
S Africa. BM, CGE, E, LTR, RNG. 7, 14, 20*, 36c*, 37c*, 207*, 212, 220,
370(17:217-245), 1029*.   *Mesembryanthemum floribundum* Haw.; incl.
D. candens (Haw.) Schwantes (*M. candens* Haw.).

EREPSIA N.E.Br.
heteropetala (Haw.) Schwantes   **Lesser Sea-fig**
• A garden escape established in an old quarry on St Mary's (Scilly).
S Africa.   BM,   CGE,   RNG.   20*,   207*,   370(17:217-245).
Mesembryanthemum heteropetalum Haw.

CARPOBROTUS N.E.Br.
acinaciformis (L.) L.Bolus   **Sally-my-handsome**
• An established garden escape in Devon, Cornwall and Scilly. S Africa.
LTR. 20, 21, 29*, 683*. Most European records are in error for *C. edulis*
var. *rubescens*, see 370(17:217-245), but not all. 473. *Mesembryanthemum
acinaciforme* L. The illustrations quoted may not refer to this taxon.

[*aequilaterus* (Haw.) N.E.Br.   In error for *C. edulis* var. *rubescens*. 370(17:217-
245).]

[*chilensis* (Molina) N.E.Br.   In error for *C. glaucescens*.   369(22:47).
Mesembryanthemum chilense Molina.]

edulis (L.) N.E.Br.   **Hottentot-fig**
•••• An established garden escape; naturalised and locally abundant on cliffs,
dunes and rocks on the coasts of southern England, Wales and Ireland,
especially in Devon, Cornwall and the Channel Is. S Africa. 2, 16, 20*(as
C. acinaciformis), 21, 22*, 25*, 35*, 370*(17:217-245). *Mesembryanthemum
edule* L.   Var. *chrysophthalmus* C.D.Preston & Sell (*C. acinaciformis* sensu
Lousley, non (L.) L.Bolus) and var. *rubescens* Druce (*C. deliciosus* auct., non

L.Bolus; *M. aequilaterum* auct., non Haw.; *M. virescens* auct., non Haw.) have
been recorded.

glaucescens (Haw.) Schwantes   Angular Sea-fig
  •• An established garden escape; naturalised in the Channel Is, on cliffs at
  Bawdsey (E Suffolk) and at Port Logan (Wigtowns). Australia. CGE. 20*,
  117*, 201, 220, 369(22:47), 370(17:217-245), 569*, 590*. *C. virescens*
  (Haw.) Schwantes; *Mesembryanthemum glaucescens* Haw.

TETRAGONIA L.
crystallina L'Hér.
  Pre-1930 only. Peru. 313(13:302), 639*.

tetragonoides (Pallas) Kuntze   New Zealand Spinach
  ••• (but confused with *Spinacia oleracea* and *Beta vulgaris*). A persistent
  garden escape and bird-seed alien, mainly in southern England; established on
  sandy beaches in Scilly. SE Asia, Australasia, S America. BM, CGE, E, LIV,
  OXF, RNG. 7, 19, 20, 37*, 51*, 201, 520*, 569*, 666*. *T. expansa* Murray.

CHENOPODIACEAE
CHENOPODIUM L.
acuminatum Willd.
  • A casual recorded from Bristol, vector unknown. No modern records.
  N Asia. 9, 617*, 630*, 636*, 1115*.

album L. **alien variants**
  Many variants have been recorded but, as the variation is continuous, none
  seem worthy of separation; some may be hybrids with other species, but this
  remains unconfirmed. 3, 1020, 1311.

[*ambiguum* R.Br. (*C. glaucum* subsp. *ambiguum* (R.Br.) Murr & Thell.)
  Dubious records. E. 3, 7, 354(7:892; 8:415).]

ambrosioides L.   Mexican-tea
  ••• An established wool, bird-seed and oil-seed alien, perhaps also a garden
  escape; abundant along the River Lea Canal towpaths from Hackney Wick to
  Bromley-by-Bow (S Essex). Tropical and subtropical America; (widespread).
  BM, E, LTN, NMW, OXF, RNG. 1, 14, 18, 19, 20, 38*, 44*, 45*,
  303*(24:14-15), 370(15:432), 595*, 617*, 1242*. *C. chilensis* Schrader;
  C. vagans Standley; *Teloxys ambrosioides* (L.) W.A.Weber; *T. vagans*
  (Standley) W.A.Weber; incl. *C. anthelminticum* L. Var. *suffruticosum*
  (Willd.) Thell. (*C. suffruticosum* Willd.) has been recorded. E. 3.

antarcticum (Hook.f.) Hook.f.
  • A wool casual. Southern S America. E, RNG. 665*.

aristatum L.
  • A wool and oil-seed casual. No modern records. N & C Asia. K, SLBI.
  1, 14, 21, 43, 134, 139, 617*, 623a*, 649*, 683*, 695*. *C. virginicum* L.;
  Teloxys aristata (L.) Moq. Contrary to many authors, this species is only a
  casual in N America. 1042.

atriplicinum (F.Muell.) F.Muell.   Starry Goosefoot
- A wool casual. Australia. K, RNG. 14, 37*, 143, 590*, 1229*. *Scleroblitum atriplicinum* (F.Muell.) Ulbr.

auricomiforme Murr & Thell.
- A wool casual. No modern records. Eastern Australia. BM, E, LIV, LTN, OXF, RNG. 3*, 14, 37*, 143, 576, 590*.

berlandieri Moq.   Pitseed Goosefoot
- •• A grain and oil-seed casual. Formerly a wool alien. USA, Mexico. BM, E, LANC, LIV, OXF, RNG. 3, 12, 20*, 354(9:368), 589*, 607*, 652, 836*(20:310), 1242*. *C. platyphyllum* Issler; *C. zschackei* Murr. Putative hybrids with *C. album* (*C.* × *variabile* Aellen, BM, LANC, LIV, OXF; *C.* × *subcuneatum* Murr) and with *C. strictum* (*C.* × *pulchellum* Murr) have been recorded.

[*bonus-henricus* L. Accepted, with reservations, as native.]

borbasioides A.Ludwig ex Asch. & Graebner
Pre-1930 only. Argentina. OXF. 354(9:35; 576), 595*, 836(20:310). *C. zobelii* A.Ludwig & Aellen, non Murr.

botrys L.   Sticky Goosefoot
- •• A wool casual or garden escape. Eurasia, N Africa; (widespread). BM, E, LIV, NMW, OXF, RNG. 1, 14, 21, 29*, 38*, 50, 245, 331(43:19), 335(108:33), 370(18:422), 536*, 1190*, 1242*. *Teloxys botrys* (L.) W.A.Weber.

capitatum (L.) Asch.   Strawberry-blite
- ••• (but confused with *C. foliosum*). A wool, bird-seed and esparto alien persistent on cultivated land; decreasing. Formerly recurrent for over a century on a farm at Farnaght (Fermanagh). Origin obscure; (widespread). BM, DBN, LTN, NMW, OXF, RNG. 2, 15, 19, 20, 22*, 38*, 303(28:15), 318(12:271), 453, 810*(283-286), 1242*. *C. blitum* F.Muell.; *Blitum capitatum* L. Records for Caerns are in error for *C. foliosum*. 16, 156.

carinatum R.Br.   Keeled Goosefoot
- •• (but confused with *C. pumilio*). A wool casual. Australia; (widespread). BFT, BM, E, LTN, OXF, RNG. 3*, 14, 20*, 37, 354(13:272), 514*, 520*. Incl. *C. holopterum* (Thell.) Thell. & Aellen. Var. *holopterum* (Thell.) Aellen; var. *melanocarpum* (J.M.Black) Aellen; and hybrids with *C. cristatum* and *C. pumilio* have been recorded.

cristatum (F.Muell.) F.Muell.   Crested Goosefoot
- •• A wool casual. Australia. BFT, BM, CGE, E, LTN, RNG. 12, 14, 20*, 37*, 133, 143, 370(1:242), 520*. A hybrid with *C. carinatum* (*C.* × *bontei* Aellen) has been recorded. RNG. 303:(20:9).

detestans Kirk   Fish-guts
Pre-1930 only. New Zealand; (Australia). 7, 36, 36c, 212, 354(7:780), 590*, 906(4:135-262).

foliosum (Moench) Asch.   Strawberry Goosefoot
- • (but confused with *C. capitatum*). A grain and esparto casual. Eurasia, N Africa. BM, CGE, K, NMW, OXF, SLBI. 1, 15, 21, 22*, 29*, 50,

354(13:167), 810*(283-286), 1190*, 1242*. *C. virgatum* (L.) Ambr., non Thunb.; *Blitum virgatum* L.

giganteum D.Don   Tree Spinach
   ••• A wool and bird-seed casual; possibly also from oil-seed or a garden escape. N India. BM, E, K, LTN, OXF, RNG. 14, 19, 20*, 21, 50, 126, 354(8:637), 594*. *C. amaranticolor* (H.Coste & A.Reynier) H.Coste & A.Reynier; *C. album* subsp. *amaranticolor* H.Coste & A.Reynier.
[*glaucum* L. Accepted, with reservations, as native.]

graveolens Lagasca & Rodriguez
   Pre-1930 only (and confused with *C. schraderianum*). SW USA, Mexico, C America; (S America, Africa). NMW. 40, 156, 1042, 1162*. *C. foetidum* Moq. pro parte, non Schrader; *C. incisum* Poiret.
[*graveolens* Willd. In error for *C. schraderianum* and *C. ambrosioides*. 3, 9, 354(5:676; 9:577). *Teloxys graveolens* (Willd.) W.A.Weber.]

hircinum Schrader   Foetid Goosefoot
   ••• A wool casual. Eastern S America. BM, CGE, E, NMW, OXF, RNG. 3, 14, 20*, 653*, 695*, 810*(126:395-397), 1042, 1242*. Putative hybrids with *C. album* and with *C. strictum* (*C.* × *haywardiae* Murr) have been recorded. E. 3*.
[*hybridum* L. Accepted, with reservations, as native. The N American *C. gigantospermum* Aellen may have been overlooked as this species. 825(2:90).]

leptophyllum group   Slimleaf Goosefoots
   ••• Grain and wool casuals. N America. BM, CGE, LANC, NMW, OXF, RNG. 1, 14, 20*, 38*, 41*, 182, 1242*. Incl. *C. dessicatum* A.Nelson; *C. leptophyllum* (Nutt. ex Moq.) S.Watson (*C. album* var. *leptophyllum* Nutt. ex Moq.); *C. pratericola* Rydb.

macrospermum Hook.f.
   • A wool casual. Temperate S America. E, LTN. 145, 354(12:291), 665*, 738*. Incl. *C. halophilum* Philippi.

missouriense Aellen   Soya-bean Goosefoot
   • An oil-seed and bird-seed casual. Texas, Missouri. BM, K. 12, 18, 20*, 126, 236, 354(12:54), 371(346:28), 545, 595*, 726, 917*(5:255). *C. paganum* Standley, non Reichb.; incl. *C. bushianum* Aellen. Possibly conspecific with *C. album*. 816(60:603).

mucronatum Thunb.
   • A wool casual. No modern records. S Africa. RNG. 14, 554, 592, 865*(Ser.2, 4:t.5). *C. pseudoauricomum* Murr.

multifidum L.   Scented Goosefoot
   •• A wool and animal-feed casual, rarely persistent. C and S America; (widespread). BM, CGE, E, NMW, OXF, RNG. 1, 14, 20, 38*, 258, 546*, 1242*. *Roubieva multifida* (L.) Moq.; *Teloxys multifida* (L.) W.A.Weber.
[*murale* L. Accepted, with reservations, as native.]

nitrariaceum (F.Muell.) Benth.   **Nitre Goosefoot**
•• A wool casual. Australia. BFT, BM, CGE, E, LTN, RNG. 14. 20, 37*,
133, 146, 147, 303(47:34), 520*, 590*, 666*, 825*(1:20-24). *Rhagodia*
nitrariacea F. Muell.

opulifolium Schrader ex Koch & Ziz   **Grey Goosefoot**
•••• A persistent grain, bird-seed, oil-seed and wool casual; decreasing.
Eurasia, N Africa. 1, 18, 19, 20*, 22*, 44*, 126*, 182, 1242*. Putative
hybrids with *C. album* or *C. strictum* (*C.* × *borbasii* Murr; *C.* × *preissmannii*
Murr; *C.* × *treasuricum* Scholz; *C.* × *tridentatum* Murr; *C.* × *vachelliae*
Druce; *C.* × *wheldonii* Murr) and with native *C. ficifolium* have been recorded.
LANC, OXF. 1311. Recent opinion is that *C.* × *borbasii* is not a hybrid but
an overlooked species. 822(102:351-373).

pallidicaule Aellen
Pre-1930 only.   Bolivia, Peru, Argentina.   E, OXF.   354(8:637; 9:576),
820*(19:548), 822(26:126), 912*(5:225; 10:302-304,354). The Selkirk plant
was originally misnamed *C. hircinum* var. *subtrilobum* Issler. 1264.

[*paniculatum* Hook.   In error for *C. hircinum* and *Einadia trigonos*. E. 3,
354(5:51; 6:145,395; 9:35,577).]

phillipsianum Aellen
• A wool casual. No modern records. S Africa. RNG. 14, 595*.

probstii Aellen   **Probst's Goosefoot**
•••• A wool, esparto, cotton, grain, bird-seed and oil-seed casual sometimes
persisting, as at Stonehill Green (W Kent). N America. 14, 18, 19, 20*,
126*, 316(20:161-3), 595*. Possibly conspecific with *C. album*.

procerum Hochst. ex Moq.
• A wool casual. No modern records. E Africa. K. 14, 618, 676, 736*,
1277*, 1335*. *C. botrys* var. *procerum* (Hochst. ex Moq.) C.B.Clarke.

pumilio R.Br.   **Clammy Goosefoot**
••• A persistent wool alien sometimes occurring in abundance. Australia;
(widespread). BM, CGE, E, LTR, OXF, RNG. 14, 20*, 21a, 37*, 143,
303*(20:8-9), 354(13:272-3), 370(8:336). *C. carinatum* auct., non R.Br.;
Teloxys pumilio (R.Br.) W.A.Weber.   A hybrid with *C. carinatum*
(*C.* × *christii* Aellen) has been recorded.

purpurascens Jacq.   **Purple Goosefoot**
• A casual escape, just outside Oxford Botanic Garden (Oxon). China. 43,
50, 51, 51b, 312*(t.5231), 371(428:7), 824*(1887:640).   *C. album* var.
purpurascens (Jacq.) Makino; *C. atriplicis* L.f.; *C. elegantissimum* Koidz.

quinoa Willd.   **Quinoa**
• An escape from cultivation recently recorded from a field edge at Shamley
Green (Surrey).   Formerly a wool alien.   Andes.   OXF.   7, 20, 50, 51,
303(54:52), 312*(t.3641), 325*(117:511), 342(Feb.1991), 354(9;134,577),
820*(19:545), 912*(10:309-317), 1364*. Now grown on farms to provide food
for game birds, so more records are to be expected. 1330.

rostratum Aellen ex Blom
Pre-1930 only. ?Sweden. 354(10:481).

rugosum Aellen
- A casual at Hanwell (Middlesex), vector unknown. No modern records. N Asia. K. 12, 630, 822(25:215). Probably conspecific with *C. album*.

schraderianum Schultes
- •• A wool casual. No modern records. E Africa; (widespread). BM, E, K, LTN, OXF, RNG. 3(as *C. graveolens*), 14, 21, 143, 354(9:577), 617*, 618*, 736*. *C. foetidum* Schrader, non Lam.; *Teloxys schraderiana* (Schultes) W.A.Weber.

solitarium Murr
Pre-1930 only. ?Europe. 7, 354(7:451), 873*(19:50).

strictum Roth   Striped Goosefoot
- •• A wool, oil-seed, bird-seed and grain casual. Europe. BM, LIV, LTN, MNE, NMW, RNG. 3, 8, 14, 18, 19, 20*, 21, 22*(fig. atypical), 589*, 822*(104:7-8), 1190*. *C. striatum* (Krašan) Murr. Putative hybrids with *C. berlandieri* and *C. hircinum* have been recorded.

subpalmatum Murr ex Druce
Pre-1930 only. Asia. OXF(holotype). 7, 354(7:780).

suecicum Murr   Swedish Goosefoot
- •• (but confused with *C. album*). A wool and grain casual; possibly also a bird-seed alien. Eurasia. BM, LIV, NMW, OXF. 3, 19, 20*, 22*, 802*(108:1-35). *C. album* forma *pseudopulifolium* Scholz; *C. pseudopulifolium* (Scholz) Nyár.; *C. viride* auct., non L. A putative hybrid with *C. album* (*C. × fursajewii* Aellen & Iljin) has been recorded. OXF. 1311.

[*urbicum* L. Accepted, with reservations, as native.]

[*virgatum* Thunb. (*C. album* subsp. *virgatum* (Thunb.) Blom)   In error for *C. foliosum*. 156.]

DYSPHANIA R.Br.

glomulifera (Nees) Paul G.Wilson   Australian Pigweed
- A wool casual. No modern records. Australia. E, K, RNG. 14, 37*, 520*, 576*, 743, 1074*(as *D. littoralis*).   *D. myriocephala* Benth.; *Chenopodium myriocephalum* (Benth.) Aellen.

littoralis R.Br.   Red Crumbweed
- A wool casual. No modern records. Australia. BM, RNG. 14, 37*, 514*, 520*, 576*, [1074*]. *Chenopodium blackianum* Aellen.

EINADIA Raf.

[*hastata* (R.Br.) A.J.Scott   In error for *E. trigonos*. 14, 236. *Rhagodia hastata* R.Br.]

polygonoides (Murr) Paul G.Wilson   Knotted Goosefoot
- A wool casual. No modern records. Australia. K. 14, 37c*, 514, 576, 590*. *Chenopodium polygonoides* (Murr) Aellen; *C. triangulare* R.Br. var. *polygonoides* Murr.

trigonos (Schultes) Paul G.Wilson subsp. **stellulata** (Benth.) Paul G.Wilson
- A wool casual. Australia. BM, RNG. 36c, 371(334:22), 576, 590*. *Chenopodium stellulatum* Aellen.

trigonos (Schultes) Paul G.Wilson subsp. **trigonos**   Fishweed
Pre-1930 only. Australia. OXF. 7(as *Chenopodium paniculatum*), 354(9:577),
567*, 743, 1010*, 1229*. *Chenopodium triangulare* R.Br., nom. illeg., non
Forsskål; *C. trigonon* Schultes.

SCLEROLAENA R.Br.
anisacanthoides (F.Muell.) Domin
• A wool casual. No modern records. Australia. K. 14, 37*. *Bassia
echinopsila* (F.Muell.) F.Muell.; *Echinopsilon anisacanthoides* F.Muell.
birchii (F.Muell.) Domin   Galvanized Bur
• A wool casual. Australia. K, RNG. 14, 37*, 145, 590*. *Bassia birchii*
(F.Muell.) F.Muell.
divaricata (R.Br.) Domin   Tangled Bassia
• A wool casual. Australia. RNG. 17, 37*, 520*, 590*, 743. *Bassia
divaricata* (R.Br.) F.Muell.
muricata (Moq.) Domin   Five-spined Bassia
• A wool casual. No modern records. Australia. BM, E, K, OXF, RNG,
SLBI. 3, 14, 37*, 354(5:51,677), 514*, 590*, 743. *Bassia quinquecuspis*
(F.Muell.) F.Muell.; *Chenolea quinquecuspis* F.Muell.

BASSIA All.
hyssopifolia (Pallas) Kuntze
• A wool casual. Eurasia; (N America). E. 21, 38a*, 41*, 50, 536*.
cf. **muricata** (L.) Asch.
• An esparto casual; also collected in 1942 from a gun-site on Southsea
Common (S Hants). No modern records. N Africa, Middle East. RNG. 44*,
46, 648*, 660*, 729*.
cf. **prostrata** (L.) A.J.Scott
• A casual, vector unknown. Eurasia, N Africa. RNG. 21, 39*, 46*.
Kochia prostrata (L.) Schrader.
scoparia (L.) Voss   Summer-cypress
••• A wool, grain, oil-seed and carrot seed casual; ornamental forms also
occur as garden escapes. E Europe, Asia; (widespread). BM, CGE, E, NMW,
OXF, RNG. 2, 14, 20, 22*, 39*, 50, 54*, 236, 264, 825(2:33-35).
Chenopodium scoparia L.; *Kochia scoparia* (L.) Schrader; incl. *K. densiflora*
Turcz.; *K. sieversiana* (Pallas) C.Meyer; *K. trichophylla* Schinz & Thell.

SPINACIA L.
oleracea L.   Spinach
••• (but confused with *Tetragonia tetragonoides* and *Beta vulgaris*). A casual
garden escape or relic of cultivation, and a bird-seed alien. Origin obscure,
perhaps W Asia. BM, CGE, E, LIV, OXF, RNG. 1, 19, 20, 29*, 80*, 536*.
S. glabra Miller; *S. inermis* Moench; *S. spinosa* Moench.

ATRIPLEX L.
australasica Moq.
- A wool casual. Australia. RNG. 37c*, 590*. *A. patula* L. var. *australasica* (Moq.) ined.

calotheca (Rafn) Fries
Pre-1930 only. Confirmed only from Co Durham. Scandinavia, Baltic. SUN. 21, 22*, 166, 300(1899:33,101,119,121), 354(9:279-280). Records from Scotland remain dubious.

eardleyae Aellen   Bell Saltbush
- A wool casual. No modern records. Australia. LTN, OXF, RNG. 7, 14, 37*, 143, 354(6:621), 520*, 590*, 743. *A. campanulata* Benth., nom. illeg., non Woods.

glauca L.
Pre-1930 only. Portugal, Spain, N Africa, Middle East. BM. 21, 46*, 288(p.lxii). Incl. *A. bocconei* Guss.

halimus L.   Shrubby Orache
- ●●● Introduced and persisting for many years near the sea in southern England; perhaps established by layering in Jersey (Channel Is). Medit. BM, CGE, JSY, LANC, OXF, RNG. 2, 20*, 39*, 44*, 201, 258.

hortensis L.   Garden Orache
- ●●●● A garden escape and bird-seed alien, sometimes persistent. Asia. 1, 20*, 22*, 29*, 39*, 283, 1242*. Var. *rubra* (Crantz) Roth (var. *atrosanguinea* hort.) has been recorded.

limbata Benth.   Spreading Saltbush
- A wool casual. Australia. RNG. 17, 37*, 590*, 666*, 743.

micrantha Ledeb.
- ●● A wool and cotton casual. Russia; (N America). BM, E, K, RNG. 17, 21, 133, 236, 303(18:11), 595*, 825*(4:103-108), 1017*. *A. heterosperma* Bunge; *A. hortensis* subsp. *heterosperma* (Bunge) Meijden.

muelleri Benth.   Annual Saltbush, incl. Australian Orache
- ●● A casual wool alien, rarely persistent. Australia; (S Africa). ABD, BM, CGE, E, OXF, RNG. 7, 14, 20*, 303*(21:14-15), 520*, 595*, 825*(4:103-108). Incl. *A. suberecta* Verdoorn.

oblongifolia Waldst. & Kit.
Pre-1930 only. Europe, N Africa, SW Asia; (N America). CGE, LIV, OXF. 21, 22*, 354(10:540), 1017*.

aff. **platensis** Speg.
- A casual on arable land, vector unknown. Argentina. BM.

rosea L.   Redscale
- A casual on tips, vector unknown. Eurasia, N Africa; (Australia, N America). BM, LANC, OXF, RNG. 1, 21, 22*, 29*, 44*, 126, 151, 192, 201, 320(15:336), 825*(4:103-108), 1017*. *A. sinuata* Hoffm. Early records were confused with native *A. glabriuscula*.

sagittata Borkh.   Purple Orache
  •• A persistent garden escape and wool alien. Eurasia. BM, RNG, SLBI. 1,
  14, 20, 21, 39*, 143, 191, 565*, 623*, 695*, 741*. *A. acuminata* Waldst. &
  Kit.; *A. nitens* Schk., nom. illeg.
cf. **semibaccata** R.Br.   Berry Saltbush
  • A wool casual. Australia; (widespread). BM, LTN, RNG. 14, 37*, 44*,
  143, 520*, 1017*, 1260*.
semilunaris Aellen
  • A wool casual. Australia. RNG. 17, 520*.
spongiosa F.Muell.   Pop Saltbush
  Pre-1930 only. Australia. BM, CGE, E, OXF, RNG, SLBI. 3*, 37*, 144,
  354(1908:396), 520*, 743.   *Senniella spongiosa* (F.Muell.) Aellen; incl.
  A. holocarpa F.Muell.
cf. **stylosa** Viv.
  Pre-1930 only. N Africa, SW Asia. 7, 44*, 354(8:35). *A. palaestina* Boiss.
tatarica L.
  • A wool casual; recently reported from West Ardsley (SW Yorks) and the
  Crown Wallpaper tip at Darwen (S Lancs). Eurasia. BM, BON, LIV, NMW,
  OXF, RNG. 1, 21, 22*, 39*, 44*, 354(12:197,502), 448, 455, 825*(4:105),
  1242*. *A. veneta* Willd. Early records were confused with *A. oblongifolia*.

AXYRIS L.
amaranthoides L.   Russian Pigweed
  •• A grain casual. Russia, N Asia; (N America). BM, CGE, LANC, NMW,
  OXF, RNG. 7, 20, 38*, 182, 303(21:14), 1242*. *Atriplex amarantoides*
  Gmelin ex Moq.; *Oxyria amaranthoides* sphalm.

BETA L.
trigyna Waldst. & Kit.   Caucasian Beet
  •• An agricultural seed alien established for more than 40 years by the
  Cambridge to Babraham road (Cambs) and long persistent as a relic of
  cultivation on the site of Warley Place gardens (S Essex); modern records are
  mostly from southern England. SE Europe, SW Asia. BM, LANC, MNE,
  OXF, RNG, SLBI. 1, 20, 31*, 126, 234, 623*, 683*, 741*, 825(5:37-38),
  1317*.
vulgaris L. **alien variants**   Cultivated Beets
  Cultivated forms, of obscure origin, occurring as escapes are subsp. *cicla*
  (L.) Arcang. (Foliage Beet; incl. Spinach Beet, Seakale Beet, Swiss Chard) and
  subsp. *vulgaris* (Root Beet; incl. Beetroot, Mangel-wurzel, Mangold, Sugar
  Beet and forms with decorative foliage). 20.

CYCLOLOMA Moq.
atriplicifolium (Sprengel) Coulter   Winged Pigweed
  Pre-1930 only. N America. 7, 21, 38*, 40*, 354(2:507). *C. platyphyllum*
  Moq.; *Kochia atriplicifolia* Sprengel.

MONOLEPIS Schrader

nuttalliana (Schultes) E.Greene   **Povertyweed**
> • A wool casual. No modern records. N America, Siberia. E, LTN, OXF, RNG. 3, 14, 38*, 40, 41*, 42, 143, 354(12:292), 1143*. *M. chenopodioides* (Nutt.) Moq.; *M. nuttallii* Engelm.; incl. *M. trifida* Schrader.

CORISPERMUM L.

hyssopifolium L.   **Hyssop-leaved Bug-seed**
> Pre-1930 only, perhaps in error for *C. leptopterum*. Eurasia; (N America). 1, 21, 38*, 40*, 41*, 247, 354(8:756; 9:759).

leptopterum (Asch.) Iljin   **Corispermum**
> • Persisted for a few years in abundance on sea-sand deposited on railway sidings at Emscote (Warks); also recently collected from a sandy foreshore at Margam Burrows (Glam) and at Bathside Bay (N Essex); the vector unknown in all cases, but a spread from NW Europe seems likely. S Europe. CGE, CLR, MNE, NMW, WARMS. 21, 22*, 129, 303*(62:41-42), 305*(54:1,25), 371(340:26). *C. hyssopifolium* auct. eur. centr., non L.

SUAEDA Forsskål ex J.Gmelin

altissima (L.) Pallas
> • A casual on Port Meadow at Oxford, vector unknown. No modern records. SE Europe, Asia. SLBI. 1, 21, 244, 247, 300(1909:43), 354(12:292), 630, 636*, 695*, 1110*, 1172*, 1245*, 1282*. *Dondia altissima* (Pallas) Druce. All plants from Spain belong to the closely allied species *S. spicata* (Willd.) Moq., which may have been overlooked. 536*.

SALSOLA L.

collina Pallas
> • A casual, vector unknown. No modern records. Russia, Asia. 21, 354(12:292), 617*, 1242*.

kali L. subsp. **ruthenica** (Iljin) Soó   **Spineless Saltwort**
> •••• A persistent wool, grain, agricultural seed and bird-seed alien. Eurasia, N Africa; (widespread). 7, 14, 19, 20, 38*, 39*, 235, 598*, 1279*. *S. australis* R.Br.; *S. calvescens* Grenier; *S. kali* var. *tenuifolia* Tausch; *S. pestifer* Nelson; *S. ruthenica* Iljin; *S. tragus* auct., non L.

kali L. subsp. **tragus** (L.) Nyman
> • A grain casual. S Europe, SW Asia. 21, 39*, 264. *S. tragus* L. Early records were confused with subsp. *ruthenica*.

soda L.
> Pre-1930 only. Medit. 2, 21, 29*, 44*, 151, 300(1905:102), 318(15:148), 546*, 683*.

POLYCNEMUM L.

arvense L.
> Pre-1930 only. Eurasia. 9, 21, 22*, 39, 354(8:637), 546*, 683*.

AMARANTHACEAE

CELOSIA L.

cristata L. Cockscomb

• A casual garden escape; possibly also a bird-seed alien. Tropics. 7, 19, 21, 50, 54*, 56*, 198, 220, 300(1909:43), 593*, 595*, 683*. *C. argentea* L. var. *cristata* (L.) Kuntze; incl. *C. plumosa* hort.

trigyna L.

Pre-1930 only. Tropical Africa. 9, 354(8:634), 517*, 577*, 676, 677*, 740*.

AMARANTHUS L.

acutilobus Uline & W.Bray

Pre-1930 only. Mexico. BM, NMW, OXF, RNG. 156, 354(13:271), 370(4:261-279); 14:173), 595*. *Euxolus emarginatus* Salzm., non A.Braun & Bouché. The specimens labelled as from St Helier's, Jersey (Channel Is) are of dubious origin.

albus L. White Pigweed

••• A persistent wool, grain, agricultural seed, oil-seed, bird-seed, esparto, tan-bark and cotton alien; known since 1934 on ash tips behind the Ford Motor Company works at Dagenham (S Essex). N America; (widespread). BM, E, K, NMW, OXF, RNG. 1, 14, 15, 18, 19, 20*, 22*, 126, 303(22*:17-20; 35:10).

aff. **arenicola** I.M.Johnston

• An oil-seed casual. N America. 18, 20, 44*, 236, 834*(13:5-46).

blitoides S.Watson Prostrate Pigweed

••• A wool, oil-seed, bird-seed, cotton and carrot seed alien, persistent in a few fields in East Anglia; increasing. N America; (Europe, S Africa). BM, CGE, E, LTN, OXF, RNG. 12, 14, 18, 19, 20*, 22*, 44*, 303*(22:17-20), 354(13:269-271), 370(2:51), 737*.

blitum L. Livid Amaranth

••• (but over-recorded for other species. A wool, bird-seed and coir-fibre alien, persistent in the Channel Is; decreasing. Eurasia, N Africa, Tropics. BM, CGE, E, NMW, OXF, RNG. 1, 19, 20, 21, 38*, 40*, 192, 370(4:261-279; 17:186), 371(334:22), 595*, 1247*. *A. ascendens* Loisel; *A. lividus* L.; *A. polygamus* auct., non L.; *A. polygonoides* L. The resurrection of the name *A. blitum* is confusing, as it has been widely applied to *A. graecizans*.

bouchonii Thell. Indehiscent Amaranth

••• (but probably overlooked as *A. hybridus*). An agricultural seed alien established and locally abundant on arable land in East Anglia; also a wool, spice and bird-seed casual. Origin obscure, perhaps derived from *A. hybridus*. BM, E, K, RNG. 19, 20*, 235, 236, 303(19:12; 22*:17-20; 58:42), 370(11:192), 595*, 825*(1:20-24), 826*(12:3-14). *A. hybridus* subsp. *bouchonii* (Thell.) O.Bolòs & Vigo; *A. buckmani* sphalm. A hybrid with *A. retroflexus* (*A.* × *ralletii* Contré ex Alliez. & Loiseaux) may have been overlooked. 874(34-35:146; 36:26-27).

capensis Thell.   Cape Pigweed
- •• A wool casual. S Africa. BM, CGE, E, LTR, OXF, RNG. 3*, 14, 20*, 303(33:10), 370(4:272; 14:241), 595*, 826*(7:75-80). *A. dinteri* auct., non Schinz. Probably all records are referable to subsp. *uncinatus* (Thell.) Brenan (*A. dinteri* var. *uncinatus* Thell.).

caudatus L.   Love-lies-bleeding
- ••• A garden escape and bird-seed casual. Tropics. BM, CGE, E, NWH, OXF, RNG. 2, 19, 20*, 54*, 303*(22:17-20), 600*. Described in 13 under the name *A. cruentus* L.

clementii Domin
- • A wool casual. Australia. RNG. 14, 370(4:277), 595*. *A. pallidiflorus* sensu Aellen, non F.Muell.

crassipes Schldl.
- • A wool casual. Tropical America. CGE. 303(10:13), 595*.

crispus (Lesp. & Théven.) Terracc.   Crisp-leaved Amaranth
- •• A wool casual. S America; (S Europe). BM, E, RNG. 14, 21, 38*, 251, 370(4:276), 595*, 683*.

deflexus L.   Perennial Pigweed
- ••• A wool, tan-bark and granite alien; established in a few places in the Channel Is and since 1981 in a plant nursery at Ufford (E Suffolk). Temperate S America; (widespread). BM, CGE, E, JSY, OXF, RNG. 1, 14, 20*, 38*, 201, 220, 303(22*:17-20; 35:10), 323(29:22), 369*(29:70-72). *A. prostratus* Balb.; *Euxolus deflexus* (L.) Raf.

[*dinteri* Schinz   In error for *A. capensis*. 303(33:10).]

graecizans L.   Short-tepalled Pigweed
- ••• (but over-recorded for *A. blitoides*). A wool and bird-seed casual; decreasing. Medit., SW Asia; (widespread). BM, CGE, E, NMW, OXF, RNG. 3*, 14, 19, 20*, 44*, 303*(22:17-20). *A. angustifolius* Lam.; *A. blitum* auct., non L.; *A. sylvestris* Villars.

hybridus L. sensu lato   Green Amaranth, incl. Purple Amaranth
- •••• A persistent wool, grain, agricultural seed, bird-seed, oil-seed, esparto, spice and pet-food alien; established on a fruit farm at Hersham (Surrey), in fields at Northfleet Green (W Kent) and undoubtedly elsewhere. Tropical & N America; (widespread). 1, 3*, 14, 15, 18, 19, 20*, 44*, 199, 303*(22:17-20), 458. Incl. *A. chlorostachys* Willd.; *A. cruentus* L.; *A. hybridus* subsp. *incurvatus* Timeroy ex Gren. & Godron; *A. hypochondriacus* L.; *A. paniculatus* L.; *A patulus* Bertol.; *A. powellii* S.Watson; *A. sanguineus* hort.; *A. speciosus* Sims. A hybrid with *A. retroflexus* (*A.* × *ozanonii* (Thell.) C.Schuster ex Goldschm.; *A.* × *adulterinus* Thell.) has been recorded. CGE, E, RNG. 19, 20. There is little agreement among authors as to how the segregate names should be applied and which are worthy of specific rank.

macrocarpus Benth.   Desert Amaranth
- • A wool casual. Australia. BM, E, OXF, RNG. 14, 37, 145, 303(10:13), 370(4:274), 514, 590*, 595*.

mitchellii Benth. Mitchell's Amaranth
- A wool casual. Australia. BM, K, RNG. 14, 37*, 145, 303(10:13), 370(4:276), 590*, 595*.

muricatus (Moq.) Hieron. Rough-fruited Amaranth
- A bird-seed casual. Temperate S America. BM, RNG. 19, 21, 303(10:13), 364(27:83), 595*, 648*, 737*, 820*(5:329-368), 826*(12:4).

palmeri S.Watson Dioecious Amaranth
- • A grain and oil-seed casual. SW USA, Mexico. BM, E, K, MNE, RNG. 14, 18, 20*, 38*, 44*, 236, 303(21:14), 370(4:278), 825*(1:49-53).

quitensis Kunth Mucronate Amaranth
- • • A wool, oil-seed and bird-seed casual. S America. BM, HTN, LTN, NWH, OXF, RNG. 7, 14, 18, 19, 20*, 21, 354(5:50), 590*, 595*, 820*(5:329-368). Possibly conspecific with *A. caudatus*.

retroflexus L. Common Amaranth
- • • • An established wool, cotton, grain, carrot seed, oil-seed and bird-seed alien; by railways, roads and canals, especially in the south; sporadic and often abundant on cultivated land. Tropical & N America; (widespread). 1, 14, 19, 20*, 22*, 26*, 44*, 235, 303*(22:17-20). Incl. *A. delilei* Richter & Loret; *A. reflexus* sphalm. A hybrid with *A. hybridus*, q.v., has been recorded.

rudis J.Sauer Western Water-hemp
- A casual, possibly from oil-seed. N America. RNG. 38*, 370(4:277), 834(13*:5-46; 21:426-434). *A. tamariscinus* auct, non Nutt.; *Acnida tamariscina* auct., non A.Gray, nec (Nutt.) Wood. True *A.* × *tamariscinus* Nutt. is a sterile hybrid, probably of *A. hybridus* × *A. rudis* parentage. 834(13:5 46; 21:426-434).

schinzianus Thell.
- A wool casual. S Africa. K. 302(12:37), 830(47:451-492).

scleropoides Uline & W.Bray
- A wool casual; possibly also from bird-seed. Texas. BM, K, RNG. 19, 303(10:13), 595*.

spinosus L. Spiny Amaranth
- • A wool, grain, coir-fibre and oil-seed casual. Tropics. BM, CGE, E, LIV, NMW, RNG. 2, 14, 36c, 38*, 44*, 236, 303(42:18), 370(4:271), 595*, 1247*. Plants without leaf-axil spines may have been overlooked; they would be referable to the tropical American *A. dubius* Thell. 736*.

standleyanus Parodi ex Covas Indehiscent Pigweed
- • A wool, oil-seed and bird-seed casual. Argentina; (S Africa). BM, K, NMW, OXF, RNG. 9, 14, 18, 19, 20*, 370(4:276), 595*, 826*(12:3-14). *A. vulgatissimus* auct., non Speg.

thunbergii Moq. Thunberg's Pigweed
- • • A wool, hides, bird-seed and esparto casual. Tropical & S Africa. ABD, BM, CGE, E, OXF, RNG. 3, 14, 15, 19, 20*, 349(3:323), 370(4:271), 595*, 830*(47:451-492).

[*torreyi* (Gray) Benth. ex S.Watson In error for *A. rudis*. RNG. In N America
this name has been misapplied to *A. arenicola*, *A. bigelovii* Uline & Bray and
A. watsonii.]

tricolor L.   Tampala
- An oil-seed casual. Tropics. 50, 51, 55*, 303(47:36), 350(47:p.xxxi), 595*.
A. gangeticus L.; *A. tristis* L.

tuberculatus (Moq.) J.Sauer   Rough-fruited Water-hemp
- A casual at Avonmouth docks (W Gloucs), vector unknown. No modern
records. N America. BM, BRIST, RNG. 38*, 370(4:277), 834*(13:5-46).
Acnida tuberculata Moq.

viridis L.
- •• A wool, hides, bird-seed and oil-seed casual. Tropical America;
(widespread). BM, K, MNE, OXF, RNG. 9, 14, 18, 19, 37*, 44*,
349(3:323), 370(4:275), 520*, 546a*, 595*, 737*, 1247*. *A. gracilis* Desf.
[*vulgatissimus* Speg. In error for *A. standleyanus*. 9.]

watsonii Standley   Watson's Amaranth
- A casual at Avonmouth docks (W Gloucs), vector unknown. California,
Arizona, Mexico. K, RNG. 42, 370(4:278), 834*(13:5-46), 1042, 1303,
1357.

DIGERA Forsskål
muricata (L.) Martius
- A casual at Dagenham (S Essex), vector unknown. No modern records.
Tropical Africa & Asia. 44*, 47*, 48, 354(12:291). *D. alternifolia* (L.)
Asch.; *D. arvensis* Forsskål; *Achyranthes alternifolia* L.

ACHYRANTHES L.
aspera L.   Chaff-flower
- A wool and bird-seed casual. W Medit., S Africa; (widespread). E, K. 14,
19, 21, 44*, 47*, 48*, 370(9:188), 595*, 737*. Incl. *A. argentea* Lam.;
A. sicula (L.) All.

ALTERNANTHERA Forsskål
caracasana Kunth
- A wool casual. C America, West Indies; (widespread). BM, E, LTN, RNG.
21, 49, 145, 536*, 572, 737*(fig. atypical), 825*(1:2-6; 5:54-60).
A. achyrantha auct., non (L.) R.Br. ex Sweet; *A. peploides* (Kunth) Urban;
A. repens auct., non (L.) Link.

ficoidea (L.) Roemer & Schultes   Parrot-leaf
Pre-1930 only. Tropical & S America. OXF. 7, 9, 49, 50, 51b, 53, 66*,
354(8:35), 696*, 825*(5:54-60), 1174*. *Telanthera ficoidea* (L.) Moq.

nodiflora R.Br.   Joyweed
- A wool casual. Tropics; (Australia). OXF, RNG. 7, 14, 21, 37*,
354(6:620), 520*, 590*.

paronychioides A.St.Hil.
- A casual at Dagenham (S Essex), vector unknown. No modern records.
Tropical America. 49, 354(11:275), 572, 696*.

pungens Kunth
- A wool casual. S America; (widespread). HbEJC. 21, 536*, 825*(1:2-6), 1027. *A. achyrantha* (L.) R.Br. ex Sweet; *A. repens* auct., non (L.) Steudel.

sessilis (L.) R.Br. ex DC.   Rabbit-meat
- A wool casual. Tropics. E. 43, 44*, 49, 87*, 729*, 825*(1:2-6), 1109*, 1247*.

PORTULACACEAE

PORTULACA L.

grandiflora hort., non Hook.   Rose-moss
- A casual garden escape and horticultural seed impurity. Brazil, Uruguay, Argentina; (C Europe). BM, RNG. 12, 16, 21, 50, 51*, 55*, 66*, 312*(t.2885, t.3084), 354(8:610), 369(21:46), 371(407:40), 373(6:23), 511*. Most plants in cultivation appear to be of hybrid origin, involving *P. grandiflora* Hook. and other species.

oleracea L.   Common Purslane
- •••• A seed impurity, long persistent and sometimes abundant as a weed of cultivated land, mainly in Scilly and the Channel Is; a bird-seed, wool, cotton and tan-bark alien elsewhere; recently recorded as growing from the peat-like packing of imported Egyptian potatoes. Origin obscure; (widespread). 1, 14, 20, 22*, 29*, 44*, 201, 207, 303(35:10), 737*. Most records are probably referable to subsp. *oleracea*, but subsp. *nitida* Danin & H.G.Baker has been recorded as a wool casual. 302(10:51). *P. papulosa* Schldl. and *P. pilosa* L. may have been overlooked. 825*(4:189-190).

MONTIA L.

.parvifolia (Moçiño) E.Greene   Small-leaved Blinks
- Probably a garden escape, established on a steep bank of the river Cart to the south of Busby (Lanarks). Western N America. BM, E, GL, OXF. 2, 20, 41*, 42, 303(44:17-19), 698*. *Claytonia filicaulis* Douglas ex Hook.; *C. parvifolia* Moçiño.

CLAYTONIA L.

perfoliata Donn ex Willd.   Springbeauty
- ••••• An established alien; probably originally imported with garden plants and escaping from botanic and nursery gardens; now widely naturalised and locally abundant, especially on sandy soils in E and SE England; known at Spurn Head (SE Yorks) since 1902. Western N America. 1, 12, 20, 22*, 35*, 50, 52*, 134, 192. *C. parviflora* Douglas; *Montia perfoliata* (Donn ex Willd.) Howell.

sibirica L.   Pink Purslane
- ••••• An established garden escape; widely naturalised in damp woods and by streamsides, especially in the north and west; a persistent weed in gardens; increasing. Western N America, E Siberia. 1, 20, 22*, 25*, 35*, 50, 130, 230. *C. alsinoides* Sims; *Montia sibirica* (L.) Howell.

virginica L.   Virginia Springbeauty
   • A garden escape. No modern records. Eastern N America. 2, 16, 38*, 50,
   300(1907:38), 698*, 1319*.

CALANDRINIA Kunth
ciliata (Ruíz Lopez & Pavón) DC.   Red-maids
   •• A grain and seed alien; persistent on cultivated sandy ground in S Devon
   (last record 1955) and the Channel Is. Western N America, S America;
   (New Zealand). BM, E, K, RNG, TOR. 13, 20, 41*, 42, 50, 201, 217, 277,
   301*(1833:t.1598), 349(3:324), 500*, 519*. Incl. *C. menziesii* (Hook.) Torrey
   & A.Gray. *C. compressa* Schrader ex DC. and *C. elegans* Spach may have
   been overlooked, although the latter may perhaps be best considered
   conspecific. 825*(3:87-89).

umbellata (Ruíz Lopez & Pavón) DC.   Rock Purslane
   • A weed of cultivated ground on St Mary's (Scilly). Peru, Chile. 50, 51,
   54*, 208a, 594*, 674*, 824*(29:271), 1072*, 1152*.

BASELLACEAE
ANREDERA Juss.
cordifolia (Ten.) Steenis   Madeira-vine
   • A persistent or established garden escape on a roadside at St Peter Port,
   Guernsey (Channel Is). Subtropical S America. 20, 21, 50, 370(14:456),
   371(397:36), 546a*, 590*, 600*, 650*, 1218. *Boussingaultia baselloides* auct.,
   non Kunth; *B. cordifolia* Ten., non (Moq.) Volkens; *B. gracilis* Miers var.
   pseudobaselloides (Hauman) L.Bailey.

CARYOPHYLLACEAE
ARENARIA L.
balearica L.   Mossy Sandwort
   •••• An established garden escape; on walls, rocks and in turf in widely
   scattered localities; known on walls at Holywood (Co Down) since 1922.
   W Medit. 1, 20, 21, 22*, 30*, 50, 319(1:180), 429, 546*, 683*.
montana L.   Mountain Sandwort
   • A garden escape which persisted for several years at Stone (W Kent);
   recently reported from Guernsey (Channel Is). SW Europe. BM, OXF. 1,
   21, 29*, 30*, 50, 126, 461, 468, 546*. A 1954 record for Dartmoor remains
   unconfirmed. 302(4:7), 349(1:192).

LEPYRODICLIS Fenzl
holosteoides (C.Meyer) Fenzl ex Fischer & C.Meyer
   • A casual at Sharpness docks (W Gloucs), vector unknown. No modern
   records. Asia. BM, OXF, RNG. 7, 39, 182, 354(4:406), 604, 617*,
   826*(12:1-2; 13:18-19), 1125*. *Arenaria holosteoides* (C.Meyer) Edgew.;
   Gouffeia holostoides C.Meyer.
stellarioides (Willd. ex Schldl.) Fischer & C.Meyer
   Pre-1930 only. Caucasus, C Asia, Iran. 7, 136, 300(1910:43), 557, 730*.
   L. cerastioides Kar. & Kir.; *Arenaria stellarioides* Willd. ex Schldl.

MINUARTIA L.
laricifolia (L.) Schinz & Thell.
> Pre-1930 only. Europe. 21, 27*, 28*, 29*, 50, 187, 195, 354(5:750). *M. striata* (L.) Mattf.; *Alsine laricifolia* (L.) Crantz; *A. striata* (L.) Gren.; *Arenaria laricifolia* L. Confirmed as a garden escape; other old records were in error for native *Minuartia verna*.

[*rubra* (Scop.) McNeill   A dubious record. 187, 354(5:750). *Alsine fastigiata* auct., non (L.) Wahlenb.; *A. jacquinii* Koch; *Arenaria fastigiata* Smith; *A. fasciculata* sensu Jacq., non L., nec Gouan.]

[HOLOSTEUM L.
umbellatum L.   Accepted, with reservations, as native.]

CERASTIUM L.
[*brachypetalum* Pers.   Accepted, with reservations, as native.]
dichotomum L.
> • A grain and seed casual. Medit., W Asia. BM, NMW, OXF. 1, 21, 44*, 303(32:18), 354(13:108), 371(367:29), 825*(1:113-117), 826*(12:66-67).

[*latifolium* L.   In error for native *C. alpinum* (*C. latifolium* Lightf., non L.). 159, 187.]
ligusticum Viv.
> Pre-1930 only. C and E Medit. BM. 21, 39, 683*. Incl. *C. campanulatum* Viv.

scaranii Ten.   Italian Mouse-ear
> Pre-1930 only. Italy, Sicily. 7, 21, 23*, 267, 300(1907:232). *C. hirsutum* auct., non Crantz.

tomentosum group   Snow-in-summer
> ••••• Established garden escapes; naturalised on dunes, roadsides, railway banks and heathland in widely scattered localities throughout the British Is. Italy, Sicily. 7, 20, 21, 22*, 23*, 27*, 199, 236. A much-confused group variously recorded as *C. biebersteinii* DC.; *C. candidissimum* Correns; *C. decalvens* Schlosser Klek & Vukot.; *C. tomentosum* L.; *C. arvense* L. × *C. biebersteinii*; and *C. arvense* × *C. tomentosum* (*C.* × *maureri* Schulze, nom. nud.). According to 20, probably all records are referable to the very variable *C. tomentosum* L. or its hybrid with native *C. arvense*. 303(57:52).

SAGINA L.
subulata (Sw.) C.Presl 'Aurea'   Cemetery Pearlwort
> •• Introduced and established in turf in cemeteries and gardens, especially in the London area. Probably S Europe, or originated in cultivation. BM, RNG. 2, 50, 52*, 192, 209, 303(49:52). Incorrectly recorded as *S. subulata* var. *glabrata* Gillot, *S. glabra* (Willd.) Fenzl and *S. pilifera* (DC.) Fenzl.

CORRIGIOLA L.
telephiifolia Pourret
> • A persistent granite alien at Gloucester Docks (W Gloucs), now gone; also recorded as a casual in the Royal Botanic Gardens at Kew (Surrey). W Medit. HbEJC. 20, 21, 45*, 182*, 328(2:231-238), 370*(13:55-57).

[*TELEPHIUM* L.
imperati L.   A dubious record.   354(6:42).   Contrary to 7, 16 & 20, we can locate no correct record.]

PARONYCHIA Miller
argentea Lam.
- An esparto casual. Medit. NMW. 15, 21, 24*, 29*, 32*, 44*, 50, 156, 160, 737*.

argyrocoma (Michaux) Nutt.   Silverling
     Pre-1930 only. Eastern N America. BM. 38*, 40, 354(9:279), 1319*.
brasiliana DC.   Brasilian Nailwort
- A wool casual. S America; (Australasia). BM, CGE, E, K, OXF, RNG. 3, 14, 36c, 300(1909:43), 533*, 590*, 653*. *P. bonariensis* DC.

[*chilensis* DC. In error for *P. brasiliana*. BM. 7, 354(4:81).]
polygonifolia (Villars) DC.
     Pre-1930 only. S Europe. 3, 9, 16, 21, 354(4:428), 546*, 683*. Recent records claimed by 20 appear to be unfounded.

HERNIARIA L.
hirsuta L.   Hairy Rupturewort
     ••• A grain alien established on waste ground and railway sidings at Burton-on-Trent (Staffs); a wool, esparto and tan-bark casual elsewhere. Eurasia, N Africa; (Australia). BM, E, LANC, NMW, OXF, RNG. 2, 15, 20*, 22*, 44*, 303(35:10), 370(1:319), 826*(11:27-29), 1279*. Incl. *H. cinerea* DC.

LOEFLINGIA L.
hispanica L.
- An esparto casual. W Medit., Canary Is, SW Asia. K, RNG. 15, 21, 44*, 45*, 46*, 737*.

SPERGULA L.
morisonii Boreau   Pearlwort Spurrey
- Established since 1943 as a weed among ericaceous plants in a shrub nursery in Broadwater Forest (E Sussex); recently recorded as an abundant weed on an experimental farm at Derrybrennan, near Lullymore (Co Kildare), probably imported with blueberry plants; also reported from a nursery in W Kent and a kitchen garden at Kirklington (Notts). Europe. BM, DBN, RNG. 16, 20*, 22*, 26*, 174, 370(15:388), 374(1950:90), 432, 737*, 1242*. *S. pentandra* auct., non L.; *S. vernalis* auct., non Willd.

[*pentandra* L. Dubious records. 354(5:753).]

SPERGULARIA (Pers.) J.S.Presl & C.Presl
[*capillacea* (Kindb. & Lange) Willk. In error for *S. purpurea*. 303(10:16), 182.]

diandra (Guss.) Boiss.   Lesser Sand-spurrey
   • A wool and possibly an esparto casual. Eurasia, N Africa; (Australia).
   RNG. 14, 21, 44*, 45*, 46*.
nicaeensis Sarato ex Burnat
   • A casual on Crown Wallpaper tip, Darwen (N Lancs), probably an esparto
   alien. Medit. OXF, RNG. 7, 21, 354(7:996; 8:285), 546*, 683*.
purpurea (Pers.) Don
   • Persistent at Gloucester Docks (W Gloucs), now gone; also casual in 1931
   at Avonmouth Docks (both W Gloucs). SW Europe. BM, RNG. 9, 21, 30*,
   323(28:415), 546*, 683. *S. longipes* Rouy.

LYCHNIS L.
chalcedonica L.   Maltese-cross
   •• A persistent garden escape. Russia. BM, NMW, OXF. 7, 20, 21, 50,
   52*, 54*, 55*, 126, 331(41:11). *Silene chalcedonica* (L.) E.H.L.Krause.
coronaria (L.) Murray   Rose Campion
   •••• A persistent garden escape, sometimes abundant for a few years; on
   heaths, railway banks, sand dunes and waste ground on light soils in widely
   scattered localities. SE Europe, N Africa, SW Asia. 1, 20, 21, 31*, 50, 52*,
   55*, 331(54:64). *Agrostemma coronaria* L.; *Silene coronaria* (L.) Clairv.
flos-jovis (L.) Desr.   Flower-of-Jove
   • A casual garden escape. Alps. 16a, 21, 23*, 27*, 50, 52*, 54*.
   Agrostemma flos-jovis L.; *Silene flos-jovis* (L.) Greuter & Burdet.
fulgens Fischer ex Sims
   Pre-1930 only. Siberia, E Asia. 43, 51, 116, 312*(t.2104). *Silene fulgens*
   (Fischer ex Sims) E.H.L.Krause.
× **haageana** Lemaire
   • Recently reported from a beach at Shorne Marshes (W Kent). Originated in
   cultivation. 50, 54*, 312*(n.s. t.314), 458, 634*. Probably *L. coronata*
   Thunb. × *L. fulgens*.
[*preslii* Sekera   Pre-1930 only. Best considered to be a variety of native
   Silene dioica (*Melandrium rubrum* (Weigel) Garke var. *glaberrimum* Rohrb.).
   RNG. 354(3:13; 8:390; 9:107), 1311. A hybrid with *S. dioica* (*Lychnis* ×
   troweriae Druce) has also been recorded. 7, 354(7:377).]

AGROSTEMMA L.
githago L.   Corncockle
   •••• A grain alien, formerly locally common as a cornfield weed throughout
   much of the British Is, now rarely seen except as casuals from bird-seed or
   wild-flower seed mixtures; also a spice casual. Reported in 1991 as occurring
   in thousands on chalky arable ground at Turnworth (Dorset). E Medit. 1, 19,
   20, 21, 22*,26*, 35*, 1242*. *Githago segetum* Link; *Lychnis githago* (L.)
   Scop.
gracilis Boiss.
   • (but perhaps overlooked as *A. githago*). A large colony persistent near
   Swanley (W Kent) and a solitary plant at Castling's Heath, Girton (W Suffolk),
   probably introduced with wild-flower seed. Greece, Turkey. 21, 39, 44*,

303(55:33), 369*(41-42,t.4). The records may be referable to *A. githago* 'Milas'.

PETROCOPTIS A.Braun
pyrenaica (Bergeret) A.Braun
Pre-1930 only. Pyrenees. JSY, OXF. 7, 21, 23*, 27*, 30*, 50, 201, 354(8:302). *Lychnis pyrenaica* Bergeret.

SILENE L.
[*amoena* L. In error for native *S. uniflora*. 218(p.58), 383(1911:309).]
antirrhina L.   Sleepy Catchfly
• A grain casual. No modern records. Formerly a wool alien. N and S America. NMW, OXF. 3, 10, 38*, 40, 41*, 354(5:17,644), 1319*.
apetala Willd.
Pre-1930 only. S Europe. BM. 21, 683*, 737*. Incl. *S. viscaginoides* Hornem.
armeria L.   Sweet-William Catchfly
••• A casual garden escape and persistent garden weed. Eurasia. BM, E, LANC, OXF, RNG. 1, 20, 22*, 25*, 29*, 34*, 199, 303(48:36), 1242*.
behen L.
•• A grain and bird-seed casual. C and E Medit. BM, OXF. 2, 21, 32*, 39*, 44*, 46*, 182, 191.
bellidifolia Juss. ex Jacq.
Pre-1930 only. Medit. 21, 39*, 354(1:40), 737*. *S. hispida* Desf.; *S. vespertina* Retz.
catholica (L.) Aiton f.
Pre-1930 only. C Medit. BM. 1, 21, 180, 354(8:729), 683*.
coeli-rosa (L.) Godron   Rose-of-heaven
•• A casual or persistent garden escape. Medit. BM, E, NMW. 16, 20, 46*, 54*, 143, 192, 331(57:71), 354(11:468), 683*. *Agrostemma coeli-rosa* L.; *Eudianthe coeli-rosa* (L.) Reichb.; *Lychnis coeli-rosa* (L.) Desr.
conica L. subsp. **subconica** (Friv.) Gavioli
• A casual. No modern records. SE Europe. 2, 21, 39*, 354(12:270; 4:475). *S. subconica* Friv.; incl. *S. juvenalis* Del.
conoidea L.
• A grass seed casual. Eurasia; (N America). BM, CGE, NMW, OXF, RNG. 1, 20, 22*, 39*, 44*, 156, 220, 303(11:10), 826*(10:91-94; 13:20-21). Early records were confused with native *S. conica* (*S. conoidea* Hudson, non L.).
cordifolia All.   Heart-leaved Catchfly
• A casual garden escape. No modern records. Maritime Alps. OXF. 21, 23*, 354(9:554), 683*.
cretica L.
• A casual, probably from agricultural seed. Formerly a weed of *Trifolium incarnatum* crops. Medit. BM, CGE, LANC, OXF, RNG, SLBI. 1, 20, 21, 31*, 39*, 201, 235, 354(4:329), 683*. Incl. *S. annulata* Thore.

cserei Baumg.
Pre-1930 only. SE Europe, Turkey. BM, CGE, LIV, OXF, RNG. 7, 12, 20, 21, 39, 354(4:9), 630*, 715*, 1172*, 1242*. *S. fabaria* subsp. *cserei* (Baumg.) Nyman. Recent records implied by 20 appear to be unfounded.

dichotoma Ehrh.   Forked Catchfly
••• A persistent agricultural seed, bird-seed and wool alien; decreasing. Eurasia. BM, JSY, LIV, NMW, OXF, RNG. 1, 8, 13, 19, 20, 21, 22*, 25*, 209, 1242*. Incl. *S. racemosa* Otth.

fabaria (L.) Sibth. & Smith
Pre-1930 only. SE Europe. 1, 21, 39*, 139, 722*.

fimbriata Sims
• A persistent garden escape or relic; known for nearly 40 years in woodland at Ardrishaig (Kintyre), now gone; a relic of cultivation in woods near Port Ellen (S Ebudes). Caucasus. BM. 7, 20, 50, 137, 169, 224, 312*(t.908), 354(3:155). *S. multifida* (Adams) Rohrb., non Edgew. These records may be in error for the allied species *S. edgeworthii* Bocqet (*Lychnis fimbriata* Wallich).

aff. **holzmannii** Heldr. ex Boiss.
Pre-1930 only. Greece, Crete. BM. 21.

inaperta L.
• An oil-seed and bird-seed casual. Formerly a wool casual. W Medit. E, RNG. 3, 19, 21, 181, 285, 354(3:155), 546*, 683*, 695*, 737*.

integripetala Bory & Chaub.
Pre-1930 only. Greece. BM, CGE, RNG, SLBI. 21, 31*, 312*(n.s. t.328).

italica (L.) Pers.   Italian Catchfly
•• Established for over 100 years on roadsides and in chalkpits in the Dartford to Greenhithe area (W Kent); casual elsewhere. Medit. BM, CGE, LCN, NMW, OXF, RNG. 1, 20*, 22*, 29*, 33*, 126, 176. *S. patens* Peete; *Cucubalus italicus* L. Early records were much confused with native *S. nutans* (*S. dubia* Herbich).

laeta (Aiton) Godron
Pre-1930 only. W Medit. OXF, RNG. 7, 21, 267, 354(4:475; 6:16), 546*, 683*, 737*.

latifolia Poiret subsp. **latifolia**
• A casual at docks, vector unknown. Medit. CLE, NMW, OXF. 7, 16, 21, 156, 354(4:405; 11:431), 374(1951:48), 411, 694*. *S. alba* (Miller) E.H.Krause subsp. *divaricata* (Reichb.) Walters; *S. macrocarpa* (Boiss. & Reuter) E.H.Krause; *S. pratensis* (Rafn) Godron & Gren. subsp. *divaricata* (Reichb.) McNeill & Prent.; *Lychnis macrocarpa* Boiss. & Reuter; *Melandrium macrocarpum* (Boiss. & Reuter) Willk.

lydia Boiss.
Pre-1930 only. E Medit. 9, 21, 39*.

muscipula L.
•• A grain and bird-seed casual. Medit. BM, CGE, LIV, NMW, OXF, RNG. 1, 8, 19, 20, 21, 44*, 45*, 46*, 258, 336*(51-52).

nocturna L.   Mediterranean Catchfly
 • A wool and bird-seed casual. Medit.; (Australia). BM, E, K, LTN, RNG.
 3, 14, 19, 21, 39*, 44*, 45*, 46*, 145. Incl. *S. brachypetala* Robill. &
 Castagne ex DC.
pendula L.   Nodding Catchfly
 •• A casual or persistent garden escape. Medit. BM, E, K, LIV, OXF, RNG.
 1, 20, 56*, 143, 369(19:363), 506*, 650*.
portensis L.
 • An esparto casual. No modern records. SW Europe. 15, 21, 245, 546*,
 683*, 737*.
pseudatocion Desf.
 Pre-1930 only. S Spain, Balearic Is, NW Africa. CGE, OXF. 7, 21, 46*,
 354(7:434), 737*.
pusilla group   Alpine Catchflies
 • Introduced and persistent since before 1974 by Lochan na Lairige, below Ben
 Lawers (Mid Perth). Europe. BM, E. 20, 21, 23*, 27*, 50, 303(10:14;
 45:26). Incl. *S. alpestris* Jacq., *S. pusilla* Waldst. & Kit.; *S. quadrifida* L.;
 Heliosperma alpestre (Jacq.) Reichb.; *H. quadrifolium* sphalm. A pre-1930
 record from Angus is dubious. 354(5:747).
rubella L.
 • A casual, probably from grain or wool. No modern records. Medit. OXF,
 RNG. 1, 3, 21, 44*, 45*, 251, 300(1907:38), 354(5:643), 370(2:193).
schafta C.Gmelin ex Hohen.
 •• An established garden escape; long naturalised on a sandstone bank between
 High and Low Dalby, near Pickering (NE Yorks); casual or persistent
 elsewhere, occurring on old walls. Caucasus. BM, OXF. 7, 20, 50, 53*,
 54*, 303(46:28), 312*(n.s. t.336), 371(328:19; 388:36), 584*.
sedoides Poiret
 • A bird-seed casual. No modern records. Medit. 19, 21, 39*, 44*, 46*,
 160. The record for Newcastle (Staffs) appears to be an error for native
 S. gallica. RNG.
stricta L.
 Pre-1930 only. W Medit. OXF. 7, 9, 21, 354(6:275), 737*, 1361*.
 S. pteropleura Boiss. & Reuter.
tatarica (L.) Pers.
 • A casual. No modern records. N Europe, Siberia. 21, 22*, 25, 251,
 1242*.
tridentata Desf.
 Pre-1930 only. W Medit., Canary Is. RNG. 21, 44*, 45*, 737*.
trinervia Sebast. & Mauri
 Pre-1930 only. Balkans. 21, 137(p.79), 184, 354(8:325), 683*.
vulgaris Garcke subsp. **angustifolia** (Miller) Hayek
 Pre-1930 only. Medit., Portugal. 9, 313(22:34). *S. angustifolia* Miller.
vulgaris Garcke subsp. **commutata** (Guss.) Hayek
 • A casual on a roadside at Kew Bridge (Middlesex). S Europe. 21, 176,
 331(54:65), 354(9:337), 683. *S. commutata* Guss.

vulgaris Garcke subsp. **macrocarpa** Turrill   Plymouth Catchfly
  • Long established on Plymouth Hoe (S Devon). Cyprus. K, RNG. 16, 20,
  21, 189, 317*, 349(2:236), 1279*. *S. angustifolia* auct., non (DC.) Guss.;
  S. linearis auct., non Sweet.

CUCUBALUS L.
baccifer L.   Berry Catchfly
  •• Established or possibly even native in woods in the Merton to Great
  Hockham area (W Norfolk), known since before 1914; recently recorded as a
  garden escape from lane hedges near Langport (N Somerset); casual elsewhere.
  Eurasia, N Africa. BM, K, LANC, LINN, OXF, RNG. 1, 20, 22*, 33*, 34*,
  235, 303(34:34), 354(4:188), 370(14:421).

GYPSOPHILA L.
acutifolia Steven ex Sprengel
  • A casual, probably a garden escape. W Ukraine, Caucasus; (China). RNG.
  21, 51, 248, 623a*, 630, 695*, 1115*, 1267*.
elegans M.Bieb.   Annual Baby's-breath
  •• A casual garden escape. Turkey, Caucasus. BM, NMW, OXF, RNG. 7,
  20, 50, 53*, 191, 650*, 683*, 826*(14:15-17), 1242*.
muralis L.   Annual Gypsophila
  • A casual, probably a garden escape. Eurasia. BM. 1,9,16, 21, 22*, 25*,
  26*, 165, 320(15:78), 331(70:159), 546*, 1242*.
paniculata L.   Baby's-breath
  •• A persistent garden escape; known for 30 years at Farnham near
  Saxmundham (E Suffolk). C Europe, W Asia. BM, CGE, LCN, LIV, OXF,
  RNG. 1, 20, 22*, 31*, 50, 310(13:20), 369(22:48).
pilosa Hudson
  Pre-1930 only. SW Asia. BM, LCN, LIV, OXF. 1, 9, 16, 21, 39, 44*,
  826*(14:12-14), 921*(9:1-203). *G. porrigens* (L.) Boiss.
repens L.   Alpine Gypsophila
  • A casual garden escape. C and S Europe. 21, 22*, 50, 54*, 458, 546*.
scorzonerifolia Ser.
  • A casual garden escape. No modern records. Russia, Asia. 21, 51,
  350(28:386), 354(12:270), 826*(11:21-26). *G. acutifolia* var. *latifolia* Fenzl.
viscosa Murray
  • A casual at Bristol, vector unknown. Middle East. 1, 39, 44*, 182, 192,
  245, 856(1:16-18), 921*(9:1-203).

SAPONARIA L.
calabrica Guss.
  • A casual, probably a garden escape. No modern records. S Europe. HTN,
  OXF. 21, 31*, 32*, 50, 354(4:475; 10:468).
glutinosa M.Bieb.
  Pre-1930 only. Medit. LSR. 2, 21, 31*, 184, 737*.

ocymoides L.   Rock Soapwort
•• A persistent garden escape; mainly on walls in southern England.
S Europe. BM, NMW, OXF, RNG. 7, 20, 22*, 23*, 27*, 29*, 303(33:12),
370(12:350).
[*officinalis* L.   Accepted, with reservations, as native.]
orientalis L.
Pre-1930 only. SW Asia. 7, 21, 354(3:13; 9:337), 550*, 695*, 718*.
Proteina orientalis (L.) Ser.

VACCARIA Wolf
hispanica (Miller) Rauschert   Cowherb
•••• A grain, agricultural seed, bird-seed, spice, esparto and wool casual, and
a garden escape. Eurasia, N Africa; (Australia). 1, 9, 15, 19, 20, 22*, 25*,
44*, 354(13:107), 546*, 1242*.   *V. parviflora* Moench; *V. pyramidata*
Medikus; *V. segetalis* (Necker) Garcke, nom. illeg.; *V. vaccaria* (L.) Druce;
Gypsophila vaccaria Sibth. & Smith;   *Saponaria segetalis* Necker;
S. vaccaria L.; incl. *V. oxyodonta* Boiss. (*S. oxyodonta* (Boiss.) Boiss.).

PETRORHAGIA (Ser. ex DC.) Link
prolifera (L.) P.Ball & Heyw.   Proliferous Pink
•• (but early records did not distinguish this species from native *P. nanteuilii*).
Established at Potton (Beds). Eurasia, N Africa; (widespread). BM, LANC,
LTN, OXF. 20, 21, 22*, 25*, 35*, 147, 235, 370(5:113-116).   *Dianthus
prolifer* L.; *Kohlrauschia prolifera* (L.) Kunth; *Tunica prolifera* (L.) Scop.
saxifraga (L.) Link   Tunic-flower
•• A persistent or established garden escape; known since 1909 at the foot of
a cliff at Tenby (Pembs), perhaps now gone; on walls, railway banks, roadsides
and in chalk pits in a few places in southern England. Eurasia. BM, NMW,
OXF, RNG. 13, 20, 22*, 25*, 26*, 156, 283, 354(2:440), 371(382:24).
Kohlrauschia saxifraga (L.) Dandy; *Tunica saxifraga* (L.) Scop.
velutina (Guss.) P.Ball & Heyw.   Hairy Pink
• A wool casual. Medit.; (Australasia). K. 1, 21, 24*, 32*, 36c*, 44*, 133,
281, 370(5:116).   *Dianthus velutinus* Guss.; *Kohlrauschia velutina* (Guss.)
Reichb.; *Tunica velutina* (Guss.) Fischer & C.Meyer.

DIANTHUS L.
arenarius L.   Stone Pink
Pre-1930 only. E Europe. 21, 22*, 50, 187, 227, 280.
barbatus L.   Sweet-William
••• A persistent garden escape; on chalk banks, railway banks, dunes and
sandy ground. S Europe. BM, DUE, LANC, NMW, OXF, RNG. 1, 20,
22*, 23*, 27*, 29*, 126, 236. A hybrid with *D. allwoodii* hort. (*D.*'Sweet
Wivelsfield') has been recorded. 371(361:26).
carthusianorum L.   Carthusian Pink
Pre-1930 only. Eurasia. BM, OXF. 1, 22*, 25*, 26*, 27*, 209, 354(4:475).
Recent records claimed by 20 appear to be unfounded.

caryophyllus L.   Clove Pink
●●● (but some records may refer to hybrids). An established garden escape;
on old walls, rocks and shingle; known for more than three centuries on the
walls of Rochester Castle (E Kent). Medit. BM, DOR, LANC, NMW, OXF,
RNG. 1, 20, 22*, 25*, 33*, 156, 176, 236. Carnations, which are cultivars
derived from this species, native D. *gratianopolitanus* and D. *plumarius*, have
been recorded as casuals.
[*chinensis* L. Unconfirmed records are claimed by 20.]
gallicus Pers.   Jersey Pink
• A garden escape, or perhaps introduced; naturalised since 1892 on sand dunes
in St Ouen's Bay, Jersey (Channel Is). SW Europe. BM, JSY, OXF, RNG.
2, 20, 22*, 25*, 200, 201*, 354(8:681,889).
guttatus M.Bieb.
Pre-1930 only. S Russia. OXF. 7, 21, 354(2:545), 394(19:8), 630*, 694,
715*.
plumarius L.   Pink
●●● An established garden escape; naturalised on roadside banks and railway
cuttings, mainly in chalk and limestone areas; also on ruins and old walls in
widely scattered localities and known for over a century on walls at Beaulieu
Abbey (S Hants). E Europe. BM, E, LANC, NMW, OXF, RNG. 1, 20,
22*, 23*, 25*, 33*, 50, 174, 1333. Putative hybrids with D. *caryophyllus* and
native D. *gratianopolitanus* have been recorded.
[*superbus* L. Unconfirmed records are claimed by 20.]
sylvestris Wulfen   Wood Pink
• An established garden escape, abundant on rocks near Whitby harbour
(NE Yorks). Europe. 21, 23*, 26*, 27*, 31*, 50, 335(920:42).

VELEZIA L.
rigida L.
Pre-1930 only. Medit., Asia. RNG. 2, 21, 29*, 44*, 300(1907:38), 737*.

[*COLOBANTHUS* Bartling
quitensis (Kunth) Bartling   Discovered in 1971 on the SS *Great Britain* docked
at Bristol after returning from the Falkland Is, but the species did not reach
land. 350(32:215), 370(9:146).]

POLYGONACEAE
PERSICARIA Miller
affinis (D.Don) Ronse Decraene   Himalayan Bistort
• A casual or persistent garden escape. No modern records. Himalayas. LIV,
OXF, SLBI. 7, 50, 51*, 51a, 52*, 54*, 354(4:208). *Bistorta affinis* (D.Don)
E.Greene; *Polygonum affine* D.Don.
alpina (All.) Gross   Alpine Knotweed
●● A persistent or established garden escape; on a shingle island at Ballater
(S Aberdeen), on a roadside near Huddersfield (SW Yorks), and on waste
ground at Montrose and Ninewells, Dundee (both Angus); also a garden weed
near Kirkintilloch (Stirlings). Eurasia, N America. BM, DEE, E, RNG. 7,

20, 22*, 23*, 83*, 187, 303(46:29), 370(11:144; 14:450), 371(374:8), 595*, 683*. *P. angustifolia* (Pallas) Ronse Decraene, nom. illeg.; *Aconogonum alpinum* (All.) Schur; *Polygonum alpinum* All.; *P. angustifolium* Pallas; *P. polymorphum* Ledeb.

amplexicaulis (D.Don) Ronse Decraene   **Red Bistort**
  •••• An established garden escape; naturalised on roadsides and railway banks, in quarries and thickets and by streams, in scattered localities, especially in Ireland. Himalayas. 2, 20, 22*, 52*, 83*. *Bistorta amplexicaulis* (D.Don) E.Greene; *Polygonum amplexicaule* D.Don.

arifolia (L.) ined.   **Halberd-leaved Tear-thumb**
Pre-1930 only. Eastern N America. 1, 38*, 40, 129, 1319*. *Polygonum arifolium* L.; *Tracaulon arifolium* (L.) Raf.   Confirmed from Leamington (Warks), but the Castlecove (S Kerry) records were in error for *P. sagittata*.

bungeana (Turcz.) Nakai
  • An oil-seed casual. No modern records. E Asia; (USA). BM, LANC, OXF, SLBI.   43, 83*, 354(10:481), 638*, 867*(104:526-533), 1157. *Polygonum bungeanum* Turcz.

campanulata (Hook.f.) Ronse Decraene   **Lesser Knotweed**
  •••• An established garden escape; naturalised by streams and on marshy ground in scattered localities, especially in Scotland and Ireland. Himalayas, W China. BM. 12, 20, 52*, 83*, 370(11:291-311), 846*(3(2):1-118). *Aconogonum campanulatum* (Hook.f.) H.Hara; *A. lichiangense* (W.Smith) Soják; *Polygonum campanulatum* Hook.f.   *Polygonum campanulatum* var. *lichiangense* (W.Smith) Steward (*P. lichiangense* W.Smith) was recorded in error for the hairy variant of *Persicaria wallichii*. 83, 156, 370(18:351-358).

capitata (Buch.-Ham. ex D.Don) ined.   **Pink-headed Knotweed**
  •• A casual garden escape. Pakistan, Himalayas, China. RNG, HbEJC. 20, 50, 52*, 303(34*:29-31; 36:28), 648*, 825*(1:49-53), 1100*. *Polygonum capitatum* Buch.-Ham. ex D.Don.; *P. nepalense* auct., non Meissner. The closely related *P. alata* (D.Don) Gross (*Polygonum alatum* D.Don), sold for ground cover as Devon Vine, may have been overlooked. 50.

glabra (Willd.) M.Gómez
  • (but confused with *P. senegalensis*). A wool and bird-seed casual. Tropics. BM. 83, 303*(41:11). *Polygonum glabrum* Willd.

mollis (D.Don) Gross   **Soft Knotweed**
  • A garden escape established for some years at Coylet (Main Argyll) and at Tunbridge Wells (W Kent). C & SE Asia. BM, RNG. 20, 83*, 236, 370(11:145,306), 589*, 724*, 846*(3(2):1-118). *Aconogonum molle* (D.Don) H.Hara; *Polygonum molle* D.Don; *P. molle* var. *rude* (Meissner) A.J.Li; *P. paniculatum* Blume; *P. frondosum* Meissner; *P. rude* Meissner. The pre-1930 record from Woodhall Spa (N Lincs) was in error for *P. wallichii*.

nepalensis (Meissner) Gross   **Nepal Persicaria**
  •• A persistent bird-seed alien; also an abundant weed for a time in a tree nursery at Wimborne (Dorset) and a weed in other nurseries in Dorset, S Somerset and S Hants, and in Edinburgh Botanic Garden. Africa, India. BM, E, K. 19, 20, 83*, 248, 303(34:29; 41:14), 1247*. *Polygonum alatum*

Buch.-Ham. ex Sprengel, nom. illeg.; *P. nepalense* Meissner. The record from
Chichester (W Sussex) was in error for *Persicaria capitata*. 371(355:26).
orientale (L.) Vilm.
- A casual garden escape. No modern records. E & SE Asia. K, RNG. 12,
  21, 38*, 50, 188, 192, 303(34:29), 648*, 650*, 725*, 1100*, 1319*.
  Polygonum orientale L.
pensylvanica (L.) M.Gómez   **Pinkweed**
- • An oil-seed casual. Eastern N America. BM, LANC, MNE, NMW, OXF,
  RNG.   12, 18, 20, 38*, 83*, 236, 354(10:481), 1319*.   *Polygonum
  pensylvanicum* L.   Most records are referable to var. *laevigatum* (Fern.)
  W.C.Ferg.
sagittata (L.) Gross ex Nakai   **American Tear-thumb**
- Possibly a grain alien, long-naturalised in abundance in wet ground near
  Castlecove (S Kerry), now much reduced or extinct there; also recorded at
  Widecombe-in-the-Moor (S Devon). E Asia, N America. BM, DBN, LANC,
  OXF, RNG, SLBI. 2, 20, 22*, 83*, 189, 319(18:331; 19:168), 320(79:103),
  1319*. *Polygonum sagittatum* L.; *Tracaulon sagittatum* (L.) Small. The Irish
  records are referable to var. *americanum* (Meissner) Miyabe. Early records
  were confused with *P. arifolia*.
senegalensis (Meissner) Soják
- • (but confused with *P. glabra* and native *P. lapathifolia*). A wool casual.
  Africa. BM, RNG. 14, 44*, 83*, 303(35:23), 349(1:493), 648*. *Polygonum
  senegalense* Meissner.
wallichii Greuter & Burdet   **Himalayan Knotweed**
- • • • An established garden escape; naturalised on roadsides, railway banks
  and river banks in widely scattered localities, commonly so along the disused
  railway near Barnstaple (N Devon); increasing. Himalayas. 7, 20, 22*, 33*,
  83*, 189, 370(11:291-311). *P. polystachya* (Wallich ex Meissner) Gross, non
  Opiz; *Aconogonum polystachyum* (Wallich ex Meissner) M.Král; *Polygonum
  polystachyum* Wallich ex Meissner. A hairy variant has been recorded, as
  Polygonum polystachyum var. *pubescens* Meissner, from Wales (Pembs, Cards
  and Merioneth), Scotland (Banff) and pre-1930 from England (N Lincs).
  370(11:306; 18:351-358).
weyrichii (F.Schmidt ex Maxim.) Ronse Decraene   **Chinese Knotweed**
- A garden escape established near Wastwater (Cumberland). E Asia. LANC.
  20, 50, 83, 370(13:138,165), 630*, 649*, 1324*. *Aconogonum weyrichii*
  (F.Schmidt ex Maxim.) H.Hara; *Polygonum weyrichii* F.Schmidt ex Maxim.

FAGOPYRUM Miller
dibotrys (D.Don) H.Hara   **Tall Buckwheat**
- An escape from cultivation, established by a roadside at Dale (Pembs).
  Himalayas; (W Europe, Azores). BM, RNG. 20, 21a, 83*, 370(11:145),
  649*, 1247*. *F. cymosum* (Trevir.) Meissner; *Polygonum dibotrys* D.Don;
  P. cymosum Trevir.

esculentum Moench   Buckwheat
●●●●● A persistent bird-seed alien; often sown for feeding game birds and frequently occurring as a relic of cultivation, mainly in S and E England; also from hamster food. C Asia; (widespread). 1, 19, 20, 22*, 33*, 83*, 191, 215, 1242*. *F. fagopyrum* (L.) Karsten; *F. sagittatum* Gilib., nom. inval.; *F. vulgaris* T.Nees; *Polygonum fagopyrum* L.

tataricum (L.) Gaertner   Green Buckwheat
●● A wool, grain and bird-seed casual; also an impurity in crops of *F. esculentum*. N & C Asia; (widespread). BM, LANC, LTN, NMW, OXF, RNG. 1, 14, 19, 20, 29*, 83*, 303(26:19), 331(60:87), 617*, 826*(17:164-167), 1242*.

POLYGONUM L.
arenarium Waldst. & Kit.   Lesser Red-knotgrass
●● (but confused with *P. patulum*). A grain and bird-seed casual. Eurasia, N Africa. BM, E, NMW, OXF, RNG, SLBI. 1, 19, 20, 83*(as *P. patulum*), 349(1:216). Incl. *P. pulchellum* Lois. All post-1930 records are referable to subsp. *pulchellum* (Lois.) D.Webb & Chater; but subsp. *arenarium* was recorded pre-1930.

bellardii All. sensu lato   Red-knotgrass
●●● (but confused with *P. arenarium*). An agricultural seed, grain, bird-seed and wool casual. Europe, N Africa, SW Asia; (Australia). BM, E, LIV, NMW, OXF, RNG. 1, 10, 14, 19, 20, 44*, 83*(as *P. arenarium),* 264, 1242*. The two segregates have been much confused in the literature; *P. bellardii* All. sensu stricto (*P. kitaibelianum* Sadler) is widespread in the Mediterranean region, but *P. patulum* M.Bieb. sensu stricto (*P. spectabile* Lehm.) is absent. 1123.

cognatum Meissner   Indian Knotgrass
●● A long-persistent grain alien at ports, railway sidings and breweries; possibly also from timber. Now gone from all its sites and probably no longer occurring as a casual. SW Asia. BM, E, K, LANC, OXF, RNG. 2, 11, 20, 22*, 83*, 370(1:319; 2:414). *P. alpestre* C.Meyer; *P. ammanioides* Jaub. & Spach.

corrigioloides Jaub. & Spach
Pre-1930 only. Iran. 1, 9, 667*.

equisetiforme Sibth. & Smith
● A wool and grain casual. Medit., SW Asia. BM, OXF, RNG. 2, 21, 30*, 83*, 267, 354(7:893; 8:756), 648*, 683*, 737*.

graminifolium Wierzb. ex Heuffel
Pre-1930 only. Danube valley. OXF. 2, 21, 150, 151, 623*.

oxyspermum C.Meyer & Bunge ex Ledeb. subsp. **oxyspermum**
● (but perhaps overlooked). Persistent on Pathhead shore, Kirkcaldy (Fife), and very rare and sporadic along the coast of E Lothian. Baltic, S Norway. 21, 22*, 113, 1242*. Various records of native subsp. *raii* from the east coast of Scotland are perhaps referable to this subspecies, suggesting that it might be an overlooked native plant. 113, 1304.

plebejum R.Br.   Small Knotweed
• A wool casual. Africa, Asia, Australasia. BM, E, LTN, OXF, RNG. 3*,
14, 83*, 145, 590*, 1247*. Incl. *P. effusum* Meissner.

FALLOPIA Adans.
baldschuanica (Regel) Holub sensu lato   Russian-vine
•••• A long-persistent or established garden escape and relic; rampant on
hedges, in thickets and on cliffs in widely scattered localities. C Asia. 12, 20,
22*, 34*, 52*, 83*, 648*, 683*. *Bilderdykia baldschuanica* (Regel) D.Webb;
Polygonum baldschuanicum Regel; *Reynoutria baldschuanica* (Regel) Mold.;
incl. *F. aubertii* (L.Henry) Holub; *B. aubertii* (L.Henry) Mold.; *P. aubertii*
L.Henry. A hybrid with *F. japonica* has been recorded. 20, 370(17:163-
181).
japonica (Houtt.) Ronse Decraene   Japanese Knotweed
••••• An established garden escape; widely and abundantly naturalised on
railway banks, roadsides and waste ground throughout the British Is; increasing.
Japan. 2, 20, 22*, 34*, 83*, 303(53:22), 370(11:291-311), 649*. *Polygonum
cuspidatum* Siebold & Zucc.; *P. sieboldii* hort. ex Meissner; *Reynoutria
japonica* Houtt.; incl. *P. compactum* Hook.f.; *R. japonica* Houtt. var. *compacta*
(Hook.f.) Mold.; *F. japonica* var. *compacta* (Hook.f.) J.Bailey. Hybrids with
F. sachalinensis and *F. baldschuanica* have been recorded.
sachalinensis (F.Schmidt ex Maxim.) Ronse Decraene   Giant Knotweed
•••• An established garden escape; naturalised, sometimes abundantly, in
widely scattered localities; increasing. Formerly grown as a forage crop.
Sakhalin, Japan. 7, 20, 22*, 83*, 370(11:291-311), 649*. *Polygonum
sachalinensis* F.Schmidt ex Maxim.; *Reynoutria sachalinensis* (F.Schmidt ex
Maxim.) Nakai; *R. vivax* J.Schmitz & Strank. A hybrid with *F. japonica*
(*F.* × *bohemica* (Chrtek & Chrtková) J.Bailey; *Reynoutria* × *bohemica* Chrtek
& Chrtková; *R.* × *vivax* auct., non Schmitz & Strank) has been recorded.
LANC, LTR. 20, 370(15:270), 826*(19:17-25).

MUEHLENBECKIA Meissner
complexa (Cunn.) Meissner   Wireplant
••• An established garden escape; long-naturalised on rocky slopes, in quarries
and along hedgebanks in Scilly and the Channel Is; recently recorded from
Bournemouth East cliffs (S Hants), Bawdsey cliffs (E Suffolk) and several
places in Cornwall. New Zealand. ABD, BM, E, OXF, RNG, SLBI. 7, 20,
83*, 207*, 212, 213, 220, 258, 648*. *Polygonum complexum* Cunn.;
Sarcogonum complexum (Cunn.) Kunze.

RHEUM L.
× **hybridum** group   Rhubarbs
•••• Garden escapes or relics of cultivation persistent in widely scattered
localities, especially in the north. Originated in cultivation. 20, 80*, 83, 187,
368(28:110). *R.* × *hybridum* Murray (*R. palmatum* × R.rhaponticum L.);
R. × *cultorum* Thorsrud & Reis., nom. nud.; *R. rhabarbum* auct., non L.;

R. rhaponticum auct., non L.   Modern culinary rhubarbs, now considered to
be complex hybrids between several Asian species, are included here.
officinale Baillon   Tibetan Rhubarb
•• (but confused with *R. palmatum*).   Introduced and persistent in a wood at
Bolton (S Lancs) and near Borve on Harris (Outer Hedrides), an escape from
cultivation on a tip at Chorley (S Lancs); also collected in 1937 from Middleton
in Teesdale (Co Durham). Tibet, China. RNG. 50, 52*, 56*(as *R.
palmatum*), 371(349:29), 467.
palmatum L.   Chinese Rhubarb
•• A persistent garden escape in a roadside ditch at Liston, near Sudbury
(N Essex); also collected from Wilcote Wood (Oxon) and Clayton Green (S
Lancs) and reported from Sandling (E Kent). NE Asia. RNG. 20, 52*, 54*,
57*, 83*, 191, 458, 683*. A hybrid with *R. kialense* Franchet is grown in
gardens and may have been overlooked.

RUMEX L.
acetosa L. subsp. **ambiguus** (Gren.) Á.Löve   Garden Sorrel
• (but confused with native subspecies). Recently recorded from Cheshunt
(Herts) and as introduced with wild-flower seed on landscaped ground at
Landguard Common, near Felixstowe (E Suffolk). Origin obscure, known only
in cultivation. RNG, HbAMB, HbEJC. 20, 21, 50, 83, 331(70:161),
369(26:64-75), 370(17:473), 810*(1944:251). *R. acetosa* hort., non L.;
R. ambiguus Gren.; *R. rugosus* Campderá.
altissimus Alph.Wood
Pre-1930 only. N America. BM, LIV, OXF. 7, 38*, 83, 251, 354(12:582),
1157*.
[*arifolius* All. Dubious records. LANC. 177, 354(7:58).]
bequaertii De Wild.
• A wool casual. Africa. BM, E, RNG. 83*, 371(334:22). *R. altissimus*
auct., non Alph.Wood; *R. camptodon* Rech.f.; *R. quarrei* De Wild.
brownii Campderá   Hooked Dock
••• A persistent wool and cotton alien; possibly also from bird-seed.
Australia; (New Zealand). ABD, BM, CGE, E, OXF, RNG. 3*, 14, 19, 20*,
83*, 251, 354(12:573), 725*, 1157*. *R. brownianus* Schultes; *R. fimbriatus*
R.Br., nom. illeg.
bucephalophorus L.   Horned Dock
• A casual. No modern records. Medit., Canary Is, Azores. NMW, OXF,
RNG. 2, 21, 39*, 44*, 45*, 83, 156, 267, 354(5:126; 12:578).
confertus Willd.   Russian Dock
• Persistent on a roadside at Headcorn and at Aylesford (both E Kent); also
known from 1942 to 1960 at Old Coulsdon (Surrey). Formerly at Marston
(Oxon) as an escape from Oxford Botanic Garden. Russia, Siberia. BM,
MNE, OXF, RNG. 7(as *R. giganteus*), 20*, 83, 209, 354(12:149).
R. pseudoalpinus auct., non Hoefft. Hybrids with native *R. crispus*
(*R. × skofitzii* Błocki) and *R. obtusifolius* (*R. × borbasii* Błocki) have been
recorded. 209, 1311.

cristatus DC.   Greek Dock
●●● (but perhaps confused with *R. obtusifolius* × *R. patientia*). Established,
possibly from grass and clover seed, in S Wales and SE England; well
naturalised by rivers, especially by the Rhymney (Mons & Glam), the Medway
and Swale (E & W Kent) and the Thames in the London area; increasing.
SE Europe, Turkey. BM, LANC, NMW, RNG, SLBI. 12, 20*, 22*, 83*,
126, 156, 236, 302(9:59), 331(65:196). *R. graecus* Boiss. & Heldr. Hybrids
with native *R. crispus* (*R.* × *dimidiatus* Hausskn.; LANC, RNG.
303*(58:1,35)), *R. conglomeratus* (331(65:196)), *R. obtusifolius* (*R.* × *lousleyi*
Kent; BM, NMW, RNG.) and *R. palustris* (331(71:179)) have been recorded.
crystallinus Lange   Glistening Dock
● A wool casual. Australia. BM, RNG. 14, 37*, 83*. *R. halophilus* F.Muell.
Early records were in error for *R. tenax*. E. 3, 197, 354(12:579).
dentatus L.   Aegean Dock
●● A wool and grain casual. SE Europe, N Africa, Asia. BM, E, LIV,
NMW, OXF, RNG. 1, 14, 20*, 83*, 144, 648*, 1157*. Most records are
referable to subsp. *halacsyi* (Rech.) Rech.f., but subsp. *dentatus*
(*R. callosissimus* Meissner), subsp. *klotzschianus* (Meissner) Rech.f. and
subsp. *mesopotamicus* Rech.f. have also been recorded. 83*. A hybrid with
native *R. maritimus* (*R.* × *kloosii* Danser) has been recorded. 1311.
[*flexuosiformis* Rech.f.   A dubious record. 83, 354(12:582).]
[*flexuosus* Sol. ex Hook.f.   Dubious records. E. 3, 83, 354(12:582).]
frutescens Thouars   Argentine Dock
●● Established; long naturalised on coastal dunes in SW England and S Wales;
also a wool casual. Temperate S America. BM, K, LANC, NMW, OXF,
RNG. 7, 20*, 22*, 83*, 354(12:579-582). *R. cuneifolius* Campderá;
R. magellanicus auct. angl., non Campderá, nec Griseb. A hybrid with native
R. conglomeratus (*R.* × *wrightii* Lousley) has been recorded. 370(2:394-397).
[*giganteus* Aiton f.   In error for *R. confertus*. 83, 354(5:307; 12:149).]
[*magellanicus* Griseb.   In error for *R. frutescens*. 354(12:583).]
maritimus L. subsp. **fueginus** (Philippi) Hultén
● A casual recorded from the Glasgow area and Tweedside. Formerly a wool
alien. Southern S America. E. 83, 607*, 665. *R. fueginus* Philippi.
nepalensis Sprengel
Pre-1930 only. SE Europe, N Africa, Asia. E. 3, 21, 39*, 83*, 354(12:583),
724*, 1247*.
obovatus Danser   Obovate-leaved Dock
● A wool, grain and bird-seed casual. Argentina, Paraguay. BM, NMW,
RNG. 6, 19, 20*, 83*, 144, 248, 331(34:246), 349(1:61). *R. paraguayensis*
Thell., non Parodi. A hybrid with native *R. crispus* (*R.* × *bontei* Danser) has
been recorded. 1157*, 1311.

obtusifolius L. subsp. **sylvestris** (Wallr.) Čelak.

• Established, vector unknown; naturalised in the Lea Valley (Herts and S Essex); casual elsewhere. E Europe. OXF, RNG. 12, 21, 83*, 120, 176, 209, 241, 267, 331(65:197), 1157*, 1242*. *R. obtusifolius* var. *microcarpus* Dierb.; *R. sylvestris* Wallr.

obtusifolius L. subsp. **transiens** (Simonkai) Rech.f.

•• Established, vector unknown; naturalised by rivers and lakes, mainly by the river Thames and its tributaries within Surrey and Middlesex; also reported from W Kent, Berks and Lanarks. C Europe. BM, RNG. 12, 21, 83*, 458. *R. obtusifolius* var. *transiens* (Simonkai) Kubát.

[*paraguayensis* Parodi   In error for *R. obovatus*. 354(6:258; 9:135).]

patientia L.   Patience Dock

•••• Established, probably as an escape from cultivation, or from grass and clover seed; naturalised on waste ground in widely scattered localities in England, especially in the London area. Formerly a grain alien. SE Europe, SW Asia. 1, 20*, 22*, 83*, 126*, 191, 192, 302(9:59), 595*, 1157*. Incl. *R. orientalis* Bernh., non Campderá. Hybrids with native *R. conglomeratus, R. crispus* (*R.* × *confusus* Simonkai) and *R. obtusifolius* (*R.* × *erubescens* Simonkai) have been recorded. 1311.

pseudoalpinus Hoefft   Monk's-rhubarb

•••• A long-persisting relic or escape from cultivation; naturalised by roadsides and streams, especially on high ground in northern England and Scotland. Europe, SW Asia. 1, 20*, 22*, 83*, 354(12:566). *R. alpinus* L. (1759), non L. (1753); *R. cordifolius* Hornem. ex Reichb.

pseudonatronatus (Borbás) Murb.

• A casual at Sharpness docks (W Gloucs), vector unknown. N Europe. BM, RNG. 21, 83, 182, 192, 589*, 641*, 1157*. *R. fennicus* (Murb.) Murb. A hybrid with native *R. crispus* has been recorded. The record from Billinge Hill (Cheshire) was in error for native *R. longifolius*. 349(7:25,561).

pulcher L. subsp. **anodontus** (Hausskn.) Rech.f.

• A casual at Avonmouth docks (W Gloucs), vector unknown. No modern records. N Africa, SW Asia. BM, BRISTM, OXF. 2, 83(p.148), 350(30:307), 354(2:418; 12:138), 648, 815(12:102-107). The subspecies of *R. pulcher* are poorly defined and are not recognized by 39.

pulcher L. subsp. **divaricatus** (L.) Murb.

•• A wool and bird-seed casual. Medit. E, RNG. 3, 14, 19, 83*, 354(12:138), 815(12:102-107), 1157*. *R. divaricatus* L.

salicifolius J.A.Weinm.   Willow-leaved Dock

••• A persistent grain and bird-seed alien. Western N America. BM, CGE, MNE, NMW, OXF, RNG . 7, 19, 20*, 22*, 83*, 354(12:575-577), 1138*, 1157*, 1242*. *R. triangulivalvis* (Danser) Rech.f.

sanguineus L. **alien variants**

Var. *sanguineus*, Blood-veined Dock, has occurred as a garden escape. 20, 83.

scutatus L.   French Sorrel, incl. Buckler-leaved Sorrel

••• An established escape from cultivation; on old walls, roadsides and railway banks in widely scattered localities, mainly in the north; known for several

centuries at Craigmillan Castle in Edinburgh (Midlothian) and Aberdour Old
Castle (Fife). Europe, N Africa, SW Asia. BM, DBN, E, NMW, OXF,
RNG. 1, 20*, 22*, 23*, 83*.

stenophyllus Ledeb.
- A persistent alien at Avonmouth docks (W Gloucs), vector unknown; also
collected from the site of a former pig-field at Hordle (S Hants). Eurasia;
(N America). BM, LIV, OXF, RNG, SLBI. 20, 22*, 83, 350(8:389-393),
354(12:150), 595*, 630*, 826*(21:80-85), 843*(60:54-57).

steudelii Hochst. ex A.Rich.
- A wool casual. Arabia, Ethiopia. BM, CGE, E, K, RNG. 17, 83*, 1054*.

tenax Rech.f.
- A wool casual. No modern records. Australia. E, K, OXF, RNG. 3, 14,
83*, 354(12:579), 725*.

cf. **thyrsiflorus** Fingerh.
- Recorded from Brechou (Channel Is). E Europe. 21, 22*, 83, 221, 1242*.

[*OXYRIA* Hill
amarantoides sphalm.   In error for *Axyris amaranthoides*. 354(4:296; 8:130).]

EMEX Campderá
australis Steinh.   Double Gee
- A wool casual. S Africa; (Australasia). E, K, RNG. 17, 37*, 83*, 725*.

spinosa (L.) Campderá
- A wool, grain and possibly a bird-seed casual. Medit.; (widespread). E, K,
OXF, RNG. 2, 19, 21, 44*, 83*, 197, 313(22;45), 648*, 737*, 1095*, 1316*.
Rumex spinosa L.; *Vibo spinosa* (L.) Medikus.

OXYGONUM Burchell ex Campderá
sinuatum (Meissner) Dammer
- A casual on waste ground of timber yard at Seaforth (S Lancs), vector
unknown. E & C Africa. LIV. 48*, 303(34:29), 618*, 1112*.

PLUMBAGINACEAE

LIMONIUM Miller
bonduellii (Lestib.) Kuntze   Algerian Statice
- A casual, perhaps bird-sown, in a garden. W Medit. RNG. 2, 21, 46*,
51*, 52*, 288. *Statice bonduellii* Lestib.

hyblaeum Brullo   Rottingdean Sea-lavender
- An established garden escape, well naturalised on chalk cliffs at Rottingdean
(E Sussex); also reported in 1993 from cliffs at West Bay (Dorset). At both
localities it grows with native or garden-escaped *Frankenia laevis*. Sicily.
LTR, HbEJC. 20*, 122, 303(61:39), 370(13*:181-184; 14:228), 403, 683,
1263*. *L. companyonis* auct., non (Gren. & Billot) Kuntze. A segregate of
L. panormitanum (Tod.) Pign.

latifolium (Smith) Kuntze   Broad-leaved Statice
   • An escape from cultivation in fields in Cambs; also reported as naturalised
   on the cliffs at Clacton (N Essex), and in grassland at Lane End, Darenth and
   at Hextable (both W Kent). E Europe. 20, 21, 55*, 303(28:16; 59:46), 458,
   1330. *Statice latifolia* Smith.
sinuatum (L.) Miller   Statice
   • A casual escape from cultivation. Medit., SW Asia. SLBI. 7, 21, 24*,
   30*, 44*, 52*, 331(70:160), 371(349:29), 520a. *Statice sinuata* L.

PSYLLIOSTACHYS (Jaub. & Spach) Nevski
leptostachya (Boiss.) Roshk.
   Pre-1930 only. Iran, Afghanistan, C Asia. 7, 136, 524, 630*, 1249.
   Limonium leptostachyum (Boiss.) Kuntze; *Statice leptostachya* Boiss.
spicata (Willd.) Nevski
   Pre-1930 only. SW Asia. OXF, SLBI. 7, 21, 44*, 51, 136, 354(4:70), 622*,
   730*. *Limonium spicatum* (Willd.) Kuntze; *Statice sisymbriifolia* Jaub. &
   Spach; *S. spicata* Willd.
suworowii (Regel) Roshk.   Rat's-tail Statice
   • A casual garden escape. Iran, C Asia. BON. 7, 51*, 52*, 53*, 66*, 143,
   251, 354(12:736), 630*. *Limonium suworowii* (Regel) Kuntze.

ARMERIA Willd.
[*maritima* (Miller) Willd. subsp. *alpina* (Willd.) Pirajá   In error for native
   subsp. *maritima*. 255.]
pseudoarmeria (Murray) Mansf.   Estoril Thrift
   • A persistent garden relic on cliffs at Lee-on-the-Solent (S Hants), now gone;
   a similar plant has been reported recently from cliffs at Boscombe (S Hants).
   Portugal. RNG. 13, 20, 21, 30*, 53*, 312*(t.7313), 404, 473, 536*.
   A. cephalotes (Aiton) Hoffsgg. & Link, nom. illeg.; *A. formosa* hort. ex Vilm.,
   non Heynh.; *A. latifolia* Willd., nom. illeg.; *Statice pseudoarmeria* Murray.

ACANTHOLIMON Boiss.
glumaceum (Jaub. & Spach) Boiss.
   • Introduced or a garden escape on Dartmoor (S Devon). SW Asia. RNG.
   39*, 51, 53*, 54*, 371(334:22) .

CERATOSTIGMA Bunge
plumbaginoides Bunge
   • A casual garden escape.   China.   RNG.   21, 51*, 56*, 63*,
   312*(n.s. t.210), 350(39:49-64).

PAEONIACEAE
PAEONIA L.
lactiflora Pallas   Chinese Peony
   •• A persistent garden escape. NE Asia. OXF. 16, 51, 52*, 54*, 56*, 119,
   126, 236, 454.

aff. **lutea** Franchet   Yellow Tree-peony
- A garden relic. China, Tibet. 54*, 61*, 62*, 63*, 340(2:4). *P. delavayi* Franchet var. *lutea* (Franchet) Finet & Gagnepain.

mascula (L.) Miller   Peony
- A garden escape established since 1803 on Steep Holm (N Somerset) and since the 1980s on Flat Holm (Glam), both in the Bristol Channel, now much reduced. Europe, N Africa, SW Asia. BM, OXF, RNG. 1, 21, 22*, 29*, 44*, 192, 245, 248, 1279*. *P. corallina* Retz.; *P. officinalis* var. *mascula* L.; incl. *P. arietina* G.Anderson; *P. russii* Biv. Confused with *P. officinalis* and *P. lactiflora*; the Shiels (Lanarks) record in 20 is in error for *P. officinalis*. 370(18:215).

officinalis L.
- • A persistent garden escape. S Europe. LANC, OXF. 1, 21, 26*, 27*, 29*, 30*, 113, 199, 370(18:215,420).

CLUSIACEAE
Guttiferae, Hypericaceae

HYPERICUM L.

aegypticum L.
- A garden escape on the rocky seawall of the estuary at Lynmouth (N Devon). Islands of C & E Medit., N Africa. RNG. 21, 51, 312*(t.6481), 502, 683*, 1299*. *Triadenia maritima* (Sieber) Boiss.

[*barbatum* Jacq.   Unconfirmed records. LIV. 1, 157, 354(5:754).]

calycinum L.   Rose-of-Sharon
- • • • • An established garden escape; naturalised, sometimes abundantly, in woods and on roadsides throughout the British Is, especially in southern England; known since 1835 at Mickleham (Surrey). SE Europe, Turkey. 1, 20, 22*, 63*, 126, 371*(398:22-25). A hybrid with *H. patulum* (*H.* × *moserianum* André) has been collected from Westmorland. LANC.

[*canadense* L.   Accepted, with reservations, as native.]

forrestii (Chitt.) N.Robson
- A garden escape persistent or established on a footbridge at Nunnery (N Somerset), on Ranmore Common (Surrey) and on limestone debris at Whitbarrow (Westmorland). China, Assam, Burma. LANC. 20, 51*, 61, 63*, 135, 199, 370(17:466), 454. *H. patulum* var. *forrestii* Chitt. Reports of *H. patulum* from Dunbarton and Cambs may refer to this species or *H.*'Hidcote'.   *H.*'Hidcote', the popular garden plant often listed under *H. forrestii* or *H. patulum*, is now considered to be a hybrid between *H. calycinum* and *H.* × *cyathiflorum* 'Gold Cup' and may have been overlooked, as have several related species and hybrids. 20, 61a, 63*.

hircinum L.   Stinking Tutsan
- • • • • An established garden escape; naturalised in woods, on river and railway banks and in quarries in widely scattered localities, mainly in southern England; known for over 100 years on the banks of the river Wey at Upway (Dorset).

Medit, SW Asia. 1, 20, 22*, 44*, 63*, 165, 371(398:22-25). *Androsaemum foetidum* Spach. All records are referable to subsp. *majus* (Aiton) N.Robson.

inodorum group　　　　　　　　　　　　　　　　　　Tall Tutsan

●●● Established garden escapes; in woods, hedges, on waste ground and commons, sometimes abundant, as at Bynea (Carms) and Helensburgh (Dunbarton). BM, E, NMW, OXF, RNG, SLBI. 1, 20, 22*, 61, 305(45:19), 312*(n.s. t.376), 370(17:466), 371*(398:22-25). *H. inodorum* Miller (*H. elatum* Aiton; *H. grandifolium* Choisy; *Androsaemum webbianum* Spach) appears to be an endemic species of the Canary Is and Madeira, and not a hybrid; it is probably naturalised in France and Britain. Very similar plants, sometimes partially sterile, occurring as garden escapes, have originated from *H. androsaemum* L. × *H. hircinum* and possibly other species; they include *H. anglicum* sensu Bab., non Bertol,; *H. elatum* auct., non Aiton; *H.*'Elstead'; *H. persistens* F.Schneider; *H. multiflorum* hort.

nummularium L.　　　　　　　　　　　　　　　Round-leaved St John's-wort

● A garden escape established on rock ledges in a quarry near Thornton Rust (NW Yorks). SW Europe. BM. 20, 23*, 27*, 30*, 371(367:29), 432.

olympicum L.

● A garden escape established on Kinnoull Hill (E Perth); casual elsewhere. Balkans, Greece, Turkey. BM, K, OXF, RNG. 21, 31*, 39*, 57*, 303(32:19; 39:9), 347*(1:192-200), 1307. *H. fragile* auct., non Heldr. & Sart. ex Boiss.; *H. polyphyllum* hort., non Boiss. & Bal.

[*patulum* Thunb. No confirmed record. This species is very rare in cultivation, having been superseded by *H. beanii* N.Robson, *H. forrestii*, *H. pseudohenryi* and *H.*'Hidcote'. Modern records of garden escapes under this name are probably referable to *H. forrestii* or *H.*'Hidcote'.]

[*polyphyllum* Boiss. & Bal. In error for *H. olympicum*. 119, 347(1:192-200).]

pseudohenryi N.Robson　　　　　　　　　　　　　　　　　Irish Tutsan

● A persistent garden escape in woodland at Glengarriff (W Cork). China. HbRCS. 61, 63*, 303(28:16). *H. patulum* var. *henryi* sensu Rehder pro parte, non Bean.

richeri Villars　　　　　　　　　　　　　　　　　　Alpine St John's-wort

Pre-1930 only. S Europe. 21, 23*, 29*, 354(3:78). *H. androsaemifolium* Villars.

xylosteifolium (Spach) N.Robson　　　　　　　　　　　　　Turkish Tutsan

● (but perhaps overlooked as *H. inodorum*). A garden escape persisting on a railway embankment near Monkton Moor (MW Yorks); also recorded from Eaves Wood at Silverdale (W Lancs). NE Turkey, SW Georgia. BM. 39*, 61, 303(24:16; 26:16). *H. inodorum* Willd., non Miller.

TILIACEAE

TILIA L.

× **euchlora** K.Koch　　　　　　　　　　　　　　　　　Caucasian Lime

● Introduced. Originated in cultivation. 20, 21, 59*, 60*, 61, 166, 350(47:35), 623a*. Probably *T. cordata* Miller × *T. dasystyla* Steven.

tomentosa Moench   Silver Lime
●●● Introduced; seedlings have been recorded at Stockwell (Surrey). SE Europe, Turkey. LANC, LIV, OXF, SLBI. 1, 20, 21, 59*, 60*, 61*, 199. *T. argentea* Desf. ex DC.; incl. *T.*'Petiolaris'.

CORCHORUS L.
olitorius L.   Tossa Jute
● A casual at Avonmouth docks (W Gloucs), probably from seeds amongst imported jute fibre. No modern records. Tropical Asia; (widespread). 9, 21, 44*, 68*, 1035*.

TRIUMFETTA L.
annua L.
● A wool casual. No modern records. Tropical Africa & Asia. K. 14, 48, 143, 312*(t.2296), 327*(39:268), 604, 676.

MALVACEAE

MALOPE L.
malacoides L.
Pre-1930 only. Medit. OXF. 7, 21, 31*, 32*, 46*, 212, 303*(53:1), 354(2:413; 5:17; 6:117). Incl. *M. hispida* Boiss. & Reut.
trifida Cav.   Mallow-wort
●●● A bird-seed and wool casual; perhaps also a garden escape. Spain, N Africa. CGE, LIV, LTN, RNG, SLBI. 7, 21, 126, 303*(53:1), 349(3:442), 370(11:71), 600*, 657*. *M. grandiflora* Dietr.

KITAIBELA Willd.
vitifolia Willd.
Pre-1930 only. Yugoslavia. BM. 1, 21, 31*, 192, 623a*.

SIDA L.
cordifolia L.   Flannel Weed
●● A wool casual. Tropics. LTN, RNG. 7, 14, 47*, 48, 49, 145, 236, 598*, 642*.
glomerata Cav.
● A wool casual. No modern records. C America. LTN. 14, 49, 143, 537*, 572, 727.
rhombifolia L.   Queensland-hemp
● A wool and oil-seed casual. Tropics; (Australasia). BM, MNE, RNG. 17, 18, 20*, 36c*, 48*, 354(11:25), 370(11:71), 371(400:36), 1109*, 1247*.
spinosa L.   Prickly Mallow
●● An oil-seed, bird-seed and wool casual. Tropics; (N America). BM, LTN, MNE, RNG. 2, 19, 20*, 38*, 147, 236, 303*(47:1,36). Incl. *S. alba* L.; *S. alnifolia* L. The somewhat similar *S. glutinosa* Comm. ex Cav. from tropical America may have been overlooked. 826*(15:45-53).

MALVASTRUM A.Gray
americanum (L.) Torrey
- A wool casual. No modern records. Tropical America; (tropics, subtropics). BM, LTN, RNG. 14, 37*, 49, 303(7:530), 370(1:242), 514, 520*, 805*(52:544-548), 1358*. *M. spicatum* (L.) A.Gray; *Malva americana* L.

campanulatum (Paxton) Nicholson
- A casual found in a ditch in Denbs, perhaps a garden escape. Chile. NMW. 51*, 156. *Malva campanulata* Paxton.

coromandelianum (L.) Garcke   Broomweed
- A wool and cotton casual. Tropical America; (tropics). K, LTN. 14, 49, 145, 236, 303(18:11), 621*, 646, 676, 1035*, 1109*.

multicaule (Schldl.) Britton
- A wool casual. S America. BM, RNG. 17, 303(7:530), 370(11:71), 371(407:41).

UROCARPIDIUM Ulbr.
peruvianum (L.) Krapov.
- A wool casual. Mexico, Venezuela, Colombia, Ecuador, Peru. LTN. 14, 143, 199, 370(11:71), 707*, 820*(10:619). *Malvastrum peruvianum* (L.) A.Gray.

shepardae (I.M.Johnston) Krapov.
- A wool casual. Peru, Bolivia, Argentina. HbEJC. 236, 370(11:71), 820*(10:619).

ANISODONTEA C.Presl
scabrosa (L.) D.Bates   Scabrous Mallow
- A casual on a refuse tip near Stone (W Kent), possibly a greenhouse escape. S Africa. HbCGH, HbEJC. 66, 236, 511, 562*, 655*, 882*(10:215-283), 1274*. *Malva scabrosa* L.; *Malvastrum scabrosum* (L.) Stapf.

MALVA L.
aegyptia L.
   Pre-1930 only. Medit., SW Asia. LCN, SLBI. 7, 21, 44*, 46*, 160, 300(1904:237; 1906:102), 1259*.

alcea L.   Greater Musk-mallow
   •• A persistent garden escape; perhaps established in a chalk pit on the Gogs (Cambs) and by Coppetts Wood (Middlesex). S Europe. CGE, E, LTN, NMW, RNG, SLBI. 1, 20, 22*, 23*, 26*, 303*(23:24-25), 331(64:123), 683*. Intermediates between *M. alcea* and native *M. moschata* have been collected. CGE.

brasiliensis Desr.
   1930 only. Brazil. 354(9:259), 633.

cretica Cav.
   Pre-1930 only. Medit.; (Australia). 21, 24*, 247, 683*, 737*. Incl. *M. althaeoides* Cav.

hispanica L.
Pre-1930 only. W Medit. NMW, OXF. 7, 10, 21, 30*, 45*, 46*, 354(6:16), 737*.

nicaeensis All.   French Mallow
●●● A persistent wool, grain, tan-bark and horticultural seed casual; probably decreasing. Medit, SW Asia; (Australasia). BM, CGE, E, NMW, OXF, RNG. 1, 14, 20*, 29*, 44*, 212, 277, 354(12:272), 737*.

oxyloba Boiss.
Pre-1930 only. E Medit. K. 44*, 192.

parviflora L.   Least Mallow
●●●● A persistent wool, grain, bird-seed and esparto alien; probably decreasing. Medit., SW Asia; (widespread). 1, 8, 14, 15, 19, 20*, 22*, 44*, 126. Incl. *M. microcarpa* Desf. A hybrid with native *M. neglecta* has been recorded. 1311.

pusilla Smith   Small Mallow
●●●● (but probably over-recorded for *M. parviflora*). A persistent wool, grain and bird-seed alien; decreasing.. Europe, SW Asia. 1, 8, 14, 19, 20*, 22*, 126*. *M. borealis* Wallr.; *M. parviflora* Hudson, non L.; *M. rotundifolia* L. pro parte.

sylvestris L. **alien variants**
Var. *ambigua* (Guss.) ined. (*M. ambigua* Guss.), var. *dasycarpa* Beck and var. *eriocarpa* Boiss. have occurred as casuals. No modern records. BM. 9, 121, 354(7:996; 12:272).

verticillata L.   Chinese Mallow
●●● A casual or persistent wool alien, decreasing. Formerly a garden escape and cotton alien. Asia; (Australia). BM, CGE, E, LCN, OXF, RNG. 1, 14, 20*, 22*, 48*, 151, 290, 683*. Incl. *M. crispa* L.

LAVATERA L.

kashmiriana Cambess.   Wild Hollyhock
Pre-1930 only. Himalayas. 7, 51, 53*, 354(7:765), 685*, 732*, 1300*. *L. cashmiriana* sphalm.

maritima Gouan   Sea Mallow
Pre-1930 only. W Medit. 21, 24*, 30*, 333(1:91), 737*.

plebeia Sims   Australian Hollyhock
●● A wool casual. Australia. E, LIV, LTN, RNG. 14, 20, 37*, 143, 354(12:271), 371(364:29), 520*, 590*, 666*.

punctata All.   Spotted-stalked Tree-mallow
● A grain and wool casual. Medit., SW Asia. LTN, NMW, OXF, RNG. 1, 21, 44*, 119, 147, 350(28:173), 354(12:37).

thuringiaca group   Garden Tree-mallows
●●● Persistent or established garden escapes in southern England; also probably introduced on newly-made road verges, sometimes abundant as in 1982 at Falmer (E Sussex). Eurasia. LTN, OXF, RNG, SLBI. 1, 20, 22*, 29*, 52*, 191, 325*(114:23-27), 409, 415, 458. A much confused group incl.

L. ambigua DC.; *L.* 'Barnsley'; *L. olbia* L.; *L.* 'Rosea'; *L. thuringiaca* L. Most
garden plants are *L. olbia* × *L. thuringiaca*, but both parents are also grown.
triloba L.
Pre-1930 only. W Medit. OXF. 7, 21, 30*, 354(5:17), 737*.
trimestris L.   Royal Mallow
●●● A wool, grain and bird-seed casual and garden escape. Medit.;
(Australia). LCN, LTN, LSR, OXF, RNG. 1, 19, 20, 29*, 44*, 53*, 146,
160, 303*(23:1,10), 737*. *L. rosea* Medikus; *Stegia trimestris* (L.) Luque &
Devesa.

ALTHAEA L.
cannabina L.
● A casual garden escape. Eurasia. HbACL. 21, 29*, 199, 615*, 683*.
[*hirsuta* L.   Accepted, with reservations, as native.]

ALCEA L.
acaulis (Cav.) Alef.
Pre-1930 only. E Medit. 39, 44*, 354(9:259). *Althaea acaulis* Cav.
rosea L.   Hollyhock
●●●● A persistent garden escape. Origin obscure. 2, 20*, 22*, 53*, 56*.
Althaea rosea (L.) Cav.; incl. *Alcea ficifolia* L. (*Althaea ficifolia* (L.) Cav.);
Althaea × *cultorum* Bergmans (*A. ficifolia* × *A. rosea*).

SIDALCEA A.Gray ex Benth.
malviflora group   Greek Mallows
●●● Long-persistent or established garden escapes; known since 1955 by the
river Wey, south of Guildford (Surrey). N America. BM, E, LANC, NMW,
OXF, RNG. 20*, 51*, 52*, 53*, 209, 805*(18:117-244). Incl. *S. candida*
A.Gray; *S. hendersonii* S.Watson; *S. malviflora* (DC.) A.Gray; *S. oregana*
(Nutt.) A.Gray; *S. spicata* E.Greene and hybrid cultivars.

ABUTILON Miller
malvifolium (Benth.) J.Black
● A wool casual. Australia. HbEJC. 37, 370(11:71), 590*.
pictum (Gillies ex Hook. & Arn.) Walp.   Chinese-lantern
● A casual on refuse tips at Romsey (Cambs) and Maidstone (W Kent),
probably a greenhouse escape. Brazil, Uruguay, Argentina. RNG. 51, 53*,
55*, 303(34:33). *A. striatum* G.Dickson ex Lindley; *Sida picta* Gillies ex
Hook. & Arn.
theophrasti Medikus   Velvetleaf
●●●● A wool, bird-seed and oil-seed casual. SE Europe, SW Asia;
(widespread). 7, 14, 19, 20*, 29*, 303*(11:8-9), 374(1949 suppl.:77).
A. avicennae Gaertner, nom. illeg.; *Sida abutilon* L.

BRIQUETIA Hochr.
spicata (Kunth) Fryxell
   Pre-1930 only. Tropical America. BM. 1, 354(4:190), 574*, 653*. *Abutilon spicatum* Kunth; *Sida spiciflora* DC.; *Wissadula spicata* (Kunth) J.S.Presl & C.Presl; *W. spiciflora* (DC.) Druce.

ALLOWISSADULA D.Bates
sessei (Lagasca) D.Bates
   1930 only. Mexico. LIV, OXF. 574*, 354(9:339), 707, 882*(11:329-354). *A. trilobata* (Hemsley) Rose; *Abutilon trilobatum* Hemsley; *Sida sessei* Lagasca; *Wissadula trilobata* (Hemsley) Rose.

MODIOLA Moench
caroliniana (L.) Don   Bristly-fruited Mallow
   • A wool casual. Tropical America; (widespread). E, LTN, RNG. 3, 21, 37*, 38*, 147, 370(11:71), 371(391:35), 519*, 533*. *M. multifida* Moench.

ANODA Cav.
cristata (L.) Schldl.   Spurred Anoda
   •• A wool, bird-seed and oil-seed casual, an impurity in horticultural seed, and possibly a garden escape. SW USA, Mexico, S America. BM, K, LTN, RNG.   14, 19, 40, 51, 303*(47:36-37), 312*(n.s. t.288), 354(12:272), 371(400:36), 650*, 652*, 826*(25:51), 848*(11:485-522), 1248*. *Sida cristata* L.; incl. *A. brachyantha* Reichb.; *A. hastata* Cav.; *A. lavateroides* Medikus. Most records are referable to var. *brachyantha* (Reichb.) Hochr.

PAVONIA Cav.
urens Cav.
   • A wool casual. Tropical Africa. BM, RNG. 17, 47*, 48*, 145, 236, 370(11:71), 371(388:37), 825*(1:113-117). Incl. *P. schimperiana* Hochst. ex A.Rich.; *P. tomentosa* Hochst.

GOSSYPIUM L.
herbaceum L.   Levant Cotton
   • A casual, from cotton imported for fibre or oil.   Asia; (widespread as a crop). 21, 29*, 68*, 303(18:11).
hirsutum L.   Upland Cotton
   • A casual, from cotton imported for fibre or oil.   No modern records. Tropical America; (widespread as a crop).   21, 24*, 68*, 354(10:468). *G. mexicanum* Tod.; *G. punctatum* Schum. & Thonn.

HIBISCUS L.
syriacus L.   Syrian Ketmia
   • A garden relic. Asia. 21, 54*, 62*, 63*, 458, 482. Self-sown seedlings appear to be restricted to gardens only.
trionum L.   Bladder Ketmia
   ••• A wool, bird-seed and oil-seed casual; also a garden escape; self-seeding since 1977 on Jersey (Channel Is). SE Europe, SW Asia, Africa; (Australasia).

BM, CGE, LANC, NMW, OXF, RNG.   1, 14, 19, 20*, 29*, 52*, 285, 303*(32:1,23), 1218.   *H. vesicarius* Cav.

SARRACENIACEAE
SARRACENIA L.
flava L.   Trumpets
  • Introduced and persistent on boggy ground at Vales Moor in the New Forest (S Hants) and on Chobham Common (Surrey). Formerly introduced in a bog in Co Roscommon, but died out during the 1930s.   SE USA.   20, 38*, 325(92:31 33; 115*:375), 342(Feb89), 352(45B:239), 370(16:173), 644*.
leucophylla Raf.   Purple Trumpet-leaf
  Pre-1930 only.   SE USA.   51, 66*, 325(92:31-33; 115*:372), 698*.
  S. drummondii Croom.
purpurea L.   Pitcherplant
  •• Introduced, now abundantly naturalised and spreading by wind-blown seed in several bogs in Ireland scattered over six vice-counties; introduced in several localities in England, persistent on Chobham Common (Surrey).   Eastern N America. BM, CGE, OXF.   16, 20, 22*, 199, 209, 252, 325(92:31-33,224; 115*:374), 644*.

DROSERACEAE
DROSERA L.
binata Labill.   Forked Sundew
  • Introduced and persistent in a ditch on wet heathland at Bisley Ranges (Surrey).   Australasia.   20, 51, 342(Feb89), 471, 519*, 569*, 591*.
capensis L.   Cape Sundew
  • Introduced and persistent in a ditch on wet heathland at Bisley Ranges (Surrey).   S Africa.   20, 51, 600*, 610*, 644*.

CISTACEAE
CISTUS L.
albidus L.   Grey-leaved Cistus
  • Reported from Sandbanks (Dorset), probably a garden escape.   SW Europe. 21, 24*, 29*, 30*, 63*, 403.
× **corbariensis** Pourret   Corbières Rock-rose
  • Reported from a steep overgrown bank above a disused nursery at Crockenhill (W Kent), probably a garden escape.   Spain, Portugal.   53*, 61, 63*, 458.   *C. populifolius* L. × *C. salviifolius*.
[*hirsutus*   An ambiguous name. 218(p.104).]
incanus L.   Hoary Rock-rose
  • A garden escape persistent for a few years on the side of a chalk pit below Mount Caburn (E Sussex) and on Great Ormes Head (Caerns); introduced on roadsides elsewhere.   Medit.   LIV, RNG, SLBI.   20, 21, 31*, 61*, 63*, 303*(34:22-23), 370(2;412).   *C. villosus* auct.; incl. *C. corsicus* Lois.; *C. creticus* L.

ladanifer L.   Gum Cistus
- Introduced and now well naturalised on a sandy warren in the grounds of Rosehill House at Farnham (E Suffolk); also a persistent garden escape at Bentley (E Suffolk), now gone. W Medit. 21, 29*, 30*, 63*, 258, 369(22:49).

laurifolius L.   Laurel-leaved Cistus
- Introduced and persistent in a roadside cutting at Barsham (E Suffolk); also a wool casual. Medit. E, HbEJC. 20, 21, 29*, 30*, 63*, 258, 369(18:159).

× **laxus** Aiton f.
- Introduced, or a persistent garden escape, among native vegetation in Silwood Park, Reading (Berks). SW Europe. 61, 325*(55:1-52), 353(25:41), 1325*. *C. populifolius* L. × *C. psilosepalus* Sweet. Not to be confused with *C. laxus* Brotero, a synonym of *C. psilosepalus*. 1043*. According to 564*, the parentage is *C. monspeliensis* L. × *C. populifolius*, and the correct binomial is *C.* × *nigricans* Pourret; in practice, the two hybrids are virtually indistinguishable. 61.

× **pulverulentus** Pourret
- A casual or persistent garden escape at Wisley (Surrey). SW Europe. 51, 63*, 199. *C. albidus* × *C. crispus* L.; *C. crispus* hort, non L.

salviifolius L.   Sage-leaved Rock-rose
- A casual or persistent garden escape on Tresco (Scilly). Medit, SW Asia. BM. 21, 29*, 30*, 31*, 63*, 354(12:36).

TUBERARIA (Dunal) Spach
lignosa (Sweet) Samp.
- Introduced and now well naturalised on a warren in the grounds of Rosehill House at Farnham (E Suffolk). W Medit. 21, 30*, 51*, 369(22:49), 737*. *T. vulgaris* Willk.; *Helianthemum lignosum* Sweet; *H. tuberaria* (L.) Miller.

HELIANTHEMUM Miller
[*ledifolium* (L.) Miller   In error for native *H. apenninum*. 188, 248.]
nummularium (L.) Miller subsp. **grandiflorum** (Scop.) Schinz & Thell.
- A persistent garden escape on a roadside wall near an old nursery at Busby (Lanarks). Europe, Caucasus. NMW, HbEJC. 21, 23*, 27*, 303(50:25). *H. grandiflorum* (Scop.) Lam.
pilosum (L.) Pers.
- A wool casual. W Medit. RNG. 17, 21, 46*, 683*.
[*salicifolium* (L.) Miller (*Cistus salicifolius* L.) In error for native *H. apenninum*. 281.]

VIOLACEAE
VIOLA L.
biflora L.   Alpine Yellow-violet
- A casual garden escape. No modern records. N temperate regions. BM. 21, 22*, 26*, 27*, 28*, 354(5:746; 9:553). Possibly recorded in error for *V. pensylvanica*, a similar but more vigorous species.

cornuta L.   Horned Pansy
●●● An established garden escape; naturalised on river banks, railway banks
and roadsides, especially in Scotland; known for over 40 years on the roadside
near Dunphail (Moray) and in Savernake Forest (N Wilts). Pyrenees. BM,
CGE, E, NMW, OXF, RNG. 1, 20, 22*, 23*, 27*, 28*, 51*, 172, 277.

declinata Waldst. & Kit.
Pre-1930 only. Carpathians. 21, 51, 354(3:12), 650*.

[*epipsila* Ledeb. In error for native *V. palustris*. 7, 380.]

labradorica Schrank   Labrador Violet
• A garden escape, recorded without locality from Yorkshire. N America.
38*, 51, 52*, 53*, 371(374:9).

pensylvanica Michaux   Smooth Yellow-violet
• A garden escape found in 1988 on a roadside bank at Great Bealings
(E Suffolk). Eastern N America. HbEJC. 38*, 40, 51, 369(27:25), 1319*.
V. eriocarpa Schwein.; *V. pubescens* Aiton var. *eriocarpa* (Schwein.)
N.Russell.

× **wittrockiana** Gams ex Kappert   Garden Pansy
●●●● A casual or persistent garden escape. Originated in cultivation. 16, 21,
51, 52*, 54*, 56*, 191. *V. altaica* Ker Gawler × *V. lutea* Hudson ×
V. tricolor L.; *V. hortensis* auct., non Schur. Hybrids and backcrosses with
native *V. arvensis* and *V. tricolor* have been recorded. 1311.

TAMARICACEAE
TAMARIX L.

africana Poiret   African Tamarisk
• A garden escape established in hedgerows at Old Grimsby, Tresco (Scilly).
W Medit. BM. 20, 21, 24*, 29*, 207, 512*.

chinensis Lour.   Chinese Tamarisk
Pre-1930 only. Mongolia, China, Japan. BM. 21, 36c*, 44*, 51, 61, 512*.

gallica L.   Tamarisk
●●●● Introduced and long persistent on many stretches of the coasts of Wales
and S and E England, established by suckering in some places; a garden escape
elsewhere. W Medit. 1, 20, 22*, 33*, 191, 207, 512*. *T. anglica* Webb;
Tamarindus gallica sphalm.

parviflora DC.
Pre-1930 only. E Medit. BM. 21, 44*, 512*.

tetrandra Pallas ex M.Bieb.
Pre-1930 only. E Medit., SW Asia. BM. 21, 248, 512*, 1181*.

FRANKENIACEAE
FRANKENIA L.
pulverulenta L.
• A wool casual. No modern records. Eurasia, N Africa. OXF, RNG.
1, 14, 21, 44*, 354(6:275; 8:728), 683*, 737*.

PASSIFLORACEAE

PASSIFLORA L.

caerulea L. Blue Passion-flower
- A casual garden escape, reported as being self-sown in Whiteknights Park at Reading (Berks). Brazil, Peru, Argentina. RNG. 52*, 61*, 63*, 64*, 126, 303(32:19), 1344*.

CUCURBITACEAE

ECBALLIUM A.Rich.

elaterium (L.) A.Rich. Squirting Cucumber
- ●● A garden escape established for some years on a cliff edge at Peacehaven (E Sussex) and as a weed in gardens at Pontac in Guernsey (Channel Is); persistent in other coastal areas in the south; a wool casual elsewhere. Medit., SW Asia; (Australia). JSY, LTN, NMW, OXF. 2, 20, 24*, 32*, 145, 201, 303(38*:17-18; 40:24), 683*, 737*. *Momordica elaterium* L.

SICYOS L.

angulatus L. Bur Cucumber
- ●● A casual or persistent oil-seed and bird-seed alien. Eastern N America; (Europe, Australasia). BM, CGE, K, MNE, RNG. 19, 20, 21, 38*, 236, 261, 303*(38:17-18), 371(367:29), 683*. Incl. *S. australis* Endl. The somewhat similar NE American *Echinocystis lobata* (Michaux) Torrey & A.Gray occurs in NW Europe and may be overlooked. 21.

CUCUMIS L.

melo L. Melon
- ●●● A food refuse casual; possibly also a wool casual. Tropical Africa, Asia; (widespread as a crop). BM, RNG. 7, 14, 20, 68*, 80*, 126, 303*(38:14,17).

myriocarpus Naudin Gooseberry Cucumber
- ●● A wool casual; perhaps also from bird-seed. S Africa; (S Europe, Australia). BM, CGE, E, LTN, OXF, RNG. 7, 14, 21, 37*, 303(38:14), 520*, 533*, 737*.

sativus L. Cucumber
- ●● A food refuse casual or garden escape. Origin obscure; (widespread as a crop). HbALG, HbTGE. 7, 20, 68*, 71*, 80*, 126, 303*(38:15,17).

CITRULLUS Schrader

colocynthis (L.) Schrader Bitter Cucumber
Pre-1930 only. N Africa, Asia; (Australia). BM. 7, 21, 24*, 29*, 44a*, 47*, 354(7:876), 683*.

lanatus (Thunb.) Matsum. & Nakai Water Melon
- ●●● A food refuse casual on tips; possibly also a wool alien. Africa; (widespread as a crop). BM, E, NMW, OXF, RNG, SLBI. 7, 20, 68*, 71*, 80*, 126, 146, 303*(38:14,16), 683*. *C. citrullus* (L.) Karsten; *C. vulgaris* Schrader; *Colocynthis citrullus* (L.) Kuntze.

CUCURBITA L.

ficifolia Bouché   **Fig-leaved Gourd**
    Pre-1930 only. Tropical America. 36c, 197, 347*(8:86-102). *Cucumis anguria* L. var. *ficifolia* Bouché.

maxima Duchesne ex Lam.   **Pumpkin**
    •• (but confused with *C. pepo*). A food refuse casual. C America; (widespread as a crop). 7, 20, 71*, 80*, 236, 303(38:13-18), 347*(8:86-102), 683*.

pepo L.   **Marrow, incl. Courgette, Ornamental Gourd**
    •••• A food refuse casual and garden escape. C America, Mexico; (widespread as a crop). 7, 20, 80*, 303*(38:15,17), 347*(8:86-102), 683*. Incl. *C. ovifera* L.

LAGENARIA Ser.

siceraria (Molina) Standley   **White-flowered Gourd**
    • A casual on a rubbish tips, perhaps a greenhouse escape. Tropics. NMW, RNG. 5, 21, 66*, 68*, 354(8:26), 371(391:34), 683*. *L. lagenaria* (L.) Druce; *L. vulgaris* Ser.

MOMORDICA L.

charantia L.   **Balsam-pear**
    • A casual on tips, perhaps from food refuse, or a greenhouse escape. Tropics. RNG, HbEJC. 51a, 56*, 69*, 126*, 303*(38:17-18), 331(56:88), 370(9:188), 1174*.

BEGONIACEAE

BEGONIA L.

grandis Dryander   **Hardy Begonia**
    • A garden escape persistent by the river Thames at Staines (Middlesex), now gone. E Asia. 51, 54*, 57*, 192. *B. evansiana* Andrews.

rex Putz.   **Painted-leaf Begonia**
    • A casual garden escape. Assam. 51, 54*, 303(37:31), 331(62:109), 593*, 600*, 719*.

× **semperflorens** hort.   **Wax Begonia**
    • A casual garden escape. Originated in cultivation. 51, 54*, 56*, 66*, 199. *B. cucullata* Willd. × *B. schmidtiana* Regel. Not to be confused with *B. semperflorens* Link & Otto, a synonym of *B. cucullata* var. *hookeri* (A.DC.) L.B.Smith & Schubert.

LOASACEAE

BLUMENBACHIA Schrader

hieronymii Urban
    • A casual garden escape; also on shingle at Loch Ranza on Arran (Clyde Is). Argentina. BM, OXF, SLBI. 7(as *B. insignis*), 303(42:18; 51*:32-33), 354(12:789), 511.

insignis Schrader
- A casual on dumped refuse outside Wisley Gardens (Surrey). Brazil, Argentina. 51, 371(382:24), 511*(p.674), 650*. Early records were in error for *B. hieronymii*. 7, 354(2:502; 12:789).

MENTZELIA L.
albicaulis (Douglas ex Hook.) Douglas ex Hook.   White-stemmed Stickleaf
Pre-1930 only. Western N America. BM, CLE, OXF. 1, 12, 41*, 42, 300(1906:42), 354(5:108), 500*. *Bartonia albicaulis* Douglas ex Hook.
lindleyi Torrey & A.Gray   Golden Bartonia
- A casual garden escape. No modern records. California. OXF. 12, 42, 51*, 52*, 55*, 301*(t.1831). *M. aurea* hort., non Nutt.; *Bartonia aurea* Lindley.

SALICACEAE
POPULUS L.
alba L.   White Poplar
●●●●● Introduced; sometimes established by suckering, especially in coastal areas. Europe. 1, 20*, 22*, 59*, 86*, 236, 1208.
balsamifera L.   Eastern Balsam-poplar
●● (but confused with *P. trichocarpa* and hybrids). Introduced; sometimes spreading by suckering. Northern N America. BM, LANC, RNG. 1, 12, 20, 50, 59*, 199, 606*. *P. tacamahacca* Miller.
× **berolinensis** K.Koch   Berlin Poplar
- Introduced. Originated in cultivation. 13, 20, 21, 50, 58*, 60*, 166, 613*. *P. laurifolia* × *P. nigra* 'Italica'.
× **canadensis** Moench   Black-poplar hybrids
●●●●● Introduced. Originated in cultivation. 2, 20*, 22*, 59*, 86*. *P. deltoides* × *P. nigra*; *P.* × *euamericana* Gruinier ex Piccarolo, nom. illeg.; incl. *P. eugenei* Simon-Louis; *P. gelrica* Houtz.; *P. lloydii* A.Henry; *P. marilandica* Bosc ex Poiret; *P.* × *regenerata* A.Henry; *P.* × *robusta* C.Schneider; *P.* × *serotina* Hartig; *P. virginiana* Foug. Putative hybrids with *P. candicans* and *P. laurifolia* have been recorded. 20, 331(65:197).
candicans Aiton   Balm-of-Gilead
●●●● (but confused with *P. balsamifera* and *P. trichocarpa*). Introduced; sometimes spreading by suckering. Origin obscure. 2, 20*, 22*, 50, 86*, 230. Possibly *P. balsamifera* × *P. deltoides*; *P. balsamifera* auct., non L.; *P. gileadensis* Roul.; *P. tacamahacca* auct., non Miller. A putative hybrid with *P. nigra* has been recorded. 458, 370(19:188-190).
× **canescens** (Aiton) Smith   Grey Poplar
●●●●● Introduced and established by suckering, especially in SE England. Europe. 2, 20*, 22*, 86*. *P. alba* × *P. tremula* L.; *P.* × *hybrida* M.Bieb.
deltoides Bartram ex Marshall   Necklace Poplar
- Introduced. Eastern N America. LANC, LIV. 1, 13, 21, 22*, 50, 59*, 86, 184, 196, 606*, 684*. Incl. *P. monilifera* Aiton. 1205*. A hybrid

with *P. trichocarpa* (*P.* × *generosa* A.Henry) has been recorded. OXF. 13, 166.

× **jackii** Sarg.
- Introduced. N America. 61, 350(35:21), 510*, 631*. *P. balsamifera* × *P. deltoides*.

laurifolia Ledeb.   Laurel-leaved Poplar
- Introduced. C Asia. BM, OXF. 7, 13, 20, 50, 61, 354(4:281; 7:212), 510*, 631*, 636*. A putative hybrid with *P.* × *canadensis* has been recorded. 331(65:197).

maximowiczii A.Henry   Japanese Balsam-poplar
- Introduced. NE Asia. 13, 50, 53*, 61. A hybrid with *P. trichocarpa* (*P.*'Androscoggin') has been recorded. 13.

nigra L. **'Gigantea'**   Giant Lombardy-poplar
- ●●●● (but overlooked as *P.*'Italica'). Introduced. Originated in cultivation. 20, 51b, 61.

nigra L. **'Italica'**   Lombardy-poplar
- ●●●●● Introduced. Originated in cultivation. 2, 20, 50, 59*, 60*, 479. *P. nigra* var. *italica* Muenchh.; *P. dilatata* Aiton; *P. fastigiata* Desf.; *P. italica* (Duroi) Moench; *P. pyramidalis* (Rozier ex Lam.) Dode. Seedlings have been erroneously reported from this male clone in Salop.

nigra L. **'Plantierensis'**
- ●●●●● (but overlooked as *P.*'Italica'). Introduced. Originated in cultivation. CGE. 20, 50, 58*, 60, 143, 245, 283. *P. nigra* subsp. *betulifolia* (Pursh) W.Wettst. × *P.*'Italica'; *P. nigra* var. *plantierensis* (Simon-Louis) C.Schneider; *P. plantierensis* Dode.

tremuloides Michaux   American Aspen
- Introduced. N America. DZS. 38*, 50, 59*, 61, 172, 505*.

trichocarpa Torrey & A.Gray ex Hook.   Western Balsam-poplar
- ●●● (but confused with *P. balsamifera* and hybrids). Introduced. Western N America. BM, LANC, NMW, OXF. 20*, 58*, 59*, 60*, 86*, 354(13:309). *P. balsamifera* subsp. *trichocarpa* (Torrey & A.Gray ex Hook.) Brayshaw. Superseded in cultivation by hybrids with *P. balsamifera*. 20*. A hybrid with *P. deltoides* has been recorded. 13.

[*wilsonii* Schneider   In error for native *P. nigra* subsp. *betulifolia* (*P. wilsonii* hort., non Schneider). 185.]

yunnanensis Dode   Yunnan Poplar
- Introduced. China. 13, 36c*, 50, 631*.

SALIX L.
alba L. **'Britzensis'**   Coral-bark Willow
- Introduced. Originated in cultivation. 61, 58*, 86, 185, 235. A cultivar of native *S. alba* var. *vitellina*.

[*appendiculata* Villars   A dubious record. 354(5:786). *S. grandiflora* Ser.]

babylonica L.   True Weeping-willow
- ●●●● (but mostly in error for hybrids). Formerly introduced, now superseded by its hardier hybrids *S.* × *pendulina* and *S.* × *sepulcralis*. China. 2, 21,

50*, 59*, 86(p.59).   *S*.'Tortuosa', Corkscrew Willow, a cultivar of
S. babylonica var. *pekinensis* Henry (*S. matsudana* Koidz.), has been reported
from Box Hill (Surrey). HbEJC. 50, 52*, 59*, 60*, 61*, 482.

candida Fluegge
- Introduced. Eastern N America. 38*, 40*, 303(58:42), 631*.

daphnoides Villars sensu lato   Violet Willow
- ●●● Introduced. Scandinavia. BM, E, LANC, NMW, OXF, RNG. 2, 20*,
22*, 58*, 86*, 251. Incl. *S. acutifolia* Willd.; *S. caspica* hort., non Pallas; *S.
pomeranica* Willd.; *S. pruinosa* Wendl. ex Reichb.; *S. violacea* Andrews.

discolor Muhlenb.   Glaucous Willow
- A garden escape persistent on the site of the Glasgow Garden Festival
(Lanarks). Eastern N America. 38*, 50*, 51b, 61, 63*, 370(19:174), 631*,
918*(9:95), 1348*, 1356*.

[× *ehrhartiana* Smith   Alien status dubious.   Both parents (*S. alba* and
S. pentandra) are native, but the hybrid is possibly of garden or European
origin. 20, 1311.]

elaeagnos Scop.   Olive Willow
- ●● Introduced or a persistent garden escape. C & S Europe, NW Africa,
Turkey. LIV. 7, 20, 22*, 50*, 86*, 174, 201, 303(21:19), 370(12:355).
S. incana Schrank.

eriocephala group   Heart-leaved Willows
- Introduced; perhaps established in boggy ground in Sutton Park (Warks) and
Heysham Reserve (W Lancs); also recorded from Chailey Common (E Sussex)
and at Llanfarian (Cards). N America. NMW, OXF, WAR. 20*, 38*, 40*,
50*, 174, 303(23:12; 56:25; 58:41), 370(13:139; 19:282), 371(419:15).
Incl. *S. acutidens* Rydb.; *S. cordata* Michaux; *S. eriocephala* Michaux;
S. missouriensis Bebb; *S. rigida* Muhlenb.

[*hastata* L.   Dubious records. BM, K. 187, 268, 354(5:786). *S. malifolia*
Smith.]

[*helvetica* Villars   A dubious record. 354(1:270; 5:787).]

hookeriana J.Barratt
- Introduced and perhaps established in swamp at Heysham (W Lancs).
Western N America. 41*, 42, 50*, 371(419:15).

irrorata Andersson   Blue-stem Willow
- Introduced in a gravel pit at Thorpe (Surrey).   SW USA.   50*, 62,
342(Feb90), 511, 1345*.

lucida Muhlenb.   Shining Willow
- Introduced and perhaps established in a swamp at Heysham (W Lancs).
Eastern N America. 38*, 40*, 50*, 371(419:15).

microstachya Turcz. ex Trautv.
- Introduced and established in a swamp at Heysham (W Lancs). NE Asia.
50*, 458, 617*.

× **pendulina** Wender.   Weeping Crack-willow
   •• (but probably under-recorded). Introduced. Origin obscure. BM, LSR.
   16, 20*, 50*, 86*, 119, 134, 135, 209. *S. babylonica* × *S. fragilis* L.;
   S. babylonica auct., non L.; *S.* × *blanda* Andersson.
petiolaris Smith   Slender Willow
   Pre-1930 only. N America. LIV. 38*, 40*, 51, 61, 159, 185, 1345*.
[*retusa* L.   In error for native *S. myrsinites*. 354(5:787).]
[*rosmarinifolia* L.   Dubious records, probably in error for native *S. repens* and
   its hybrids. 7, 86, 354(7:62).]
schraderiana Willd.
   • A garden escape reported from a steep overgrown bank above a disused
   nursery at Crockenhill (W Kent). Europe. 21, 50*, 458, 546*, 1089*.
   S. bicolor Willd., non Ehrh.; *S. phylicifolia* auct., non L.
× **sepulcralis** Simonkai   Hybrid Weeping-willow
   ••• (but under-recorded as *S. babylonica*). Introduced. Originated in
   cultivation. 16, 20*, 22*, 59*, 86*, 1311. *S. alba* × *S. babylonica*; incl.
   S. alba 'Tristis'; *S. alba* 'Vitellina Pendula'; *S.* × *chrysocoma* Dode.
× **seringeana** Gaudin
   Pre-1930 only. Origin obscure. 50, 61, 108, 695*. *S. caprea* L. ×
   S. eleagnos.
udensis Trautv. & Meyer
   • A garden escape established since the 1960s in a woodland bog at Camis
   Eskan (Dunbarton); also collected from ?Viol Moor (Cumberland). NE Asia.
   LANC. 20, 50*, 62*, 316(21:556). *S. sachalinensis* Schmidt; incl. *S.* 'Sekka'.

CAPPARACEAE
Capparidaceae

CLEOME L.
gynandra L.
   • A casual at Bristol (N Somerset), vector unknown. Formerly a grain casual.
   No modern records. Tropics. BM, OXF. 7, 47*, 48*, 51, 354(7:765; 8:107;
   11:23), 1035*. *Gynandropsis gynandra* (L.) Briq.; *G. pentaphylla* (L.) DC.
serrulata Pursh   Rocky Mountain Bee-plant
   Pre-1930 only. Western N America. 7, 41*, 42, 319(3:21), 354(3:10), 500*.
sesquiorygalis Naudin ex C.Huber   Spiderflower
   •• A casual garden escape. S America. HbEJC. 51*, 54*, 55*, 56*(all as
   C. spinosa), 20, 49, 126, 199, 303(26:14). *C. hassleriana* Chodat; *C. spinosa*
   auct., non Jacq.
[*spinosa* Jacq.   In error for *C. sesquiorygalis*. 303(26:14).]

BRASSICACEAE
Cruciferae

SISYMBRIUM L.

altissimum L.   **Tall Rocket**

•••• An established wool, bird-seed and grass seed alien; widespread throughout the British Is, locally plentiful on waste ground especially in the south and east. Europe, SW Asia; (widespread). 1, 14, 19, 20*, 22*, 90*, 303*(30:11). *S. pannonicum* Jacq.; *S. sinapistrum* Crantz; incl. *S. rigidulum* Decne., non Tchich.; *S. ucranium* Błocki.

austriacum Jacq.   **Austrian Rocket**

• Recently recorded as a casual from Whiteknights Park at Reading (Berks) and reported from a re-seeded area of Danson Park at Welling (W Kent). Europe; (S America). BM, CGE, LANC, LNHS, NMW, OXF. 2, 16, 21, 22*, 23*, 303*(30:11), 353(31:22), 458. *S. pyrenaicum* (L.) Villars, non L.; incl. *S. hispanicum* Jacq. Early records were confused with *S. volgense*. 370(12:311-314).

burchellii DC.

• A wool casual. No modern records. S Africa. E, OXF. 3*, 7, 133, 354(4:189), 541*. Incl. *S. turczaninowii* Sonder.

capense Thunb.

• A wool casual. S Africa. RNG. 541*, 592.

erysimoides Desf.   **French Rocket**

•• A wool casual. Medit.; (Australia). BM, CGE, E, LIV, OXF, RNG. 7, 14, 20*, 44*, 45*, 90.

irio L.   **London Rocket**

•••• An established grain and wool alien, mostly in C and E England; known in London since about 1660, and still on walls and in shrubberies about the City (Middlesex). An early colonist which has been claimed as native. Eurasia, N Africa; (widespread). 1, 20*, 22*, 33*, 35*, 90*.

loeselii L.   **False London-rocket**

•••• Established, probably as a grain and bird-seed alien; naturalised on waste ground and roadsides, locally in abundance, especially about London north of the river Thames; also a wool casual. Eurasia; (N America). 1, 8, 20*, 22*, 90*, 303*(30:11), 310(25:41). *S. irio* auct., non L.

[*obtusangulatum* Schleicher ex Willd. In error for *S. volgense*. 370(12:311-314).]

orientale L.   **Eastern Rocket**

••••• An established bird-seed, grain and wool alien; widespread and locally abundantly naturalised on waste ground in town centres, in quarries and gravel pits, especially in southern England. Medit.; (widespread). 1, 19, 20*, 22*, 90*, 143, 303*(30:11). *S. columnae* Jacq.; *S. irio* auct., non L.; incl. *S. subhastatum* Willd.

polyceratium L.   **Many-podded Hedge-mustard**
   •• A casual, vector unknown. Formerly a grain and ballast alien long persistent at Bury St Edmunds (W Suffolk). Europe, N Africa. BM, CGE, E, LANC, NMW, OXF. 1, 16, 21, 39*, 90, 235, 303*(30:11), 354(2:37).

polymorphum (Murray) Roth
   Pre-1930 only. Eurasia. 1, 21, 195, 623*, 630*, 715*, 1048*. *S. junceum* M.Bieb.

runcinatum Lagasca ex DC.
   • An esparto casual. SW Europe, N Africa. CGE, K, RNG. 5, 21, 44*, 45*, 90, 300(1909:41), 303(29:12; 30*:11).

septulatum DC.
   • A grain casual. SW & C Asia. BM, RNG. 39, 182, 354(11:465), 350(29:23), 734*. *S. grandiflorum* Post; *S. pannonicum* Boiss. pro parte, non Jacq.; *S. rigidulum* Tchich., non Decne.; *Sinapis oliveriana* DC.

strictissimum L.   **Perennial Rocket**
   •• Established about the churchyard on Kew Green (Surrey), in the Marine Park at South Shields (Durham) and in and near Chelsea Physic Garden (Middlesex); casual elsewhere. Recently refound by the river Mersey at Heaton Mersey (S Lancs) where it had been recorded in 1890. Europe. BM, CGE, K, OXF, RNG, SLBI. 1, 20*, 22*, 90*, 126, 166, 303*(30:10-11), 320(36:32), 455.

thellungii O.Schulz
   • A wool casual. S Africa. OXF, RNG. 7, 14, 354(5:277), 541*, 1035*.

volgense M.Bieb. ex Fourn.   **Russian Mustard**
   •• An established grain alien; naturalised on a roadside bank at Caistor St Edmunds (E Norfolk), by disused railways at Hadleigh (W Suffolk) and Alnwick (Cheviot), and on waste ground at ports. SW Russia; (Europe). BM, BRIST, LPL, NMW, OXF, RNG. 7, 9, 20*, 22*, 90*, 303*(30:10-11), 370(12:311-314), 1138*, 1242*, 1342*. *S. hispanicum* auct., non Jacq.

LYCOCARPUS O.Schulz
fugax (Lagasca) O.Schulz
   Pre-1930 only. SE Spain. 2, 16, 21, 300(1904:175), 564*, 1114*. *Sisymbrium fugax* Lagasca.

DESCURAINIA Webb & Berth.
[*myriophylla* (Willd.) R.E.Fries (*Sisymbrium myriophyllum* (Willd.) DC.)   In error for *D. pulcherrimum*. E. 354(6:15).]

pinnata (Walter) Britton   **Tansy-mustard**
   • A casual at Gloucester docks (W Gloucs), vector unknown. Formerly a wool casual. No modern records. N America. BM, E, OXF. 1, 3, 9, 38*, 40, 41*, 42, 182, 354(5:15; 7:28,556). *D. brachycarpa* O.Schulz; *D. multifida* (Pursh) O.Schulz; *Sisymbrium brachycarpon* Richardson; *S. brachyphyllum* sphalm.; *S. canescens* Nutt.; *S. multifidum* (Pursh) MacMillan; *S. pinnatum* E.Greene.

pulcherrima Muschler
 Pre-1930 only. Bolivia, Argentina. E, OXF. 7, 354(6:15), 861(49:200).
 Sisymbrium pulcherrima (Muschler) Druce.
richardsonii (Sweet) O.Schulz   Western Tansy-mustard
 Pre-1930 only. N America. OXF. 7, 16, 38*, 40, 41*, 42, 318(19:238),
 354(7:433; 9:258). *Sisymbrium richardsonii* Sweet; incl. *S. incisum* Engelm.
 (*Sophia incisa* (Engelm.) E.Greene); *Sisymbrium procerum* Schumann (*Sophia
 procera* (Schumann) E.Greene).
 [*sophia* (L.) Webb ex Prantl   Accepted, with reservations, as native.]

HUGUENINIA Reichb.
tanacetifolia (L.) Reichb.   Tansy-leaved Rocket
 Pre-1930 only. SW Europe. 7, 21, 23*, 27*, 267, 1242*. *Sisymbrium
 tanacetifolium* L.

MYAGRUM L.
perfoliatum L.   Mitre Cress
 •• A grain and bird-seed casual. Europe, SW Asia. BM, CLE, LIV, OXF.
 1, 11, 19, 21, 22*, 25*, 90*, 303*(47:34-35).

ISATIS L.
glauca Aucher-Éloy ex Boiss.
 • A casual weed at Helensburgh (Dunbarton), perhaps a garden escape.
 SW Asia. 7, 39*, 51, 223.
lusitanica L.
 Pre-1930 only. E Medit., SW Asia. 7, 16, 21, 39*, 44*, 300(1910:43).
 I. aleppica Scop., incl. var. *pamphylica* Boiss.
 [*tinctoria* L.   Accepted, with reservations, as native.]

BUNIAS L.
erucago L.   Southern Warty-cabbage
 • A grain casual. No modern records. Medit. BM, LIV, OXF. 7, 8, 22*,
 29*, 32*, 90*, 143, 234. The 1987 record from Winchbottom (Bucks) is in
 error for *B. orientalis*. 353(40:29), 403.
orientalis L.   Warty-cabbage
 •••• An established grain, bird-seed and ballast alien; widespread and locally
 plentiful in chalk and gravel pits, on roadsides and by railways, mainly in
 southern England. Europe, SW Asia. 1, 8, 19, 20*, 22*, 90*, 191.

GOLDBACHIA DC.
laevigata (M.Bieb.) DC.
 Pre-1930 only. Asia. 1, 16*, 21, 39*, 300(1910:43).

BOREAVA Jaub. & Spach
orientalis Jaub. & Spach
 Pre-1930 only. Turkey, Syria. 1, 16*, 39*, 300(1905:98), 354(8:726).

ERYSIMUM L.

allionii group   Siberian Wallflowers
  ••• Casual garden escapes. Originated in cultivation. BM, OXF, RNG.
  1, 20*, 52*, 54*, 90*, 191, 192, 371(343:30). Incl. *E. allionii* hort., non
  (DC.) Kuntze (*Cheiranthus allionii* hort.); *E. perovskianum* Fischer &
  C.Meyer; *E. asperum* × *E. hieracifolium*; *E. decumbens* × *E. perofskianum*;
  and *E.* × *marshallii* Stark. The correct name for this group may be
  E. × *marshallii* (Heslop-Harrison) Bois.

asperum (Nutt.) DC.   Western Wallflower
  • A persistent garden escape. N America. 38*, 40, 41*, 51, 371(343:30).
  E. arkansanum Nutt. ex Torrey & A.Gray; *Cheiranthus asper* Nutt.

[*cheiranthoides* L.   Accepted, with reservations, as native.]

cheiri (L.) Crantz   Wallflower
  ••••• An established garden escape; known for centuries on old walls and
  ruins; widely naturalised on coastal cliffs, rock outcrops, railway cuttings
  and in quarries. SE Europe. 1, 20*, 22*, 34*, 90*. *Cheiranthus cheiri* L.;
  incl. *C. fruticulosus* Smith. A hybrid with *E. bicolor* DC. (*E.* × *kewensis*
  hort.; *C.* × *kewensis* hort.) has been reported. 458.

decumbens (Schleicher ex Willd.) Dennst.   Decumbent Treacle-mustard
  Pre-1930 only. SW Europe. 21, 23*, 27*, 156. *E. ochroleucum* DC., nom.
  illeg.

diffusum Ehrh.
  Pre-1930 only. Eurasia. 7, 16, 21, 354(2:495), 695*, 715*. *E. canescens*
  Roth.

grandiflorum Desf.
  Pre-1930 only. W Medit. BM. 21, 30, 546*. *E. longifolium* DC.

helveticum (Jacq.) DC.
  •• Introduced in a bomb crater on Brockham Hill (Surrey); a casual garden
  escape elsewhere. Origin obscure. BM, CGE, LANC, OXF, RNG, SLBI. 7,
  12, 51, 53*, 312*(n.s. t.535), 371(385:35). Possibly of *E. cheiranthoides*
  × *E. cheiri* origin. *E. arkansanum* hort., non Nutt. ex Torrey & A.Gray;
  E. helveticum Jacq.; *E. pumilum* auct.; *E. suffruticosum* Sprengel; *E.* 'Golden
  Gem'. The name of this plant has, unfortunately, been widely misapplied to
  both *E. rhaeticum* (Schleicher ex Hornem.) DC. from the Alps, as in 21, and
  to *E. drenowskii* Degen from the Balkans. 21a.

hieraciifolium group
  • Casuals on a refuse tip at Sundon (Beds) and at Seaford (E Sussex), vector
  unknown. Eurasia. K, LCN, LIV. 1, 21, 22*, 23*, 27*, 122, 143, 160, 248,
  1242*. Incl. *E. durum* J.S.Presl & C.Presl; *E. hieraciifolium* L.; *E. strictum*
  Gaertner, Meyer & Scherb.; *E. virgatum* Roth.

incanum Kunze   Hoary Treacle-mustard
  Pre-1930 only. SW Europe, NW Africa. LSR. 7, 16, 21, 23*, 184, 737*.
  E. aurigeranum Jeanb. & Timb.-Lagr.

linifolium (Pers.) Gay
  • A casual garden escape. SW Europe. ABD, BM, CGE, E, LIV. 21, 30*,
  52*, 65*, 277, 303(16:18), 475.

[*orientale* Miller   Unconfirmed.   Probably in error for *Conringia orientalis*.
178, 215.]

repandum L.   Spreading Treacle-mustard
●● A persistent grain and agricultural seed alien.   Formerly a wool casual.
Europe, SW Asia; (widespread).   BM, CGE, E, LANC, LIV, OXF. 1, 3, 21,
22*, 90*, 182, 303*(42:18-19), 354(12:480), 1242*.   *Sisymbrium repandum*
sphalm.

scoparium (Brouss. ex Willd.) Wettst.
● A casual in Guernsey (Channel Is), probably a garden escape. No modern
records. Canary Is. 51a, 220, 301*(t.1431 as *Cheiranthus mutabilis*), 530*.
Cheiranthus scoparius Brouss. ex Willd.

SYRENIA Andrz. ex Besser
cana (Piller & Mitterp.) Simonk.
Pre-1930 only. C Europe, Caucasus, W Siberia. SLBI. 21, 623a*, 630,
715*, 741*.   *S. angustifolia* (Ehrh.) Reichb.; *Cheiranthus bocconei* All.;
C. canus Piller & Mitterp.; *Erysimum canum* (Piller & Mitterp.) Polatschek;
E. cinereum Moench.

HESPERIS L.
laciniata All.
● A casual, vector unknown.   Medit.   2, 21, 27*, 30*, 31*, 90, 220,
300(1907:38).

matronalis L.   Dame's-violet
●●●●● An established garden escape; widely naturalised on roadsides and by
streams throughout the British Is. Eurasia. 1, 20*, 22*, 23*, 90*. Incl.
H. candida Kit.; *H. cladotricha* Borbás; *H. nivea* auct.; *H. voronovii* N.Busch.

MALCOLMIA R.Br.
africana (L.) R.Br.   African Stock
● A casual in 1951 and 1990 at Askham Pier (Westmorland),and in newly sown
grass in 1932 at Kew Gardens (Surrey).   Formerly mainly at ports. Medit.
BM, CGE, E, LANC, NMW, OXF. 1, 9, 21, 22*, 39*, 44*, 90, 303(58:38;
60*:41-44).   *Wilckia africana* (L.) F.Muell.

[*chia* (L.) DC.   A dubious record. 303*(60:41-44).]

crenulata (DC.) Boiss.
Pre-1930 only. Middle East. CLE, E. 1, 39, 44*, 303(58:38; 60*:41-44).
Wilckia crenulata (DC.) Druce.

[*littorea* (L.) R.Br.   A dubious record. 199, 303(60:44).   *Wilckia littorea* (L.)
Druce.]

maritima (L.) R.Br.   Virginia Stock
●●●● A garden escape and esparto casual. SE Europe. 1, 15, 20*, 32*, 90*,
303*(60:41-44).   *Cheiranthus maritimus* L.; *Wilckia maritima* (L.) Halácsy.

[*ramosissima* (Desf.) Thell. (*M. parviflora* (DC.) DC.; *Wilckia parviflora* (DC.)
Druce)   In error for *M. africana* and *M. maritima*. 303(60:44).]

MATTHIOLA R.Br.
fruticulosa (L.) Maire   Sad Stock
- An esparto casual. No modern records. Medit. 1, 15, 16, 21, 27*, 30*, 90*, 267, 354(5:367). *M. tristis* (L.) R.Br.

[*incana* (L.) R.Br. (*Cheiranthus incanus* L.) Accepted, with reservations, as native.]

longipetala (Vent.) DC.   Night-scented Stock
- ••• A casual garden escape. SE Europe, SW Asia. BM, CLE, CGE, E, LANC, OXF. 1, 20*, 32*, 90*. Incl. *M. bicornis* (Sibth. & Smith) DC.; *M. oxyceras* DC.

tricuspidata (L.) R.Br.
- A casual on a tip near Cheltenham (E Gloucs), vector unknown. No modern records. Medit. NMW. 2, 16, 21, 31*, 44*, 90*, 182, 263.

TETRACME Bunge
quadricornis (Stephan) Bunge
Pre-1930 only. Russia, C Asia. 7, 21, 136, 617*.

CHORISPORA R.Br. ex DC.
syriaca Boiss.
- A casual at Langstone Harbour (S Hants), vector unknown. SW Asia. BM, CLE. 1, 39*, 44*, 300(1906:101), 371(425:8). *C. purpurascens* (Banks & Sol.) Eig; *Chorispermum syriacum* (Boiss.) Druce.

tenella (Pallas) DC.
- A grain and grass seed casual. E Europe, Asia. BM. 1, 9, 16*, 21, 303(60:35), 349(2:369), 589*, 1242*. *Chorispermum tenellum* (Pallas) R.Br.

EUCLIDIUM R.Br.
syriacum (L.) R.Br.
- A casual, probably from grain. No modern records. SW Asia. BM, CGE, LSR, OXF, SLBI. 1, 16*, 21, 39*, 248, 354(12:36), 589*. *Anastatica syriaca* L.; *Soria syriaca* (L.) Desv. ex Steudel.

BARBAREA R.Br.
intermedia Boreau   Medium-flowered Winter-cress
- •••• Locally established or persistent on arable and waste land in widely scattered localities throughout the British Is; increasing. C & S Europe, N Africa. 2, 20*, 22*, 90*, 370*(16:389-396).

[*sicula* C.Presl A dubious record, perhaps in error for *B. intermedia* or *B. verna*. 300(1909:41), 370(16:395), 1075.]

[*stricta* Andrz. Accepted, with reservations, as native.]

[*taurica* DC. Recorded pre-1930, as a native of E Europe and W Asia, but it is not separable, even as a variety, from native *B. vulgaris*. 354(2:273), 394(25:174-175).]

verna (Miller) Asch.   American Winter-cress
- •••• An established escape from, or relic of, cultivation mainly in southern England and Wales; known since 1877 at Old Grimsby (Scilly). SW Europe.

1, 20*, 22*, 90*, 207, 370*(16:389-396). *B. patula* Fries; *B. praecox* (Smith) R.Br.; *Erysimum praecox* Smith; *E. vernum* Miller.

RORIPPA Scop.

austriaca (Crantz) Besser   Austrian Yellow-cress

●●● Established; originally a grain alien, now locally naturalised on railway banks, roadsides, riversides, in meadows, orchards and pits in scattered localities in England and Wales; the largest colony, the only one in Scotland, may be at Possilpark in Glasgow. Europe, SW Asia. BM, CGE, E, NMW, OXF, RNG.   1, 20*, 22*, 90*, 192, 209, 350(37:27), 1083*, 1242*. *Nasturtium austriacum* Crantz; *Radicula austriaca* (Crantz) Druce, nom. illeg. Hybrids with native *R. amphibia* (*R.* × *hungarica* Borbás) and *R. sylvestris* (*R.* × *armoracioides* (Tausch) Fuss) have been recorded.   GL.   90*, 126, 303*(64:1,39), 370(17:174-176).

lippizensis (Wulfen) Reichb.

Pre-1930 only.   SE Europe.   NMW.   21, 156, 683*.

nana (Wedd.) Thell.

Pre-1930 only.   Bolivia, Peru.   E, OXF.   3, 7, 354(5:15).   *Nasturtium nanum* Wedd.; *Radicula nana* (Wedd.) Druce, nom. illeg.

pyrenaica (Lam.) Reichb.

Pre 1930 only.   Europe.   BM.   21, 22*, 683*.   *Nasturtium pyrenaicum* (Lam.) R.Br.

ARMORACIA P.Gaertner, Meyer & Scherb.

macrocarpa (Waldst. & Kit.) Kit. ex Baumg.

Pre-1930 only.   Danube basin.   21, 197, 354(9:258), 715*.   *Cochlearia armoracia* L. var. *macrocarpa* Waldst. & Kit.

rusticana P.Gaertner, Meyer & Scherb.   Horse-radish

●●●●● An established garden escape; widely naturalised and locally plentiful on roadsides, railway banks, river banks and waste ground, especially in England.   Origin obscure, probably W Asia.   1, 20*, 22*, 34*, 90*. *A. lapathifolia* Gilib., nom. inval.; *Cochlearia armoracia* L.

CARDAMINE L.

[*bellidifolia* L.   In error.   248, 281, 354(5:742).   Incl. *C. alpina* Willd.]

chelidonia L.

● A garden escape persistent for several years in a wood near Edge (E Gloucs); also collected from Adwell, near Watlington (Oxon). Italy, Yugoslavia. BM. 21, 182, 303(33:12), 371(394:35), 546*, 683*.

corymbosa Hook.f. sensu lato

● A persistent weed in the Royal Botanic Garden at Edinburgh (Midlothian); recently recorded from two gardens in Berwicks and one at Bracebridge Heath (S Lincs).   New Zealand.   36, 36c, 303(58*:38-39; 59:49; 64:16).   Incl. *C. uniflora* (Hook.f.) Allan, nom. illeg.

glanduligera O.Schwarz
- • A garden relic long established on the site of Warley Place gardens (S Essex). E Europe. 21, 303(43:17), 630*, 695*, 715*, 741*. *Dentaria glandulosa* Waldst. & Kit.

heptaphylla (Villars) O.Schulz   Pinnate Coralroot
- •• A garden relic long established on the site of Warley Place gardens (S Essex); introduced and persistent at Castle Eden Dene (Co Durham), Peasholme Glen at Scarborough (NE Yorks) and in Newark Park at Wortley (W Gloucs); also recorded from Crawley (W Sussex). Europe. BM, SUN. 20, 22*, 23*, 90*, 122, 303(19:14), 335(114:53), 370(9:411), 371(400:36). *C. pinnata* (Lam.) R.Br.; *Dentaria pinnata* Lam.

kitaibelii Bech.   Kitaibel's Bittercress
- • Introduced at Forres (Moray). C Europe. K. 21, 23*, 27*, 277. *C. polyphylla* (Waldst. & Kit.) O.Schulz, non D.Don; *Dentaria polyphylla* Waldst. & Kit.

laciniata (Muhlenb.) Alph.Wood   Cut-leaved Bittercress
- • A garden relic long established on the site of Warley Place gardens (S Essex). Eastern N America. 38*, 40*, 371(361:27). *Dentaria laciniata* Muhlenb.

pentaphyllos (L.) Crantz   Five-leaflet Bittercress
- • A casual garden escape. W & C Europe. 21, 22*, 26*, 27*, 29*, 236, 371(346:26-27). *Dentaria digitata* Lam.; *D. pentaphyllos* L. The record for Castle Eden Dene (Co Durham) was in error for *C. heptaphylla*. 167, 303(19:14).

raphanifolia Pourret   Greater Cuckooflower
- ••• An established garden escape; naturalised by streamsides and lakes, sometimes in abundance, especially in the north and west; known since 1932 by a stream on the lower slopes of Loughrigg, near Ambleside (Westmorland); increasing. S Europe, Turkey, Caucasia. BM, CGE, E, NMW, OXF, RNG. 13, 20, 23*, 90*, 303*(28:1,15), 354(13:257), 370(2:37), 371(376:25). *C. chelidonia* Lam., non L.; *C. latifolia* Vahl, non Lej.

trifolia L.   Trefoil Cress
- •• A garden escape and relic established in a few places in the north and west; known for many years in a churchyard at Trentishoe (N Devon), at Casterton (Westmorland) and at Craigmillar quarry (Midlothian). C Europe. BM, E, OXF, RNG, SLBI. 2, 20. 22*, 23*, 26*, 90, 349(3:396), 354(10:516; 13:52).

ARABIS L.

arenosa (L.) Scop.   Sand Rock-cress
- • A casual at gas-works and on colliery spoil heaps, persisting for several years, perhaps imported with ores. Formerly a grain alien. Europe. BM, LANC, RNG. 1, 20, 22*, 23*, 25*, 26*, 191, 258, 303(32:18), 715*, 1242*. *Cardaminopsis arenosa* (L.) Hayek.

[*borealis* Andrz. ex Ledeb. In error for *Camelina* sp. 370(9:369).]

caucasica Willd. ex Schldl.   Garden Arabis

•••• An established garden escape; in quarries, on walls, railway banks and riversides, mainly in England; known for over 100 years on limestone cliffs near Matlock (Derbys), now extensively naturalised there. S Europe, SW Asia. 1, 20, 22*, 31*, 90*, 130. *A. albida* Steven ex M.Bieb.; *A. alpina* L. subsp. *caucasica* (Willd. ex Schldl.) Briq.

collina Ten. sensu lato   Rosy Cress

•• A garden escape established on walls and rocks in a few widely scattered localities in the north and west. C & S Europe. ABD, BM, OXF, RNG. 7, 20, 90, 303(34:22), 312*(t.3246), 371(346:27), 683*, 825(1:113-117). Incl. *A. muralis* Bertol., non Salisb.; *A. muralis* subsp. *collina* (Ten.) Thell. var. *rosea* (DC.) Thell.; *A. muricola* Jordan; *A. rosea* DC.

nova Villars

Pre-1930 only. S Europe, SW Asia. LSR. 7, 21, 44*, 354(5:15), 366(1905):108). *A. auriculata* Lam.; *A. saxatilis* All.

[*saggitata* (Bertol.) DC. In error for native *A. hirsuta*. 2, 1008.]

turrita L.   Tower Cress

•• An early garden escape, established since 1722 on old walls of St John's College grounds at Cambridge (Cambs); also a grain casual. C Europe, Medit., SW Asia. BM, LANC, LCN, OXF, RNG. 1, 20, 22*, 29*, 35*, 90, 160, 234, 354(10:960), 432.

verna (L.) R.Br.

Pre-1930 only. Medit. 2, 21, 24*, 29*, 44*, 136, 156.

AUBRIETA Adans.

deltoidea (L.) DC.   Purple Rock-cress

•••• An established garden escape; long naturalised on limestone cliffs in the Matlock area (Derbys); elsewhere in chalk pits and on walls in widely scattered localities. SE Europe, Turkey. 7, 20*, 22*, 90*, 126, 354(12:702), 431. *Alyssum deltoideum* L.

LUNARIA L.

annua L.   Honesty

•••• An established garden escape; naturalised in chalk pits, on dunes, roadsides and woodland edges, especially in southern England; increasing. SE Europe. 1, 20*, 22*, 34*, 90*, 126, 209, 256. *L. biennis* Moench, nom. illeg.

rediviva L.   Perennial Honesty

• A garden relic long established on the site of Warley Place gardens (S Essex); casual elsewhere. Europe, W Asia. BM, LANC, LIV. 2, 9, 20, 22*, 23*, 27*, 90*, 126, 331(29:11), 332(12:31).

ALYSSUM L.

alyssoides (L.) L.   Small Alison

••• An agricultural seed and wool alien persistent as a weed of cultivated land, mainly in S and E England; few modern records away from the Breckland (W Suffolk and W Norfolk). Europe, N Africa; (New Zealand). BM, CGE,

E, LANC, OXF, RNG.   1, 20*, 22*, 26*, 90*, 235, 354(12:35).
A. calycinum L., nom. illeg.; *Clypeola alyssoides* L.
dasycarpum Stephan ex Willd.
Pre-1930 only. SW Asia. 2, 21, 39*, 44*.
desertorum Stapf
Pre-1930 only. Europe, SW Asia. GLR.   7, 21, 39*, 320(1911:227), 354(3:9), 1342*. *A. vindobonense* G.Beck.
hirsutum M.Bieb.
Pre-1930 only. Europe, SW Asia. 1, 21, 39*, 248. The record from Oulton Broad (W Suffolk) was in error for *A. strigosum*. K.
montanum L.   Mountain Alison
Pre-1930 only. Europe. BM. 2, 8, 21, 22*, 25*, 27*, 30*, 151.
murale Waldst. & Kit. sensu lato
• A casual on a railway bank near Dinton Station (S Wilts) and on bombed sites in the City of London (Middlesex). No modern records. C & S Europe, SW Asia. RNG. 21, 31*, 172. Incl. *A. argenteum* Vitman.
saxatile L.   Golden Alison
••• A garden escape sometimes established, as on the shingle of the river Garry, near Killicrankie (Mid Perth), in a chalk pit at Cherry Hinton (Cambs) and on walls and waste ground in SE England; casual elsewhere. C & S Europe, SW Asia. BM, E, LANC, NMW, OXF, RNG. 7, 20*, 22*, 26*, 90, 135, 349(1966:237), 370(12:286). *Aurinia saxatilis* (L.) Desv.
serpyllifolium Desf.
Pre-1930 only. W Medit. 2, 21, 51, 181, 525*, 546.
simplex Rudolfi
Pre-1930 only. Europe, NW Africa, SW Asia. K. 1, 9, 21, 44*, 139, 300(1905:98), 737*. *A. campestre* auct. pro max. parte, non (L.) L.; *A. minus* Rothm., nom. illeg.; *A. parviflorum* M.Bieb.
strigosum Banks & Sol.
• A casual, vector unknown. No modern records. SE Europe, Cyprus, SW Asia. K, OXF. 21, 39, 44*, 354(8:104), 658, 683*. *A. minus* subsp. *strigosum* (Banks & Sol.) Stoj.; *A. campestre* auct. pro parte.

BERTEROA DC.
incana (L.) DC.   Hoary Alison
•••• A grain alien established by railways, on field edges and in pits in a few places in S and E England; more often a casual on tips and arable land. Europe, W Asia; (N America). 1, 11, 20*, 22*, 26*, 90*, 126, 209, 236. *Alyssum incanum* L.(incl. var. *viride* Tausch); *Farsetia incana* (L.) R.Br.

LOBULARIA Desv.
maritima (L.) Desv.   Sweet Alison
••••• An established garden escape and bird-seed alien; naturalised in scattered localities, mainly in coastal areas in the south. Medit., SW Asia. 1, 19, 20*, 22*, 34*, 90*, 207, 220. *Alyssum maritimum* (L.) Lam.; *Clypeola maritima* L.; *Koniga maritima* (L.) R.Br.

CLYPEOLA L.
jonthlaspi L.   Disk Cress
Pre-1930 only. Medit. 21, 29, 39*, 197, 683*. The record from Ireland is
dubious. 132.

EROPHILA DC.
conferta Wilm.
• A persistent garden escape at Layer Marney (N Essex) and Aberfeldy (Perth),
not seen since the 1970s. Rhodes. BM(holotype). 191, 370(1:137; 2:309),
1330. This species is not mentioned in any modern works.

COCHLEARIA L.
glastifolia L.   Upright Scurvy-grass
• Introduced in a bomb crater on Brockham Hill (Surrey), now gone, but the
progeny still persist in a few gardens in southern England. SW Europe.
BM, LANC, LIV, OXF, RNG, SLBI. 7, 21, 126, 300(1911:100), 546*, 683*,
1108*.

KERNERA Medikus
saxatilis (L.) Reichb.   Kernera
Pre-1930 only. C & S Europe. 2, 21, 22*, 23*, 156, 244. *Cochlearia
saxatilis* L.

CAMELINA Crantz
alyssum (Miller) Thell.
•• (but perhaps over-recorded in error for *C. macrocarpa*). A casual weed in
flax fields. No modern records. Europe. BM, E, LANC, LIV, OXF, RNG.
2, 16, 21, 22*, 25*, 248, 354(9:552), 1242*. *C. dentata* (Willd.) Pers.;
C. foetida (Schk.) Fries; *C. linicola* Schimper & Spenner; *C. sativa*
subsp. *alyssum* (Miller) Hegi & E.Schmid; *Myagrum alyssum* Miller;
M. foetidum Bergeret. The record from Pusey (Berks) was in error for
C. macrocarpa. 119.
hispida Boiss.
Pre-1930 only. SW Asia. 39, 44*, 300(1904:175). Incl. *C. grandiflora* Boiss.
macrocarpa Wierzb. ex Reichb.
• (but possibly overlooked as *C. alyssum* or *C. sativa*). A casual weed in flax
fields. No modern records. Europe, Russia. K. 21, 22*, 25*, 90*.
C. alyssum subsp. *integerrima* (Čelak) Smejkal. Doubtfully distinct at specific
rank from *C. alyssum*.
microcarpa Andrz. ex DC.   Lesser Gold-of-pleasure
••• (but confused with *C. sativa*). A grain alien sometimes persisting a few
years, as at Flitwick (Beds). Eurasia, N Africa; (N America). BM, CGE, E,
LANC, OXF, RNG. 1, 20*, 22*, 25*, 90*, 143, 323(29:22), 1242*.
C. sativa subsp. *microcarpa* (Andrz. ex DC.) Thell.; *C. sylvestris* Wallr.

rumelica Velen.
- A grain casual at Gloucester and Sharpness docks (W Gloucs). Europe, SW Asia. E, RNG. 21, 90*, 182*, 303(25:14).

sativa (L.) Crantz   Gold-of-pleasure
•••• A widespread persistent weed of corn, lucerne and flax fields, now almost exterminated; but now frequent and increasing as a bird-seed casual. Formerly also a wool casual. C & SE Europe, SW Asia; (widespread). 1, 3, 19, 20*, 22*, 26*, 90*. *C. friesii* Nobre; *Alyssum sativum* (L.) Smith; *Myagrum sativum* L.; incl. *C. pilosa* (DC.) Vassilcz. (*C. sativa* subsp. *pilosa* (DC.) Thell.).

NESLIA Desv.

paniculata (L.) Desv. sensu lato   Ball Mustard
••• A grain and bird-seed casual; decreasing. Formerly a wool casual. Eurasia, N Africa. BM, CGE, E, NMW, OXF, RNG. 1, 3, 9, 19, 20*, 22*, 26*, 90*, 1242*. *Myagrum paniculatum* L.; *Vogelia paniculata* (L.) Hornem.; incl. *N. apiculata* Fischer, C.Meyer & Avé-Lall.; *N. thracica* Velen.; *Camelina sagittata* Crantz; *Vogelia apiculata* Vierhapper; *V. sagittata* Medikus.

CAPSELLA Medikus

rubella Reuter   Pink Shepherd's-purse
••• Possibly a grain alien, sometimes persisting in abundance for a few years. Medit. BM, CGE, E, NMW, RNG, WAR. 16, 20*, 22*, 23*, 25*, 277, 303(16:29). *Bursa rubella* (Reuter) Druce. Perhaps best regarded as conspecific with native *C. bursa-pastoris*. 21a. A hybrid with *C. bursa-pastoris* (*C.* × *gracilis* Gren.) has been recorded. E, EXR, K, RNG. 20.

[*PRITZELAGO* Kuntze
alpina (L.) Kuntze (*Hutchinsia alpina* (L.) R.Br.; *Noccaea alpina* (L.) Reichb.) A dubious record. 196, 197, 320(1:359), 354(5:744).]

IONOPSIDIUM Reichb.

acaule (Desf.) Reichb.   Violet Cress
•• A casual garden escape and persistent garden weed. SW Europe. BM, LIV. 2, 20*, 51*, 52*, 207, 236. *Cochlearia acaulis* Desf.

BIVONAEA DC.
lutea (Biv.) DC.
Pre-1930 only. Sicily, Algeria. 21, 175, 181, 683*. *Thlaspi luteum* Biv.

[*TEESDALIA* R.Br.
coronopifolia (Bergeret) Thell. (*T. lepidium* DC.)   In error for native *T. nudicaulis*. 20, 349(12:23).]

THELYPODIUM Endl.
lasiophyllum (Hook. & Arn.) E.Greene   Cutleaf Thelypody
- A grain casual. No modern records. Western N America. OXF. 7, 16, 41*, 42, 169, 354(12:267), 500*. *Microsisymbrium lasiophyllum* (Hook. & Arn.) O.Schulz; *Turritis lasiophylla* Hook. & Arn.

THLASPI L.

alliaceum L.   Garlic Penny-cress

•• Established as a weed of arable land in a few places in southern England; known since 1923 at Rippers Cross, near Hothfield (E Kent), and about Beeleigh Abbey at Maldon (S Essex) since 1951. Europe, Turkey. BM, E, NMW, OXF, RNG, SLBI. 7, 20*, 22*, 90, 191, 353(31:21), 354(7:28), 374(1953:33). Records from Denbs and Co Antrim are in error for native *T. arvense*. 156, 319(23:121), 339(13:48).

macrophyllum Hoffm.   Caucasian Penny-cress

• An established garden escape or introduction; naturalised in the grounds of Bancroft Castle, near Craven Arms (Salop), in Belmont Wood at Failand (N Somerset) and on the disused site of the John Innes Institution at Bayfordbury (Herts). Turkey, Caucasia. BM, BRIST, CGE, OXF, TCD. 20*, 53*, 90*, 303(24:18; 26:18), 327*(87:77-82), 350(42:105). *T. latifolium* M.Bieb.; *Pachyphragma macrophyllum* (Hoffm.) N.Busch.

AETHIONEMA R.Br.

grandiflorum Boiss. & Hohen.   Persian Candytuft

• A garden escape established on an old wall of Sidney Sussex College in Cambridge (Cambs). Middle East, Caucasus. CGE. 39, 51a, 53*, 370(16:184), 371(349:27), 543*, 650*, 734*. *A. pulchellum* Boiss. & Huet.

saxatile (L.) R.Br.   Burnt Candytuft

Pre-1930 only. S Europe. BM. 7, 21, 23*, 29*, 354(4:403), 923*(13:3-42). Incl. *A. graecum* Boiss. & Heldr.

IBERIS L.

crenata Lam.

Pre-1930 only. SW Europe. OXF. 7, 21, 30*, 354(5:16). *I. bourgaei* Boiss.; *I. pectinata* Boiss.

intermedia Guers.

Pre-1930 only. Europe. 7, 21, 300(1910:43), 683*.

odorata L.

Pre-1930 only. E Medit. OXF. 7, 21, 44*, 354(7:29; 9:20), 528*, 658, 1262*. According to 39* the correct name is *I. acutiloba* Bertol. (*I. odorata* Boiss., non L.) but other authors disagree.

sempervirens L.   Perennial Candytuft

•• A garden escape established on old walls and long-persistent as garden relics; known for 30 years at Sand Point, Kewstoke (N Somerset). Medit. E, SLBI. 7, 20*, 27*, 31*, 90*, 199, 370(15:62).

umbellata L.   Garden Candytuft

•••• A garden escape sometimes persistent in southern England; established for some years in a quarry at Luffenham (Rutland). Medit., or originated in cultivation. 1, 20*, 55*, 90*, 222.

BISCUTELLA L.
auriculata L.
- A spice casual. W Medit. BM. 7, 21, 169, 199, 354(7:765), 459, 737*, 825*(5:44-46). *Iondraba auriculata* (L.) Webb & Berth.

cichoriifolia Lois.   Chicory-leaved Buckler-mustard
Pre-1930 only. S Europe. OXF. 21, 23*, 169, 354(8:609), 546*.

laevigata L.   Buckler-mustard
- An esparto casual. No modern records. Europe. BM. 2, 15, 22*, 23*, 27*, 28*, 260. *B. didyma* L. may have been overlooked. 825*(5:44-46).

LEPIDIUM L.
africanum (Burm.f.) DC.   African Pepperwort
••• A wool casual. S Africa; (tropical Africa, Australasia). BM, CGE, E, K, OXF, RNG. 3*, 14, 20*, 36c*, 90, 143, 370(11:367-372), 520*, 533*. *L. divaricatum* subsp. *linoides* (Thunb.) Thell.; *L. hyssopifolium* auct., non Desv.; *L. linoides* Thunb.; incl. *L. iberioides* Desv.

aletes Macbr.
- A wool casual. S America. E, RNG. 3, 14, 354(5:16), 370(11:367-372), 1066*. *L. calycinum* Godron.

aucheri Boiss.
Pre-1930 only. Egypt, SW Asia. E, OXF. 3*, 44*, 370(11:367-372), 729, 743, 1005*. Formerly a weed of flax in Australia, but now extinct there.

bipinnatifidum Desv.   Wayside Pepperwort
- A wool casual. S America; (N America). E, OXF. 3, 133, 370(11:367-372), 500*.

bonariense L.   Argentine Pepperwort
••• A wool and bird-seed casual. S America; (Australasia). BM, CGE, E, LANC, OXF, RNG. 3*, 14, 19, 20*, 22*, 36c*, 90, 370(11:367-372), 1066*, 1242*.

capense Thunb.
- A wool casual. S Africa. E, OXF. 3, 370(11:367-372), 541*. *L. africanum* var. *capense* (Thunb.) Thell.

desertorum Ecklon & C.Zeyher
- A wool casual. S Africa. E, K, RNG. 14, 370(11:367-372), 541*.

desvauxii Thell.   Bushy Pepperwort
Pre-1930 only. Australia. OXF. 7, 36c*, 354(6:603), 576*, 811*(4:217-308).

divaricatum Sol.
•• (but over-recorded for *L. africanum*). A wool casual; possibly also a bird-seed alien. S Africa. BM, E, K, LTN, OXF, RNG. 3, 14, 19, 21, 22*, 36c*, 370(11:367-372). Incl. *L. ecklonii* Schrader ex Regel; *L. trifurcum* Sonder.

draba L. sensu lato   Hoary Cress
••••• Originally a grain, straw and ballast alien at ports, long established on waste ground and as a colonist of arable land throughout the British Is, especially in the south and east; increasing. Formerly also a wool casual. Medit., SW Asia; (N America). 1, 3, 16, 20*, 21, 22*, 90*. *Cardaria draba*

(L.) Desv.; incl. *C. chalepensis* (L.) Hand-Mazz.; *L. chalepense* L. Most records are referable to subsp. *draba*; subsp. *chalepense* (L.) Thell. is rare. 303*(48:12-14).

fallax Ryves
- A wool casual. Probably S America. K(holotype). 370(11:367-372). Known only from the type gathering of 1971 from Blackmoor (N Hants).

fasciculatum Thell.   Bundled Pepperwort
- •• A wool casual. Australia. CGE, E, K, OXF, RNG. 3, 14, 37, 370(11:367-372), 743, 811*(4:217-308), 1193*.

graminifolium L.   Tall Pepperwort
- •• A wool and grain alien sometimes persistent or established, as at Cardiff and Barry docks (both Glam) and Avonmouth docks (W Gloucs). Europe, N Africa, SW Asia. BM, E, LANC, NMW, OXF, RNG. 1, 13, 20*, 22*, 90*, 126*, 350(28:173), 370(11:367-372). Incl. *L. iberis* L.

hirtum (L.) Smith
- Reported from a newly made road bank at Mateley Passage, near Lyndhurst (S Hants), vector unknown. Medit. HbEJC. 16, 21, 22*, 90, 370(11:367-372), 404. *Thlaspi hirtum* L. Early records were mostly in error for native *L. heterophyllum*. 354(5:743).

[*hyssopifolium* Desv. In error for *L. africanum*. Contrary to 20 and most other literature, the true plant has no tendency to weediness and is an endangered endemic of Australia and Tasmania. 576, 811*(4:217-308).]

[*incisum* Roth   This name has been misapplied to at least four *Lepidium* species, including *L. pinnatifidum*, *L. sagittulatum* and *L. sativum* 'Curled'; according to 595b it is a synonym of *L. virginicum* sensu stricto. BM. 2, 300(1904:175; 1907:38; 1908:58), 318(19:238). The illustration in 1103* portrays the New Zealand *L. flexicaule* Kirk.]

lasiocarpum Nutt.
Pre-1930 only. Western N America. OXF. 42, 354(9:553), 500*, 671*. Recorded as subsp. *georginum* (Rydb.) Thell.

lyratum L. sensu lato
Pre-1930 only. Russia, SW Asia. BM. 1, 21, 195(782,Errata), 630*. Incl. *L. coronopifolium* Fischer; *L. lacerum* C.Meyer.

monoplocoides F.Muell.   Winged Pepperwort
- A wool casual. Australia. 37, 133, 303*(29:8-9), 725*, 811*(4:217-308).

montanum Nutt. ex Torrey & A.Gray
Pre-1930 only. N America. 7, 354(2:496), 652*. Incl. *L. alyssoides* A.Gray; *L. eastwoodiae* Wooton.

oxytrichum Sprague
Pre-1930 only. Australia. E, OXF. 3*, 37, 354(3:152 as *L. papillosum*; 5:16), 370(11:367-372), 520*, 811*(4:217-308).

papillosum F.Muell.   Warty Pepperwort
Pre-1930 only. Australia. OXF. 3, 37*, 354(5:16), 370(11:367-372), 520*, 743, 811*(4:217-308).

peregrinum Thell.
   Pre-1930 only.   Probably Australia.   E.   3*, 354*(3:153), 576*, 811*(217-
   308).   Curiously, this species has been found in Australia on only two
   occasions, and may not be native there, or is now extinct.
perfoliatum L.   **Perfoliate Pepperwort**
   ●●● A persistent grain and grass seed alien. Eurasia; (N & S America). BM,
   E, LANC, NMW, OXF, RNG. 1, 20*, 22*, 26*, 38a*, 90*, 113, 336(16:41),
   370(11:367-372), 500*, 1066*.
pinnatifidum Ledeb.
   ● A casual, vector unknown. S & SE Russia; (California). BM, RNG. 21,
   303*(29:8-9), 1066*, 1143.   *L. incisum* auct. pro parte, non Roth;
   L. micranthum Ledeb. Since publication of the Cambs record, the identity has
   been verified.
pseudodidymum Thell.
   ● A wool casual. S America. E, K, OXF. 3*, 14, 42, 354*(3:308-309),
   370(11:367-372), 665*, 820*(13:506-528), 1066*. *L. inclusum* O.Schulz.
pseudohyssopifolium Hewson
   Pre-1930 only. Australia. E, OXF. 3, 354(5:278), 576*. *L. peregrinum* var.
   glabripes Thell.
pseudoruderale Thell.
   Pre-1930 only. Australia. 37, 370(11:367-372), 520*.
[*ruderale* L.   Accepted, with reservations, as native.]
sagittulatum Thell.
   Pre-1930 only. Australia. E, OXF, RNG. 3*, 354(6:374), 370(11:367-372),
   576*, 725*, 840*(31:318, as *L. incisum*). *L. incisum* auct. pro parte, non
   Roth.
sativum L.   **Garden Cress**
   ●●●●● A persistent bird-seed, spice, grain and wool alien and escape from
   cultivation. Egypt, W Asia. 1, 19, 20*, 22*, 90*, 303*(29:8-9), 370(11:367-
   372), 1066*. Incl. *L.*'Curled'.
schinzii Thell.
   ● A wool casual. S Africa. BM, CGE, E, K, OXF, RNG. 3, 21, 354(5:97;
   6:374), 370(11:367-372), 541*. *L. lacerum* Syme nom. nud., non C.Meyer.
spicatum Desv.
   ● A wool casual. Chile, Patagonia. E, RNG. 3, 354(5:278), 370(11:367-
   372), 820*(13:506-528), 1066*. *L. racemosum* Griseb.
spinosum Ard.
   Pre-1930 only. E Medit. K, SLBI. 21, 39, 44*, 667*, 722*. *L. cornutum*
   Sibth. & Smith.
strictum (S.Watson) Rattan
   Pre-1930 only. S America; (western N America). 7, 41*, 42, 354(4:8).
   L. oxycarpum Torrey & A.Gray var. *strictum* S.Watson; *L. pubescens* auct.,
   non Desv.; *L. reticulatum* Howell, non sensu Thell.
villarsii Gren. & Godron
   Pre-1930 only.   SW Europe. 16, 21, 23*, 370(11:367-372).   Incl.
   L. heterophyllum var. *pratense* (Serres) F.Schultz; *L. pratense* Serres.

virginicum L. sensu lato   Least Pepperwort

●●●● A persistent wool, timber, grain and bird-seed alien. N America;
(widespread). 1, 11, 14, 20*, 22*, 90*, 182, 303*(29:8-9; 45:21), 370(11:367-
372), 1247*. Incl. *L. apetalum* Asch.; *L. densiflorum* Schrader; *L. neglectum*
Thell.; *L. ramosissimum* Nelson; *L. texanum* Buckley.

CORONOPUS Zinn

didymus (L.) Smith   Lesser Swine-cress

●●●●● Originally a ballast alien at ports, now an established weed of arable
and waste land in many places, especially in southern England and Ireland; also
a wool alien; increasing. Probably S America; (widespread). 1, 20*, 22*,
90*. *C. pinnatifidus* Dulac; *Carara didyma* (L.) Britton; *Lepidium anglicum*
Hudson; *L. didymum* L.; *Senebiera didyma* (L.) Pers.; *S. pinnatifida* DC.

OCHTHODIUM DC.

aegyptiacum (L.) DC.

Pre-1930 only. E Medit. RNG. 39*, 44*. *Bunias aegyptiaca* L.

CONRINGIA Heister ex Fabr.

austriaca (Jacq.) Sweet

• A casual, possibly a grain alien. SE Europe, SW Asia. OXF. 1, 8, 16*,
20, 90, 258, 349(1:163). *Erysimum austriacum* (Jacq.) Roth.

orientalis (L.) Dumort.   Hare's-ear Mustard

●●●● A grain, clover seed and bird-seed casual; decreasing. Formerly a wool
casual. Medit. 1, 3, 9, 19, 20*, 22*, 26*, 90*, 1242*. *C. perfoliata* Link;
Brassica orientalis L.; *Erysimum orientale* (L.) Crantz, non Miller;
E. perfoliatum (Link) Crantz.

MORICANDIA DC.

arvensis (L.) DC.   Violet Cabbage

• An esparto casual. No modern records. Medit. CLE, OXF. 1, 15, 21,
24*, 29*, 45*, 121, 303(58:38), 354(5:640), 600*. *Brassica arvensis* L.

DIPLOTAXIS DC.

catholica (L.) DC.

Pre-1930 only. W Medit. OXF. 7, 16, 21, 45*, 354(7:28), 737*.

crassifolia (Raf.) DC.

Pre-1930 only. W Medit. 7, 21, 354(4:8), 648*. Incl. *D. lagascana* DC.

erucoides (L.) DC.   White Wall-rocket

●● A grain and wool casual; apparently increasing after an absence for some
years. Medit. BM, K, LANC, NMW, OXF, RNG. 1, 16, 21, 22*, 44*,
303(25:14; 58:36), 370(18:421), 737*.

harra (Forsskål) Boiss.

• An esparto casual. No modern records. N Africa, SW Asia. K, RNG.
44*, 45*, 46*.

muralis (L.) DC.   Annual Wall-rocket
●●●●● Originally a ballast alien at ports, now widely established especially by railways and on limestone cliffs; also a bird-seed alien. Formerly a wool casual. Medit.; (Australasia). 1, 3, 20*, 22*, 25*, 90*. *Brassica brevipes* Syme, nom. illeg., pro parte; *B. muralis* (L.) Hudson pro parte; *Sinapis muralis* (L.) R.Br.; *Sisymbrium murale* L.

tenuisiliqua Del.
• A casual at Sharpness docks (W Gloucs), vector unknown. No modern records. N Africa. 7, 45*, 46, 182, 354(8:301). *D. auriculata* Durieu.

viminea (L.) DC.   Vineyard Wall-rocket
Pre-1930 only (but confused with *D. muralis*). Medit. TOR. 7, 9, 16, 21, 22*, 44*, 217, 220, 225, 245, 737*.

virgata (Cav.) DC.
• An esparto casual. W Medit. RNG. 21, 682*, 737*.

BRASSICA L.
[*balearica* Pers. In error for *B. juncea*. 354(4:120; 5:96).]

barrelieri (L.) Janka
Pre-1930 only. W Medit. OXF. 2, 21, 30*, 45*, 300(1906:101; 1907:231). *B. laevigata* Lagasca; *B. oxyrrhina* Cosson; *B. parra* Durieu & Schinz; *B. sabularia* Brot. The 1907 record from Leith docks (Midlothian) was in error for *B. tournefortii*. SLBI.

carinata A.Braun   Ethiopian Rape
●● A bird-seed or oil-seed casual, sometimes appearing to be persistent, but probably repeatedly introduced; on tips and waste ground especially in the London area and S Lancs. Origin obscure, probably E Africa. BM, CGE, OXF, RNG. 12, 19, 20, 90*, 192, 251, 303*(47:25-26), 349(2:269), 354(12:467), 370(1:39), 825*(1:20-24). *B. integrifolia* var. *carinata* (A.Braun) O. Schulz.

elongata Ehrh.   Long-stalked Rape
●● A persistent grain alien; known for 20 years in Luffenham quarry (Rutland), now probably gone. C & SE Europe, SW Asia; (USA). BM, DUE, E, LANC, LIV, OXF. 1, 16, 20, 21, 90*, 222, 1242*. Incl. *B. persica* Boiss. & Hohen. All records are referable to subsp. *integrifolia* (Boiss.) Breistr.

fruticulosa Cirillo   Twiggy Turnip
• A casual in a pit at Beetley (W Norfolk) and in Dublin port (Co Dublin), vector unknown. Medit. DBN, LANC, NMW, OXF. 4, 21, 264, 303*(57:36-37), 354(7:433), 514*, 525*, 737*.

juncea (L.) Czernj.   Chinese Mustard
●●●● A grain, bird-seed, agricultural seed, spice and wool casual, and a relic of cultivation. E Africa, Asia; (Australasia). 1, 14, 19, 20, 22*, 90*, 235, 303*(47:25-26), 1242*. Possibly of *B. nigra* (L.) Koch × *B. rapa* origin; *B. integrifolia* (West) O.Schulz; *B. lanceolata* Lange; *Sinapis juncea* L.

napus L.   Rape, incl. Oil-seed Rape, Swede, Chinese Cabbage
●●●●● A widespead relic of, or escape from, cultivation and bird-seed casual, sometimes abundant on roadsides; increasing. Origin obscure; (widespread).

1, 19, 20*, 22*, 90*, 303*(45:6-8). Possibly of *B. oleracea* L. × *B. rapa* origin; incl. *B. chinensis* L.; *B. oleifera* Moench; *B. napobrassica* (L.) Miller; *B. rutabaga* (DC.) Briggs. A hybrid with native *B. rapa* (*B.* × *harmsiana* O.Schulz) has been recorded in crops. 20.

[*rapa* L. subsp. *oleifera* (DC.) Metzger (Turnip-rape), a bird-seed and oil-seed casual, and subsp. *rapa* (Turnip), a relic of cultivation, may be of alien origin.]

repanda (Willd.) DC.
    Pre-1930 only. S Europe, N Africa. LANC. 21, 27*, 303(34:22), 737*, 1290*. *Diplotaxis saxatilis* DC.

tournefortii Gouan   **Pale Cabbage**
    ●●● A wool, grain, esparto and spice casual. Medit.; (Australasia). BM, CGE, E, OXF, RNG, SLBI. 2, 14, 15, 20*, 44*, 90, 277, 737*.

SINAPIS L.

alba L.   **White Mustard**
    ●●●●● A relic or escape from cultivation established as a weed of arable land on calcareous soils; also a bird-seed alien. Formerly a wool casual. Medit., SW Asia; (widespread). 1, 3, 19, 20*, 22*, 90*, 143. *S. dissecta* Lagasca; *Brassica alba* (L.) Rabenh.; *B. dissecta* (Lagasca) Boiss.; *B. hirta* Moench.

allionii Jacq.
    Pre-1930 only. Egypt. 7, 251, 354(5:549), 729*, 1337*. *S. arvensis* subsp. *allionii* (Jacq.) Baillarg.; *S. arvensis* var. *turgida* Asch. & Schweinf.; *S. nigra* L. var. *turgida* (Del.) Asch. & Schweinf.; *S. turgida* (Pers.) Del., non Pers.; *Brassica turgida* Pers.

arvensis L. **alien variants**
    Var. *orientalis* (L.) Koch & Ziz (*S. orientalis* L.) has occurred as a casual. Var. *schkuhriana* (Reichb.) Hagenb. (*S. schkuhriana* Reichb.; *Brassica kaber* (DC.) Wheeler var. *schkuhriana* (Reichb.) Wheeler) was recorded in error for *Rapistrum rugosum*. 166.]

flexuosa Poiret
    ● A wool casual. W Medit. RNG. 1, 20, 21, 45*, 278, 354(9:683), 648*, 850*(35:287,321), 1127*. *S. hispida* Schousboe; *Brassica hispida* (Schousboe) Boiss.

ERUCA Miller

vesicaria (L.) Cav.   **Garden Rocket**
    ●●● A persistent grain, bird-seed, oil-seed, lucerne seed, spice and esparto alien and garden escape. Europe, N Africa, SW Asia; (widespread). BM, CGE, E, NMW, OXF, RNG. 1, 15, 19, 20*, 22*, 90*, 182, 236, 303*(50:28), 1242*. *E. cappadocica* Reuter; *E. eruca* (L.) Asch. & Graebn.; *E. longirostris* Uechtr.; *E. sativa* Miller; *Brassica eruca* L. Most records are referable to subsp. *sativa* (Miller) Thell.; subsp. *vesicaria* was recorded pre-1930 only. CLE. 300(1901:41), 303(58:38).

ERUCASTRUM C.Presl
gallicum (Willd.) O.Schulz   **Hairy Rocket**
●●●● An established grain and bird-seed alien; naturalised on a few railway
banks and in quarries, mainly in southern England; naturalised since 1885 on
the Devil's Dyke (Cambs). Europe. 2, 19, 20*, 22*, 26*, 90*, 234, 683*.
E. inodorum Reichb.; *E. ochroleucum* Beck; *E. pollichii* Schimper & Spenner;
Brassica gallica (Willd.) Druce; *B. pollichii* Shuttl.; *Diplotaxis bracteata* Gren.
& Godron; *Sisymbrium erucastrum* Pollich, non Gouan.
griquensis (N.E.Br.) O.Schulz
● A wool casual. S Africa. LTN, RNG. 14, 143, 541*. *Brassica griquana*
N.E.Br.
nasturtiifolium (Poiret) O.Schulz   **Watercress-leaved Rocket**
● A casual at Gloucester docks (E Gloucs) and in a gravel pit near Mitcham
Junction station (Surrey), vector unknown. No modern records. SW Europe.
BM, LANC, OXF. 1, 16*, 20, 22*, 90, 245, 348(20:29), 354(12:480), 546*,
683*. *E. obtusangulum* (Willd.) Reichb.; *Brassica erucastrum* Villars, vix L.
cf. **varium** (Durieu) Durieu
● A spice casual. N Africa. RNG. 46, 303(51:8), 648*. *Brassica varia*
Durieu.

COINCYA Rouy
hispida (Cav.) Greuter & Burdet   **Wallflower Cabbage**
Pre-1930 only. SW Europe. NMW. 21, 156, 737*. *Brassica valentina* DC.;
Eruca hispida Cav.; *Hutera hispida* (Cav.) Gómez Campo; *Rhynchosinapis*
hispida (Cav.) Heyw.
longirostra (Boiss.) Greuter & Burdet
Pre-1930 only. S Spain. 1, 21, 522*, 737*. *Brassica longirostra* Boiss.;
Erucastrum longirostre (Boiss.) Nyman; *Rhynchosinapis longirostra* (Boiss.)
Heyw.
monensis (L.) Greuter & Burdet subsp. **recurvata** (All.) Leadlay
●●● An established ballast alien; by roadsides and railways, on shingle and
dunes; known for over 50 years on a roadside in Waterworks Valley in Jersey
(Channel Is). W Europe. BM, CGE, E, NMW, OXF, RNG. 1, 20, 22*, 25*,
90*, 201, 370(9:376). *C. cheiranthos* (Villars) Greuter & Burdet; *Brassica*
cheiranthos Villars; *Brassicella erucastrum* O.Schulz pro parte; *Hutera*
cheiranthos (Villars) Gómez-Campo; *Rhynchosinapis cheiranthos* (Villars)
Dandy; *R. erucastrum* sensu Clapham, non (L.) Dandy.

HIRSCHFELDIA Moench
incana (L.) Lagr.-Fossat   **Hoary Mustard**
●●●●● An established wool, grain and bird-seed alien; locally abundant on
waste ground, roadsides, shingle beaches, sand dunes and by railways,
especially along the coasts of SE England and S Wales, and in the Channel Is;
increasing. Medit.; (Australasia). 1, 14, 19, 20*, 22*, 90*, 220, 236.
Brassica adpressa Boiss.; *B. geniculata* (Desf.) Ball; *B. incana* (L.) Meigen,
non Ten.; *Erucastrum incanum* (L.) Koch; *Sinapis incana* L.

CARRICHTERA DC.
annua (L.) DC.   **Cress Rocket**
•• A wool, grain and esparto casual. Medit.; (Australia). BM, CGE, E, NMW, OXF, RNG. 1, 14, 15, 16*, 21, 44*, 90*, 277. *C. vella* DC.; *Vella annua* L.

SUCCOWIA Medikus
balearica (L.) Medikus
• A wool casual. W Medit. RNG. 16*, 17, 21, 46*, 354(11:23), 737*. *Bunias balearica* L.

ERUCARIA Gaertner
hispanica (L.) Druce
•• A grain and wool casual. Formerly a cotton casual. E Medit. BM, CGE, NMW, OXF, RNG, SLBI. 1, 8, 16*, 21, 39*, 44*, 145, 354(4:120). *E. aleppica* Gaertner; *E. latifolia* DC.; *E. myagroides* (L.) Hal.; *Sinapis hispanica* L.; incl. *E. linearifolia* sphalm.; *E. lineariloba* Boiss.

[*CAKILE* Miller
edentula (Bigelow) Hook. In error for native *C. maritima*. 253.]

RAPISTRUM Crantz
perenne (L.) All.   **Steppe Cabbage**
••• (but over-recorded for *R. rugosum*). A persistent grain and bird-seed alien. E Europe, W Asia. BM, E, LANC, NMW, OXF, RNG. 1, 19, 20*, 22*, 29*, 90, 264, 1242*.

rugosum (L.) Bergeret   **Bastard Cabbage**
•••• An established wool, grain and bird-seed alien; locally abundant on waste ground, roadsides, arable land and by railways; increasing. Medit.; (widespread). 1, 14, 19, 20*, 22*, 29*, 90*, 303(1:70). *R. glabrum* Host; *R. hispanicum* (L.) Crantz; *R. linnaeanum* Boiss. & Reuter, nom. illeg.; *R. orientale* (L.) Crantz; *R. tenuifolium* (Sibth. & Smith) Benth. & Hook.f.; *Cakile rugosa* (L.) L'Hér. ex DC.; *Didesmus tenuifolius* DC.; *Myagrum orientale* L. According to 20, most records are referable to subsp. *linnaeanum* (Cosson) Rouy & Fouc., whereas 39 claims that the subspecies are unworkable.

DIDESMUS Desv.
aegyptius (L.) Desv.
Pre-1930 only. E Medit. OXF. 21, 564*, 648*, 658.

CRAMBE L.
cordifolia Steven   **Greater Sea-kale**
•• A garden escape established and increasing on waste ground near the canal at Boston Manor (Middlesex); persisting elsewhere on roadsides and waste ground in a few localities in C and S England. Caucasus. BM, LANC, RNG. 20, 53*, 55*, 57*, 90, 303(49:30; 52:30), 371(349:28).
orientalis L.   **Oriental Sea-kale**
Pre-1930 only. SW Asia. 1, 39*, 44*, 51, 125, 354(4:296).

CALEPINA Adans.
irregularis (Asso) Thell.   White Ball-mustard
 • A casual, vector unknown. Medit. 2, 16a*, 21, 22*, 25*, 29*, 44*,
300(1905:98), 546*, 1059. *C. corvini* (All.) Desv.; *Myagrum irregulare* Asso.

ENARTHROCARPUS Labill.
lyratus (Forsskål) DC.
 • A casual at Gloucester docks (W Gloucs), vector unknown. No modern
records. E Medit. BM, E, LANC, LIV, OXF. 1, 16*, 21, 112, 182.

RAPHANUS L.
raphanistrum L.
    subsp. **landra** (Moretti ex DC.) Bonnier & Layens   Mediterranean Radish
 •• A grain casual. Medit. BRISTM, DZS, E, K, OXF. 2, 20*, 90,
303(59:48). *R. landra* Moretti ex DC.
[*raphanistrum* L. subsp. *raphanistrum* Accepted, with reservations, as native.]
raphanistrum L. subsp. **rostratus** (DC.) Thell.   Beaked Radish
 • A casual at Sharpness docks (W Gloucs), vector unknown. No modern
records. E Medit. RNG. 44*. *R. rostratus* DC.
sativus L.   Garden Radish
 ••••• A persistent garden escape or relic of cultivation and a bird-seed alien.
Origin obscure, probably Medit. 1, 19, 20*, 22*, 90*, 303*(51:13-15), 1242*.
Incl. *R. caudatus* L. A hybrid with *R. raphanistrum* (*R.* × *micranthus*
(Uechtr.) O.Schulz) has been recorded. 20, 1311.

RESEDACEAE

RESEDA L.
alba L.   White Mignonette
 •••• An established or persistent garden escape, grain, wool and bird-seed
alien in a few places in southern England, especially near the sea; known since
1826 on sandy ground at Hugh Town on St Mary's (Scilly), perhaps now gone.
Medit., SW Asia; (Australasia). 1, 19, 20, 22*, 24*, 33*, 34*, 207. Incl.
R. fruticulosa L.; *R. hookeri* Guss.; *R. suffruticulosa* L.
inodora Reichb.   Scentless Mignonette
 • A casual, probably a grain alien. SE Europe. BM, OXF, SLBI. 7, 21, 39*,
303(17:16), 354(11:467), 695*.
lutea L. **alien variants**
    Var. *gracilis* (Ten.) ined. (*R. gracilis* Ten.), var. *laxa* Lange and var. *pulchella*
J.Mueller have been recorded.
odorata L.   Garden Mignonette
 •••• A casual garden escape. Origin obscure. 1, 20, 52*, 54*, 55*.
phyteuma L.   Corn Mignonette
 •• A persistent alien for more than 30 years in a cornfield near Westwell, and
recently recorded from a cornfield near Faversham (both E Kent); a wool and
grain casual elsewhere. Medit. BM, CGE, NMW, OXF, RNG. 7, 20, 22*,
23*, 29*, 236.
[*stricta* Pers. In error for native *R. lutea*. 349(4:273).]

[*SESAMOIDES* Ortega
pygmaea (Scheele) Kuntze   (*Astrocarpus sesamoides* (L.) DC.;   *Reseda sesamoides* L.)   In error for native *Reseda luteola*. 9, 354(6:15).   Contrary to 16 and 20, we cannot locate a correct record.]

CAYLUSEA A.St.Hil.
hexagyna (Forsskål) Green
   Pre-1930 only. N Africa. 44*, 46*, 354(8:301).   *C. canescens* A.St.Hil.

CLETHRACEAE
CLETHRA L.
alnifolia L.   Sweet Pepperbush
   • A persistent garden relic collected from Surrey. Eastern N America. CGE. 7, 38*, 54*, 61, 62*, 63*, 354(7:776).   *C. paniculata* Aiton.
arborea Aiton   Lily-of-the-valley-tree
   • A garden escape self-sown on walls in Guernsey (Channel Is); introduced and reproducing abundantly in woodland on Tresco (Scilly). Madeira. 20, 51, 61, 66*, 220, 312*(t.1057), 325(107:89).

[*EMPETRACEAE*
EMPETRUM L.
rubrum Vahl   Discovered in 1971 on the SS *Great Britain* docked at Bristol after returning from the Falkland Is, but the species did not reach land. 350(32:215), 370(9:146).   The name has also been misapplied to a wild plum (*Prunus* sp.) in Wales. 263(p.2).]

ERICACEAE
LEDUM L.
palustre L.   Labrador-tea
   ••• An established garden escape, tenuously claimed as native on Flanders Moss (W Perth); well naturalised on boggy ground in Surrey and a few northern counties. N America, Greenland. ABD, BM, LANC, OXF, RNG. 7, 20*(fig. as *Vaccinium uliginosum*), 33*, 38*, 41*, 316(19:219-233; 22:41-46), 1253.   *L. groenlandicum* Oeder; *L. latifolium* Jacq.

RHODODENDRON L.
arboreum W.Smith   Tree Rhododendron
   • Introduced for game cover in Yorkshire; self-sown from gardens in Argyllshire. Himalayas, Sri Lanka. 61, 61a, 63*, 197, 371(349:28), 505*, 685*, 724*.
ferrugineum L.   Alpenrose
   Pre-1930 only. C & S Europe. OXF. 21, 22*, 26*, 27*, 28*, 354(8:747).
luteum Sweet   Yellow Azalea
   ••• An established garden escape, mainly in woodland on acid soils; long naturalised in abundance on heathland at Burnham Beeches (Bucks); increasing.

C & E Europe, SW Asia. BM, LANC, NMW, OXF, RNG. 13, 20, 22*, 31*, 54*, 354(12:285). *R. flavum* G.Don; *Azalea pontica* L.

macrophyllum D.Don ex G.Don   California Rosebay
  • Introduced or a garden relic near Preston (S Lancs). Western N America. 41*, 51, 61, 371(346:28), 631*. *R. californicum* Hook.

minus Michaux
  • Introduced for game cover and self-sown in Yorkshire. SE USA. 51, 61, 63*, 197, 631*. *R. punctatum* Andrews.

ponticum L.   Rhododendron
  ••••• A well-established garden escape; naturalised, sometimes abundantly, on heaths and in woodlands on acid soils throughout the British Is, now a major threat to native vegetation in some areas; increasing. S Europe, SW Asia. 1, 20, 22*, 29*, 308(73), 312*(t.650), 322(63:345-364), 737*. Incl. *R. baeticum* Boiss. & Reuter (*R. ponticum* subsp. *baeticum* (Boiss. & Reuter) Hand.-Mazz.). Naturalised plants may be partly of hybrid origin; *R. catawbiense* Michaux × *R. ponticum* is recorded from Dunbarton, and some plants on the Surrey-Sussex border appear to be *R. catawbiense* × *R. maximum* L. × *R. ponticum*, whereas others nearby match pure subsp. *baeticum*. 61, 467, 1218.

sutchuenense Franchet
  • A garden escape or relic reported in 1992 from woods at Whorlton (Co Durham). China. 51*, 61, 312*(t.8362), 371(428:6), 379*(84:115), 594*, 617*, 631*, 1295*, 1341*. Incl. *R. praevernum* Hutch.

KALMIA L.

angustifolia L.   Sheep-laurel
  ••• A garden escape established in boggy areas on moorland and commons; known since before 1913 at Rixton Moss (S Lancs) and persistent for at least 60 years on the moorland site of an abandoned nursery at Whitesprings, near Darley Dale (Derbys). Eastern N America. BM, OXF, RNG. 7, 20*, 22*, 63*, 199*, 209, 251, 370(10:87).

latifolia L.   Calico-bush
  • A garden escape or relic persistent at Bayham Abbey (W Kent) and at Brookwood (Surrey). Eastern N America. RNG. 51*, 52*, 61*, 63*, 342(Feb89).

polifolia Wangenh.   Bog-laurel
  • An established garden escape in a few widely scattered boggy areas; naturalised for at least 80 years on Chobham Common (Surrey). N America. BM, OXF, RNG. 7, 20*, 22*, 38*, 62*, 209, 228. *K. glauca* L'Hér. ex Aiton.

ENKIANTHUS Lour.

deflexus (Griffith) Schneider
  • A garden escape or relic at Aros Park on Mull (Mid Ebudes). Himalayas, W China, Burma. BM. 51, 61*, 190, 312*(t.6460), 685*. *E. himalaicus* Hook.f. & Thomson.

PIERIS D.Don
floribunda (Pursh ex Sims) Benth. & Hook.   Mountain Fetter-bush
 • A garden escape, possibly established in Glencoyne wood, Ullswater (Cumberland/Westmorland). Eastern N America. LIV. 38*, 51*, 61, 63*, 183, 347*(4:65-75). *Andromeda floribunda* Pursh ex Sims.
formosa (Wallich) D.Don
 • Introduced or a garden escape at Rempstone, near Wareham (Dorset). Himalayas, China. RNG. 61, 61a, 63*, 347*(4:65-75), 403. Incl. *P. forrestii* Harrow ex W.Smith.
japonica (Thunb.) D.Don
 • Reported as self-seeding plentifully in Brookwood Cemetery (Surrey). Japan, China. HbKWP. 20, 51*, 61*, 61a, 62*, 63*, 342(Feb89), 347*(4:65-75). Incl. *P. taiwanensis* Hayata.

ZENOBIA D.Don
pulverulenta (Bartram ex Willd.) Pollard   Zenobia
 • Introduced or a garden escape in a plantation at Tower Hill (Berks). Eastern N America. K. 40, 52*, 61*, 62*, 63*, 119. *Z. speciosa* (Michaux) D.Don; *Andromeda cassinefolia* (Vent.) Pollard; *A. pulverulenta* Bartram ex Willd.

GAULTHERIA L.
mucronata (L.f.) Hook. & Arn.   Prickly Heath
 •••• An established garden escape; naturalised on moorland and in woods on acid soils, especially in Scotland and Ireland. Chile, Argentina. 7, 20*, 21, 22*, 54*, 61*, 63*. *Pernettya mucronata* (L.f.) Gaudich. ex Sprengel. A hybrid with *G. shallon* (× *Gaulnettya wisleyensis* W.Marchant, nom. nud.) has been recorded. 20.
procumbens L.   Checkerberry
 • A garden escape established in a pinewood in Glen Garry (Westerness), in woodland behind a plant nursery at Seal (W Kent) and extensively spreading in Brookwood Cemetery (Surrey). Eastern N America. 20, 38*, 52*, 61*, 341(98:3), 354(12:47), 432.
shallon Pursh   Shallon
 •••• Well established and widespread on acid soils; introduced as cover and food for pheasants and locally abundantly naturalised; also a garden escape. Western N America. 7, 20*, 22*, 52*, 63*, 631*. A hybrid with *G. mucronata* has been recorded.

ARBUTUS L.
[*andrachne* L. In error for *Gaultheria shallon*. 199.]
× **andrachnoides** Link   Hybrid Strawberry-tree
 • A persistent garden escape, bird-sown in a chalkpit at Stone (W Kent). Greece. 21, 58*, 61, 63*, 371(414:13). *A. andrachne* × *A. unedo* L.; *A.* × *hybrida* Ker Gawler.

ERICA L.
arborea L.   Tree Heath
- A garden escape recently recorded as self-sown on a hedgebank in Budleigh Salterton (S Devon), on a roadside at Brook (Surrey) and on Portelet Common, Jersey (Channel Is). Medit. HbKWP, HbLJM. 21, 22*, 30*, 63*, 127, 342(Feb93:5), 375(24:455), 390(120:208).

carnea L.   Winter Heath
- A garden escape persistent on Apple Tree Banks on Tresco (Scilly), on Dartford Heath (W Kent), and in Brookwood Cemetery and at Churt (both Surrey). C & S Europe. 1, 21, 22*, 23*, 27*, 29*, 236, 342(Feb89), 458. *E. herbacea* L.

× **darleyensis** Bean   Darley Dale Heath
- Introduced and persistent since the 1930s on grassy banks by the A12 road near Woodbridge (E Suffolk); also recently reported from Brookwood Cemetery (Surrey). Originated in cultivation. 20*, 61, 342(Feb89), 371(343:28), 597*. *E. carnea* × *E. erigena* R.Ross.

lusitanica Rudolphi   Portuguese Heath
- •• A well-established garden escape; naturalised in quantity along the main railway line between Lostwithiel and St Germans (E and W Cornwall); introduced and long naturalised in abundance on Lytchett Heath (Dorset), now much reduced or gone. SW Europe. BM, CGE, E, LANC, OXF, RNG. 2, 20*, 22*, 29*, 30*, 163, 212, 519*.

[*multiflora* L.   Unconfirmed record.   Probably in error for native *E. vagans*. 218(p.44), 287, 354(5:777).]

terminalis Salisb.   Corsican Heath
- An established garden escape or introduction; long naturalised on Magilligan dunes (Co Londonderry); also recorded from Youlbury (Berks) and Onich (Westerness). W Medit. BM, K, OXF, RNG. 7, 13, 20*, 22*, 30*, 119, 371(374:8). *E. stricta* Donn ex Willd.

VACCINIUM L.
corymbosum L. sensu lato   Blueberry
- An established escape from cultivation; naturalised on the sandy ground of Ashley Heath, near Ringwood (S Hants) and in Wareham Forest (Dorset). Eastern N America. RNG, HbEJC. 20*, 62*, 63*, 80*, 303(37:20; 40:7), 370(16:190), 597*, 825*(13:65-69), 1280*. The records are probably referable to a modern crop plant of complex hybrid origin involving several species.

macrocarpon Aiton   American Cranberry
- •• An established escape from cultivation; naturalised on boggy ground in widely scattered localities; spreading on the Isle of Arran (Clyde Is), on Kingsley Common (N Hants) and in the New Forest (S Hants).   Eastern N America. CLE, E, NMW, OXF, RNG, SLBI. 1, 20. 22*, 38*, 80*, 370(10:456), 1128*. *Oxycoccus macrocarpos* (Aiton) Pursh.

MYRSINACEAE
MYRSINE L.

africana L.   African Boxwood

- A persistent garden escape and relic of cultivation on Tresco (Scilly) and under the cliffs of Fort George in Guernsey (Channel Is). Azores, Africa, Asia. BM, HbEJC. 20, 21, 51*, 220, 303(36:23), 600*, 680*, 1112*, 1294*.

PRIMULACEAE
PRIMULA L.

auricula L.   Auricula

- Introduced and established on rock ledges in Caenlochan Glen (Angus); still there in small quantity (contrary to 187) more than 100 years after it was planted. This or its hybrid with *P. hirsuta* All. (*P.* × *pubescens* Jacq.) has occurred as a short-lived escape in Dunbarton and W Kent. C & S Europe. 13, 20, 22*, 27*, 28*, 52*, 304(1:6).

bulleyana Forrest

- Introduced or a garden escape, established around a pool in Rothiemurchus Forest (Moray). China. 51, 55*, 76, 325*(112:273), 371(337:29).

florindae Kingdon-Ward   Tibetan Cowslip

- ••• An established garden escape, mainly in Scotland; plentifully naturalised along a stream at Coldingham (Berwicks) and on the raised beach near Druim House (Nairns); increasing. Tibet. E, RNG. 20*, 52*, 54*, 55*, 277, 370(12:355), 371(361:26).

japonica A.Gray   Japanese Cowslip

- •• Introduced or a garden escape, established in wet soils in a few places in the south and west; naturalised in quantity in a field at Salruck (W Galway). SE Asia. NMW, RNG. 20*, 52*, 55*, 199, 212, 275, 371(349:28).

× **polyantha** group   Polyanthus

- •• Casual or persistent garden escapes. Originated in cultivation. BM. 20, 53*, 54*, 199, 220, 305(42:20). A complex of hybrids involving native and alien races of *P. veris*, *P. vulgaris* and other species; *P.* × *polyantha* Miller; *P.* × *tommasinii* Gren. & Godron; *P.* × *variabilis* Goupil, non Bast.; incl. *P.* Pacific Hybrids.

× **pruhonicensis** Zemann ex Bergmans   Hybrid Primulas

- Casual garden escapes. Originated in cultivation. OXF. 51*, 52*, 54*, 253, 350(47:42), 371(374:9). A complex of hybrids involving *P. juliae* Krecz. and native *P. elatior* and *P. vulgaris*; *P.* × *juliana* hort.; incl. *P.*'Wanda'.

pulverulenta Duthie

- A garden escape established at Pusey (Berks) and near Linn Mill at Alloa (W Perth). China. OXF. 20, 51*, 52*, 54*, 76*, 119, 1307.

sikkimensis Hook.f.   Sikkim Cowslip

- Introduced in 1921 at high altitude on Snowdon (Caerns) and still surviving. Himalayas. 20, 51*, 54*, 55*, 76*, 371(374:8).

spectabilis Tratt.   Spectacular Primrose
• Introduced or a garden escape on a field bank at Woldingham (Surrey).
Alps. 21, 23*, 27*, 56*, 209.

ANDROSACE L.
maxima L.   Annual Androsace
Pre-1930 only. Eurasia, N Africa. 1, 9, 21, 22*, 23*, 27*, 44*, 169, 683*.

CYCLAMEN L.
coum Miller   Eastern Cyclamen
•• Introduced and established since 1972 in private woodland near Bradley
(N Somerset), in grassland in Saumarez Park in Guernsey (Channel Is), on a
roadside verge at Effingham and at Ewhurst (both Surrey); also recorded
without locality from S Wales. SE Europe, SW Asia. HbKWP. 20, 21, 54*,
78*, 220, 342(Feb90), 371(357:23), 700*, 713*. Incl. *C. ibericum* Steven ex
Boiss.; *C. orbiculatum* Miller; *C. vernum* Sweet.
graecum Link   Greek Cyclamen
• A garden escape persistent for a few years on Dartford Heath, now gone, and
introduced by the Nepicar roundabout (both W Kent). E Medit. 20, 21, 29*,
78*, 303(62:45), 370(15:429), 371(403:39), 700*, 713*.
hederifolium Aiton   Cyclamen
•••• An established garden escape or relic sometimes naturalised in woodland,
mainly in the south and west. S Europe. 1, 20, 22*, 23*, 78*, 713*.
C. europaeum sensu Miller, non L.; *C. neapolitanum* Ten.
persicum Miller   Florist's Cyclamen
• A casual greenhouse escape. E Medit. NMW. 21, 31*, 54*, 78*, 156,
342(Feb89), 458, 713*. *C. latifolium* Smith.
repandum Smith
• Introduced or a garden escape persistent or established in Tehidy Woods,
Camborne and at Trelowarren (both W Cornwall), in woodland at Torquay
(S Devon) and at West Porlock (N Somerset). Medit. BM, OXF. 7, 20, 21,
29*, 31*, 78*, 156, 212, 371(382:24), 423, 713*.

LYSIMACHIA L.
ciliata L.   Fringed Loosestrife
••• (but confused with *L. punctata*). An established garden escape; naturalised
and locally abundant in northern England and Scotland. N America. BM, E,
LANC, NMW, OXF, RNG. 2, 20*, 22*, 41*, 126*, 303(31:23). *Steironema
ciliatum* (L.) Raf.
ephemerum L.
Pre-1930 only. SW Europe. 21, 30*, 54*, 192, 525*.
[*lanceolata* Walter   In error for native *L. nemorum*. E. 303*(58:48-49).]
punctata L.   Dotted Loosestrife
•••• A garden escape established by roadsides and rivers in scattered localities
throughout the British Is; increasing, in spite of appearing to be sterile.
Europe, SW Asia. 1, 20*, 22*, 52*, 126*. *L. westphalica* hort., non Weihe.
L. westphalica Weihe appears to be a synonym of native *L. vulgaris*.

quadrifolia L.   Whorled Loosestrife
Pre-1930 only. Eastern N America. 2, 38*, 40, 193(p.199), 217, 1268*.

terrestris (L.) Britton, Sterns & Pogg.   Lake Loosestrife
• An established garden escape naturalised since 1885 by Lake Windermere
(Westmorland and Furness); also at Leonardslee and South Lodge (both
W Sussex). N America. BM, K, LANC, OXF, RNG, SLBI. 1, 20*, 22*,
38*, 41*, 174, 284, 349(1:375), 354(11:488; 12:737). *L. racemosa* Lam.;
L. stricta Aiton.

TRIENTALIS L.
borealis Raf.   Starflower
• A persistent weed in plant nurseries at Seal (W Kent) and Winchester
(S Hants). Eastern N America. RNG. 20, 38*, 40, 51*, 236, 310(31:25),
349(7:191), 825*(2:138-143). *T. americana* Pursh; *T. europaeus* Michaux,
non L.

ANAGALLIS L.
monelli L.   Garden Pimpernel
• A casual garden escape. W Medit. RNG. 7, 21, 24*, 30*, 57*,
303(29:10), 342(Feb89), 354(6:297), 584*, 683*, 737*. *A. collina* Schousboe;
A. grandiflora hort.; *A. linifolia* L.

PITTOSPORACEAE
PITTOSPORUM Banks ex Gaertner
crassifolium Banks & Sol. ex Cunn.   Karo
•• A garden escape and hedging plant established and spreading by bird-sown
seed in Scilly; single plants have also been recorded from St Ives
(W Cornwall), Tutshill (W Gloucs) and Jersey (Channel Is). New Zealand.
BM, RNG. 13, 20*, 36, 52*, 55*, 201, 207, 212, 323(33:171), 600*.

ralphii T.Kirk
• (but possibly overlooked as *P. crassifolium*). A garden escape reported in
1990 from Tresco (Scilly). New Zealand. 51, 371(422:9), 686*, 901*(4:446),
1286*.

tenuifolium Gaertner   Kohuhu
•• Reported as self-seeding in woods, churchyards and on walls in several
places in W Cornwall and in Talbot Valley in Guernsey (Channel Is); also
recorded from Scilly and from Jethou (Channel Is). New Zealand. OXF.
20*, 36, 58*, 59*, 63*, 202, 213, 303(36:23), 1181*, 1218. *P. mayi* hort.

HYDRANGEACEAE
Incl. *Philadelphaceae*

HYDRANGEA L.
anomala D.Don   Japanese Climbing-hydrangea
• A garden relic on an old wall at South Darenth (W Kent). China, Japan. 20,
51*, 61*, 62*, 65*, 236. Incl. *H. petiolaris* Sieb. & Zucc.; *H. scandens*
Maxim.

macrophylla (Thunb.) Ser.   Common Hydrangea
•• A persistent garden relic or escape recorded from the Channel Is,
E and W Cornwall, W Lancs and Dunbarton. Japan. 51a, 52*, 62*, 63*, 66*,
201, 202, 204, 213. Incl. *H. hortensis* Smith.

sargentiana Rehder   Sargent's Hydrangea
• Introduced and naturalised on Valencia Is (S Kerry). E Asia. 20, 51*, 61*,
62*, 63*, 303(32:19). *H. aspera* Buch.-Ham. ex D.Don subsp. *sargentiana*
(Rehder) McClintock.

PHILADELPHUS L.
coronarius L.   Mock-orange
••• (but confused with hybrids and other species). Introduced and long
persistent in woodlands; also a garden escape or relic reproducing in a few
places. SE Europe, Turkey. BM, LANC, OXF, RNG, SLBI. 7, 20, 61, 62*,
63*, 126, 347(2:104-116). *P. pallidus* Hayek.

garden hybrids   Hybrid Mock-oranges
• (but probably under-recorded). A persistent garden escape on Bookham
Common (Surrey) and introduced on Chislehurst Common (W Kent). 20*,
53*, 54*, 61, 63*, 331(60:77), 371(337:28), 347(2:104-116). Incl.
P. × *cymosus* Rehder; *P.* × *lemoinei* Lemoine; *P.* × *polyanthus* Rehder;
P. × *purpureomaculatus* Lemoine; *P.* × *virginalis* Rehder.

pubescens Lois.   Broad-leaved Mock-orange
• (but probably under-recorded). A garden escape or relic recorded in 1990
from Curzon Bridge (Surrey). SE USA. 40, 51, 61, 342(Feb91), 347(2:104-
116), 507*, 726*. *P. grandiflorus* hort.

subcanus Koehne
• A garden escape or relic c.1984 at Glasgow (Lanarks). China. 61, 63*,
316(21:206), 347(2:104-116).

DEUTZIA Thunb.
× **elegantissima** (Lemoine) Rehder
• Reported as a garden relic persisting in abundance on the site of a demolished
hospital at Stone (W Kent). Originated in cultivation. 52*, 55*, 61, 62, 63*,
458. *D. purpurascens* (L.Henry) Rehder × *D. sieboldiana* Maxim.

scabra Thunb.
•• Introduced or a persistent garden escape. China, Japan. LANC, OXF,
RNG. 21, 56*, 61, 63*, 119, 371(340:28; 397:35), 445, 613*, 856(19:133-
147). Some records are referable to *D. scabra* 'Flore Pleno'.

GROSSULARIACEAE
Incl. *Escalloniaceae*

ESCALLONIA Mutis ex L.f.
macrantha group   Escallonias
••• Established garden escapes and relics; naturalised on cliffs, shores and
railway banks, especially in Ireland, Wales and Cornwall. Chile, Argentina.
BM, LANC, NMW, OXF, RNG, SLBI. 1, 20*, 61, 63*, 156, 207*, 213,
252. Incl. *E. macrantha* Hook. & Arn.; *E. rubra* (Ruíz Lopez & Pavón) Pers.

The hybrid *E. macrantha* × *E. virgata* Pers. (*E.* × *langleyensis* Veitch) has been recorded from Dunbarton. 467.

RIBES L.
americanum Miller   American Black-currant
Pre-1930 only. Eastern N America. 38*, 40, 51, 318(21:237).

[*aureum* Pursh   Unconfirmed. Perhaps always in error for *R. odoratum* but, according to 51b, this less attractive species is still grown in gardens; it may have been overlooked.]

[*nigrum* L. Accepted, with reservations, as native.]

odoratum Wendl.f.   Buffalo Currant
•• A garden escape persistent in a few places in SE England; perhaps established by the river Brent at Boston Manor (Middlesex) and in a chalk pit at Houghton Regis (Beds); also recorded from the bank of the river Clyde at Hutchesontown (Lanarks). N America. E, OXF, SLBI. 20*, 38*, 61*, 63*, 331(55:46), 370(19:147), 371(376:25), 415. *R. aureum* Lindley, non Pursh.

[*rubrum* L.   Accepted, with reservations, as native.]

sanguineum Pursh   Flowering Currant
•••• An established garden escape; naturalised, sometimes abundantly, in woodland and on grassy banks and cliffs; bird-sown in hedges and on wall tops; in scattered localities throughout the British Is, especially in the north and west. Western N America. 3, 20*, 61, 62*, 63*.

speciosum Pursh   Fuchsia-flowered Gooseberry
• Introduced in field hedges in Ireland. Western N America. 42, 51, 61*, 62*, 63*, 65*, 274. The related *R. divaricatum* Douglas occurs as naturalised from hedges in Scandinavia and may have been overlooked. 915*(86:275-279).

FRANCOA Cav.
ramosa D.Don   Bridal Wreath
• A garden escape reported in 1993 from Scilly.   Chile.   312*(t.3824), 379*(90:423), 480, 565*, 728*(6:t.223), 1257*. *F. glabrata* DC.

CRASSULACEAE
CRASSULA L.
alata (Viv.) Berger
        subsp. **pharnaceoides** (Fischer & C.Meyer) Wickens & Bywater
Pre-1930 only. Tropical Africa. 3, 300(1911:100), 328(34:629-638), 604, 676, 736. *C. campestris* subsp. *pharnaceoides* (Fischer & C.Meyer) Tölken pro parte; *C. pharnaceoides* Fischer & C.Meyer; *Tillaea pharnaceoides* (Fischer & C.Meyer) Hochst. ex Britten.

aff. **campestris** (Ecklon & C.Zeyher) Endl. ex Walp.
• A wool casual. S Africa. K, HbEJC. 328(34:629-638), 370(10:391).

colorata (Nees) Ostenf.   Dense Pigmyweed
• A wool casual. Australia. K, HbEJC. 36c, 37*, 370(10:393), 520*, 533, 590*. *Tillaea acuminata* Reader.

decumbens Thunb.   Scilly Pigmyweed
•• Established in bulbfields and by tracks on St Mary's (Scilly), perhaps
imported with cultivated plants; also a wool casual. S Africa, Australasia.
BM, CGE, E, K, RNG. 14, 20*, 36c, 207*, 349(4:42), 370(10:391), 501,
520*, 549*, 590*. Incl. *C. macrantha* (Hook.f.) Diels & Pritzel; *Tillaea
macrantha* Hook.f.

helmsii (Kirk) Cockayne   New Zealand Pigmyweed
•••• Introduced as an oxygenating plant, or accidentally with ornamental
aquatic plants, now well established and abundant in ponds, reservoirs and
canals in widely scattered localities throughout the British Is; increasing rapidly,
already a threat to native vegetation in some areas. Australasia. 13, 20*, 87*,
303*(19:10; 22:20; 49:43), 370(5:59-63), 590*. *C. recurva* (Hook.f.) Ostenf.,
non N.E.Br.; *Tillaea helmsii* Kirk; *T. recurva* (Hook.f.) Hook.f., non N.E.Br.;
T. verticillaris Hook., non DC.

intricata (Nees) Ostenf.
• A wool casual. Australia. K, HbEJC. 317*(t.295), 370(10:393). *Tillaea
intricata* Nees.

natans Thunb.
• A wool casual. S Africa, Australia. K, RNG, HbEJC. 370(10:392), 501,
520*, 628*, 655*, 818(8:1-595), 863*(10:70). *Tillaea capensis* L.f.; incl.
C. levynsiae Adamson.

peduncularis (Smith) Meigen   Purple Pigmyweed
• A wool casual. Australasia, S America. K, HbEJC. 21, 37c*, 370(10:392),
514, 520*, 533, 590*. *C. bonariensis* DC. ex Cambess., nom. illeg.;
C. purpurata (Hook.f.) Domin; *Tillaea purpurata* Hook.f.

pubescens Thunb.   Jersey Pigmyweed
• A persistent garden escape at the foot of a granite boulder on a sandy
hillside at Les Quennevais in Jersey (Channel Is). S Africa. HbEJC. 20*,
201, 303*(29:26-27). Incl. *C. radicans* (Haw.) D.Dietr. (*C. pubescens* subsp.
radicans (Haw.) Tölken).

[*recurva* N.E.Br., non (Hook.f.) Hook.f.   In error for *C. helmsii*. 303(23:13),
370(11:395).]

sarmentosa Harvey   Jade-tree
• A casual greenhouse escape. S Africa. 53*, 122, 503*, 593*, 818(8:1-595).
C. argentea Thunb.; *C. ovata* E.Meyer ex Drège, nom. nud.

setulosa Harvey
• A wool casual. S Africa. ABD, BM, CGE, E, K, RNG. 17, 51a, 592,
602*, 818(8:1-595). Incl. *C. milfordae* Byles.

sieberiana (Schultes & Schultes f.) Druce   Australian Pigmyweed
• A wool casual. Australasia. ABD, E, K, RNG. 3, 14, 36c, 37c*,
370(10:392), 533*, 569*, 666*. *Tillaea sieberiana* Schultes & Schultes f.;
T. verticillaris DC.

thunbergiana Schultes
• A wool casual. S Africa. HbEJC. 370(10:392), 501, 628*, 818(8:1-595).
C. debilis Thunb.; *C. decumbens* Harvey, non Thunb.; *C. zeyheriana* Schönl.;
Purgosea debilis (Thunb) G.Don; *Tillaea decumbens* Willd.

vaillantii (Willd.) Roth
Pre-1930 only. S Europe, N Africa. E, OXF. 3*, 21, 22*, 45*, 46*, 354(2:414), 501. *Tillaea vaillantii* Willd.

KALANCHOE Adans.
blossfeldiana Poelln.   Flaming Katy
• A casual greenhouse escape. Madagascar. 51*, 54*, 67*, 199, 503*, 719*. Plants grown under this name are often of complex hybrid origin; they may also escape.

[*UMBILICUS* DC.
erectus DC. (*Cotyledon lusitanica* Lam.; *C. lutea* Hudson; *C. umbilicus-veneris* L.) In error for native *U. rupestris*. 373(6:20).]

CHIASTOPHYLLUM (Ledeb.) Stapf ex A.Berger
oppositifolium (Ledeb. ex Nordm.) A.Berger   Lamb's-tail
• An established garden escape on an old wall·at Bodnant (Denbs); probably casual elsewhere. HbEJC. Caucasus. 51*, 53*, 55*, 57*, 354(10:471). *Cotyledon oppositifolia* Ledeb. ex Nordm.; *C. simplicifolia* hort.; *Umbilicus oppositifolius* (Ledeb. ex Nordm.) Ledeb. Easily misidentified as *Crassula* sp.

PISTORINIA DC.
hispanica (L.) DC.
Pre-1930 only. SW Europe. 2, 21, 30*, 737*. *Cotyledon hispanica* L.

SEMPERVIVUM L.
andreanum Wale
• Reported in 1989 from old walls at Hextable (W Kent). Pyrenees. 21, 306*(9:104), 458.
montanum L.   Mountain House-leek
Pre-1930 only. C & S Europe. 21, 26*, 27*, 28*, 104.
tectorum L.   House-leek
••••• Introduced and long-established on old walls and cottage roofs in many places; decreasing and becoming rare. S Europe. 1, 20, 22*, 23*, 26*, 27*.

JOVIBARBA Opiz
sobolifera (Sims) Opiz   Hens-and-chickens House-leek
• Introduced or a garden escape on a stone wall near Hallaton (Leics). N Eurasia. 21, 22*, 27*, 392. *Sempervivum soboliferum* Sims.

AEONIUM Webb & Berth.
arboreum (L.) Webb & Berth.   Tree Aeonium
• A persistent garden escape on walls of Abbey Gardens on Tresco (Scilly). Morocco. 20, 36c*, 303(36:23), 593*, 681*, 687*, 702*. *Sempervivum arboreum* L. The record from Jersey (Channel Is) in 1920 was in error for *Sedum praealtum*. 354(6:22), 201.

canariense (L.) Webb & Berth.
- A persistent garden escape on the shores of St Agnes and Bryher (both Scilly). Canary Is. HbEJC. 51, 56*, 207, 530, 687*.

cf. **cuneatum** Webb & Berth.   Aeonium
- A garden escape self-sown and established on walls on St Mary's and St Martin's and persistent on the shore of Bryher (all Scilly). Canary Is. BM, HbEJC. 20, 51, 207, 371(374:9), 458, 530*, 687*, 1117*(as *Aeonium* sp.).

SEDUM L.
[*album* L.   Accepted, with reservations, as native.]

anacampseros L.   Love-restoring Stonecrop
- A garden escape established on a disused airfield at Ashbourne (Derbys). S Europe. 20*, 21, 23*, 27*, 51b, 72*, 183, 303(17:19). *Hylotelephium anacampseros* (L.) H.Ohba.

cepaea L.   Pink Stonecrop
- A casual garden escape. C & S Europe, Turkey. BM, LIV, OXF. 1, 20, 21, 23*, 31*, 72*, 119, 150, 199, 354(5:379).

confusum Hemsley   Lesser Mexican-stonecrop
- A persistent garden escape; recorded from Newlyn Harbour, the side of a quarry car park at Falmouth, a hedgebank near Garras (all W Cornwall) and a roadside bank at Ruettes Brayes in Guernsey (Channel Is). Mexico. 20, 72*, 212, 213, 325*(46:1-314), 371(391:34).

dasyphyllum L.   Thick-leaved Stonecrop
- ●●● A garden escape established in quarries, on old walls and rocks in scattered localities, especially in Wales and SW England; long naturalised on limestone rocks near Cork. C Europe, Medit. BM, LANC, LTN, NMW, OXF, RNG. 1, 20, 22*, 27*, 33*, 72*, 238. *S. glaucum* Lam., non Waldst. & Kit.; *Oreosedum dasyphyllum* (L.) Grulich.

[*dendroideum* Sessé & Moçiño ex DC.   In error for *S. praealtum*. 201, 220.]

hispanicum L.   Spanish Stonecrop
- ●● A garden escape long established on walls at Garford (Berks); also recorded from a gravel pit at Beckford (Worcs), a disused airfield at Ashbourne (Derbys), and from Bury St Edmunds (W Suffolk) and Eastbourne (E Sussex). SE Europe, SW Asia. BM, OXF. 7, 20*, 21, 22*, 31*, 44*, 119, 122, 303(17:19; 45:20), 369(26:63). *S. glaucum* Waldst. & Kit., non Lam.; incl. *S. bithynicum* Boiss.

hybridum L.   Siberian Stonecrop
- A garden escape naturalised in Holton Park (E Suffolk) and short-lived on the slipway at Fort Saumarez, Guernsey (Channel Is). Siberia, Mongolia. 1, 20, 21, 22*, 72*, 325*(46:1-314), 369(21:38), 371(385:34).

kamtschaticum Fischer & C.Meyer sensu lato
- A garden escape persistent on the old tramway at Slepe Farm, near Corfe Castle and near Stoborough (both Dorset); also on waste ground outside the RHS Gardens, Wisley (Surrey). NE Asia. BM. 20, 51*, 52*, 57*, 72*, 371(394:35; 416:7). Incl. *S. ellacombianum* Praeger; *S. middendorffianum* (Maxim.) Boiss.

lydium Boiss.   **Least Stonecrop**
- A garden escape recorded from Dartford and Gravesend (both W Kent), Littlestone (E Kent) and Geddes House (Nairn). Turkey. BM, DUE, E, OXF, RNG. 3, 20, 39, 72*, 236, 277, 325*(46:1-314). Recorded in error from Shanes Castle (Co Antrim) and Alport (Derbys). 319(17:427), 1250.

mexicanum Britton   **Mexican-stonecrop**
- A garden escape persistent on waste ground, Granton, Edinburgh. C America; (Mexico). 36c, 51, 72*, 303(32:19), 1060*. *S. aureum* hort., non Wirtgen ex F.Schultz.

nicaeense All.   **Pale Stonecrop**
- Recorded without locality from Salop and reported as abundant on a grassy bank at Eltham (W Kent). Medit. RNG. 7, 20, 21, 24*, 44*, 72*, 206, 354(4:198), 458. *S. altissimum* Poiret; *S. sediforme* (Jacq.) Pau.

[*ochroleucum* Chaix (*S. anopetalum* DC.)   A dubious record. 354(5:762).]

pallidum M.Bieb.
    Pre-1930 only.   SE Europe, SW Asia.   7, 21, 44*, 72*, 288(p.lix). *S. intermedia* hort.

populifolium Pallas
- A casual garden escape recorded without locality. Siberia. 51b, 72*, 347*(8:1-20), 371(352:27). *Hylotelephium populifolium* (Pallas) H.Ohba.

praealtum A.DC.   **Greater Mexican-stonecrop**
- An established garden escape; naturalised since 1920 on rocks at St Catherine's Bay in Jersey; also found in a few places in Guernsey (both Channel Is). Mexico. BM, JSY, OXF, RNG. 20, 72*, 117*, 201, 220*, 371(391:34), 1060*. *S. dendroideum* subsp. *praealtum* (A.DC.) R.T.Clausen.

[*pruinatum* Link ex Brot.   In error for native *S. forsterianum*. 149.]

rubrotinctum R.T.Clausen   **Jelly-beans**
- A persistent garden or greenhouse escape at the base of a wall in Cambridge. Originated in cultivation, possibly of hybrid origin. 36c, 54*, 72*, 371(379:25), 593*, 719*. *S. guatemalense* hort., non Hemsley.

rupestre L.   **Reflexed Stonecrop**
••••• An established garden escape; widespread on walls, rocks and banks, especially in southern England; also a wool casual. Europe, Turkey. 2, 20*, 22*, 72*, 825*(6:67-72). *S. albescens* Haw.; *S. glaucum* Haw., non Lam., nec Waldst. & Kit.; *Petrosedum rupestre* (L.) P.V.Heath.   *S. reflexum* L. (*P. reflexum* (L.) Grulich; *P. rupestre* subsp. *reflexum* (L.) Velayos) does not seem to be worthy of even subspecific rank.

sexangulare L.   **Tasteless Stonecrop**
••• An established garden escape on walls, cliffs, rocky slopes and in churchyards in widely scattered localities in England and Wales; naturalised since 1912 on Wick (or Wyck) Rocks near Bristol (W Gloucs). Europe. BM, LANC, NMW, OXF, RNG, TOR. 1, 20, 22*, 23*, 72*, 245, 303(3:11-12). *S. boloniensis* Lois.; *S. mite* Gilib., nom. illeg.

sieboldii Sweet ex Hook.
Pre-1930 only. Japan. BM. 43, 51b, 52*, 72*, 593*, 702*. *Hylotelephium sieboldii* (Sweet ex Hook.) H.Ohba.

spathulifolium Hook.   Colorado Stonecrop
• A garden escape persistent on walls; recently recorded from Hastings (E Sussex) and Dunbarton. Western N America. 13, 20, 52*, 53*, 72*, 122, 156, 347*(8:1-20).

spectabile Boreau   Butterfly Stonecrop
••• An established garden escape; on roadsides, commons and waste ground in widely scattered localities; known since 1930 in woodland at Nettleton (N Wilts). SE Asia. OXF, RNG. 20*, 52*, 54*, 72*, 199, 303(17:19; 64*:40), 347(8:1-20), 477. *Hylotelephium spectabile* (Boreau) H.Ohba. A hybrid with native *S. telephium* (*S.*'Autumn Joy') may have been overlooked. 303*(64:40).

spurium M.Bieb.   Caucasian-stonecrop
•••• An established garden escape; on cliffs, rocks, railway embankments, walls, roadsides and in sand and gravel pits; long-naturalised on the cliff at Pettycur (Fife). Turkey, Caucasus. 2(as *S. stoloniferum*), 20*, 22*, 52*, 72*, 113, 347(8:1-20), 908(56:29-45). *S. stoloniferum* auct., non S.Gmelin; *Asterosedum spurium* (M.Bieb.) Grulich; *Spathulata spuria* (M.Bieb.) A.Löve & D.Löve.

stellatum L.
Pre-1930 only. Medit. 1, 21, 32*, 72*, 149, 281, 320(44:17).

stoloniferum S.Gmelin   Lesser Caucasian-stonecrop
• (but confused with *S. spurium*). A persistent or established garden escape; modern records are from Ruxley gravel pits (W Kent), near the reservoir in Guernsey (Channel Is), and on rocks at Old Tongland Bridge (Kirkcudbrights). SW Asia. JSY, SLBI. 13, 20, 39, 72*, 220, 325*(46:197), 365(3:63), 371(331:19). *Asterosedum stoloniferum* (S.Gmelin) Grulich. Early records, including 2, were in error for *S. spurium*.

telephium L. subsp. **maximum** (L.) Krocker
• Recorded as a garden escape. Europe, N Turkey, Caucasus. 21a, 39, 192, 312*(n.s. t.429), 595b*, 1242*. *S. haematodes* Miller; *S. maximum* (L.) Suter; *S. stapposum* Boiss.; *Hylotelephium maximum* (L.) J.Holub; *H. telephium* (L.) H.Ohba subsp. *maximum* (L.) H.Ohba. Alien variants of the two native subspecies may also occur, as well as hybrids with *S. spectabile*. 51b.

ECHEVERIA DC.
derenbergii J.Purpus   Painted-lady
• A casual garden escape. Mexico. 51, 54*, 371(376:25), 702*.

SAXIFRAGACEAE

ASTILBE Buch.-Ham. ex D.Don

× **arendsii** Arends   **Red False-buck's-beard**
•• A persistent garden escape and relic. Originated in cultivation. OXF, RNG. 20*, 54*, 66*, 303(48:35), 316(21:206). *A. chinensis* (Maxim.) Franchet & P.A.L.Savat. × *A. japonica*; *A.* × *rosea* hort.

japonica (Morren & Decne.) A.Gray   **False-buck's-beard**
• A persistent garden escape. Japan. NMW. 2, 20, 43, 156, 371(343:29), 650*.

rivularis Buch.-Ham. ex D.Don   **Tall False-buck's-beard**
• Introduced or a garden escape, naturalised in a ravine and on the sides of a forest road near Meall Mhor House (Kintyre). Nepal. 20, 370(17:472), 685*.

RODGERSIA A.Gray

aesculifolia Batalin
• A garden relic long persistent on the site of Warley Place gardens (S Essex) and on a stream bank at Wellingborough (Northants). China. 51, 52*, 53*, 57*, 371(352:28), 435.

podophylla A.Gray   **Rodgersia**
•• Introduced and persistent by streams and ponds at Carbery Towers (E Lothian), Dumbarton golf course (Dunbarton), Ilam (Staffs), Warley Place gardens (S Essex) and Gregynog Hall, near Newtown (Monts). E Asia. BM. 20, 43, 52*, 53*, 54*, 156, 313(31:370), 370(17:189), 371(349:28; 385:34), 1324*.

BERGENIA Moench

ciliata (Haw.) Sternb.   **Hairy Bergenia**
• A casual garden escape. Himalayas. HbPM. 51a, 53*, 66*, 198, 325*(108:483), 685*. *B. ligulata* (Wallich) Engl., nom. illeg.

cordifolia (Haw.) Sternb.   **Heart-leaved Bergenia**
•• A casual or persistent garden escape. Siberia. 20, 51a, 53, 54*, 55*, 156, 173, 199, 371(349:29).

crassifolia (L.) Fritsch   **Elephant-ears**
•• A garden escape established or persistent on tips and waste ground; also long persistent in derelict gardens. Siberia. 20, 51a, 119, 191, 236, 506*, 511*(p.1014), 600*.

purpurascens (Hook.f. & Thomson) Engl.
• Reported from a roadside bank at Hastings (E Sussex). Himalayas, China, Burma. 51a, 53, 54*, 417, 685, 724*. *B. delavayi* (Franchet) Engl.

× **schmidtii** (Regel) Silva-Tar.
• Introduced and established in woodland between Leatherhead by-pass and Daymerslea Ridge (Surrey); also persistent in Guernsey (Channel Is). Originated in cultivation. 51a, 53, 199, 220, 584*. *B. ciliata* × *B. crassifolia*.

DARMERA Voss ex Post & Kuntze
peltata (Torrey ex Benth.) Voss ex Post & Kuntze   Indian-rhubarb
   ••• Introduced and a garden escape, long-persistent or established; naturalised
   in damp woodland and by streams and lakes, especially in the north and west.
   California, Oregon. BM, CGE, E, LANC, OXF, RNG. 13, 20*, 41*, 51*,
   52*, 220, 354(8:616,799), 370(14:192).   *Peltiphyllum peltatum* (Torrey ex
   Benth.) Engl., nom. illeg.; *Saxifraga peltata* Torrey ex Benth.

SAXIFRAGA L.
× **andrewsii** Harvey
   Pre-1930 only. Some records are dubious. Originated in cultivation. BM.
   255, 1311. *S.* × *geum* × *S. paniculata*; *S. guthriana* hort.
× **arendsii** group   Garden Mossy-saxifrages
   • A garden escape persistent on a grassy bank at Llanwrda (Carms).
   Originated in cultivation.   51b, 96, 305(53:13), 900*(46:257), 1004,
   1183*(t.6,7). Hybrids between *S. exarata* Villars, *S. rosacea* Moench and
   other species; incl. *S.* × *arendsii* Engl.
[*cotyledon* L.   A dubious record. 354(5:760).]
cuneifolia L.   Lesser Londonpride
   • A garden escape persistent on an old wall at Clapham (MW Yorks); casual
   elsewhere. S Europe. 20*, 23*, 27*, 96*, 156, 348(20:31), 371(409:40).
cymbalaria L.   Celandine Saxifrage
   ••• Persistent as a garden weed and on damp walls near gardens. Medit.
   BM, LANC, LSR, NMW, OXF, RNG. 2, 20, 46*, 51*, 96*, 303*(25:10-12),
   546*. Incl. *S. huetiana* Boiss.; *S. sibthorpii* auct. angl., non Boiss. Records
   are probably all referable to var. *huetiana* (Boiss.) Engl. & Irmscher.
× **geum** L.   Scarce Londonpride
   • (but confused with its congeners). Introduced or a garden escape; modern
   records are from Ravenscraig Park (Fife), Hunmanby Hall grounds (SE Yorks),
   and a woodland bank near Polawyn, Lizard (W Cornwall). Pyrenees. BM, K,
   LANC. 20*, 22*, 96*, 113, 134, 371(403:40). *S. hirsuta* × *S. umbrosa*. The
   Aros Park (Mid Ebudes) record was in error for *S. hirsuta* L. 303(47:20).
hirsuta L. **alien variants**
   Unnamed varieties have been recorded as garden escapes. 155, 316(21:557),
   319(19:206), 1328.
hypnoides L. **alien variants**
   Alien varieties probably occur as garden escapes. A hybrid with *S. trifurcata*
   has been collected. RNG.
[*moschata* Wulfen (*S. muscoides* Wulfen, non All.; *S. pygmaea* Haw.) A dubious
   record. 188, 354(5:760), 361(7:222), 1149.]
paniculata Miller   Live-long Saxifrage
   • Recently recorded from rocks at Beckermonds in Upper Wharfedale
   (MW Yorks), and reported in 1992 to be still on a path at the Garden Festival
   site in Glasgow (Lanarks). Europe. 21, 22*, 29*, 30*, 96*, 335(114:106),
   445. *S. aizoon* Jacq.

[*pedemontana* All. (incl. *S. pedatifida* auct., vix Smith)   Dubious records.   159,
187, 354(5:759).]

rotundifolia L.   Round-leaved Saxifrage
•• A garden escape established in a few places in northern England and
Scotland; known since the 1920s, now naturalised in abundance, by Fin Burn
and Glazert Burn at Campsie (Stirlings); also naturalised in Holden Clough,
near Bolton-by-Bowland (MW Yorks) and by the river Kelvin near Lennoxtown
(Dunbarton).   C & S Europe, SW Asia.   CGE, CLE, E, OXF, RNG.   1, 20,
22*, 23*, 27*, 96*, 349(7:189-191), 1083*.

[*sibthorpii* Boiss.   In error for *S. cymbalaria*.   303(25:12).]

stolonifera Curtis   Strawberry Saxifrage
• A garden or greenhouse escape established on a north-facing wall NE of
Garras, Lizard (W Cornwall) and in the churchyard at St Just-in-Roseland
(E Cornwall).   E Asia.   BM.   7, 20, 56*, 368(1876:249), 371(403:40), 593*,
650*.   *S. sarmentosa* L.f.

trifurcata Schrader
Pre-1930 only.   N Spain.   OXF.   7, 21, 96*, 354(5:27,759), 1149.

umbrosa L.   Pyrenean Saxifrage
•• (but confused with *S.* × *urbium*).   An established garden escape; known
since 1792 in Heselden Gill (MW Yorks).   Pyrenees.   BIRM, BM, OXF, RNG.
2(as native), 20*, 22*, 29*, 96*.

× **urbium** D.Webb   London Pride
•••• An established garden escape; widely naturalised in woods, on commons
and roadsides, abundantly so in several glens in the Isle of Man.   Originated in
cultivation.   12, 20*, 22*, 34*, 96*, 103.   *S. spathularis* Brot. × *S. umbrosa*;
S. umbrosa var. *crenatoserrata* Bab.

HEUCHERA L.
× **brizoides** hort. ex Lemoine
Pre-1930 only.   Originated in cultivation.   LANC.   51b, 53*, 55*, 612*,
1163*.   A complex of hybrids involving *H. americana* L., *H. micrantha*
Douglas ex Lindley and *H. sanguinea*.   Incl. *H.*'Red Spangles'.

sanguinea Engelm.   Coral-bells
•• A persistent garden escape.   Southern USA, Mexico.   20*, 55*, 126, 192,
236, 371(337:29), 657*, 680*.

× HEUCHERELLA Wehrh.
tiarelloides (Lemoine) Wehrh. ex Stearn
• A casual garden escape.   Originated in cultivation.   51, 57*, 312*(n.s. t.31),
371(346:29), 1163*.   *Heuchera* × *brizoides* × *Tiarella cordifolia*; *Heuchera*
× *tiarelloides* hort. ex Lemoine.

TIARELLA L.
cordifolia L.   Foamflower
• An established garden escape at Virginia Water (Berks).   Eastern N America.
LIV, RNG.   38*, 51, 53*, 54*, 312*(t.1589), 1257*.

wherryi Lakela   Tufted Foamflower
- A garden escape on an old wall at Stradey Woods, Llanelli (Glam). Eastern N America. 40, 51, 54*, 156, 303(36:29), 315*(124:85), 1257*.

TOLMIEA Torrey & A.Gray
menziesii (Pursh) Torrey & A.Gray   Pick-a-back-plant
●●●● An established garden escape; naturalised, sometimes abundantly, on shaded or damp ground, especially by streams in the north. Western N America. 7, 20*, 25*, 41*, 52*, 500*. *Leptaxis menziesii* (Pursh) Raf.; *Tiarella menziesii* Pursh.

TELLIMA R.Br.
grandiflora (Pursh) Douglas ex Lindley   Fringe-cups
●●●● An established garden escape; naturalised in shaded places and by streams in widely scattered localities throughout the British Is. Western N America. 2, 20*, 25*, 41*, 55*, 500*, 826*(19:26-28). *Mitella grandiflora* Pursh.

MITELLA L.
caulescens Nutt. ex Torrey & A.Gray   Star-shaped Mitrewort
- Recorded from Bagley Wood (Berks), perhaps introduced. No modern records. Western N America. 41*, 42, 354(11:477), 500*. *Mitellastra caulescens* (Nutt. ex Torrey & A.Gray) Howell.
diphylla L.   Two-leaved Mitrewort
Pre-1930 only. Eastern N America. 1, 38*, 40, 51, 650*, 1349.

BOYKINIA Nutt.
major A.Gray
- A casual at Newport (Mons), possibly a garden escape. Western N America. NMW. 42, 51, 156, 500*, 511*(p.1014).

ROSACEAE
SORBARIA (Ser. ex DC.) A.Braun
grandiflora (Sweet) Maxim.
- An established garden escape, found in 1986 in Oxleas Wood (W Kent). E Siberia. HbEJC. 20, 51b, 631, 678*, 824*(9:t.295c). *S. alpina* (Pallas) Dippel; *S. sorbifolia* var. *alpina* Pallas; *Spiraea grandiflora* Sweet.
kirilowii (Regel) Maxim.   Chinese Sorbaria
●● A garden escape established by suckering; recorded from a roadside hedge at Milton Bryan (Beds), several localities in W Kent, and Jersey (Channel Is). China. 20, 54*, 61*, 62*, 63*, 331(56:85), 371(409:39; 414:13). Incl. *S. arborea* C.Schneider; *S. assurgens* A.Vilm. & Bois; *Spiraea arborea* (C.Schneider) Bean.
sorbifolia (L.) A.Braun   Sorbaria
●● A garden escape established by suckering; on roadsides and by railways in a few widely scattered localities in England. N Asia. BM, RNG. 20*, 53*, 61, 192, 350(38:43), 371(343:29), 1242*. *Spiraea sorbifolia* L.; incl. *S. stellipila* Maxim.

tomentosa (Lindley) Rehder   Himalayan Sorbaria
•• An established garden escape; on walls and railway banks in a few widely
scattered localities. Himalayas, China. LSR, RNG. 20*, 51*, 199, 236,
371(361:27), 685*. Incl. *S. aitchisonii* (Hemsley) Hemsley ex Rehder.

PHYSOCARPUS (Cambess.) Maxim.
amurensis (Maxim.) Maxim.
Pre-1930 only. Manchuria, Korea. 51, 61, 394(3:293), 617*, 631*, 638*,
824*(t.489). *Neillia amurensis* (Maxim.) Bean; *Spiraea amurensis* Maxim.
opulifolius (L.) Maxim.   Ninebark
•• An established garden escape and introduction; naturalised in woodlands,
by rivers and on railway banks, mainly in the north; known since 1909 near the
river Conon (E Ross). Eastern N America. BM, CGE, E, OXF, RNG, SLBI.
7, 20*, 31*, 38*, 63*, 153, 312*(n.s. t.459), 546*. *Neillia opulifolia* (L.)
Brewer & Watson; *Spiraea opulifolia* L. Possibly conspecific with the western
American *P. capitatus* (Pursh) Kuntze.

NEILLIA D.Don
sinensis Oliver
• Introduced at Ness Castle (Easterness) and recorded from south of the bridge
between Lamlash and Brodick, on Arran (Kintyre). China. E. 51, 53, 61*,
277, 303(29:12), 617*, 631*.

STEPHANANDRA Siebold & Zucc.
incisa (Thunb.) Zabel   Lace-shrub
• A persistent garden escape or introduction reported from a shrubby roadside
at New Ash Green and on chalky banks at High Elms, Downe (both W Kent).
Japan, Korea. HbEJC. 51, 61*, 62*, 303(59:46), 371(414:14), 458.
[*tanakae* (Franchet & P.A.L.Savat.) Franchet & P.A.L.Savat. The syn. is *Neillia
tanakae* Franchet & P.A.L.Savat. In error for *Holodiscus discolor*.
371(379:25; 382:25).]

SPIRAEA L.
alba Duroi   Pale Bridewort
••• (but confused with *S. salicifolia* and hybrids). An established garden
escape and relic; in scattered localities throughout Britain, naturalised in
numerous places in the Isle of Man. N America. BM, CGE, E, LANC, RNG.
13, 20*, 61, 103, 370(13:135; 18*:147-151). Incl. *S. latifolia* (Aiton) Borkh.;
S. salicifolia var. *paniculata* Aiton. Most records are referable to var. *latifolia*
(Aiton) Dippel.
× **arguta** Zabel   Bridal-spray
• A persistent garden escape in Darenth Wood (W Kent); also recorded from
Helensburgh and Craigendoran (both Dunbarton) and Siston Common
(W Gloucs). Originated in cultivation. HbMARK. 20, 61*, 62*, 63*, 223,
236, 454. *S.* × *multiflora* Zabel × *S. thunbergii* Siebold ex Blume.

× **billardii** Hérincq   Billard's Bridewort
●●●● (but much over-recorded for *S.* × *pseudosalicifolia*). A persistent garden
escape and relic. Originated in cultivation. 20, 370(14:423; 15:395; 18*:147-
151). *S. alba* × *S. douglasii*; incl. *S.* × *macrothyrsa* Dippel.

× **brachybotrys** Lange   Lange's Spiraea
● A persistent garden escape recorded for Scotland. Originated in cultivation.
20, 51, 691, 862(13:26-28), 1295. *S. canescens* × *S. douglasii.*

canescens D.Don   Himalayan Spiraea
● A persistent garden escape recorded from Longfield (W Kent), Hayling Island
golf-course (S Hants), Goodwick (Pembs) and Cwrtnewydd (Cards).
Himalayas. BM, CGE, NMW, OXF. 7, 20*, 61, 63*, 156, 236, 303(10:14),
354(6:21), 370(19:284), 1063*, 1300*.

chamaedryfolia L.   Elm-leaved Spiraea
● A garden escape established since 1966 by Isla Water (Angus). Eurasia.
OXF, RNG. 1, 20*, 23*, 31*, 349(7:189). Incl. *S. ulmifolia* Scop. Early
records were confused with *S. hypericifolia.*

douglasii Hook.   Steeplebush
●●●● (but confused with its hybrids). An established garden escape or relic;
naturalised in woods, hedgerows and on commons in widely scattered localities.
Western N America. 7, 20*, 63*, 126*, 370(15:130; 18*:147-151). Incl.
S. menziesii Hook.

[× *fulvescens* Dippel (*S. douglasii* × *S. tomentosa*)   Dubious records. 209,
371(328:19-20), 1275.]

hypericifolia L.
● A garden escape recorded from Redlynch (S Wilts). No modern records.
S Europe, Asia. BM, CGE, OXF, SLBI. 1, 21, 29*, 61, 151, 172, 181,
354(5:559), 617*. Incl. *S. obovata* Waldst. & Kit. ex Willd.
Var. *plukenetiana* DC. has been recorded. Early records were confused with
S. chamaedryfolia.

japonica L.f.   Japanese Spiraea
●● An established garden escape; naturalised on Milton Heath (Surrey) and on
sandy ground at Merthyr Mawr (Glam). E Asia. NMW, OXF, RNG, SLBI.
7, 20*, 56*, 61a, 62*, 63*, 199, 370(15:130). Incl. *S.* × *bumalda* Burv.;
S.'Anthony Waterer'.

media Schmidt
● Recently recorded from a limestone quarry above Row, Whitbarrow
(Westmorland). C Europe, Russia, Asia. LANC. 20, 21, 61*, 370(18:423),
617*. *S. confusa* Regel & Koern.; *S. oblongifolia* Waldst. & Kit.

× **pseudosalicifolia** Silverside   Confused Bridewort
●●●● A long-persistent garden escape; on commons, roadsides, railway banks
and lakesides in widely scattered localities. Originated in cultivation.
20, 370*(18:147-151). *S. douglasii* × *S. salicifolia*; *S.* × *billardii* hort.,
non Hérincq.

× **rosalba** Dippel   Intermediate Bridewort
●●●● An established garden escape by roadsides and on river banks, especially
in Wales and central Scotland; abundantly naturalised on rough marshy ground

near Rosneath Castle (Dunbarton). Originated in cultivation. 20, 153,
370(17:187; 18*:147-151). *S. alba* × *S. salicifolia*; incl. *S.* × *rubella* Dippel.
Most records are referable to nothovar. *rubella* (Dippel) Silverside.
[*salicifolia* L. In the past *S. salicifolia* was interpreted in a broad sense to include
S. alba Duroi (incl. *S. latifolia* (Aiton) Borkh.) and the hybrids between them.
According to 370*(18:147-151), the occurrence of true *S. salicifolia* in the
British Is must be regarded as highly doubtful, a statement contradicted by 20.]

tomentosa L.   Hardhack
- (but confused with *S. douglasii* and hybrids). A persistent garden escape.
Eastern N America. BM, RNG. 1, 20, 38*, 61, 209, 354(1:67; 10:523),
371(334:21), 631*, 698*.

trilobata L.
- A garden relic. N Asia. 51, 61, 126, 617*, 631*, 645*(t.1272).

× **vanhouttei** (Briot) Carrière   Van Houtte's Spiraea
- • A long-persistent garden escape or relic in widely scattered localities.
Originated in cultivation. RNG. 7, 20*, 52*, 61*, 199, 370(14:423; 15:62),
371(391:33). *S. cantoniensis* Lour. × *S. trilobata*.

ARUNCUS L.
dioicus (Walter) Fern.   Buck's-beard
- • • An established garden escape; reproducing and abundantly naturalised in
a glen near Moulin (E Perth) and self-seeding very freely in several places in
Dunbarton; elsewhere usually single-sex groups or individual plants persisting
in widely scattered localities, mainly in northern England and Scotland.
N temperate regions. BM, E, LANC, RNG. 12, 20, 22*, 23*, 25*, 27*,
316(21:557), 371(352:26). Incl. *A. sylvester* Kostel; *A. vulgaris* Raf.;
Spiraea aruncus L.

HOLODISCUS (K.Koch) Maxim.
discolor (Pursh) Maxim.   Ocean-spray
- • • A garden escape established on railway banks, roadsides and old walls;
abundant on cliffs on disused railway land at Wetherby (MW Yorks);
naturalised, perhaps for over 100 years, on the outside of a garden wall
and on a nearby rock-face at Kilmacolm (Renfrews). Western N America.
BM, CGE, LSR, RNG. 20*, 52*, 53*, 62*, 349(5:121), 371(382:23; 394:34).
Spiraea ariifolia Smith; *S. discolor* Pursh.

FILIPENDULA Miller
camtschatica (Pallas) Maxim.   Giant Meadowsweet
- • (but confused with other species). A persistent garden escape and relic; in
damp soils in a few scattered localities in the north and west. E Asia. LSR,
OXF, RNG. 20, 43, 54*, 55*, 316(21:558), 371(346:29), 430, 432, 649*,
1324*. *Spiraea camtschatica* Pallas; *S. gigantea* hort.; *S. kamtschatica* sphalm.

purpurea Maxim.
- • (but confused with *F. camtschatica* 'Rosea'). A long-persistent garden
escape. Probably originated in cultivation. 20, 43, 220, 312*(t.5726),
370(15:395), 371(346:29; 409:40), 594*. Perhaps *F. auriculata* (Ohwi) Kitam.

× *F. multijuga* Maxim.; *Spiraea palmata* hort., non Murray.   Contrary to
much literature, this taxon is not known as a wild plant in Japan.

rubra (Hill) Robinson   Queen-of-the-prairie
  • A persistent garden escape at Helensburgh (Dunbarton). Eastern N America.
  38*, 51, 54*, 223. *Spiraea lobata* Gronov. ex Jacq.; *S. palmata* Murray, non
  Pallas, nec Thunb.; *S. venusta* hort. ex Otto & A.Diels. The record for
  Guernsey (Channel Is) was in error for *S. purpurea*. 220.

vestita (Wallich ex Don) Maxim.
  Pre-1930 only.   Afghanistan, Himalayas, SW China.   OXF.   7, 51,
  301*(1841:t.4), 354(4:407), 604, 685*, 1063. *Spiraea vestita* Wallich ex Don;
  S. camtschatica var. *himalensis* Lindley.

KERRIA DC.

japonica (L.) DC.   Kerria
  ••• A persistent garden escape or relic; on roadsides, commons, and in woods
  in widely scattered localities. China. BM, NMW, OXF. 20, 52*, 62*, 63*,
  119, 190, 199, 649*. Most records are referable to *K. japonica* 'Flore Pleno'.

RUBUS L.

allegheniensis Porter   Highbush Blackberry
  • A garden escape established on Lindow Common (Cheshire) and on
  Littleworth and Wimbledon Commons (both Surrey). Eastern N America.
  BM, CGE.   12, 38*, 51, 199, 370(11:381), 586*, 882(1*:288; 2:380),
  883*(9:132), 1094. *R. villosus* Aiton, non Thunb.

armeniacus Focke   Giant Blackberry
  •••• An established garden escape ; naturalised on commons and in woods,
  especially in S and E England.   Europe, SW Asia.   21, 36c*, 41*, 119.
  R. discolor auct., non Weihe & Nees; *R. oplothyrsus* Sudre; *R. procerus*
  Muller; incl. *R.*'Himalaya Giant'.   Contrary to 21, *R. discolor* Weihe & Nees
  is a synonym of native *R. ulmifolius*. 436.

canadensis L.   American Dewberry
  • A garden escape established in thickets in Epping Forest (S Essex); also
  recorded without locality from E Ross. Eastern N America. 13, 38*, 40, 51,
  191, 371(343:28), 586*, 1094.

cockburnianus Hemsley   White-stemmed Bramble
  • A persistent garden escape or relic on Hayling Island (S Hants), in woods at
  Brodie Castle (Moray) and beside an old railway between Glasgow and Renfrew
  (Lanarks). China. E, HbAB. 20*, 61*, 277, 325*(94:511), 370(16:187),
  371(379:25), 1094. *R. giraldianus* Focke. Some records may be referable to
  R. thibetanus Franchet or other related species. 325(117:578-579).

aff. **coreanus** Miq.   Korean Bramble
  • A persistent garden escape close to riverside allotments in York. China,
  Korea. 43, 51, 61, 371(388:35), 617*, 631*, 649*. *R. tokkura* Siebold.

deliciosus Torrey   Rocky Mountain Raspberry
  • A persistent garden relic at Farnham (E Suffolk). Western N America.
  51, 55*, 61*, 63*, 369(22:49), 1181*.

elegantispinosus (A.Schum.) H.E.Weber   Slender-spined Bramble
●●● An established garden escape; widely naturalised, particularly in suburban
areas, sometimes extensively, especially in eastern Scotland; increasing.
NW Europe. BM.   21, 370(11:381; 15:70,370), 845*(3:261), 1094.
R. argenteus Weihe & Nees subsp. *elegantispinosus* A.Schum.; *R. elegans*
Utsch, non P.J.Mueller.   This taxon is apparently unknown in gardening
literature.

× **fraseri** Rehder
● Recorded from an old refuse tip at Bannerdown, near Bath (N Somerset).
Originated in cultivation.   51b, 61, 315*(73:51), 371(382:23), 691.
R. odoratus × *R. parviflorus*; *R. robustus* G.Fraser, non Presl.

illecebrosus Focke   Strawberry-raspberry
● Collected as a weed in a garden at Delph, near Oldham (S Lancs). Japan.
RNG. 21, 61, 596*, 649*. *R. commersonii* Poiret var. *illecebrosus* (Focke)
Makino.   *R. idaeus* group.

laciniatus Willd.   Cut-leaved Blackberry
●●●● An established garden escape; naturalised in widely scattered localities,
especially in SE England. Origin obscure, probably NW Europe.. 1, 21, 61,
126, 350*(47:41), 500*, 595*, 1094. *R. fruticosus* L. var. *laciniatus* Weston.
A hybrid with native *R. ulmifolius* has been recorded.

linkianus Ser.
● Recorded from a roadside near Downhill Hall (Co Londonderry).   No
modern records. Origin obscure, probably Europe. 21, 51b, 352(52B:64),
691, 804*(t.722), 880*(17:553), 882(1:197).   *R. fruticosus* 'Flore Pleno';
R. hedycarpus Focke forma *linkianus* Zabel.   *R. candicans* Weihe ex Reichb.
group.

loganobaccus L.Bailey   Loganberry
●●● An established garden escape; naturalised on commons, roadsides and
railway banks, mainly in southern England. Originated in cultivation. BM,
CGE, NMW. 12, 20*, 54*, 61, 80*, 199, 882*(5:57), 1094. Probably of
R. idaeus L. subsp. *strigosus* (Michaux) Focke × *R. ursinus* Cham. & Schldl.
subsp. *vitifolius* Cham. & Schldl. origin.   *R. idaeus* group.

odoratus L.   Purple-flowered Raspberry
● Introduced or a garden escape established on the site of a former railway
track near Muswell Hill (Middlesex); also at Shottermill and West Humble
(both Surrey). Eastern N America. BM. 7, 20, 38*, 55*, 62*, 63*, 199,
331(62:110), 1094, 1210*.

parviflorus Nutt.   Thimbleberry
●● Introduced or a garden escape, well established in a few widely scattered
localities; naturalised in quantity by the lake below Kenwood House
(Middlesex), on a railway embankment at Shalford (Surrey) and in the
grounds of Brahan Castle (E Ross). N America. BM, CGE, DUE, E, OXF,
RNG. 7, 20*, 38*, 41*, 61, 126, 209, 370(10:423), 882*(5:908), 1094.
R. nutkanus Moçiño ex Ser.

pergratus Blanchard
•• An established escape from cultivation; naturalised in a few localities in England, as in Toft Wood at Knutsford (Cheshire) and Epping Forest (S Essex). Eastern N America. BM, CGE, LSR. 38*, 40, 199, 240, 370(11:380-382; 15:361-380), 1094, 1210*.

phoenicolasius Maxim.   Japanese Wineberry
••• An established garden escape; naturalised in hedges and woodland in a few places in southern England and Wales. E Asia. BM, NMW, OXF, RNG. 7, 13, 20, 54*, 55*, 62*, 312*(t.6479), 350(42:105), 649*, 882*(5:902), 1094. A hybrid with native *R. idaeus* (*R.* × *paxii* Focke) has been recorded. 354(9:260).

spectabilis Pursh   Salmonberry
•••• An established garden escape and introduced for game-cover; naturalised in woods and on waste ground in scattered localities throughout the British Is; known for over 100 years and now abundant near Sandling Park at Hythe (E Kent) and near Dolphinton (Peebles); increasing. Western N America. 1, 20*, 41*, 61, 61a, 63*, 110, 236, 320(21:251), 349(3:400), 882*(5:898), 1094.

[*thyrsigeriformis* (Sudre) D.Allen   Accepted, with reservations, as native. 370(17:435).]

tricolor Focke   Chinese Bramble
•• A garden escape, frequently planted in recent years as ornamental ground cover, established at Drakelow (Derbys); also recorded from elsewhere in Derbys and as introduced and increasing in N Devon and Middlesex. China. HbEJC. 20*, 61*, 62*, 63*, 183, 302(18:13), 312*(t.9534), 436, 1094. *R. polytrichus* Franchet, non Progel.

POTENTILLA L.
alba L.   White Cinquefoil
Pre-1930 only. Europe. BM, LIV, OXF. 2, 21, 22*, 26*, 29*, 192, 354(6:381; 9:114).

alchemilloides Lapeyr.   Alchemilla-leaved Cinquefoil
• Introduced and persistent near Appletreewick (MW Yorks). Pyrenees. RNG. 21, 23*, 27*, 28*, 371(331:20; 352:27), 546*.

arguta Pursh   Tall Cinquefoil
Pre-1930 only. N America. 38*, 40*, 51, 113, 169.

aurea L.   Golden Cinquefoil
Pre-1930 only. C & S Europe. OXF. 2, 21, 22*, 23*, 27*, 354(5:15).

bifurca L.
Pre-1930 only. SE Europe, Russia. 21, 143, 589*, 630*, 715*, 1197*. *P. orientalis* Juz.

canadensis L.
Pre-1930 only. Eastern N America. 7, 38*, 40*, 354(4:12).

chrysantha group   Thuringian Cinquefoils
• Established for many years on railway embankments between Leysmill and Friockheim (Angus), now gone, vector unknown. Europe, SW Asia. BM,

CGE, DUE, E, OXF, RNG. 7, 20, 21, 22*, 23*, 187, 251, 354(2:342), 1242*. Incl. *P. chrysantha* Trev.; *P. nestleriana* Tratt.; *P. parviflora* Gaudin, non Desf.; *P. thuringiaca* Bernh. ex Link.

cinerea Chaix ex Villars   Ashy Cinquefoil
• Casual or persistent on a stone wall at Hastings (E Sussex), vector unknown. Europe. 21, 22*, 23*, 25*, 27*, 303(29:12).

collina Wibel sensu lato
Pre-1930 only. Europe. 1, 21, 22*, 683*. A variable taxon which has been segregated into twelve or more species.

crassinervia Viv.
• Recorded from Edington (Surrey), probably as a garden escape. Corsica, Sardinia. 21, 371(355:26), 546*, 683*.

cuneata Wallich ex Lehm.
A garden escape found in 1992 on a sandstone bank by the old railway at Harpford Woods (S Devon). Himalayas, SW China. 51, 312*(t.4613), 325*(94:214), 390(125:?), 511.

fruticosa L. **alien variants**
Var. *parvifolia* (Fischer ex Lehm.) ined. (*P. parvifolia* Fischer ex Lehm.), *P.* 'Vilmoriniana' and other variants have been recorded as garden escapes.
[*grandiflora* L. In error, probably for *P. intermedia*. 394(3:15).]
[*heptaphylla* L. (*P. opaca* L., non sensu Smith) In error for *P. intermedia* or the *P. chrysantha* group. 354(3:316-322).]

hirta L.
• A casual at Kensington Gardens (Middlesex), vector unknown. No modern records. W Medit. BM, LIV, OXF, SLBI, TOR. 1, 21, 192, 354(2:500), 546*, 683*. *P. pedata* Willd.

inclinata Villars   Grey Cinquefoil
•• A persistent garden escape. Eurasia. ABD, BM, E, OXF, RNG, SLBI. 1, 20*, 22*, 40*, 371(337:30), 546*. *P. canescens* Besser.

intermedia L.   Russian Cinquefoil
••• Probably a grain alien, persistent or established in widely scattered localities, mainly in England. Russia. BM, CGE, E, NMW, OXF, RNG. 1, 20*, 22*, 209, 1242*. *P. opaca* sensu Smith, non L.

montana Brot.   Mountain Cinquefoil
• A garden escape, near Fincastle House, NW of Pitlochry (Mid Perth). Portugal, Spain, France. 21, 51, 525*, 546*, 1288*, 1290*, 1307. *P. splendens* Ramond ex DC., non Wallich.

nepalensis Hook.
• A casual garden escape. Himalayas. 51, 135, 312*(t.9182), 371(391:34), 728*. *P. formosa* D.Don.

norvegica L.   Ternate-leaved Cinquefoil
•••• An established garden escape and grain, bird-seed and iron ore alien; naturalised in widely scattered localities, mainly in England. Eurasia, N America. 1, 19, 20*, 22*, 29*, 126, 1242*. *P. monspeliensis* L.

recta L.   Sulphur Cinquefoil
●●●● An established garden escape and grass seed alien; naturalised in widely
scattered localities, mainly in S and E England. Eurasia, NW Africa;
(widespread). 1, 20*, 22*, 25*, 26*, 185. Incl. *P. obscura* Willd.;
P. sulfurea Lam.

rivalis Nutt. ex Torrey & A.Gray   Brook Cinquefoil
● A well-established alien by Barnsley Pool at Roughton, near Bridgenorth
(Salop), vector unknown. N America. BM, HbEJC. 20, 38*, 40*, 41*, 259,
370(13:49). Incl. *P. pentandra* Engelm.

supina L.
● Reported from the Crown wallpaper tip at Darwen (S Lancs) and waste
ground at West Barns (E Lothian), vector unknown. Formerly probably a grain
casual. Eurasia, N Africa. BM, BON, RNG, SLBI. 1, 21, 22*, 121, 234,
419, 475, 683*.

tridentata Aiton   Three-toothed Cinquefoil
Pre-1930 only. Eastern N America, Greenland. BM, DEE, LINN, OXF.
7, 38*, 40*, 51, 131, 187, 354(5:757).

FRAGARIA L.

ananassa (Duchesne) Duchesne   Garden Strawberry
●●●●● An established garden escape; widespread in Britain, especially on
railway banks. Originated in cultivation. 12, 20, 22*, 25*, 33*. Probably of
ancient *F. chiloensis* × *F. virginiana* (Duchesne) Duchesne origin; *F. chiloensis*
auct., non (L.) Duchesne; *F. grandiflora* Ehrh.; *F. virginica* sphalm.

chiloensis (L.) Duchesne   Beach Strawberry
●● (but confused with *F. ananassa*). An established garden escape; known
since 1933 on a bank on Sark (Channel Is). Western N & S America. BM.
1, 41*, 42, 51, 201, 220, 669*.

moschata (Duchesne) Duchesne   Hautbois Strawberry
●●● (but confused with *F. ananassa* and native *F. vesca*). An established
garden escape; naturalised in woods and on roadsides in widely scattered
localities in Britain; known since 1931 at Hascombe and W Horsley (both
Surrey). Europe. BM, CGE, LANC, NMW, OXF, RNG. 2, 20, 22*, 23*,
25*, 33*, 199, 209, 1242*. *F. elatior* Ehrh., nom. illeg.; *F. magna* Thuill.,
nom. illeg.; *F. muricata* sensu Kent, non L., vix Miller.

DUCHESNEA Smith
indica (Andrews) Focke   Yellow-flowered Strawberry
●●● An established garden escape; naturalised in woods in the south. Asia;
(widespread). BM, LANC, NMW, OXF, RNG. 7, 20, 303*(25:15), 500*,
595*. *Fragaria indica* Andrews.

WALDSTEINIA Willd.
fragarioides (Michaux) Tratt.
● Recorded in 1974 from Murhill (Wilts), probably a garden escape. Eastern
N America. E. 28*, 38*, 40, 51*, 372(35:22), 699*. *Dalibarda fragarioides*
Michaux.

ternata (Stephan) Fritsch
- A casual or persistent garden escape. Carpathians, Siberia, Japan. BM.
  21, 43, 51, 52*, 53*, 57*, 303(50:30), 313(27:121-138). *W. trifolia* Rochel
  ex W.D.J.Koch; *W. trifoliata* Steudel.

GEUM L.
aleppicum Jacq.
   Pre-1930 only. Eurasia, N America. 7, 21, 36c, 38*, 40*, 41*, 114.
   G. strictum Aiton.
macrophyllum Willd.   Large-leaved Avens
   ••• An established garden escape; naturalised on roadsides and river banks,
   mainly in Scotland. E Asia, N America. BM, CGE, E, K, LANC, RNG.
   2, 20, 38*, 40*, 41*, 153, 277, 349(5:131), 1242*. *G. japonicum* ?hort. ex
   Scheutz, non Thunb.
quellyon Sweet   Scarlet Avens
   •• A garden escape persistent or established on waste ground near Rolvendon
   (E Kent) and at Braehead quarry (Midlothian); casual elsewhere. S America.
   20, 52*, 54*, 56*, 126, 236, 371(391:34). *G. borisii* hort., non Kellerer &
   Sunderm.; *G. chilense* Lindley; *G. chiloense* Balbis ex Ser., nom. nud.;
   G. coccineum hort., non Sibth. & Smith.

AREMONIA Necker ex Nestler
agrimonioides (L.) DC.   Bastard Agrimony
   •• An established garden escape or introduction; naturalised in woods in
   eastern Scotland; known for well over 100 years near Rait, Carse of Gowrie
   (E Perth); also a survivor since 1862 from W.Borrer's private botanical
   garden at Henfield (W Sussex). Europe, SW Asia. ABD, BM, CGE, DUE,
   OXF, RNG. 1, 20, 22*, 23*, 33*, 174, 349(2:301), 361(2:263), 683*.
   Agrimonia agrimonioides L.

SANGUISORBA L.
canadensis L.   White Burnet
   ••• An established garden escape; naturalised by lakes and streams, mainly in
   Scotland; long-naturalised in the marshes of the lower reaches of the river Tay
   (E Perth). Eastern N America. BM, CGE, GL, NMW, OXF, RNG. 2, 20,
   38*, 40, 52*, 316(18:522), 594*. *Poterium canadense* (L.) A.Gray.
minor Scop. subsp. **magnolii** (Spach) Briq.
   Pre-1930 only. Medit. 2, 21, 44*, 300(1905:99), 546*. *Poterium magnolii*
   Spach; *P. verrucosum* Ehrenb.
minor Scop. subsp. **muricata** (Gremli) Briq.   Fodder Burnet
   •••• A persistent escape from cultivation; naturalised on railway banks and
   roadsides, mainly in S and E England. S Europe. 1, 16*, 20*, 33*, 546*.
   Poterium muricatum Spach, nom. illeg.; *P. polygamum* Waldst. & Kit.;
   P. sanguisorba L. subsp. *muricatum* (Gremli) Rouy & Camus; incl.
   P. platylophum Jordan; *P. stenolophum* (Spach) Jordan.

MARGYRICARPUS Ruíz Lopez & Pavón
pinnatus (Lam.) Kuntze   Pearl-fruit
* A garden escape established on moorland near the western extremity of the Dawros promontory (W Donegal), apparently now gone. S America. 51, 52*, 57*, 61, 352(51B:34). *M. setosus* Ruíz Lopez & Pavón.

ACAENA Mutis ex L.
anserinifolia (Forster & Forster f.) Druce   Bronze Pirri-pirri-bur
●● (but over-recorded for *A. novae-zelandiae*). An established garden escape and wool alien, recorded from a few places in Scotland and Ireland; abundantly established for 20 years at St Fillans, Loch Earn (Mid Perth). New Zealand. BM, CGG, RNG, ZCM. 20*, 36c, 319(19:365-367), 371(361:26; 385:34), 746*, 1275*. *A. sanguisorbae* (L.f.) Vahl pro parte; incl. *A. pusilla* (Bitter) Allan. A putative hybrid with *A. inermis* has been recorded. 20, 746*.
[*caesiiglauca* (Bitter) Bergmans (*A. glauca* hort.) A dubious record. 303(42:18), 746*.]
hieronymi Kuntze
* A wool casual. No modern records. S America. 14, 511, 807*(17.74:1-336).
magellanica (Lam.) Vahl sensu lato
* A garden escape and wool casual. New Zealand, Chile, Argentina. BM, CGE, E, OXF, RNG, TCD. 3, 20, 36c, 145, 300(1911:100), 354(7:436), 665*, 746*, 1148*. Incl. *A. adscendens* Vahl; *A. laevigata* Aiton f.
microphylla Hook.f. sensu lato
●● An established garden escape; naturalised by streams and reservoirs in a few places in the north; first recorded from the Isle of Raasay (N Ebudes) in 1937, still there; also known for over 20 years near Culter Water, Coulter (Lanarks). New Zealand. BM, CGE, CGG, E, RNG. 20*, 28*, 349(4:274), 370(14:191; 19:145), 746*, 1275*. Incl. *A. inermis* Hook.f.; *A. microphylla* var. *pallideolivacea* Bitter.
novae-zelandiae Kirk   Pirri-pirri-bur
●●●● An established garden escape and wool alien; naturalised in woods and on commons in widely scattered localities; known since 1923 on Kelling Heath (E Norfolk); increasing. Australasia. 3*(as *A. anserinifolia*), 20*, 146, 209, 235, 590*, 746*, 1275*. *A. anserinifolia* auct., non (Forster & Forster f.) Druce; *A. sanguisorbae* (L.f.) Vahl pro parte.
ovalifolia Ruíz Lopez & Pavón   Two-spined Acaena
●●●● An established garden escape; naturalised, often in abundance, in woods, on moorland and dunes, mainly in Ireland and Scotland. S America. 20*, 319(19:365-367), 349(4:415), 370(11:394), 705*, 746*, 825(2:65-66), 1275*.
ovina Cunn. sensu lato   Sheep's-bur
* Recorded without locality or vector. Australia. 37*, 51, 302(2:41), 514*, 533*, 569*, 590*.
pinnatifida Ruíz Lopez & Pavón
* A wool casual. S America. ABD, BFT, BM, E, RNG, TCD. 17, 51, 301*(t.1271), 665, 705*, 820*(13:309-313).

sericea Jacq.f.
- • A wool casual. S America. BFT, BM, CGE, E, RNG, TCD. 17, 51, 511, 665*.

splendens Hook. & Arn.
- • A wool casual. Chile. BM, LTN, RNG. 14, 145, 665, 820*(13:325-329). *A. integerrima* Gillies ex Hook. & Arn.

ALCHEMILLA L.
conjuncta Bab.   Silver Lady's-mantle
- ••• An established garden escape and introduction mainly in northern England and Scotland; long known and formerly claimed as native by rocky streams in Glen Clova (Angus) and in Glen Sannox on Arran (Clyde Is). Alps, Juras. CGE, LANC, OXF, RNG, SUN. 2, 16*, 20*, 22*, 33*, 1251. *A. argentea* (Trevelyan) Trevelyan, non Lam.

mollis (Buser) Rothm.
- ••• A persistent or established garden escape in widely scattered localities in Britain. SE Europe, SW Asia. BM, LANC, NMW, RNG. 12, 20*, 54*, 325*(113:535; 116:63), 1251.

tytthantha Juz.
- •• Established, possibly as an escape from the Edinburgh Botanic Garden (Midlothian); naturalised in a few places in southern Scotland; abundant on the Bowhill estate near Selkirk (Selkirks). Crimea. BFT, BM, CGE, DHM, E, RNG. 16, 20*, 22*, 316(18:521), 370(4*:281-2; 15:396). *A. multiflora* Buser ex Rothm.

venosa Juz.
- • Introduced and well established in a churchyard at Hampton-in-Arden (Warks) and in an old plantation near the top of Nantyffrith (Denbs). Caucasus. LIV, RNG. 51b, 129, 236, 325(113:535; 116*:63), 1122.

ROSA L.
× **alba** group   Alba Roses, incl. White-Rose-of-York
- •• Long-persistent or established garden escapes; widely naturalised in the north of the Isle of Man. Origin obscure. HbEC. 1, 20, 61, 89*, 103, 220, 335(102:151), 515*, 744*, 1121*. Incl. *R.* × *alba* L.; *R. gallica* × *R. arvensis* Hudson; *R. gallica* × *R. canina* L.; *R.* × *collina* Jacq.; *R.* × *monsoniae* (Lindley) Druce.

blanda Aiton   Smooth Rose
- • A casual garden escape. N America. BM, NMW, OXF. 2, 21, 61, 156, 515*, 586*, 613*, 744*. *R. cinnamomea* auct., non L.

carolina L.   Pasture Rose
- • A garden escape persistent or established in a quarry near Wickwar (W Gloucs) and on a roadside at Hogmoor, Woolmer Forest (N Hants); casual elsewhere. Eastern N America. NMW, HbEJC. 38*, 40, 61, 89*, 156, 350(31:356), 404, 586*, 613*, 744*, 1128*. *R. humilis* Marshall.

centifolia group   Cabbage Roses, incl. **Moss Roses**
- Long-persistent garden escapes and relics; recorded from the top of Jacob's
Ladder at Cheddar (N Somerset), a disused railway at Downton (S Wilts), a
riverside slope at Kettlewell (MW Yorks) and a hedge near Haugh of Urr
(Kirkcudbrights). Origin obscure, probably derived from *R. gallica* and other
species. 61, 89*, 312*(t.69), 371(382:23; 397:35), 432, 475, 515*, 744*.
Incl. *R. centifolia* L.; *R.* 'Muscosa'.

chinensis group   China Roses
- Persistent garden escapes or introductions; reported from below Darenth
Wood and on waste ground near Swanley (both W Kent). Originated in
cultivation. BM. 51, 52*, 53*, 54, 89*, 458, 744*. Hybrids derived from
R. chinensis Jacq.; incl. *R.* 'Semperflorens'.

damascena group   Damask Roses
- Persistent garden escapes at Hitchcopse pit, near Tubney (Berks), on the old
Bannerdown tip near Bath (N Somerset) and in a remote hedge at Stonehill
Green (W Kent). Origin obscure, probably derived from *R. gallica* and other
species. OXF. 54*, 61, 89*, 303(59:46), 371(382:23), 515*, 744*. Incl.
R. damascena Miller and its hybrids.

cf. **davurica** Pallas
- A long-persistent garden relic on the site of Warley Place gardens (S Essex).
NE Asia. 51, 371(361:27), 617*, 678, 679, 691. Possibly conspecific with
R. majalis.

[*elliptica* Tausch   In error for native *R. agrestis*, *R. micrantha* or hybrids of
them. 20, 370(18:122).]

filipes Rehder & E.Wilson
- Recorded from the golf course at Mallow (Mid Cork) and from the disused
Stony Cross airfield in the New Forest (S Hants). China. HbRPB. 51*, 61,
65*, 89*, 515*, 1351.

foetida Herrm.   Austrian Briar
Pre-1930 only. SW Asia. 22*, 61, 89*, 271. *R. lutea* Miller.

gallica group   Gallica Roses, incl. **Red-Rose-of-Lancaster**
- Introduced in the 1920s and now well established on the Fort Saumarez
peninsula in Guernsey (Channel Is); also recorded from a railway embankment
at West Tanfield (NE Yorks) and from among reeds at Dibden and by the old
railway station at Holmsley (both S Hants). Europe, SW Asia. HbRPB.
1, 20*, 22*, 23*, 89*, 220, 354(6:123), 371(379:25), 1121*. *R. gallica* L.
(*R. austriaca* Crantz) and hybrids; *R.* 'Tuscany'.

gigantea Collet ex Crépin
- A garden escape or relic reported from Swanley (W Kent). China. 53, 54,
89*, 458.

glauca Pourret   Red-leaved Rose
- •• A persistent garden escape or introduction in Beacon Woods at Bean
(W Kent), by the old railway line at Cranleigh (Surrey) and by the river Spey
at Pitcroy (Moray); seedlings have been recorded from a roadside at Oxted
(Surrey) and as abundant in a churchyard at Chislehurst (W Kent). C Europe.
LANC, HbEJC. 20*, 22*, 22*, 52*, 89*, 277, 303(46:28), 331(62:106), 458,

1121*.   *R. rubrifolia* Villars, nom. illeg.   Possibly conspecific with
R. ferruginea Villars, but not to be confused with *R.* glauca Villars ex Lois, a
synonym of the Mediterranean *R. vosagiaca* Desp.

hybrid tea group   Hybrid Tea Roses
• (but probably under-recorded). A persistent garden escape. Originated in
cultivation from many parent species. 53*, 54*, 89*, 199, 515*.

luciae group   Rambler Roses
•• Long-persistent or established introductions or garden escapes; naturalised
in abundance near Fort Saumarez, Guernsey (Channel Is). E Asia; hybrids
originated in cultivation. 20*, 61, 89*, 220, 302(9:10), 744*, 1121*, 1218.
R. luciae Franchet & Rochebr. (*R. wichuraiana* Crépin ex Déségl.) and
hybrids with *R. setigera* and other species; incl. *R.*'Albéric Barbier';
R.'American Pillar'; *R.*'Dorothy Perkins'; *R.*'May Queen'; *R.*'Red Rambler'.

majalis Herrm.   Cinnamon Rose
• A persistent garden escape; also introduced and naturalised in cemeteries.
N & C Europe, Siberia. 2, 20, 22*, 23*, 26*, 89*, 156, 242, 303(58:37),
354(12:277), 1242*. *R. cinnamomea* L.(1759), non L.(1753). Early records
were confused with *R. blanda*.

multiflora Thunb. ex Murray   Many-flowered Rose
••• An established garden escape; naturalised in hedges, on railway banks and
on waste ground in widely scattered localities. E Asia. BM, E, LANC, LSR,
OXF, RNG. 12, 20*, 61, 89*, 354(9:254), 515*, 613*, 1121*. Incl.
R. cathayensis (Rehder & E.H.Wilson) L.H.Bailey. A hybrid with native
R. rubiginosa has been recorded from Herm (Channel Is). 20, 1311.
[*nitida* Willd. Unconfirmed; no localised record found to justify the entry in 20.]

noisettiana group   Noisette Roses
• A garden escape or relic reported from Swanley (W Kent). Originated in
cultivation. 53, 54, 89*, 458, 515*. *R. chinensis* group × *R. moschata*
Herrm.; incl. *R. noisettiana* Thory.

palustris Marshall   Swamp Rose
• Introduced in the Ham Burn on Foula (Shetland). Eastern N America.
RNG. 40, 61, 89*, 115, 349(4:49), 586*, 744*. *R. carolina* L.(1762),
non L.(1753); *R. virginiana* Duroi, non Herrm., nec Miller.

pendulina L.   Alpine Rose
• Persistent in a roadside hedge far from habitation at Craigton, Milngavie
(Dunbarton). C & S Europe. OXF, SLBI. 1, 20, 22*, 23*, 27*, 89*,
354(1:106), 600*, 613*. *R. alpina* L.; *R. pyrenaica* Gouan.

rugosa Thunb. ex Murray   Japanese Rose
•••• An established garden escape; naturalised in widely scattered localities
throughout the British Is. E Asia. 7, 20*, 22*, 52*, 89*, 303(55*:24-27),
1121*, 1242*. Incl. *R. camtschatica* Vent.; *R.*'Hollandica'; *R. kamtschatica*
sphalm. Hybrids with native *R. arvensis* (*R.* × *paulii* Rehder), *R. caesia*,
R. canina (*R.* × *praegeri* Wolley-Dod) and *R. mollis* have all been recorded.
20, 303(58:36).

sempervirens L.
- A garden escape reported from Kirkcaldy (Fife). A variant was formerly naturalised at Madresfield (Worcs) and another collected at Sneem (S Kerry). S Europe, Turkey, NW Africa. LIV. 1, 20, 21, 104, 113, 354(2:169), 546*, 683*, 737*. *R. melvinii* Towndrow. A hybrid with native *R. stylosa* (*R.* × *bibracteata* Bast.) was recorded in error. 176.

sericea Lindley   **Silky Rose**
- Recently recorded without locality from a hedge in MW Yorks. Himalayas, China. 20, 51*, 61, 679*. Incl. *R. omeiensis* Rolfe.

[*setigera* Michaux   Unconfirmed; all records appear to refer to hybrids. Records under this name of *R.* 'American Pillar' (*R. luciae* × *R. setigera*) from the Channel Is are included in our *R. luciae* group. 20*, 1121*.]

villosa L.   **Apple Rose**
Pre-1930 only. Europe, SW Asia. 1, 21, 300(1898:167), 354(2:45; 6:123), 515*, 744*. *R. pomifera* Herrm.; incl. *R. dicksonii* Lindley. Not to be confused with *R. villosa* auct., non L., a synonym of native *R. mollis*.

virginiana Herrm.   **Virginian Rose**
- • A persistent garden escape. Formerly persisted for over 60 years at Clova (Angus). Eastern N America. BM, K, LIV, NMW, OXF. 1, 20*, 38*, 61, 89*, 354(9:708), 1121*. *R. lucida* auct., non Ehrh. The correct application of the name *R. virginiana* Miller remains unresolved, although it has been widely used as a synonym.

willmottiae Hemsley   **Miss Willmott's Rose**
- Recorded from Ellar Ghyll at Otley (MW Yorks). China. 20, 51*, 52*, 54*, 61, 89*, 371(382:23).

PRUNUS L.
armeniaca L.   **Apricot**
- Recorded from a bombed site in central London (Middlesex). No modern records. Asia. BM. 21, 58*, 61, 192, 505*, 606*, 684*. *Armeniaca vulgaris* Lam.

cerasifera Ehrh.   **Cherry Plum**
- • • • • Introduced, mainly in hedges, especially in S and E England; bird-sown and becoming naturalised in a few places. SE Europe, SW Asia. 7, 20, 22*, 33*, 59*, 606*. Incl. *P. divaricata* Ledeb.; *P. pissardii* Carrière; *P.* 'Atropurpurea'.

cerasus L.   **Dwarf Cherry**
- • • • • (but confused with native *P. avium* and hybrids). An established escape from cultivation; widespread in hedges and copses, mainly in S and E England. Origin obscure; (Eurasia). 2, 20, 22*, 33*, 35*, 1242*. Incl. *P. acida* Ehrh. var. *semperflorens* (Ehrh.) Koch (*P. semperflorens* Ehrh.); *P. austera* Leighton.

cornuta (Wallich ex Royle) Steudel   **Himalayan Bird-cherry**
- A garden escape. No modern records. Himalayas. BM. 61, 192, 685*. Possibly conspecific with native *P. padus*.

domestica L. sensu lato   Plum, incl. **Bullace, Damson, Greengage**
●●●●● An established escape from cultivation; widely naturalised in hedges and
scrub.   Origin obscure, probably SW Asia.   1, 20, 22*, 33*, 35*.   Incl.
P. communis Hudson; *P. insititia* L.; *P.* × *italica* Borkh. A hybrid with native
P. spinosa (*P.* × *fruticans* Weihe) has been recorded.   20, 1311.

dulcis (Miller) D.Webb   Almond
●● A persistent garden escape or food refuse alien; possibly also an oil-seed
casual.   W Asia; (widespread).   BM, RNG.   12, 20, 55*, 59*, 60*, 126.
P. amygdalus Batsch.; *P. communis* (L.) Arcang., non Hudson; *Amygdalus
communis* L.

incisa Thunb. ex Murray   Fuji Cherry
● An established garden escape freely bird-sown in wooded scrub at Chinnor
Hill, Oxford (Oxon).   Japan.   20*, 55*, 63*, 371(391:33).

laurocerasus L.   Cherry Laurel
●●●●● An established garden escape; naturalised in woods, especially in the
south; locally dominant in woodlands in Ireland; increasing.   SE Europe,
Turkey.   1, 20, 22*, 25*, 59*, 319(20:468; 21:417).   *Laurocerasus officinalis*
M.Roemer; *L. vulgaris* Carrière.

lusitanica L.   Portugal Laurel
●●●● An established garden escape; naturalised in woodlands in widely
scattered localities, forming dense stands in the Killarney district (N Kerry).
SW Europe, Atlantic Is.   16, 20, 55*, 59*, 60*, 63*, 319(21:417).
Laurocerasus lusitanica (L.) M.Roemer; incl. subsp. *azorica* (Mouill.) Franco.

mahaleb L.   St Lucie Cherry
●● An established escape from cultivation; naturalised in a wood between
Tregony and Tressilian (E Cornwall) and plentifully on railway banks in the
Longfield to Southfleet area (W Kent).   Its use as a stock for grafting
Prunus cultivars might account for some records.   Eurasia.   BM, K, OXF,
RNG.   20*, 21, 22*, 23*, 59*, 303(33:9), 331(55:44), 349(4:41), 1242*.
Cerasus mahaleb (L.) Miller.

pensylvanica L.f.   Pin Cherry
● Introduced and now well established by suckering, possibly also by seed,
in woodland and on the adjacent banks of the Basingstoke Canal at Deepcut,
near Frimley (Surrey).   N America.   HbKWP.   20, 38*, 59*, 61, 303(50:34).
Cerasus pensylvanica (L.f.) Loisel.

persica (L.) Batsch   Peach, incl. **Nectarine**
●●● A garden escape and food refuse casual sometimes persistent on tips;
introduced elsewhere.   China, or derived in cultivation from the Chinese
P. davidiana (Carr.) Franchet.   OXF.   12, 20, 21, 59*, 80*, 606*, 684*.

aff. **pumila** L.   American Dwarf-cherry
Pre-1930 only.   Eastern N America.   BM.   38*, 40, 61, 631*.

serotina Ehrh.   Rum Cherry
●●● An established garden escape; naturalised and spreading aggressively in
woods and on heaths in southern England and Wales.   Eastern N America.
BM, CGE, NMW, OXF, RNG, SLBI.   7, 20*, 22*, 55*, 59*, 61, 209,

303(50:34), 370(18:426), 546a*. Some records may be referable to
P. virginiana L. 21, 546a*.
serrulata group   Japanese Cherries
• Introduced or garden relics. E Asia. RNG. 20, 58*, 59*, 60*, 61, 61a,
119. Incl. *P. lannesiana* (Carrière) E.Wilson; *P. serrulata* Lindley;
P. speciosa (Koidz.) Nakai.

OEMLERIA Reichb.
cerasiformis (Torrey & A.Gray ex Hook. & Arn.) Landon   Oso-berry
• A garden escape established by suckering on Hampstead Heath (Middlesex)
and on Chislehurst Common (W Kent); also a garden relic at Tintern (Mons).
Western N America.   20*, 41*, 61*, 62*, 63*, 126, 303(60:35),
312*(n.s. t.582), 331(68:148), 347*(14:252-255), 370(15:131), 371(382:24).
Osmaronia cerasiformis (Torrey & A.Gray ex Hook. & Arn.) E.Greene.

CYDONIA Miller
oblonga Miller   Quince
••• A persistent garden escape and relic of cultivation in hedges and woods,
mainly in southern England. Asia; (Medit.). BM, CGE, LANC, OXF, RNG,
SLBI. 7, 20, 29*, 59*, 61*, 617*. *C. vulgaris* Delarbre; *Pyrus cydonia* L.

CHAENOMELES Lindley
× **californica** W.Clarke ex C.Weber
• Seedlings reported from a roadside at Stone (W Kent). Originated in
cultivation. 51b, 61, 458. *C. cathayensis* × *C.* × *superba*; incl. *C.*'Cardinal'.
cathayensis (Hemsley) C.Schneider
• Seedlings reported from a roadside at Stone (W Kent). C China. 51, 61,
63*, 317*(t.2657; t.2658), 458, 842*(1924:64). *C. lagenaria* var. *cathayensis*
(Hemsley) Rehder; *Cydonia cathayensis* Hemsley.
japonica (Thunb.) Lindley ex Spach   Maule's Quince
•• (but confused with *C. speciosa* and hybrids). A garden escape established
on Walton Common and on Alderstead Heath (both Surrey). Japan. 43,
54*, 61, 199, 613*, 617*.   *C. lagenaria* (Lois.) Koidz., nom. illeg.;
Cydonia maulei (Masters) T.Moore; *Pyrus maulei* Masters; *P. japonica* Thunb.,
non Sims.
speciosa (Sweet) Nakai   Flowering Quince
•• (but confused with *C. japonica* and hybrids). An established garden escape
or long-persistent relic. China; (Japan). HbLJM. 20, 54*, 56*, 61, 126, 201,
236, 613*. *C. lagenaria* auct., non Koidz.; *Cydonia japonica* auct., non
(Thunb.) Lindley ex Spach; *Pyrus japonica* Sims, non Thunb.
× **superba** (Frahm) Rehder
• A persistent garden escape at Ash Vale (Surrey), on a roadside bank at
Wrotham Hill (W Kent) and in grassland between Glasgow and Renfrew
(Lanarks). Originated in cultivation. 51b, 53*, 54, 61, 62*, 316(21:507),
342(Feb91), 458. *C. japonica* × *C. speciosa*.

PYRUS L.

amygdaliformis Villars   Almond-leaved Pear
Pre-1930 only. Medit. NMW. 21, 27*, 31*, 58*, 59*, 156. *P. nivalis* sensu
Lindley, non Jacq.; *P. parviflora* Desf.

communis L.   Pear, incl. Wild Pear
●●●●● An established garden escape, mainly in C and E England. An ancient
introduction which has been claimed as native. Eurasia. 1, 20, 22*, 33*, 35*.
Incl. *P. pyraster* (L.) Burgsd.

salicifolia Pallas   Willow-leaved Pear
● Introduced or a persistent garden escape at Broxbourne (Herts), Hounslow
Heath (Middlesex), Sandbeck Park (SW Yorks) and Darlington (Co Durham).
Turkey, Cauacasia, Iran. 22*, 58*, 60*, 61, 166, 432, 436, 458.

MALUS Miller

baccata (L.) Borkh.   Siberian Crab
● Recorded in 1981 from Wisley Common (Surrey). N Asia. 20, 58*, 59*,
60*, 61*, 199.

domestica Borkh.   Apple
●●●●● An established garden escape and food refuse alien; naturalised in
hedges and scrub throughout the British Is. Originated in cultivation. 2, 20,
22*, 684*. *M. sylvestris* Miller subsp. *mitis* (Wallr). Mansf.

floribunda Siebold ex Van Houtte   Japanese Crab
● Introduced. Japan. 20, 54*, 58*, 61, 199.

hupehensis (Pampan.) Rehder   Hupeh Crab
● A garden escape; reported in 1986 as bird-sown on sandy heathland at Hale
Purlieu, New Forest (S Hants). China, N India. HbRPB. 51b, 59*, 60*, 61,
312*(t.9667), 403. *M. theifera* (L.Bailey) Rehder; *Pyrus hupehensis* Pampan.
According to 440, the plant in cultivation does not match oriental wild material.

prunifolia (Willd.) Borkh.   Plum-leaved Apple
● Introduced or a bird-sown garden escape; persistent on a tip at Guildford
(Surrey) and in a hedge near Ringwood (S Hants). E Siberia, N China. 51*,
61*, 199, 371(346:28). Incl. *M.*'Cheal's Golden Gem'.

× **purpurea** (Barbier) Rehder   Purple Crab
●● A persistent garden escape in a sand pit at Horns Cross and at Dartford
(both W Kent); also recorded from Dorset, S Hants and Cambs. Originated in
cultivation. RNG, HbEJC, HbRPB. 20, 59*, 60*, 61, 135, 236, 458.
M. atrosanguinea (Spaeth) Schneider × *M. niedzwetzkyana* Dieck.

× **robusta** (Carrière) Rehder   Hybrid Siberian Crab
● A persistent garden escape reported from Dartford Heath (W Kent) and from
Holmwood (Surrey). Originated in cultivation. HbEJC. 51a, 58*, 61,
371(385:34), 458, 504*. *M. baccata* × *M. prunifolia.*

sieboldii (Regel) Rehder   Siebold's Crab
● Introduced by Hatchet Pond in the New Forest (S Hants). Japan. HbEJC,
HbRPB. 43, 51*, 54*, 61, 371(394:34), 649*. *M. toringo* Sieb.; incl.
M. sargentii Rehder.

spectabilis (Aiton) Borkh.   Chinese Crab
  • Introduced or a garden escape; persistent on Tunbridge Wells Common
  (W Kent). Origin obscure; (China). 51, 58*, 59*, 60*, 61, 371(340:28).

SORBUS L.
croceocarpa Sell   Orange-berried Whitebeam
  ••• (but confused with *S. latifolia*). Introduced and a garden escape;
  reproducing and well established in the Bristol area, Salop, Caerns and
  Anglesey. Origin obscure. BM, CGE, E, NMW, OXF, RNG. 16, 20*,
  35*(as *S. latifolia*), 61, 370(17:392-395). *S.* 'Theophrasta'.

decipiens (Bechst.) Irmisch   Sharp-toothed Whitebeam
  • (but confused with native *S. subcuneata*). Introduced and reproducing in the
  Avon Gorge (N Somerset) and in a railway cutting at Achnashellach (W Ross);
  also collected from Ashstead Park (Surrey). France, Germany. CGE, LANC.
  20*, 370(17:386), 631*.

domestica L.   Service-tree
  •• A garden escape; bird-sown in a thicket by the railway near Croome Court
  (Worcs), in a hedge at Kingston Bagpuize (Berks) and self-sown in the grounds
  of St Ann's Hospital, Tottenham (Middlesex); introduced on roadsides
  elsewhere. A very recent report suggests that this species is native on seacliffs
  near Cardiff (Glam). Medit. BM, E, LANC, LIV, OXF, RNG. 2, 20, 22*,
  58*, 59*, 60*, 119, 191, 303(58:37), 371(401:19).   *Pyrus domestica* (L.)
  Ehrh.; *P. sorbus* Gaertner.

hupehensis C.Schneider   Hupeh Rowan
  • A garden escape; apparently bird-sown in a hedgerow at Woburn (Beds); also
  spreading by suckers and seedlings on waste ground at Barnes Cray (W Kent).
  China. LTN. 20, 58*, 59*, 60*, 61*, 303(32:19-20), 312*(n.s. t.96), 458.

hybrida L.   Swedish Service-tree
  • (but confused with native *S.* × *thuringiaca*). A garden escape; bird-sown in
  a chalk pit at Stone (W Kent) and in semi-natural woodland at Eden
  (N Aberdeen). Scandinavia, Finland. ABD, LANC. 2, 20*, 22*, 52*, 61,
  277, 370(16:443), 371(414:13), 1242*. *S. fennica* Fries; *Pyrus pinnatifida*
  Smith; *P. semipinnata* auct., non Roth.

intermedia (Ehrh.) Pers.   Swedish Whitebeam
  •••• Introduced; sometimes reproducing and naturalised, especially in southern
  England. Baltic region. 2, 20, 22*, 33*, 59*, 199, 1242*. *S. scandica* Fries;
  S. suecica (L.) Krok; *Pyrus intermedia* Ehrh.; *P. scandica* Bab., non *Pyrus
  scandica* Fries. A hybrid with native *S. aucuparia* (*S.* × *pinnatifida* auct.) has
  been recorded. BM, HbEJC. 1083*, 1311.

latifolia (Lam.) Pers.   Fontainebleau Service-tree
  ••• (but confused with *S. croceocarpa*). Introduced or a garden escape;
  sometimes reproducing, as in Leigh Woods (N Somerset). W Alps, Pyrenees.
  ABD, BM, CGE, E, NMW, OXF. 59*, 61, 248, 370(17:396), 684*, 1217*.
  Crataegus latifolia Lam.; *Pyrus latifolia* (Lam.) Syme. Not to be confused
  with *S. latifolia* sensu lato, which includes native *S. bristoliensis, S. devoniensis*
  and *S. subcuneata*.

sargentiana Koehne   Sargent's Rowan
- Introduced. China. 58*, 59*, 60*, 61, 62*, 166.

vilmorinii Schneider   Vilmorin's Rowan
- Introduced. China. 58*, 60, 61, 62*, 166.

ARONIA Medikus
arbutifolia (L.) Pers.   Red Chokeberry
- A garden escape established on Banstead Heath (Surrey).   Eastern
N America. HbEJC. 20, 38*, 61, 126, 199, 303(16:16), 331(56:86), 586*,
698*.

melanocarpa (Michaux) Elliott   Black Chokeberry
- A garden escape established by suckering near a lake at Llyn Hafod-y-llyn
(Caerns); also recorded in 1991 on Bere Heath (Dorset). Eastern N America.
NMW. 20*, 38*, 52*, 61, 156, 303(37:14; 59:45), 586*. Probably not
reliably distinguishable from hybrids with *A. arbutifolia* (*A.* × *prunifolia*
(Marshall) Rehder), the taxon which is increasingly reported in western Europe.
61a.

ERIOBOTRYA Lindley
japonica (Thunb.) Lindley   Loquat
- A casual garden weed, vector unknown. China, Japan. 21, 36c, 199, 504*,
546a*, 683*, 716*, 1344*. *Mespilus japonica* Thunb.

AMELANCHIER Medikus
lamarckii F.-G.Schroeder   Juneberry
- •••• Introduced and now established in woods and on commons on acid soils,
especially in southern England; naturalised for over 100 years in the Hurtwood
area (Surrey). Origin obscure, probably E Canada. 1(as *A. canadensis*), 20*,
22*, 370*(8:155-162). This taxon has been variously recorded as *A. alnifolia*
Nutt.; *A. arborea* (Michaux f.) Fern.; *A. canadensis* (L.) Medikus; *A. confusa*
N.Hylander; *A. florida* Lindley; *A. grandiflora* Rehder; *A. intermedia* Spach;
A. laevis Wieg.; and as *Crataegus racemosa* Lam., non *Amelanchier racemosa*
Lindley. One or more of these species may occur as garden relics.   61a,
370(8:155-162).

ovalis Medikus   Snowy Mespilus
Pre-1930 only. Medit. 21, 22*, 23*, 26*, 27*, 354(8:397). *A. vulgaris*
Moench.

spicata (Lam.) K.Koch   Low Juneberry
- Recorded in 1971 from Frensham Great Pond (Surrey). Probably originated
in cultivation. BM. 21, 22*, 61, 199, 1242*. Possibly *A. canadensis* ×
A. stolonifera Wieg.; *A. ovalis* Borkh., non Medikus; *Crataegus spicata* Lam.

PHOTINIA Lindley
davidiana (Decne.) Cardot   Stranvaesia
- •• A garden escape; seedlings have been reported from old walls at Chester
(Cheshire), at New Ash Green (W Kent) and on Wimbledon Common (Surrey);
also introduced in quantity on forest verges in Scotland. China, Vietnam. BM,

E, RNG.   20*, 51*, 61*, 62*, 236, 277, 303(36:27), 331(69:140), 371(349:29). *Stranvaesia davidiana* Decne.

COTONEASTER Medikus
acuminatus Lindley
• A garden escape established on rough hillsides near Glendore and Union Hall in the Skibbereen district (W Cork). Himalayas. 21, 51, 61, 319(10:237), 645*(as *Mespilus acuminata*), 685, 724*. This record requires confirmation. The Ampfield Wood (S Hants) record was in error for *C. laetevirens*. 319(10:237), 458.
[*acutifolius* Turcz. In error for *C. nitens* and *C. villosulus*.]
adpressus Bois   Creeping Cotoneaster
• A garden escape recorded from W Kent, Surrey and, with some doubt, from the Glasgow area (Lanarks). W China. 20, 61*, 303*(39:20-21), 316*(22:111-114). *C. horizontalis* var. *adpressus* (Bois) C.Schneider.
affinis Lindley   Purpleberry Cotoneaster
• A garden escape recorded from a cemetery at Guildford and Wood Field station, Ashstead (both Surrey) and reported as seedlings on a roadside bank, Park Farm Road, Bickley (W Kent). Himalayas. HbACL, HbJDF. 20, 61, 199, 458, 685*. The records for Bargate pit, Godalming (Surrey) and Hurn Forest (S Hants) were in error for *C. transens*. Several other records are in error for *C. bacillaris*.
[*ambiguus* Rehder & E.Wilson In error for *C. laetevirens*. 458.]
amoenus E.Wilson   Beautiful Cotoneaster
• An established garden escape; a colony near Long Road, Cambridge (Cambs) and recorded from a chalky roadside bank north of Winchester (S Hants); seedlings reported from Lambeth (Surrey). China. HbJDF. 20, 61, 61a, 310(30:26), 315*(51:2), 336(24:36), 342(Feb90).
apiculatus Rehder & E.Wilson
• A garden escape recorded from Shawlands (Renfrews). China. 51, 61, 316*(22:239-242), 1003.
atropurpureus Flinck & Hylmö   Purple-flowered Cotoneaster
• An established garden escape at Leigh Woods, Bristol (N Somerset) and naturalised at Cambuslang (Lanarks). China. HbJDF, HbPM. 20, 316*(22:111-114), 370(18:311-313; 19:146). *C. horizontalis* 'Prostratus'.
bacillaris Wallich ex Lindley   Open-fruited Cotoneaster
•• An established garden escape; naturalised in a few places in southern England. Himalayas. BM, OXF. 7, 20, 61, 236, 303(43:30), 354(7:436), 685*, 1295*. *C. affinis* var. *bacillaris* (Wallich ex Lindley) C.Schneider. The Hawkerland Valley (S Devon) record was in error for *C. cooperi*, and the Darenth Wood (W Kent) record for *C. transens*.
bullatus Bois   Hollyberry Cotoneaster
•••• (but over-recorded in error for *C. rehderi*). An established garden escape; naturalised in woodland in widely scattered localities throughout the British Is. Tibet, W China. 13, 20, 51*, 61, 61a, 62*, 63*, 199, 303(32:18; 44*:18-19), 316*(22:111-114), 915*(87:305-330).

buxifolius Wallich ex Lindley   Box-leaved Cotoneaster
• Reported as a garden escape from the Chew Valley (N Somerset). S India.
HbJDF. 20, 388(133:234). *C. affinis* DC. pro parte, non Lindley. The record
for Northfleet (W Kent) was in error. 458.

calocarpus (Rehder & E.Wilson) Flinck & Hylmö
• An established garden escape; reported as abundant on the site of a
demolished hospital at Stone (W Kent). China. 61, 303(50:33), 458,
810*(119:449). *C. multiflorus* var. *calocarpus* Rehder & E.Wilson. This
record requires confirmation. 483.

cashmiriensis Klotz   Kashmir Cotoneaster
• A garden escape in the Gorbals, Glasgow (Lanarks) and, as seedlings,
along a bank of the river Medway at Maidstone (E Kent). Kashmir. 20,
303(50:33 as *C. cochleatus*), 316*(22:111-114), 458, 1003. *C. cochleatus*
auct., non (Franchet) Klotz.

cinerascens (Rehder) Flinck & Hylmö
• An established garden escape, naturalised about Hextable and recorded, as
seedlings, from New Ash Green (both W Kent). China. 61, 303(50:33), 458.
C. franchetii var. *cinerascens* Rehder. These records require confirmation.
483.

congestus Baker   Congested Cotoneaster
• A garden escape recorded from Hextable (W Kent). Himalayas. 20, 55, 61,
303(43:30), 370(17:471). *C. buxifolius* hort., non Wallich ex Lindley;
C. microphyllus sensu Yü, non Wallich ex Lindley; *C. microphyllus* var.
glacialis Hook.f.; *C. pyrenaicus* hort., non Chanc. The Mogshade Hill
(S Hants) record was an error for *C. sherriffii*. 404, 483.

conspicuus Marquand   Tibetan Cotoneaster
• An established garden escape; recorded from screes on Cefn-yr-Ogo (Denbs);
many bird-sown plants of varying ages on a roadside verge at Chertsey
(Surrey); seedlings at New Ash Green and Erith (W Kent). Tibet. 20, 54*,
61*, 62*, 63*, 305(45:7), 316*(22:239-242), 370(15:429; 17:471), 458.
C. microphyllus var. *conspicuus* (Marquand) Yü. A hybrid with *C. dammeri*
(*C.* × *suecicus* Klotz, q.v.) has been recorded.

cooperi Marquand
• Collected from a roadside, Hawkerland Valley (S Devon). N India. BM.
61, 312*(t.9478), 317*(32:t.3146), 1003. The Dartford Heath (W Kent) record
was in error for *C. obtusus*.

dammeri C.Schneider   Bearberry Cotoneaster
•• An established garden escape; dominant on 50m of railway embankment at
St Philips, Bristol; abundant on waste lime heaps at Carnforth (W Lancs);
known for over 20 years on a gravelly bank at Kirkfieldbank Brae (Lanarks).
China. HbJDF, HbKWP. 20, 61, 61a, 303(42:18), 316(19:340), 458, 1295*.
C. humifusus Praeger. A hybrid with *C. conspicuus* (*C.* × *suecicus* Klotz,
q.v.) has been recorded.

dielsianus E.Pritzel ex Diels   Diels' Cotoneaster
●●● (but confused with *C. franchetii*).   An established garden escape;
naturalised on downs, in quarries and by roadsides in scattered localities in
Britain.   China.   BM, LANC, NMW, RNG.   20, 61, 156, 303(43:30),
316*(22:111-114), 349(5:338), 350*(47:37), 915*(87:305-330).

divaricatus Rehder & E.Wilson   Spreading Cotoneaster
●●● A garden escape established on walls and roadsides; naturalised in
profusion on a roadside bank near Eynsford (W Kent).   China.   CGE, RNG.
16b, 20, 61, 303(34*:27-28; 43:30), 316*(22:111-114), 613*, 915*(87:305-
330).

ellipticus (Lindley) Loudon   Lindley's Cotoneaster
●● (but confused with *C. hissaricus*).   An established garden escape;
naturalised in quantity by chalk pits at Northfleet and frequent around derelict
buildings in Dartford (both W Kent).   Himalayas.   RNG.   20, 61, 197, 236,
303(50:33), 331(50:114), 458.   *C. insignis* Pojark.; *C. lindleyi* Steudel,
nom. illeg.; *C. nummularius* Lindley, non Fischer & C.Meyer.

'Firebird'
● Recorded as a garden escape at Ashford Hangers, Steep (N Hants).   Origin
obscure.   404, 1003.   Formerly considered to be *C. bullatus* × *C. franchetii*,
this plant appears to be an unnamed species.

foveolatus Rehder & E.Wilson
● A garden escape recorded from Eaves Wood, Silverdale (W Lancs);
a seedling has also been reported from Southport (S Lancs).   China.   RNG.
51, 61, 370(15:424), 371(349:28), 458, 1003.   Both records are probably in
error for *C. villosulus*.   483.

franchetii Bois   Franchet's Cotoneaster
●●● (but confused with *C. dielsianus* and *C. sternianus*).   An established
garden escape; mainly in southern England, Wales and Ireland; naturalised on
an open limestone rock-face by Dundag Bay, near Killarney (N Kerry).
Burma, Tibet, China.   OXF, LANC, RNG.   20*, 52*, 61, 63*, 303(49:36),
316*(22:111-114), 331(57:74).

frigidus Wallich ex Lindley   Tree-cotoneaster
●●● (but confused with hybrids).   An established garden escape in widely
scattered localities in Britain.   Himalayas.   ABD, BM, E, NMW, OXF, RNG.
7, 20, 59*, 60*, 156, 685*, 724*.   *C. affinis* auct., non Lindley.   For hybrids
see *C.*'Watereri' group.

[*glaucophyllus* Franchet   In error for *C. serotinus*.   483.]

henryanus (C.Schneider) Rehder & E.Wilson
● Reported as seeding freely from gardens about Coleraine (Co Londonderry).
C China.   HbJDF.   51, 61, 842*(1919:t.264), 1003.   *C. rugosus* E.Pritzel ex
Diels var. *henryanus* C.Schneider.

hissaricus Pojark.   Round-leaved Cotoneaster
•• An established garden escape; abundant in hedges, on roadsides and
tracksides around Park Gate, Lullingstone, and spreading towards Hulberry
(W Kent); many seedlings from Charing Cross to Temple, and around Hyde
Park Corner (Middlesex). Himalayas. 20, 303(50:33), 371(414:13), 458.
C. racemiflorus auct., non (Desf.) K.Koch.

hjelmqvistii Flinck & Hylmö   Hjelmqvist's Cotoneaster
•• An established garden escape; recorded from a bank of the river Clyde at
Glasgow (Lanarks); seedlings reported from Bermondsey (Surrey), Southport
(Lancs), Shawlands (Renfrews) and several places in W Kent. ?W China.
HbJDF, HbPM. 20, 316*(22:111-114), 370(18:311-313), 458. *C. horizontalis*
'Robustus'.

horizontalis Decne.   Wall Cotoneaster
•••• An established garden escape; naturalised in old chalk pits, quarries and
on rocky banks in widely scattered localities. W China. 12, 20*, 54*, 61,
63*, 349(1:156).

hummelii Flinck & Hylmö, nom. nud.
• A garden escape recorded from the Winchester by-pass (S Hants). China.
HbJDF. 1003.

hupehensis Rehder & E.Wilson
• Recorded without locality from W Kent. W China. 51, 61*, 303(62:43),
312*(n.s. t.245), 1003.

ignotus Klotz   Black-grape Cotoneaster
• A garden escape established on Witley Common (Surrey) and in rough
grassland and copses between Bowman's Heath and Old Bexley (W Kent).
Himalayas. Hb JDF, HbKWP. 20, 371(428:7), 1003. *C. hissaricus* auct.,
non Pojark.

insculptus Diels   Engraved Cotoneaster
• Naturalised in Caerns. China. 20.

[*integerrimus* Medikus   Unconfirmed records. Possibly introduced from
European stock onto Great Ormes Head (Caerns) to increase the indigenous
population that is now considered to be an endemic species, *C. cambricus* Fryer
& Hylmö (*C. integerrimus* auct. angl., non Medikus). 483.]

integrifolius (Roxb.) Klotz   Small-leaved Cotoneaster
••••• An established garden escape; naturalised in many places, especially in
the south and west. Himalayas, China. 20*. *C. microphyllus* auct., non
Wallich ex Lindley; *C. thymifolius* Wallich ex Lindley.

lacteus W.Smith   Late Cotoneaster
••• An established garden escape; bird-sown on wall-tops, roadsides and by
railways in SE England. China. RNG, HbJDF. 20*, 51*, 53*, 54*, 61, 63*,
303(14:13; 42:18), 331(64:116).

laetevirens (Rehder & E.Wilson) Klotz   Ampfield Cotoneaster
• An established garden escape; naturalised in Ampfield Wood (S Hants).
China. HbJDF. 20, 458, 915*(87:305-330). *C. ambiguus* auct., non Rehder
& E.Wilson.

linearifolius (Klotz) Klotz   Thyme-leaved Cotoneaster
- An established garden escape; naturalised in SE England and collected from Mull (S Ebudes). Nepal. BM. 20, 1003. *C. microphyllus* forma *linearifolius* Klotz; *C. thymifolius* auct., non Wallich ex Lindley. Most records of *C. thymifolius* are in error for *C. integrifolius*. 483.

lucidus Schldl.   Shiny Cotoneaster
- Reported from Pitt (S Hants), and as seedlings near bushes planted on roadsides at New Ash Green (W Kent). Siberia, Mongolia. 20, 51, 61, 303(50:33), 404, 458, 915*(87:305-330), 1295*. *C. acutifolius* Lindley, non Turcz.

mairei Léveillé
- A garden escape established on Wall's Hill, Babbacombe (S Devon). SW China. BM, RNG. 349(5:338 as *C. dielsianus*).

marginatus Lindley ex Schldl.
- Recorded from the limestone slopes of Cefn-yr-Ogo, near Abergele (Denbs) and without locality from S Wilts; reported as seedlings at Hextable and Farningham (both W Kent) and along the Victoria Embankment (Middlesex); collected from E Gloucs in 1923. Kashmir. BM, RNG. 172, 261, 303(50:33), 458, 1295*. *C. buxifolius* Baker, non Wallich ex Lindley; *C. prostratus* var. *lanatus* (Jacques) C.Schneider; *C. wheeleri* hort.

microphyllus Wallich ex Lindley
- •• (but over-recorded for *C. integrifolius*). Collected from a few places pre-1930 and from a wood in S Aberdeen in 1961. Nepal. BM. 51b, 301*(t.1114), 1003.

moupinensis Franchet   Moupin Cotoneaster
- Recorded for S Hants, W Kent, Essex and Westmorland. China. 20, 51, 55*, 61, 302(20:26), 458. *C. foveolatus* auct., non Rehder & E.Wilson.

mucronatus Franchet   Mucronate Cotoneaster
- An established garden escape; naturalised in Middlesex. China. 20, 303(50:33), 1003.

[*multiflorus* Bunge   In error for *C. veitchii*. 483.]

nanshan A.Vilm. ex Mottet   Dwarf Cotoneaster
- A garden escape; seedlings near planted bushes recorded from Crayford (W Kent). China. 20, 61, 303(50:33), 458. *C. adpressus* var. *praecox* auct., non Bois & Berthault; *C. praecox* auct., non (Bois & Berthault) A.Vilm. ex Bois & Berthault.

newryensis Lemoine
- A garden escape reported from chalk downs above Kemsing (W Kent). Probably China. 458, 691, 1003. This record requires confirmation. 483.

nitens (Rehder & E.Wilson) Flinck & Hylmö   Few-flowered Cotoneaster
- Reported from the Winchester by-pass (S Hants), and from W Kent and Middlesex as seedlings near planted bushes. China. HbJDF. 20, 61, 303(50:33), 404, 458, 1003.

nitidus Jacques   Distichous Cotoneaster
- Introduced; on a bank at Warsash (S Hants). Himalayas. 20, 483, 1003. *C. distichus* Lange; *C. rotundifolius* hort., non Wallich ex Lindley.

obtusus Wallich ex Lindley   **Dartford Cotoneaster**
• A garden escape persistent on Dartford Heath and Barnehurst golf course
(both W Kent) and on the towpath of the river Thames at Hampton Court
(Middlesex). Himalayas. HbJDF. 20, 331(71:177; 72:119), 371(419:15).
C. cooperi auct., non Marquand.

pannosus Franchet   **Silverleaf Cotoneaster**
• A garden escape on waste ground at Stone (W Kent); seedlings have recently
been recorded from central Dartford (W Kent). China. 20, 51*, 61,
303(42:18; 43:30; 60:35; 62*:43-44).

perpusillus (C.Schneider) Flinck & Hylmö
• (but perhaps overlooked as *C. horizontalis*). A garden escape reported from
Hants and from scrub near the river, Mylor Bridge (W Cornwall). China. 61,
370(16:229), 371(403:39), 1003. *C. horizontalis* var. *perpusillus* C.Schneider.

[*prostratus* Baker
• Reported without locality from W Kent and as introduced as cover for game
birds and thereby naturalised in S Wilts. Himalayas. HbJDG. 51, 61, 172,
261. The true identity of this species remains unresolved; the records are
probably referable to *C. rotundifolius*. 483.]

rehderi Pojark.   **Bullate Cotoneaster**
• (but confused with *C. bullatus* and under-recorded). An established garden
escape; recorded from Ashford Hangers, Steep (N Hants), Grubbins Wood,
Arnside (Westmorland) and Clints Quarry, Egremont (Cumberland); bird-sown
at New Ash Green (W Kent). C China. LANC, HbJDF. 20*, 303(50:33),
371(403:39; 414:14), 458. *C. bullatus* var. *macrophyllus* Rehder & E.Wilson.

rotundifolius Wallich ex Lindley
• A garden escape collected from Wall's Hill, Torquay (S Devon). Also
reported as being introduced as cover for game birds and thereby naturalised
in Princethorpe Wood in the Rugby district, but this record could refer to
C. nitidus. Himalayas. HbJDF. 51*, 101, 301*(t.1187), 312*(t.8010), 1003,
1295*. *C. microphyllus* var. *uva-ursi* Lindley.

salicifolius Franchet   **Willow-leaved Cotoneaster**
••• An established garden escape; naturalised in a few places in the south.
China. NMW, RNG. 20, 52*, 55*, 56*, 61, 63*, 303(43:30; 50:33),
371(349:28). Incl. *C. rugosus* E.Pritzel ex Diels.

sanguineus Yü
• A garden escape recorded from shrubby waste ground, Cambuslang
(Lanarks); reported from a roadbank at Knockholt, as abundantly naturalised
on roadsides at Erith, and bird-sown in a chalk pit at Stone (all W Kent) and
from the city wall of Chester (Cheshire). Himalayas. 370(18:219),
371(414:14), 458, 1236.

serotinus Hutch.
• A garden escape self-sown on a chalk bank above Woldingham (Surrey);
reported without locality from chalkpits and, as seedlings, at New Ash Green
(both W Kent). W China. 36c, 51*, 61, 303(62:43), 312*(t.8854, t.9171).
C. glaucophyllus forma *serotinus* (Hutch.) Stapf. The Woldingham (Surrey)
record needs confirmation. 199.

sherriffii Klotz
- A garden escape established at Mogshade Hill in the New Forest (S Hants). SE Tibet. HbJDF, HbRPB. 1003. *C. orbicularis* hort., non Schldl.

silvestrei Pampan.
- A garden escape; seedlings reported from waste ground at Shooters Hill (W Kent). China. 303(50:33), 458. This record requires confirmation; it has been wrongly equated to *C. hupehensis*. 483.

simonsii Baker   Himalayan Cotoneaster
- •••• An established garden escape; naturalised in woodland in widely scattered localities throughout the British Is. Himalayas. 3, 20*, 22*, 61, 63*, 303*(44:1).

splendens Flinck & Hylmö   Showy Cotoneaster
- An established garden escape; on a vertical rock face near Porth Navas (W Cornwall); also naturalised seedlings in the Horton Kirby and S Darenth area (W Kent). China. 20, 61, 303(59:46), 371(394:35), 1003.

sternianus (Turrill) Boom   Stern's Cotoneaster
- •• Recorded as a garden escape at Winchester and at Ashford Hangers, Steep (both N Hants), and at Keston Mark and Bromley (both W Kent). E, RNG. 312*(n.s. t.130), 316*(22:239-242), 342(Feb91), 371(394:35), 458. *C. franchetii* var. *sternianus* Turrill; *C. stonianus* sphalm.; *C. wardii* hort., non W.Smith. Literature records are inseparably confused with *C. franchetii* and *C. dielsianus*.

× suecicus Klotz   Swedish Cotoneaster
- •• Seedlings reported from a few places in Kent, Denbs, Cheshire and Lanarks. Originated in cultivation. HbJDF. 20, 61, 303(43:30), 458, 1236. *C. conspicuus* × *C. dammeri*; incl. *C.*'Coral Beauty'; *C.*'Skogholm'.

tengyuehensis Hylmö, nom. nud.
- One plant reported from Hextable (W Kent). China. HbJDF. 371(419:15).

transens Klotz   Godalming Cotoneaster
- (but confused with *C. affinis*). An established garden escape; naturalised in the old Bargate sandstone pit at Godalming (Surrey) and Darenth Wood (W Kent); also recorded from a conifer plantation in Hurn Forest (S Hants). China. HbCGH, HbJDF. 20, 303(24:14 as *C. affinis*; 48:36 as *C. bacillaris*), 370(19:146), 371(391:33 as *C. bacillaris*), 458.

turbinatus Craib
- Seedlings reported from near Charing Cross (Middlesex) and at Lambeth (Surrey). China. 61, 312*(t.8546), 458. The identity has recently been confirmed. 483.

veitchii (Rehder & E.Wilson) Klotz   Many-flowered Cotoneaster
- A persistent garden escape; recorded as a single shrub on Bison Hill, Whipsnade (Beds) and as seedlings in W Kent. China. E, LTN. 20, 61, 303(32:19; 62:43), 370(16:166), 915*(87:305-330). *C. racemiflorus* (Desf.) K.Koch var. *veitchii* Rehder & E.Wilson. The Beds record has been widely known as *C. multiflorus* but is an error; the W Kent record is most likely to be the same error. 483.

vestitus (W.Smith) Flinck & Hylmö
- A garden escape; apparent seedlings near planted bushes at New Ash Green (W Kent). China. 303(50:33). *C. glaucophyllus* var. *vestitus* W.Smith.

villosulus (Rehder & E.Wilson) Flinck & Hylmö   Lleyn Cotoneaster
- An established garden escape; naturalised in West Lleyn (Caerns) and Red Hills Wood, Arnside (Westmorland); several plants at Pitt (S Hants) and one seedling between Charing Cross and Temple (Middlesex). Himalayas, China. LANC, HbJDF. 20, 61, 303(50:33), 370(15:424), 458. *C. acuminatus* auct., non Lindley; *C. acutifolius* var. *villosulus* Rehder & E.Wilson.

vilmorinianus Klotz
- A garden escape by the railway at Montpelier, Bristol (W Gloucs) and on chalk downland at Portsdown Hill, Portsmouth (S Hants). HbJDF. 350(51:39), 401.

'Watereri' group   Hybrid Tree-cotoneasters
- ●●● Established garden escapes; naturalised on chalk downland in Kent and Surrey, and in a chalk hanger at Steep (N Hants); bird-sown on walls and by railways. Originated in cultivation. LANC, RNG. 20*, 54*, 61, 61a, 62*, 63*, 199, 303(43:30; 50:33), 312*(n.s. t.282), 370(18:219). A complex of hybrids between *C. frigidus, C. henryanus, C. salicifolius* and possibly other species; incl. *C.* × *watereri* Exell; *C.*'Cornubia'; *C.*'Hybridus Pendulus'; *C.*'Rothschildianus'.

zabelii C.Schneider   Cherryred Cotoneaster
- A garden escape established on walls and grassy banks at Gravesend and bird-sown on a roadside at Hextable (both W Kent); also recorded from Guildford and Banstead Downs (both Surrey). C China. 20, 61, 303(29:14; 42*:20-21), 342(Feb93), 458, 1295*.

PYRACANTHA M.Roemer

atalantoides (Hance) Stapf   Gibbs' Firethorn
- A garden escape recently recorded from Arundel (W Sussex). C China. 51, 61, 63*, 122, 597*. *P. gibbsii* A.B.Jackson.

coccinea M.Roemer   Firethorn
- ●●● An established garden escape; naturalised in several places on chalk in southern England, plentifully so in chalk pits at Grays (S Essex) and Rainham (E Kent). S Europe, SW Asia. BM, NMW, OXF, RNG. 1, 20, 31*, 54*, 61*, 63*, 191, 310(21:6). *Cotoneaster pyracantha* (L.) Spach; *Crataegus pyracantha* (L.) Medikus. A hybrid with *P. rogersiana* was recorded in 1992 from S Hants. HbRPB. 404.

crenulata (D.Don) M.Roemer   Nepalese Firethorn
Pre-1930 only. Himalayas. 7, 51, 61, 354(6:22), 590*, 685*, 724*. *Cotoneaster crenulata* Wenzig; *Crataegus crenulata* Roxb. Possibly conspecific with *P. coccinea*.

rogersiana (A.B.Jackson) Coltman-Rogers   Asian Firethorn
•• A persistent garden escape; bird-sown in an old gravel pit at Aylesford
(E Kent), on chalk grassland at Northfleet Brooks, below Darenth Woods and
at Kemsing (all W Kent); also reported as self-sown on river banks at Dartford
(W Kent).   China.   HbRPB.   20, 51, 52*, 61, 63*, 312*(n.s. t.74),
331(68:149), 371(379:25), 590*, 1327*.   *P. crenulata* var. *rogersiana*
A.B.Jackson.

MESPILUS L.
germanica L.   Medlar
•••• An established garden escape; in hedges and thickets especially in
southern England; long naturalised and widespread in Jersey (Channel Is).
SE Europe, SW Asia. 1, 20, 22*, 35*, 59*, 61*, 201. *Pyrus germanica* (L.)
Hook.f.   A hybrid with native *Crataegus laevigata* (× *Crataemespilus*
grandiflora (Smith) Camus; × *Crataegomespilus grandiflora* (Smith) Bean;
Mespilus grandiflora Smith) has been recorded. BM, OXF, RNG. 51, 62,
1311.

CRATAEGUS L.
[*azarolus* L. (*C. aronia* Bosc)   In error for *C. laciniata*.   354(5:652; 8:25).]
[*coccinea* L.   An ambiguous name covering several species. 1, 7, 126, 184,
354(1:218).]
coccinioides Ashe   Large-flowered Cockspur-thorn
• Introduced or bird-sown in an old hedge at Epsom (Surrey) and outside a car
park in York. Central N America. 20, 38*, 40*, 61*, 199, 371(388:35). The
record from Crockenhill (W Kent) was in error for *C. pedicillata*. 236.
crus-galli L.   Cockspur-thorn
••• (but confused with *C. persimilis*). A persistent garden escape in scrub,
gravel pits and hedges in the south and west, perhaps bird-sown in some places.
N America. BM, LNHS, NMW, OXF, RNG. 7, 20, 58*, 59*, 60*, 61,
371(355:27), 683*. *C. crus-corvi* sphalm.
douglasii Lindley   Black Hawthorn
• A relic of cultivation at Hob Moor, near York (MW Yorks). Western
N America. 38*, 59*, 61, 371(394:34).
elliptica Aiton
• Collected from a hedgerow opposite Givons Grove, near Leatherhead
(Surrey). Origin obscure. BM. 301*(t.1932, t.1939), 315*(27:404), 691.
C.flava Aiton var. *lobata* Lindley; *C. lobata* (Lindley) Loudon, non Bosc.
heterophylla Fluegge   Various-leaved Hawthorn
• An established garden escape in scrub on a railway bank at South Tottenham;
also introduced and now well established by seeding in woodlands at Nunhead
and Abney Park cemeteries (all Middlesex). Caucasus. OXF, RNG. 20, 61,
301*(14:t.1161; 22:t.84), 303(61:43), 354(9:269), 458. *C. neapolitanus* hort.
intricata Lange   Lange's Thorn
• Introduced or a garden escape in hedges in Glam. Eastern N America.
NMW. 21, 38*, 61, 156. *C. coccinea* auct., non L.

laciniata Ucria   Oriental Hawthorn
•• (but confused with *C. tanacetifolia*). A garden escape, probably bird-sown,
in a hedge near Leatherhead (Surrey) and in a disused sand pit at Stone
(W Kent); introduced elsewhere, persisting since 1921 on Shapwick Moor
(N Somerset). SE Europe, SW Asia. BM, LANC, OXF, RNG. 12, 20, 31*,
58*, 59*, 61, 63*, 236, 248, 350(32:19), 683*. *C. orientalis* Pallas ex
M.Bieb.; *C. tanacetifolia* auct., non (Lam.) Pers. Possibly better treated as
C. monogyna var. *lasiocarpa* (Lange) Christensen.

laevigata (Poiret) DC. subsp. **palmstruchii** (Lindman) Franco
• Introduced or a garden escape in hedges. No modern records. Sweden,
Denmark, Poland. BM, OXF, RNG. 21, 192, 327(71:127-139), 354*(12:847-
866), 826*(10:59-65). *C. palmstruchii* Lindman. Probably not worthy of
subspecific rank.

× **lavallei** Hérincq   Hybrid Cockspur-thorn
• Introduced on Chislehurst Common (W Kent) and in Avon Gorge
(W Gloucs); seedlings have been reported from Grove Park Cemetery (W Kent)
and Southfleet (S Lancs). Originated in cultivation. RNG. 54*, 59*, 60*, 61,
126, 350(47:37), 458. *C. crus-galli* × *C. pubescens* Steudel; *C.* × *carrierei*
Vauvel.

mollis (Torrey & A.Gray) Scheele
• Introduced. No modern records. Central N America. BM, K, LANC,
SLBI. 21, 38*, 59*, 61*, 172.

monogyna Jacq. subsp. **monogyna**
• Recently recorded from London parks and Nunhead Cemetery (all
Middlesex). 21, 303(61:43), 631, 683, 691. *C. monogyna* subsp. *azarella*
(Griseb.) Franco.

× **mordenensis** Boom
• Reported in 1988 from chalk scrub below Darenth Wood (W Kent).
Originated in cultivation. 61, 458, 631. *C. laevigata* × *C. succulenta*; incl.
C.'Snowbird'; *C.*'Toba'.

nigra Waldst. & Kit.   Hungarian Thorn
• Recorded in 1989 from a gully on the Downs at Avon Gorge (W Gloucs).
Hungary. 21, 61, 350*(47:37), 631*.

pedicillata Sarg.   Pear-fruited Cockspur-thorn
• A garden escape apparently self-seeding from an introduced tree, in a copse
near Crockenhill (W Kent); also reported in 1992 from Lyminster (W Sussex)
and Alderney (Channel Is). Eastern N America. RNG. 20, 53*, 61,
303(61*:1; 62:38), 370(15:429), 458. *C. coccinea* auct., non L.

persimilis Sarg. 'Macleod'   Broad-leaved Cockspur-thorn
••• (but confused with *C. crus-galli*). Introduced, and a bird-sown garden
escape, persistent in a few places in the south. Origin obscure, possibly
C. crus-galli × *C. macracantha* Lodd. BM, K, RNG. 20, 54*, 55*, 59*,
60*, 61, 126, 199, 327*(96:368), 370(15:429). *C. crus-galli* hort., non L.;
C. prunifolia Pers., nom. illeg.

punctata Jacq.   Dotted Hawthorn
  • Introduced. Eastern N America. BM, CGE, LIV, RNG. 7, 12, 38*, 40*, 61, 354(10:102), 586*.
sanguinea Pallas
  • Reported as an apparent garden escape from near Brackley (Northants). Russia, Siberia. 7, 21, 51, 61, 354(5:560), 435, 631, 691, 1295*, 1350*.
submollis Sarg.   Hairy Cockspur-thorn
  • Introduced or a garden escape in hedgerows near St Mary Cray and Goddington (both W Kent) and at Egg Buckland (S Devon). Eastern N America. 20, 21, 38*, 61, 303*(61:1), 331(61:101), 371(414:14), 390(73:63-72), 684*, 1289*. *C. coccinea* auct., non L.
succulenta Schrader   Round-fruited Cockspur-thorn
  • A garden escape, bird-sown in hedges and woods at Muddiford, Shirwell, Landkey and Umberleigh (all N Devon). Eastern N America. K, RNG. 20, 38*, 61, 303*(61:1, fig. atypical), 349(5:25), 371(331:19), 1289*. *C. coloradensis* Nelson.
tanacetifolia (Lam.) Pers.   Tansy-leaved Thorn
  • (but confused with *C. laciniata*). Introduced. Turkey. 39*, 55*, 61, 199, 354(5:379).
[*tomentosa* L.   Unconfirmed.   Perhaps in error for *C. calpodendron* Medikus (*C. tomentosa* auct., non L.). 2, 354(3:57), 376(3:57).]
viridis L.   Southern Thorn
  • A garden escape in a hedgerow by the golf course at Emmer Green (Berks). SE USA. 38*, 61, 354(14:25), 371(334:22). *C. arborescens* Elliott.

MIMOSACEAE
ACACIA Miller
falciformis DC.   Hickory Wattle
  • Introduced and reproducing in woods on Tresco (Scilly).   Australia. 303(32:18), 514, 533*, 569*, 743.
melanoxylon R.Br.   Australian Blackwood
  • Introduced, or a garden escape, reproducing on cliffs at Blackpool Sands (S Devon) and in woods on Tresco (Scilly). Australia. RNG. 20, 21, 37*, 303(32:18), 349(4:274), 569*, 606*, 684*. *Racosperma melanoxylon* (R.Br.) C.Martius.
[*retinodes* Schldl.   In error. In 21, Br (Britain) is a misprint for Bl (Balearic Is).]

CAESALPINIACEAE
CAESALPINIA L.
spinosa (Molina) Kuntze   Tara
  • A casual, probably from pods imported for tanning or dyeing; many seedlings in 1978 and 1987 at Avonmouth docks (W Gloucs). Tropical America; (Africa).   HbEJC.   350(41:88; 47:xxxi), 511, 608*, 689*, 723*, 736. *C. pectinata* Cav.; *C. tinctoria* (Dombey ex Kunth) Benth. ex Taubert; *Coulteria tinctoria* Dombey ex Kunth.

CERCIS L.
siliquastrum L.   Judas-tree
- A garden escape on waste ground at Hythe (E Kent), and reported as seedlings in Hextable Nature Reserve and at Eynsford (both W Kent). Medit. LANC. 21, 24*, 58*, 60*, 236, 458.

CERATONIA L.
siliqua L.   Carob
- A casual, probably from pods imported for food-processing or for animal feed; hundreds of seedlings in 1978 on an industrial dump at Greenhithe (W Kent). Medit. 2, 21, 29*, 32*, 44*, 55*, 236, 300(1905:235), 303(24:14,17), 331(58:63).

CHAMAECRISTA Moench
nictitans (L.) Moench   Sensitive Cassia
- Pre-1930 only. Eastern N America, W Indies. 38*, 40, 51, 354(8:305), 1319*. *Cassia nictitans* L.

SENNA Miller
obtusifolia (L.) Irwin & Barneby   American Sicklepod
- A casual, probably from oil-seed. Tropical America; (widespread). HbALG. 38*, 40 (both as *Cassia tora*), 303(36:28), 1109*. *Cassia obtusifolia* L.; *C. tora* auct., non L.

occidentalis (L.) Link   Coffee Senna
- An oil-seed casual. Tropics. RNG. 20, 38*, 47*, 48*, 126, 236, 331(54:64), 736*. *Cassia occidentalis* L.

CASSIA L.
fistula L.   Purging Cassia
- A casual on tips; possibly imported for tanning or pharmacy, or perhaps a greenhouse escape. No modern records. Tropical Asia. RNG. 12, 51, 126, 604, 680*, 681*, 685*.

FABACEAE
WISTERIA Nutt.
floribunda (Willd.) DC.
- A garden escape or relic persistent on waste ground at Hextable (W Kent). Japan. 51, 61, 63*, 458, 631*, 649*.

sinensis (Sims) Sweet   Chinese Wisteria
- A garden escape; seedlings recorded from south London, and from Guernsey (Channel Is). China. 21, 51, 54*, 61, 63*, 331(65:195), 631*, 1218.

venusta Rehder & E.Wilson
- A garden relic on waste ground at Hextable (W Kent). Japan. 51*, 61, 63*, 458, 631*, 649*. This species should probably be sunk into the synonymy of *W. brachybotrys* Siebold & Zucc., but this name has long been misapplied to *W. floribunda*.

ROBINIA L.
pseudoacacia L.   False-acacia
●●●●● An established garden escape and introduction; naturalised in woodland
and scrub, spreading by both suckering and seed, especially on sandy soils in
southern England; increasing. Eastern N America. 1, 20, 22*, 58*, 59*.

SESBANIA Scop.
[*aculeata* (Willd.) Poiret   In error for *S. bispinosa*.]
benthamiana Domin
● A wool casual. Northern Australia. K, RNG. 14, 37(as *S. aegyptiaca*),
37a, 855*(13:103-141).
bispinosa (Jacq.) W.Wight
● A wool casual. Tropics. LTN. 14, 37, 328*(13:287-288); 17:91-159),
544*, 620*. *Aeschynomene aculeata* Schreber, non *Sesbania aculeata* (Willd.)
Poiret.
exaltata (Raf.) Cory   Colorado River-hemp
● An oil-seed casual. N America. HbCGH, HbEJC. 38*, 40, 42, 126, 236,
303(22:12; 47:36), 501*, 699*. *S. macrocarpa* Muhlenb. ex Raf.

DESMODIUM Desv.
varians (Labill.) Endl.   Slender Tick-trefoil
● A wool casual. Australia. RNG. 17, 514, 533*, 549, 569*, 590*, 743.

PHASEOLUS L.
coccineus L.   Runner Bean
●●● A food refuse casual. Tropical America; (widespread as a crop). RNG.
7, 20, 21, 80*, 126, 553*. *P. multiflorus* Willd.
lunatus L.   Butter Bean
● A casual on a tip at Stone (W Kent), vector unknown. Tropical America;
(tropics). 36c*, 51, 68*, 80*, 236.
vulgaris L.   French Bean
●● A food refuse and bird-seed casual. Tropical America; (widespread as a
crop). LIV, OXF, SLBI. 2, 19, 20, 21, 80*, 192, 236, 270, 1141*.

VIGNA Savi
[*mungo* (L.) Hepper (*Phaseolus mungo* L.)   In error for *V. radiata*. 303(15:11),
370(9:189).]
radiata (L.) Wilczek   Mung-bean
●● A food refuse casual. E Asia; (widespread as a crop). 20, 43, 68*, 80*,
126, 236, 303(15*:1,11; 30:29). *V. mungo* auct., non (L.) Hepper; *Phaseolus
aureus* Roxb.

GLYCINE Willd.
max (L.) Merr.   Soya-bean
●●● An oil-seed, food refuse and bird-seed casual. Origin obscure, possibly
derived from *G. soja* Siebold & Zucc.; (widespread as a crop). BM, CGE, E,
NMW, OXF, RNG. 7, 19, 20, 21, 80*, 126, 303*(30:1,14), 683*. *G. soja*
auct., non Siebold & Zucc.; *G. hispida* (Moench) Maxim.

LABLAB Adans.
purpureus (L.) Sweet   Lablab-bean
- An oil-seed casual. Tropics. 47*, 48*, 68*, 80*, 236, 331(58:63). *L. niger* Medikus; *Dolichos lablab* L.; *D. purpureus* L.

PSORALEA L.
americana L.   Scurfy-pea
- • A bird-seed casual. W Medit. BM, E, K, NWH, RNG. 12, 19, 20*, 21, 30*, 303*(18:14), 737*. *P. dentata* DC.

bituminosa L.   Pitch Trefoil
- A wool casual. Medit. BM, CGE, LTN, NMW, OXF. 7, 21, 24*, 29*, 44*, 147, 354(6:607; 7:713), 737*. *Bituminaria bituminosa* (L.) C.H.Stirt.

cinerea Lindley   Hoary Scurfy-pea
- A wool casual. Australia. LTN, RNG. 37, 145, 303*(23:8-9), 840*(10:253).

patens Lindley   Spreading Scurfy-pea
- A wool casual. Australia. RNG. 17, 37*, 743, 1055*, 1340*.

ARACHIS L.
hypogaea L.   Peanut
- • A bird-seed and tan-bark casual. Tropical S America; (widespread as a crop). BM, HTN, K, NMW, OXF, RNG. 4, 19, 20, 21, 68*, 69*, 71*, 80*, 303(45:20), 354(11:473).

ZORNIA J.Gmelin
sp.
- A wool casual. HbMMcCW. 14, 1154, 1351.

GALEGA L.
officinalis group   Goat's-rue
- •••• Established garden escapes; locally abundant on roadsides, railway banks and waste ground, mainly in southern England; increasing. Europe, SW Asia. 1, 20*, 22*, 25*, 34*. Incl. *G.* × *hartlandii* Hartland ex Clarke (*G. officinalis* hort., non L.); *G. officinalis* L.

COLUTEA L.
arborescens L.   Bladder-senna
- •••• An established garden escape; widely naturalised, especially on railway banks in SE England. Medit. 7, 20*, 22*, 23*, 26*. A hybrid with *C. orientalis* Miller (*C.* × *media* Willd.) has been recorded. 20, 63*.

CARAGANA Fabr.
arborescens Lam.   Siberian Pea-tree
- A persistent garden escape at the side of an old chalk pit at Kemsing (W Kent); seedlings on a country roadside at Wrotham Heath (W Kent); also recorded from Middlesex. Siberia, Mongolia. RNG. 21, 55*, 61, 62*, 303(50:27), 371(346:27; 407:41), 613*, 1242*. *Robinia caragana* L.

ASTRAGALUS L.
boeticus L.
- (but confused with *A. odoratus*). A grain casual. No modern records. Formerly a wool casual. Medit. BM, LIV, NMW, OXF, RNG, SLBI. 2, 9, 20, 21, 39*, 44*, 156, 182, 212, 729*.

cicer L.   Chick-pea Milk-vetch
- An established grain alien; known for over 60 years along a hedgebank near a mill at Fushiebridge, near Edinburgh (Midlothian). Europe, W Asia. BM, CGE, E, NMW, OXF, RNG. 7, 20*, 22*, 23*, 27*, 349(3:400; 4:327), 371(331:19).

aff. **cruciatus** Link
- An esparto casual. No modern records. SW Asia. E. 15, 44*, 667*.

hamosus L.
- •• A grain and tan-bark casual. Medit., SW Asia. BM, E, NMW, OXF, SLBI. 1, 21, 39*, 44*, 182, 258, 303(39:6), 354(13:289), 615*, 729*.

hispidulus DC.
Pre-1930 only. NE Africa, SW Asia. SLBI. 7, 44*, 300(1909:41), 667*.

nuttallianus DC.
Pre-1930 only. Western N America. 2, 42, 233, 699*, 1062*.

odoratus Lam.   Lesser Milk-vetch
- •• An established grain alien; by canals, rivers and railways in England; known for over 60 years near maltings at Burton-on-Trent (Staffs). SE Europe, SW Asia. BM, LANC, NMW, RNG. 20*, 21, 126, 185, 248, 349(3:400), 354(9:259), 371(409:39), 683*, 1080*.

scorpioides Pourret ex Willd.
Pre-1930 only. Medit. BM. 21, 44*, 45*, 46*. *A. uncinatus* Bertol.

sesameus L.
Pre-1930 only. Medit. BM, SLBI. 2, 21, 39*, 300(1907:39), 546*, 683*, 695*.

stella Gouan
- A wool casual. No modern records. Formerly an esparto casual. Medit. NMW, OXF, RNG. 7, 9, 14, 21, 39*, 354(6:607), 546*.

tribuloides Del.
Pre-1930 only. E Medit., Asia. SLBI. 7, 21, 39*, 44*, 354(4:407), 667*.

BISERRULA L.
pelecinus L.
- A casual collected in 1951 from mud dredged from the river Stour at Stonar (E Kent), vector unknown. No modern records. Medit. RNG. 21, 44*, 45*, 683*. *Astragalus pelecinus* (L.) Barneby.

GLYCYRRHIZA L.
echinata L.   German Liquorice
Pre-1930 only. E Medit., Asia. 1, 21, 31*, 44*, 149, 151.

glabra L.   Liquorice
Pre-1930 only.   Medit., Asia.   LIV.   1, 21, 29*, 39, 44*, 53*, 185,
218(16,42,43,90,96,109).   *G. officinalis* Lepechin.
lepidota (Nutt.) Pursh   American Liquorice
• A wool casual.   N America.   BM, RNG.   38*, 41*, 51, 371(379:26).

ONOBRYCHIS Miller
aequidentata (Sibth. & Smith) Urv.
• A tan-bark casual.   Medit.   SLBI.   21, 31*, 32*, 39*, 303(35:10; 45:20).
alba (Waldst. & Kit.) Desv.
Pre-1930 only.   Medit.   7, 21, 31*, 46*, 354(5:281), 683*.
caput-galli (L.) Lam.   Cockscomb Sainfoin
Pre-1930 only.   Medit.   21, 24*, 39*, 44*, 46*, 300(1904:238; 1907:39).
crista-galli (L.) Lam.
• A grain casual.   No modern records.   E Medit.   OXF.   2, 21, 39*, 44*, 46*,
300(1905:99), 354(11:473; 12:39).   *O. squarrosa* Viv.
[*viciifolia* Scop.   Accepted, with reservations, as native.]

HEDYSARUM L.
coronarium L.   Italian Sainfoin
• A casual, from wool and as a garden weed.   C & W Medit.   CGE, NMW,
OXF, RNG.   1, 21, 24*, 29*, 32*, 36c*, 336(9:18,54), 371(385:35), 415.
glomeratum Dietr.
• A wool casual.   Medit.   LTN, NMW.   4, 7, 21, 29*, 30*, 147, 354(6:607).
H. capitatum Desf., non Burman f.
spinosissimum L.
• A wool casual.   Medit.   LTN, RNG.   14, 21, 39*, 44*, 46*, 145,
371(400:36).   *H. pallens* (Moris) Hal.

ANTHYLLIS L.
tetraphylla L.   Bladder Vetch
Pre-1930 only.   Medit.   NMW.   4, 7, 21, 24*, 29*, 44*, 46*, 354(6:277),
683*.   *Physanthyllis tetraphylla* (L.) Boiss.
vulneraria L. subsp. **carpatica** (Pant.) Nyman
•• (but probably overlooked).   A grass seed or agricultural seed alien in widely
scattered localities; perhaps also a garden escape.   NW & C Europe.   BM, E,
K, LIVU, RNG.   20, 135, 220, 303*(46:13,15), 370(17:187; 18:401-403),
371(427:35).   All records are referable to var. *pseudovulneraria* (Sagorski)
Cullen.   A hybrid with subsp. *polyphylla* has been reported.   458.
[*vulneraria* L. subsp. *maritima* (Schweigger) Corbière (*A. maritima* Schweigger)
In error for native subsp. *vulneraria* var. *langei*.   20.]
vulneraria L. subsp. **polyphylla** (DC.) Nyman
••• (but perhaps overlooked).   A persistent grass seed or agricultural seed
alien sometimes abundant on grassy banks; possibly also a fodder alien.
E Europe, Turkey, Caucasia.   ABD, BM, CGE, E, LIV, RNG.   20, 21, 39,
277, 303*(46:13,15), 344(35:1-38), 370(18:217,401-403), 630*.   *A. polyphylla*

(DC.) Kit. ex G.Don; *A. schiwereckii* (Ser. ex DC.) Błocki.   Plants intermediate between this and subsp. *vulneraria* have been recorded. 370(18:401-403).

vulneraria L. **alien variants**

'Amaranth Purple' has been recorded pre-1930 with a synonym of *A. dillenii* Schultes ex G.Don f., but this is insufficiently precise to equate it to a subspecies as defined in 344(35:1-38). 326(24:12-13).

DORYCNIUM Miller

hirsutum (L.) Ser.   Canary Clover

• A garden escape; many seedlings reported from a trackside on the outskirts of Faversham (E Kent). Formerly a ballast alien in the Cardiff area (Glam). Medit. BM, CGE, NMW, OXF. 4, 7, 21, 29*, 44*, 46*, 52*, 458. *Bonjeanea hirsuta* (L.) Reichb.; *Lotus hirsutus* L.

pentaphyllum Scop.   Badassi

• A persistent garden escape; recorded for some years from the north cliffs of Sheppey (E Kent), probably now gone; also on a roadside verge at Canterbury (E Kent), on downs at Meopham (W Kent) and on the Regent's Canal towpath verge near Victoria Park (Middlesex). Medit. BM, CGE, NMW, OXF, RNG, SLBI. 7, 16, 21, 22*, 26*(t.152), 46*, 236, 303(45:22), 320(72:349), 331(33:159), 354(6:607). Incl. *D. gracile* Jordan; *D. herbaceum* Villars; *D. suffruticosum* Villars.

rectum (L.) Ser.   Greater Badassi

• An introduction which persisted for 20 years in a bomb crater on Brockham Hill (Surrey); also a garden weed, probably from an impurity in flower seed. Formerly persistent at Barry docks (Glam). Medit. BM, CGE, LANC, NMW, OXF, RNG. 7, 10, 21, 29*, 44*, 46*, 191, 209, 354(7;34). *Bonjeanea recta* (L.) Reichb. Wrongly described in 13 under *D. pentaphyllum* Scop.

LOTUS L.

americanus (Nutt.) Bisch.   Prairie Trefoil

Pre-1930 only. Eastern N America. OXF. 7, 38*, 40, 354(5:554; 7:566). *Hosackia americana* (Nutt.) Piper.

aff. **borbasii** Ujh.

• A casual at Alnwick and Amber (S Northumb), probably introduced with grass seed. E Europe. 21, 265, 852*(52:185-195). *L. corniculatus* subsp. *major* auct. pro parte; incl. *L. degenii* Ujh. Possibly in error for *L. corniculatus* var. *sativus*.

conimbricensis Brot.

• A casual in a nursery at Freshfield (Cumberland), vector unknown. No modern records. Medit. NMW, RNG. 5, 7, 21, 39, 44*, 46*, 354(8:22).

corniculatus L. **alien variants**

Var. *sativus* Chrtková has been recorded from sown grass verges. E. 20, 380(56:129-138), 382(13:3-6), 825*(14:137-140), 913(83:1-94).

creticus L.   Cretan Bird's-foot-trefoil

Pre-1930 only. Medit., Azores. NMW, OXF. 4, 21, 29*, 30*, 44*, 46*, 354(7:34).

cytisoides L.
> Pre-1930 only. Medit. NMW. 21, 24*, 31*, 44*, 46, 156. *L. creticus* subsp. *cytisoides* (L.) Asch.

edulis L.
> Pre-1930 only. Medit. NMW, OXF, SLBI. 2, 21, 24*, 44*, 46*, 300(1905:99), 354(7:179).

jacobaeus L.   St James' Trefoil
> Pre-1930 only. Cape Verde Is. 7, 51, 312*(t.79), 354(3:156).

[*lathyroides* ?Greene   A dubious record; perhaps in error for native *Vicia lathyroides*. 187.]

ornithopodioides L.   Southern Bird's-foot-trefoil
> • An esparto casual. Formerly a ballast alien. Medit. BM, NMW, OXF, RNG, SLBI. 4, 6, 7, 21, 24*, 44*, 46*, 300(1906:102), 354(6:277).

palustris Willd.
> Pre-1930 only. Medit. 21, 44*, 46, 300(1910:44), 667*, 885*(19:271-292). *L. lamprocarpus* Boiss.; *L. angustissimus* L. subsp. *palustris* (Willd.) Ponert; *L. clausonis* Pomel ex Battand. *L. decumbens* Poiret is a synonym of native *L. pedunculatus* but the name has been widely misapplied to *L. preslii* Ten. (546*, 546a, 1052*) as well as to this species. 1123.

TETRAGONOLOBUS Scop.

biflorus (Desr.) Ser.
> Pre-1930 only. Medit. OXF. 7, 21, 354(6:277), 683*. *Lotus biflorus* Desr.

maritimus (L.) Roth   Dragon's-teeth
> ••• Established, perhaps as a grass seed or fodder alien; locally naturalised, or possibly even native, in grassland in a few places in southern England; known for many years at West Mersea (N Essex), Sheppey (E Kent) and Remenham (Berks). Europe, SW Asia, N Africa. ABD, BM, CGE, E, OXF, RNG. 1, 20*, 22*, 27*, 35*, 119, 191, 236, 354(8:111), 825(2:34-35). *T. siliquosus* (L.) Roth; *Lotus maritimus* L.; *L. siliquosus* L.

purpureus Moench   Asparagus-pea
> • A casual, perhaps from bird-seed. Formerly a grain alien. Medit., SW Asia. BM, K, NMW, OXF, RNG, SLBI. 1, 8, 21, 29*, 30*, 44*, 46*, 339(16:219), 371(346:28). *Lotus tetragonolobus* L.; incl. *T. palaestinus* Boiss.

HYMENOCARPUS Savi

circinnatus (L.) Savi   Disk Trefoil
> • A tan-bark casual. Medit. OXF, SLBI. 2, 21, 24*, 39*, 44*, 46*, 300(1904:238; 1905:99), 303(35:10), 354(4:480). *Circinnus circinnatus* (L.) Kuntze.

ORNITHOPUS L.

compressus L.   Yellow Serradella
> •• A tan-bark and granite alien persisting a few years; known since 1957 on a roadside bank near Ruxley (W Kent), where it may have originated from wool waste. Medit. CGE, JSY, NMW, OXF, RNG, SLBI. 1, 20*, 21, 22*, 24*, 31*, 44*, 201, 303(10:16; 35:10; 59:45), 354(11:171), 365*(9:106-107).

sativus Brot. Serradella
•• Introduced and persistent on china-clay waste in the St Austell area (E Cornwall); and on gabbro spoil at Dean Quarries near St Keverne (W Cornwall); casual elsewhere. W Medit. BM, CGE, E, NMW, OXF, RNG. 7, 20, 21, 22*, 133, 212, 213, 371(403:40), 546*, 566*, 1242*. Incl. *O. roseus* Dufour.

CORONILLA L.
cf. **coronata** L. Scorpion-vetch
• A tan-bark casual. C & S Europe, SW Asia. 21, 22*, 26*, 29*, 303(35:10).

scorpioides (L.) Koch Annual Scorpion-vetch
•• A bird-seed, tan-bark and wool casual. Medit. BM, CGE, LTN, NMW, OXF, RNG. 1, 19, 20*, 24*, 32*, 44*, 212, 370(11:428). *Arthrolobium scorpioides* (L.) DC.; *Astrolobium scorpioides* (L.) DC.; *Ornithopus scorpioides* L.

valentina L. Shrubby Scorpion-vetch
•• An established garden escape; naturalised on a cliff at Torquay (S Devon), at the southern end of the parade at Eastbourne (E Sussex) and at Porlock Weir (S Somerset); casual elsewhere. Medit. LANC, RNG. 13, 20*, 66*, 303(32:18), 370(1:317), 452, 506*, 593*. Records are probably all referable to subsp. *glauca* (L.) Battand. (*C. glauca* L.).

HIPPOCREPIS L.
emerus (L.) Lassen Scorpion Senna
•• A persistent or established garden escape in hedges and on roadsides; modern records are from Waddington (S Lincs), Gidea Park (S Essex), Hextable (W Kent) and Eastbourne (E Sussex). Europe, SW Asia. BM. 20*, 22*, 26*, 27*, 122, 126, 288, 303(59:46), 331(59:79), 371(364:28), 1242*. *Coronilla emerus* L.

multisiliquosa L.
Pre-1930 only. Medit. 7, 21, 39*, 44*, 45*, 300(1906:102).

unisiliquosa L. sensu lato
• A grain and tan-bark casual. Medit., SW Asia. NMW, OXF, SLBI. 1, 4, 21, 24*, 31*, 39*, 44*, 213, 247, 354(12:39). The common Medit. species has recently been segregated as *H. biflora* Sprengel. 894*(5:225-261).

SECURIGERA DC.
cretica (L.) Lassen
Pre-1930 only. E Medit. OXF. 7, 21, 31*, 44*, 354(5:281), 683*. *Coronilla cretica* L.

securidaca (L.) Degen & Doerfler Hatchet Vetch
• A tan-bark casual. Medit., SW Asia. BM, NMW, OXF, SLBI. 1(under Cruciferae), 4, 21, 31*, 39*, 44*, 121, 303(35:10), 349(3:442), 683*, 825*(5:38-39). *S. coronilla* DC.; *Bonaveria securidaca* (L.) Desv.; *Coronilla securidaca* L.

varia (L.) Lassen   Crown Vetch
•••• An established garden escape; naturalised in quarries, on roadsides and
railway banks in widely scattered localities, mainly in S and E England.
Europe, SW Asia. 1, 20*, 22*, 26*, 35*, 143, 1242*. *Coronilla varia* L.

SCORPIURUS L.
muricatus L.   Caterpillar-plant
•••• A wool, bird-seed, spice and tan-bark casual. Medit., SW Asia. 1, 14,
19, 20*, 24*, 29*, 44*, 303(35:10), 825(2:37-39). Incl. *S. laevigatus* Sibth.
& Smith; *S. subvillosus* L.; *S. sulcatus* L.
vermiculatus L.
• A bird-seed casual. Formerly a ballast alien. W Medit. LCN. 7, 19, 21,
45*, 46*, 166, 182, 371(340:27), 737*.

VICIA L.
altissima Desf.
Pre-1930 only. W Medit. BM, LIV, OXF. 7, 21, 354(5:554,814), 546*,
556, 1036*.
amoena Fischer
Pre-1930 only. Siberia, E Asia. LIV, OXF. 7, 313(22:37), 354(3:80,156),
638*, 1115*.
articulata Hornem.
Pre-1930 only. Medit. 1, 9, 21, 267, 354(6:405; 8:734), 546*, 683*, 737*.
V. monanthos (L.) Desf., non *V. monantha* Retz.; *Ervum monanthos* L.
benghalensis L.   Purple Vetch
•• A wool and grain casual, and an impurity in Sweet Pea seed. Medit.;
(N America). BM, NMW, OXF, RNG. 7, 11, 14, 20, 29*, 45*, 439, 546*,
683*. *V. atropurpurea* Desf.
cassubica L.   Danzig Vetch
• A persistent alien, vector unknown; naturalised for over 30 years in a chalk
pit near Greenhithe (W Kent), now gone. Europe, SW Asia. BM, LANC,
NMW, OXF, RNG. 12, 20, 21, 22*, 26*, 126, 156, 374(1949, suppl.:78),
1242*. The record from Bow (Middlesex) was in error for *V. villosa* subsp.
varia (Host) Corb. 331(65:197; 70:154).
cracca L. subsp. **gerardii** Gaudin
• Reported from a roadside near Wrotham Heath (W Kent), possibly introduced
with a grass seed mixture. C & S Europe. BM. 21, 39, 346(n.s. 3:335), 458,
546*, 683*. *V. gerardii* All.; *V. incana* Gouan.
cretica Boiss. & Heldr.
Pre-1930 only. Aegean, Cyprus, Turkey. BM, CGE. 7, 21, 39, 354(4:337),
658. Incl. *V. spruneri* Boiss.
ervilia (L.) Willd.   Ervil
• A grain casual. No modern records. Medit. BM, OXF, SLBI. 1, 21, 44*,
46*, 349(2:140), 370(1:248), 546*, 737*. *Ervum ervilia* L.

faba L.   Broad Bean
●●●● A relic or escape from cultivation, sometimes persistent; also a bird-seed
casual. Origin obscure; (widespread). 1, 19, 20, 71*, 80*, 222, 546*, 1242*.
Faba bona Medikus; *F. vulgaris* Moench.

grandiflora Scop.
● A casual at Avonmouth docks (W Gloucs), vector unknown. No modern
records. C & SE Europe, SW Asia. BM. 1, 21, 354(11:27), 683*, 1242*,
1271*.

cf. **hookeri** G.Don   Hooker's Vetch
1930 only. Chile. 9, 557. *V. micrantha* Hook. & Arn.; *V. parviflora* Hook.,
non Cav.

hybrida L.   Hairy Yellow-vetch
●● (but confused with *V. pannonica*). A grain and tan-bark casual. Formerly
established for more than 150 years on Glastonbury Tor (N Somerset). Medit.,
SW Asia. BM, CGE, E, NMW, OXF, RNG. 1, 20, 21, 24*, 32*, 44*, 160,
248, 303(39:6), 546*, 1271*.

hyrcanica Fischer & C.Meyer
Pre-1930 only. SW & C Asia. 7, 39, 281, 1173*, 1271*.

leucantha Biv.
Pre-1930 only. Medit. 1, 21, 46*, 346(n.s. 3:335). *V. agrigentina* (Guss. ex
Ser.) Link; *V. bivonea* Raf.; *Ervum agrigentinum* Guss. ex Ser.

lutea L. subsp. **vestita** (Boiss.) Rouy
● (but perhaps overlooked). A casual on a motorway embankment at Swanley
(W Kent) and at Edinburgh (Midlothian), possibly introduced with grass seed
mixtures. SW Europe, N Africa, SW Asia. K, NMW. 5, 9, 12, 21, 458,
475, 546*, 737*, 1271*. *V. lutea* var. *hirta* (Balbis) Lois.; *V. vestita* Boiss.

melanops Sibth. & Smith   Black-eyed Vetch
● A casual at Bristol (N Somerset) and Sharpness docks (W Gloucs), vector
unknown. No modern records. S Europe. BM, LIV, OXF. 2, 21, 31*, 182,
354(12:275), 546*.

monantha Retz.   Few-flowered Vetch
●● A grain casual. Medit., SW Asia. ABD, BM, CGE, E, LCN, OXF. 2,
11, 21, 44*, 556*, 737*, 1271*. *V. biflora* Desf.; *V. calcarata* Desf.;
V. triflora Ten. Most records are referable to subsp. *triflora* (Ten.) B.L.Burtt
& P.Lewis, but it is not worthy of subspecific rank.

narbonensis L.   Narbonne Vetch
●●● A persistent grain, bird-seed and tan-bark alien; known since 1976 from
the edge of a plantation at Icklingham (W Suffolk). Medit. BM, CGE, LCN,
NMW, OXF, RNG. 1, 19, 20, 22*, 44*, 303(39:6), 354(13:156), 546*,
1271*. Incl. *V. serratifolia* Jacq.

onobrychioides L.
Pre-1930 only. Medit. 21, 29*, 30*, 288(p.lviii), 546*.

palaestina Boiss.
● A casual at Bristol, date and vector unknown. No modern records. E Medit.
39, 44*, 245, 1271*.

pannonica Crantz   Hungarian Vetch
●●● Introduced with grass seed on roadside banks, now well established and
spreading in several places in Kent, especially in the Dartford to Northfleet area
(W Kent); also a grain and bird-seed casual. Medit. BM, E, MNE, NMW,
OXF, RNG. 1, 19, 20, 22*, 303(17:18; 34*:26-27), 546*, 1242*, 1271*.
Incl. *V. purpurascens* DC.; *V. striata* M.Bieb. Both subsp. *pannonica* and
subsp. *striata* (M.Bieb.) Nyman have been recorded. 20.

peregrina L.
● A grain casual. No modern records. Medit., SW & C Asia. BM, K,
NMW, OXF. 1, 8, 21, 44*, 354(7:567; 9:758), 546*, 737*, 1271*.

pubescens (DC.) Link
Pre-1930 only. Medit., Atlantic Is. OXF. 3, 8, 21, 44*, 346(n.s. 3:335),
546*, 737*. *Ervum pubescens* DC.

pyrenaica Pourret   Pyrenean Vetch
● A grain casual. SW Europe. 7, 21, 23*, 27*, 28*, 160, 354(3:18),
371(355:26), 546*.

sativa L. **alien variants**
●●●●● Established escapes from cultivation and bird-seed aliens. 2, 20, 21,
26*. Owing to varied taxonomic treatments in this very variable species, most
records of the two alien subspp. now recognised, subsp. *sativa* (subsp. *obovata*
(Ser.) Gaudin) and subsp. *segetalis* (Thuill.) Gaudin (*V. angustifolia* subsp.
segetalis (Thuill.) Archang.) cannot now be separated. Other possibly alien
forms recorded under various infraspecific ranks, mainly pre-1930, are:
"*amphicarpa*", "*canescens*", "*cordata*", "*cordifolia*", "*macrocarpa*",
"*nemoralis*" and "*triflora*". Their taxonomic rank is uncertain. 370(12:1-14).

tenuifolia Roth   Fine-leaved Vetch
●●● An established grain alien; locally naturalised on grassy banks and bushy
ground; known for over 50 years at Warren Barn near Warlingham (Surrey)
and for 40 years at Coates, near Cirencester (E Gloucs). Eurasia, NW Africa.
BM, CGE, E, LSR, OXF, RNG. 1, 11, 20, 22*, 44*, 126*, 209,
303*(34:22,24), 354(13:27), 546*, 1059. Incl. *V. dalmatica* Kerner;
V. elegans Guss.; *Cracca tenuifolia* (Roth) Gren. & Godron.

villosa Roth   Fodder Vetch
●●●● A persistent grain, bird-seed, wool and tan-bark alien and an escape from
cultivation; often locally abundant on grassy banks for several years. Eurasia,
N Africa; (Australasia). 1, 11, 14, 19, 20, 22*, 26*, 44*, 303*(50:31), 546*,
1242*. Incl. *V. boissieri* Heldr. & Sart.; *V. dasycarpa* Ten.; *V. elegantissima*
Shuttlew.; *V. eriocarpa* (Hausskn.) Hal.; *V. glabrescens* (Koch) Heimerl.;
V. godronii Rouy; *V. microphylla* Urv.; *V. pseudocracca* Bertol.; *V. salaminia*
Heldr. & Sart.; *V. varia* Host.

LENS Miller
culinaris Medikus   Lentil
●●● A bird-seed and food refuse casual. Origin obscure; (widespread as a
crop). BM, CGE, E, NMW, OXF, RNG. 2, 19, 20*, 29*, 44*, 80*,
354(12:676), 546*, 1242*. *L. esculenta* Moench; *L. lens* (L.) Huth;

Ervum lens L.; *Vicia lens* (L.) Cosson & Germ.; incl. *L. orientalis* (Boiss.)
Popov (*Ervum orientale* Boiss.)
nigricans (M.Bieb.) Godron
  •• A casual at Norwich (E Norfolk), perhaps introduced with nursery stock.
  Medit., SW Asia. 7, 9, 21, 303(50:27), 354(12:676), 546*, 683*, 737*.
  L. culinaris subsp. *nigricans* (M.Bieb.) Thell.

LATHYRUS L.
angulatus L.   Angular Pea
  • A grain casual. No modern records. Medit. BM, LCN. 2, 21, 160, 251,
  300(1905:99), 533*, 546*, 683*.
annuus L.   Fodder Pea
  •• A grain, bird-seed and tan-bark casual. Medit., SW & C Asia. BM, LCN,
  OXF, RNG. 1, 11, 19, 20, 21, 24*, 45*, 303(35:10; 18*:13), 737*.
[*aphaca* L.   Accepted, with reservations, as native.]
blepharicarpus Boiss.
  Pre-1930 only. E Medit. 2, 39, 44*, 300(1905:99), 667*.
cicera L.   Red Vetchling
  •• A tan-bark casual and possibly a bird-seed alien. Formerly a cotton alien.
  Medit., SW Asia. BM, JSY, LCN, LIV, NMW, OXF. 1, 19, 21, 24*, 32*,
  39*, 44*, 303(35:10), 349(3:442), 354(4:295). *L. erythrinus* C.Presl.
clymenum L.
  • A bird-seed and spice casual. Medit. K, NMW, OXF, WARMS. 7, 19, 21,
  24*, 29*, 45*, 156, 236, 370(13:331), 371(385:35; 388:37).   Incl.
  L. articulatus L.
gorgoni Parl.
  • A grain casual. No modern records. E Medit. BM, OXF, RNG. 2, 11,
  21, 39*, 44*, 300(1904:238; 1905:99), 667*. *L. amoenus* Fenzl.
grandiflorus Sibth. & Smith   Two-flowered Everlasting-pea
  ••• (but confused with *L. latifolius, L. odoratus* and *L. tingitanus*).   An
  established garden escape; widespread and locally plentiful, especially on
  railway banks, sometimes spreading vegetatively over hundreds of square yards.
  SE Europe. BM, CGE, E, NMW, OXF, RNG. 2, 20, 21, 31*, 36c*, 54*,
  209, 303*(16:17). *L. tingitanus* auct., non L.
heterophyllus L.   Norfolk Everlasting-pea
  • A garden escape established for many years on dunes at Burnham Overy
  Staithe (W Norfolk) and known for at least ten years in an old railway
  cutting at Kingsley (N Hants). Europe. LANC, RNG. 20, 22*, 27*, 235,
  349(2:238-240), 370(10:423), 405.   Possibly conspecific with *L. latifolius*.
  Plants from N Hants and W Norfolk are referable to var. *unijugus* Koch.
hierosolymitanus Boiss.
  • A grain casual. No modern records. Crete, Greece, SW Asia. BM, OXF.
  2, 21, 44*, 354(4:194; 7:567; 12:275,486), 667*.

hirsutus L.   Hairy Vetchling
●●● An established alien; long naturalised in several areas in S Essex and at
Caterham (Surrey); in recent years a grain and bird-seed casual. Europe,
N Africa, SW Asia. BIRM, BM, OXF, RNG. 1, 19, 20, 22*, 25*, 29*, 191,
209, 303(34:9).

inconspicuus L.
●● A bird-seed and tan-bark casual. Medit, SW & C Asia. BM, K, LIV,
OXF. 2, 21, 39*, 44*, 212, 303(10:12), 387(44:32). Incl. *L. erectus* Lagasca.

latifolius L.   Broad-leaved Everlasting-pea
●●●●● An established garden escape, possibly also a bird-seed alien; widely
naturalised, especially on railway banks; increasing. Europe, N Africa.
1, 19, 20, 22*, 23*, 34*, 36c*.

marmoratus Boiss. & Bal.
● A grain casual. No modern records. E Medit. OXF. 39*, 44*, 667*.

niger (L.) Bernh.   Black Pea
● A long-persistent or established garden escape; known for many years on a
railway bank near Tunbridge Wells (W Kent). Formerly claimed as native in
Warwicks, Cheviot, E Perth and Angus, now presumed extinct. Europe,
NW Africa, SW Asia. BM, CGE, E, LANC, OXF, RNG. 2, 16, 20, 22*,
26*, 34*, 220, 236, 354(6:19-21), 1242*. *Orobus niger* L.

ochrus (L.) DC.   Winged Vetchling
●● A casual, possibly from bird-seed. Formerly a grain and ballast alien.
Medit. BM, CGE, JSY, LIV, NMW, OXF. 1, 21, 24*, 29*, 44*, 156, 201,
354(7:567). *Pisum ochrus* L.

odoratus L.   Sweet Pea
●●● (but confused with *L. grandiflorus*). A casual or persistent garden escape.
Italy, Sicily. LIV, RNG, SLBI. 1, 20, 21, 54*, 166, 191, 192, 594*, 680*.

palustris L. subsp. **pilosus** (Cham.) Hultén
● Recorded from dunes at Berrow (N Somerset) and Towyn Burrows, Pembrey
(Carms). Asia, N America. 43, 303(20:14), 607*, 649*. *L. palustris*
var. *pilosus* (Cham.) Ledeb.; *L. pilosus* Cham.; *L. ugoensis* Matsum.

rotundifolius Willd.   Round-leaved Vetchling
● (but confused with *L. grandiflorus*). A casual in a sandpit at Blaxhall
(E Suffolk), perhaps a garden escape. No modern records. Russia, SW Asia.
21, 39, 51, 258, 312*(t.6522), 915*(86:85-87), 1125*. Incl. *L. miniatus*
M.Bieb. ex Steven.

sativus L.   Indian Pea
● A bird-seed casual. C Europe, Medit., Africa, SW Asia, India. BM, LIV,
OXF. 1, 19, 20, 29*, 39*, 199, 235, 285, 303(45:22), 658.

saxatilis (Vent.) Vis.
Pre-1930 only. Medit. 7, 21, 354(2:500), 546*, 667*, 683*. *L. ciliatus*
Guss.; *Orobus saxatilis* Vent.; *Vicia saxatilis* (Vent.) Tropea.

setifolius L.   Brown Vetchling
● A wool casual. No modern records. Medit., SW Asia. RNG. 7, 14, 21,
23*, 39*, 44*, 136, 354(12:715).

sphaericus Retz.   Round-seeded Vetchling
Pre-1930 only. Eurasia, N Africa. BM, CGE, NMW, SLBI. 1, 4, 21, 22*,
39*, 45*, 48*, 144.

tingitanus L.   Tangier Pea
• (but confused with *L. grandiflorus* and *L. odoratus*). A bird-seed casual.
W Medit., Atlantic Is. RNG. 21, 30*, 32*, 36c*, 126, 192, 519*, 737*. The
Abbotsbury (Dorset) record was in error for *L. grandiflorus*. 371(334:22;
337:30). The description in 13 refers to *L. grandiflorus*.

tuberosus L.   Tuberous Pea
•••• An established grain and bird-seed alien, and probably also a garden
escape; locally naturalised in fields and grassy places and known since about
1800 at Fyfield (N Essex). Eurasia. 1, 20, 22*, 33*, 35*, 191, 209. The
hybrid with *L. rotundifolius* may have been overlooked. 312*(n.s. t.384).

vernus (L.) Bernh.   Spring Pea
•• A persistent garden escape and relic. Eurasia. LIV, HbACL. 20, 22*,
23*, 26*, 27*, 126, 199, 217, 261, 1242*. *Orobus vernus* L.

PISUM L.

fulvum Sibth. & Smith
Pre-1930 only. E Medit. OXF. 7, 39, 44*, 354(5:555), 667*.

sativum L. sensu lato   Pea
•••• Mainly an escape or relic of cultivation, but also a grain, tan-bark and
bird-seed casual. Medit., SW Asia; (widespread). 1, 11, 19, 20, 32*, 44*,
80*, 235, 303(35:10), 1242*. Incl. *P. arvense* L.; *P. elatius* M.Bieb.;
P. humile Boiss. & Noë.

CICER L.

arietinum L.   Chick Pea
••• A grain, bird-seed and food refuse casual, sometimes abundant near docks.
SW Asia. BM, CGE, LIV, NMW, OXF, RNG. 1, 19, 20*, 29*, 80*,
303*(34:1,22).

ONONIS L.

alopecuroides L.
•• A casual, probably from bird-seed. Medit. ABS. 2, 21, 44*, 45*, 46*,
130, 154, 156, 236, 737*.

[*antiquorum* L.   In error for native *O. spinosa*. 141, 195.]

[*arvensis* L.   In error for native *O. repens*. 141.]

baetica Clemente   Salzmann's Restharrow
••• A bird-seed casual. W Medit. ABD, BM, CGE, E, RNG. 19, 20*, 21,
349(2:45).   *O. salzmanniana* Boiss. & Reuter.   According to 737*, the
synonymy given here is incorrect.

mitissima L.   Mediterranean Restharrow
•• A bird-seed casual. Medit., Atlantic Is. BM, LSR. 1, 19, 20*, 21, 44*,
45*, 371(394:36), 737*.

174 ONONIS   FABACEAE

natrix L.   Yellow Restharrow
• An established alien, vector unknown; in plenty on the old Memsbury
airfield, and persistent for several years on dumped soil on the airfield on
Crookham Common (both Berks); also recorded from Hounslow Heath tip
(Middlesex) and Charlton Forest (W Sussex). Formerly a ballast alien. Medit.
BM, CGE, E, NMW, RNG. 1, 20, 22*, 27*, 44*, 119, 331(45:20),
353(4:22), 370(2:40), 371(385:34), 737*. Incl. *O. ramosissima* Desf.
serrata Forsskål
• A wool casual. N Africa, Crete, SW Asia. K. 21, 44*, 45*, 46*, 667*.
viscosa L.
Pre-1930 only. Medit. OXF. 7, 21, 44*, 45*, 354(5:18), 546*.

MELILOTUS Miller
albus Medikus   White Melilot
••••• An established wool and bird-seed alien, and an escape or relic of
cultivation; locally abundant, especially in the south; increasing. Eurasia;
(widespread). 1, 14, 19, 20*, 22*, 26*, 33*,. 235.   *M. leucanthus* Koch
ex DC.; *M. vulgaris* Willd.; *Trifolium germanicum* Smith; *T. leucanthum* auct.,
non M.Bieb.
[*altissimus* Thuill. Accepted, with reservations, as native.]
elegans Salzm. ex Ser.
Pre-1930 only. Medit. 2, 21, 39*, 44*, 46*, 300(1905:234).
indicus (L.) All.   Small Melilot
•••• An established wool, grain, spice, bird-seed and granite alien; locally
plentiful, especially in the south. Eurasia; (Australasia). 1, 14, 19, 20*, 22*,
33*, 35*, 182. *M. parviflorus* Desf.; *M. rugulosus* Willd.; *M. tommasinii*
Jordan; *Trifolium indicum* L.
infestus Guss.
• A bird-seed and grain casual. W Medit. BM, OXF, RNG, SLBI. 2, 19,
21, 46*, 160, 300(1907:39), 354(8:732), 546*, 683*.
messanensis (L.) All.   Sicilian Melilot
• A bird-seed casual. Medit. BM, CGE, LIV, NMW, OXF, RNG. 1, 9, 19,
21, 39*, 44*, 191, 371(343:30), 683*, 737*. *M. sicula* (Turra) B.D.Jackson;
Trifolium messanense L.
neapolitanus Ten.
Pre-1930 only. Medit. 21, 318(15:146), 546*, 571*, 695*. *M. gracilis* DC.
officinalis (L.) Pallas   Ribbed Melilot
••••• An established bird-seed alien and escape from cultivation; widespread
and locally frequent, especially in the south and east. Eurasia. 1, 19, 20*,
22*, 26*, 33*. *M. arvensis* Wallr.; *M. petitpierreanus* Willd.; *Trifolium
officinalis* L.
polonicus (L.) Pallas
Pre-1930 only (but confused with *M. wolgicus*). Russia. OXF. 7, 21,
354(5:554), 630*, 861*(29:t.6). *Trifolium polonicum* L.

sulcatus Desf.   Furrowed Melilot
•• A wool and bird-seed casual.   Medit.   BM, CGE, E, LANC, OXF, RNG.
1, 14, 19, 20*, 31*, 39*, 44*, 198.   *Medicago sulcata* sphalm.
wolgicus Poiret
Pre-1930 only.   Russia.   OXF, RNG.   7, 21, 354(5:554), 595b*, 630*,
861*(29:t.6), 1242*.   *M. ruthenicus* (Bieb.) Ser.

TRIGONELLA L.
arabica Del.
• A grain casual.   No modern records.   N Africa, SW Asia.   LCN, OXF,
SLBI. 1, 11, 44*, 160, 300(1907):39), 354(9:110; 12:38), 667*, 729*, 1305*.
aurantiaca Boiss.
• A wool and grain casual.   No modern records.   SW Asia.   BM, RNG.
2, 11, 14, 39*, 300(1904:237; 1905:99), 1305*.
caelesyriaca Boiss.
• A wool casual.   Turkey, Syria, Iraq.   K, SLBI.   7, 9, 39*, 44*,
335(116:140), 350(28:22; 29:107), 354(2:499; 13:108), 448, 667*.   *T. aleppica*
Boiss. & Hausskn.
caerulea (L.) Ser.   Blue Fenugreek
•• (but confused with *T. procumbens*).   A wool, grain and bird-seed casual.
Origin obscure, probably derived from *T. procumbens*.   ABD, CGE, E, LCN,
OXF, RNG.   1, 14, 19, 20*, 22*, 29*, 371(352:27), 1242*.   *Melilotus
caerulea* (L.) Desr.; *Trifolium caeruleum* L.
caerulescens (M.Bieb.) Hal.
Pre-1930 only.   SE Europe, SW Asia.   OXF, SLBI.   2, 21, 39*,
300(1906:102), 354(5:102), 1305*.   *T. azurea* C.Meyer.
corniculata (L.) L.   Sickle-fruited Fenugreek
• A spice and bird-seed casual.   Medit.   BM, CGE, E, LIV, NMW, OXF.
1, 19, 20*, 39*, 44*, 156, 236, 370(13:134), 459.
crassipes Boiss.
Pre-1930 only.   Turkey, Iraq, Iran.   2, 39*, 300(1907:39), 667*, 1305*.
fischeriana Ser.
Pre-1930 only.   Crimea, SW Asia.   OXF.   7, 21, 39*, 354(4:223,406), 1305*.
Medicago fischeriana (Ser.) Trautv.
foenum-graecum L.   Fenugreek
••• A persistent grain, bird-seed and spice alien.   Medit.   BM, LANC, LCN,
LIV, OXF, RNG.   1, 9, 13, 19, 20*, 22*, 25*, 44*, 126, 737*.
gladiata Steven ex M.Bieb.
• A casual in a garden, vector unknown.   No modern records.   Medit.,
SW Asia.   BON, LIV.   1, 21, 39*, 45*, 46*, 195, 272, 278, 300(1907:39),
546*, 737*.   *T. prostrata* DC.   The Bristol record was in error for
T. foenum-graecum.   9.
hamosa L.   Egyptian Fenugreek
•• A grain and wool casual.   N Africa, SW Asia.   BM, CLE, NMW, OXF,
RNG, SLBI.   1, 44*, 182, 303*(33:22-23), 335(116:140), 350(29:431), 729*,
1054*.   Incl. *T. uncata* Boiss. & Noë.

incisa Benth.
- A wool casual at East Ardsley (SW Yorks), and on soil dredged from the river at Preston Crowmarsh (Oxon), vector unknown. SW Asia. OXF, HbEJC. 7, 335(108:33), 354(5:18), 371(334:23), 1305*.

kotschyi Fenzl ex Boiss.
Pre-1930 only. Turkey. BM, OXF, SLBI. 7, 39*, 354(3:17), 667*, 1305*.

laciniata L.
- A casual in a newly rotovated garden at Sunderland (Co Durham), possibly from infilled ballast. Formerly a grain and ballast alien. Egypt, Syria. NMW, RNG. 1, 4, 166, 245, 300(1906:102; 1907:39), 303(14:16), 320(4:149), 527*, 729*.

maritima Del. ex Poiret
- A wool casual. Medit., SW Asia. OXF. 1, 21, 44*, 46*, 303(9:18), 354(2:499; 6:277).

monantha C.Meyer
Pre-1930 only. SW Asia. 1, 39*, 44*, 136, 1242*. Incl. *T. noeana* Boiss.

monspeliaca L.   Star-fruited Fenugreek
- A tan-bark casual. Medit. CLE, NMW, RNG. 1, 21, 22*, 39*, 44*, 195, 300(1904:109), 303*(39:6,7), 354(6:723; 7:177), 737*.

polyceratia L.
- A grain casual. W Medit. BM, LIV, NMW, OXF, RNG, SLBI. 1, 9, 21, 45*, 46*, 371(352:27), 546*. *T. pinnatifida* Cav.

procumbens (Besser) Reichb.
- • (but confused with *T. caerulea*). A grain casual. SE Europe, SW Asia. BM, LIV, OXF, RNG, SLBI. 1, 8, 9, 21, 31*, 39*, 1242*. *T. besseriana* Ser.

stellata Forsskål
- A wool casual. N Africa, SW Asia. 44*, 46*, 335(116:140), 542*, 667*, 729*, 734*, 1271*.

tenuis Fischer ex M.Bieb.
Pre-1930 only. SE Europe, SW Asia. OXF. 7, 21, 39*, 354(6:18). *T. striata* auct., non L.f.

MEDICAGO L.

arabica (L.) Hudson **alien variants**
Var. *longispina* Rouy has been recorded pre-1930 as an alien, but seems unworthy of recognition. 2, 192, 354(8:22).

arborea L.   Tree Medick
- A persistent garden escape on a cliff at Wain's Hill, Clevedon (N Somerset). Medit., Canary Is. 20, 24*, 31*, 32*, 39*, 303(14:14), 350(33:28), 358(1874:43).

blancheana Boiss.
- A grain casual. Formerly a wool casual. No modern records. E Medit. BM. 1, 11, 21, 44*, 1142*. Incl. *M. bonarotiana* Arcang.

coronata (L.) Bartal.
   Pre-1930 only. Medit., SW Asia. 7, 21, 39*, 44*, 166, 300(1909:41), 683*.
disciformis DC.
   • A wool casual. Medit. 2, 21, 31*, 39*, 300(1907:39), 448, 683*.
doliata Carmign.
   Pre-1930 only. Medit. BM, NMW, RNG. 2, 9, 21, 44*, 45*, 46*, 156,
   300(1904:237). *M. aculeata* Gaertner; *M. turbinata* Willd., non (L.) All.
glutinosa M.Bieb.
   • Reported from grassy banks near the railway at Northfleet (W Kent).
   S Europe, Turkey, Caucasus. 21, 39, 458, 546*, 683, 1123. *M. glomerata*
   Balbis; *M. sativa* subsp. *glomerata* (Balbis) Tutin.
intertexta (L.) Miller   Hedgehog Medick
   •• A wool and bird-seed casual. Medit. BM, E, LTN, NMW, OXF, RNG.
   1, 14, 19, 21, 39*, 46*,303(14:15). Incl. *M. ciliaris* (L.) Krocker; *M. echinus*
   Lam. ex DC.
laciniata (L.) Miller   Tattered Medick
   ••• A persistent wool, cotton and grain casual; probably also a tan-bark
   casual. N Africa, SW Asia; (Australia). BM, CGE, E, LIV, OXF, RNG.
   1, 3*, 11, 14, 20*, 44*, 303*(35:16-17). Incl. *M. aschersoniana* Urban.
littoralis Rohde ex Lois.   Shore Medick
   • A casual on sand and granite heaps at Gloucester docks (W Gloucs).
   Formerly a ballast and wool alien. Medit., SW Asia. NMW, OXF. 1, 3, 21,
   44*, 45*, 46*, 182, 263, 303(10:16). *M. striata* Bast.
marina L.   Sea Medick
   Pre-1930 only. Medit., Black Sea. NMW. 1, 21, 22*, 29*, 39*, 44*,
   195(p.784), 263.
minima (L.) Bartal. **alien variants**
   Several variants have been recorded, but most records are of var. *longiseta* DC.
   (var. *longispina* Lowe), a wool casual. According to 1142*, the correct name
   is var. *minima*, and our native plant is var. *brevispina* Benth. 303*(35:16-17).
murex Willd.
   • A wool casual. No modern records. Medit. BM, NMW, OXF, RNG,
   SLBI. 1, 21, 39*, 44*, 121, 191, 354(6:605). *M. sphaerocarpos* Bertol.
noeana Boiss.
   • A casual at Bristol, vector unknown. No modern records. E Turkey, Iraq,
   Syria. 39*, 354(11:472), 1142*.
orbicularis (L.) Bartal.   Button Medick
   •• (but confused with *M. tornata*). A grain and tan-bark casual. Medit.,
   SW Asia. BM, CGE, LCN, OXF, RNG, SLBI. 1, 21, 24*, 29*, 39*, 44*,
   122, 160, 213, 354(10:93; 11:653). Incl. *M. marginata* Willd.
[*platycarpa* Trautv. A dubious record. 2, 300(1905:99).]
polymorpha L. **alien variants**
   Several variants have been recorded, some at specific level, but, since the
   variation is continuous, none seem worthy of retention even as varieties.

praecox DC. Early Medick
●●● A wool casual. Medit. BM, CGE, E, LANC, OXF, RNG. 3*, 14, 20*, 39*, 251, 303*(35:16-17).
prostrata Jacq.
Pre-1930 only. C & SE Europe. 7, 21, 166, 354(2:414), 683*.
radiata L.
Pre-1930 only. SW Asia. 7, 39*, 44*, 136, 667*. *Trigonella radiata* (L.) Boiss.; *Pocockia radiata* (L.) Trautv.
rigidula (L.) All.
• A casual at Avonmouth docks (W Gloucs), vector unknown. No modern records. Formerly a ballast alien. Medit. BM, NMW. 1, 21, 44*, 136, 166, 281, 350(29:23), 543*, 683*. *M. agrestis* Ten. ex DC.; *M. gerardii* Waldst. & Kit. ex Willd.
rotata Boiss.
Pre-1930 only. SW Asia. OXF. 7, 39*, 44*, 354(5:554), 1142*.
rugosa Desr.
Pre-1930 only. Medit. 2, 21, 24*, 39*, 44*, 300(1904:238; 1906:102), 354(4:261), 683*. *M. elegans* Jacq. ex Willd.; *Melilotus rugosa* sphalm.
sativa L. subsp. *falcata* (L.) Arcang. **alien variants**
Var. *diffusa* (Schur) ined. and var. *tenuifoliata* (Vuyck.) ined. have been recorded. 7, 9, 191, 245, 331(65:196), 354(3:17; 8:21).
sativa L. subsp. **sativa** Lucerne
●●●●● An established escape from cultivation throughout most of the British Is; widely naturalised in grassy places, especially in the south and east; introduced with grass seed on roadsides and becoming naturalised; also a bird-seed alien. Eurasia, N Africa. 1, 19, 20*, 22*, 33*, 35*. Hybrids or intermediates with native subsp. *falcata*, known as subsp. *varia* (Martyn) Arcang. (*M.* × *varia* Martyn; *M.* × *sylvestris* Fries) have been frequently recorded, especially in East Anglia.
scutellata (L.) Miller Snail Medick
• A wool casual. Medit., SW Asia; (Australia). BM, CLE, LIV, LTN, OXF, RNG. 1, 21, 31*, 39*, 44*, 45*, 121, 370(16:166). *M. scutellaris* sphalm.
soleirolii Duby
Pre-1930 only. W Medit.; (Crimea). OXF, RNG. 1, 4, 21, 300(1908:102), 546*, 683*, 1142*.
tenoreana Ser.
Pre-1930 only. S Europe. CGE. 1, 21, 121, 251, 546*, 683*, 1142*.
tornata (L.) Miller
• A casual at ports, vector unknown. Formerly a ballast alien. Medit. NMW, SLBI. 2, 9, 21, 44*, 112, 156, 546*, 683*. *M. italica* (Miller) Fiori subsp. *tornata* (L.) Emberger & Maire; *M. obscura* Retz.; *M. muricata* All.
truncatula Gaertner Strong-spined Medick
●●● A wool, tan-bark and grain casual. Medit., SW Asia; (Australia). BM, CGE, E, NMW, OXF, RNG. 2, 14, 20*, 39*, 44*, 277, 303*(35:16-17). Incl. *M. tribuloides* Desr. ex Lam.; *M. uncinata* Willd.

turbinata (L.) All.
- A grain casual. Medit. BM, NMW, OXF, SLBI. 2, 11, 21, 39*, 44*, 45*, 160. *M. tuberculata* (Retz.) Willd. Not to be confused with *M. turbinata* Willd. which is a synonym of *M. aculeata* Willd.

TRIFOLIUM L.
affine C.Presl
- A tan-bark casual. SE Europe, Turkey. BM, HbLJM. 21, 39*, 212, 303(10:12; 35:10; 45:20), 747*. *T. preslianum* Boiss., nom. illeg.

albidum Retz.
Pre-1930 only. Origin obscure. SLBI. 7, 9, 139, 192, 267, 557, 697. This taxon is not discussed in 747; it may be only a variant of native *T. ochroleucum*.

alexandrinum L.   Egyptian Clover
- • (but confused with *T. constantinopolitanum*). A grass seed or bird-seed casual. E Medit.; (widespread). BM, OXF. 1, 19, 20*, 39*, 44*, 251, 331(70:160), 371(376:25). Possibly conspecific with *T. constantinopolitanum*.

alpestre L.   Mountain Zigzag-clover
- A casual in 1974 at St Peter's on Guernsey (Channel Is), vector unknown. Europe, SW Asia. 21, 22*, 26*, 31*, 39*, 357(19:367). The early Dublin record was an error. 1346.

angulatum Waldst. & Kit.
Pre-1930 only. E Europe. OXF. 7, 21, 354(5:281), 741*, 747*.

angustifolium L.   Narrow Clover
- • • A wool, tan-bark and bird-seed casual. Medit., SW Asia; (Australasia). BM, CGE, E, NMW, OXF, RNG. 2, 14, 19, 20*, 24*, 31*, 44*, 212. Incl. *T. infamiaponertii* Greuter; *T. intermedium* Guss., nom. illeg.

arvense L. **alien variants**
Var. *longisetum* (Boiss. & Bal.) Boiss. (*T. longisetum* Boiss. & Bal.) has been recorded as a wool casual, but is not separable from native material. 143, 354(7:178).

aureum Pollich   Large Trefoil
- • • • A long-persistent or established bird-seed, wool and agricultural seed alien; widespread in grassy places throughout the British Is. Europe, SW Asia; (New Zealand). 1, 14, 20*, 22*, 33*, 209, 303*(42:7). *T. agrarium* L., nom. inval.; *Chrysaspsis aurea* (Pollich) Greene.

badium Schreber   Brown Clover
- A grain casual. No modern records. C & S Europe. 7, 21, 22*, 23*, 27*, 29*, 160, 354(4:407).

beckwithii Brewer ex S.Watson   Beckwith's Clover
Pre-1930 only. Western N America. 41*, 42, 354(7:435), 747*, 1221*.

cernuum Brot.   Nodding Clover
- • A persistent wool and granite alien. SW Europe; (Australasia). BM, CGE, E, LTN, OXF, RNG. 3, 14, 20*, 37*, 182, 303(23:8), 546*, 683*.

cherleri L.
- A wool and tan-bark casual. Formerly a ballast alien. Medit.; (Australia). NMW, OXF, HbAB. 1, 4, 21, 24*, 37*, 39*, 44*, 303(23:8; 48:37), 371(379:26).

ciliolatum Benth.   Foothill Clover
Pre-1930 only. Western N America. OXF. 7, 41*, 42, 354(5:554), 747*, 886*(n.s. 1:152). *T. ciliatum* Nutt., non Clarke.

clypeatum L.
Pre-1930 only. E Medit. 2, 21, 39*, 44*, 300(1904:238; 1906:102), 1259*.

constantinopolitanum Ser.
- •• (but confused with *T. alexandrinum*). A wool, grain and tan-bark casual. E Medit. BM, LTN, NMW, OXF. 2, 39*, 44*, 145, 303(35:10), 370(1:42), 747*. *T. alexandrinum* var. *phleoides* (Boiss.) Boiss.

dasyurum C.Presl
Pre-1930 only. E Medit. RNG. 7, 21, 39*, 44*, 300(1905:99), 734*, 1271*. *T. formosum* Urv., non Savi.

depauperatum Desv.   Poverty Clover
Pre-1930 only. Western N America, Peru, Chile. 7, 41*, 42, 747, 1221*.

diffusum Ehrh.
- A wool casual. C & S Europe, Caucasus. BM, CGE, LSR, NMW, OXF, SLBI. 7, 21, 39*, 303(18:15), 354(2:282), 371(340:27; 403:39), 546*, 695*.

echinatum M.Bieb.   Hedgehog Clover
- •• A wool and bird-seed casual. E Medit., SW Asia. BM, E, LTN, LSR, OXF, RNG. 2, 14, 19, 20*, 39*, 44*, 683*. Incl. *T. carmeli* Boiss., *T. supinum* Savi.

fucatum Lindley
Pre-1930 only. Western N America. 2, 41, 42, 300(1905:99), 747*, 1221*.

gemellum Pourret ex Willd.
- A tan-bark casual. SW Europe, NW Africa. 21, 303(39:6), 737*, 747*.

gracilentum Torrey & A.Gray   Slender Clover
- Collected from the Seed Testing Station at Cambridge (Cambs), vector unknown. No modern records. Western N America, Mexico. CGE. 41*, 42, 747*, 1221*.

grandiflorum Schreber
- A tan-bark casual. Medit. HbLJM. 21, 39*, 212, 303(10:12; 35:10; 48:37), 737*, 747*, 1271*. *T. speciosum* Willd.

hirtum All.   Rose Clover
- ••• A wool, grain and tan-bark casual. Medit. ABD, BM, CGE, E, OXF, RNG. 2, 20*, 31*, 236, 303*(39:6-7), 371(367:29), 734*.

hybridum L.   Alsike Clover
- ••••• An established escape from cultivation, and a grain, bird-seed, wool and tan-bark alien; naturalised by roadsides throughout the British Is. S Europe, SW Asia; (widespread). 1, 19, 20*, 22*, 26*, 35*, 303(39:6). *Amoria hybrida* (L.) C.Presl; incl. *T. anatolicum* Boiss.; *T. elegans* Savi.

incarnatum L. subsp. **incarnatum**   **Crimson Clover**
   ●●●●● An escape or relic of cultivation, especially in the south; also a tan-bark casual; decreasing. S Europe. 1, 20, 22*, 33*, 213, 747*.

isthmocarpum Brot.
   Pre-1930 only. W Medit. OXF. 7, 9, 21, 45*, 46*, 354(5:281; 6:379), 737*.

lappaceum L.   **Bur Clover**
   ●●● A wool, bird-seed, grain and tan-bark casual. Medit., SW Asia; (Australia). BM, CGE, E, NMW, OXF, RNG. 1, 14, 19, 20*, 37*, 44*, 303(39:6), 371*(371:25). *T. latipaceum* sphalm.

leucanthum M.Bieb.
   ● A wool and tan-bark casual. Medit. BM, CGE, LTN, RDG. 21, 44*, 147, 303(37:15), 683*, 747*.

ligusticum Balbis ex Lois.
   ● A wool casual. Medit., Atlantic Is. RNG. 21, 683*, 737*, 747*.

michelianum Savi   **Annual White-clover**
   ● A wool casual. Formerly a grain alien. Medit. BM, RNG. 1, 4, 21, 22*, 46*, 144, 354(10:699), 683*.

microcephalum Pursh   **Small-headed Clover**
   Pre-1930 only. Western N America. 7, 41*, 42, 354(5:281), 747*, 1221*, 1242*.

mutabile Portenschlag
   Pre-1930 only. S Europe, Turkey. RNG, SLBI. 2, 21, 300(1905:99; 1907:39), 683*, 747*.

cf. **nervulosum** Boiss. & Heldr.
   ● A casual, probably from bird-seed. E Medit. RNG. 21, 667*, 747*. Possibly better treated as *T. glanduliferum* Boiss. var. *nervulosum* (Boiss. & Heldr.) Zoh.

nigrescens Viv.
   ● A grain and tan-bark casual. Medit. BM, NMW, OXF. 2, 21, 39*, 44*, 46*, 303(17:14; 39:6), 354(9:341), 370(2:106). Incl. *T. petrisavii* Clementi.

[*obscurum* Savi   A dubious record. 1333.]

pallidum Waldst. & Kit.
   ● A casual, perhaps a fodder alien. No modern records. C Europe, Medit. RNG. 21, 39*, 46*, 197, 737*, 747*.

pannonicum Jacq.   **Hungarian Clover**
   ●● A possible fodder alien, or perhaps introduced, persistent or established by railways; abundant on a railway bank south of Forty Green, near Beaconsfield (Bucks). Formerly a ballast alien. S Europe. BM, NMW, RNG. 2, 20*, 23*, 31*, 51b, 251, 332(20:34), 370(17:187), 371(355:25). Incl. *T. armenium* Willd.

patens Schreber
   ● A wool casual. C & S Europe, SW Asia. BM, CGE, LTN, OXF, RNG, SLBI. 7, 14, 21, 39*, 145, 354(5:375,648), 683*, 747*.

pilulare Boiss.
   ● A tan-bark casual. E Medit. 31*, 39*, 44*, 303(48:37), 734*, 1259*, 1271*.

pratense L. **alien variants**
Var. *americanum* Harz and var. *sativum* Sturm have been recorded as relics of cultivation. BM. 21.

purpureum Lois.   Purple Clover
 • A wool casual. Medit. LTN, RNG. 2, 21, 32*, 39*, 44*, 145, 182, 300(1907:39), 354(8:111), 546*.

resupinatum L.   Reversed Clover
 •••• A wool, grain, bird-seed, esparto and tan-bark alien established in a few places in the south; known since 1929 at Combwich (S Somerset). Medit, SW Asia; (Australia). 1, 9, 14, 19, 20*, 22*, 44*, 248, 251, 303(35:10). Incl. *T. suaveolens* Willd.; *Galearia resupinata* (L.) C.Presl.

retusum L.
 • A tan-bark and grass seed casual. C & S Europe, SW Asia. CGE, E, K, OXF, RNG. 2, 21, 39*, 212, 303*(39:6,7), 371(346:28). *T. parviflorum* Ehrh.

aff. **roussaeanum** Boiss.
 • A wool casual. No modern records. S Turkey. BM. 39*, 747*.

spumosum L.
 •• A tan-bark, grain and grass seed casual. Medit., SW Asia. BM, CGE, E, LCN, OXF, RNG. 1, 21, 39*, 44*, 46*, 303(17:14; 33:12; 39*:6,7).

squarrosum L.
 Pre-1930 only. Medit., Atlantic Is. 1, 21, 46*, 180, 354(7:565), 683*, 747*.

stellatum L.   Starry Clover
 • An established alien, known since 1804 on shingle at Shoreham (W Sussex); casual elsewhere, possibly a wool alien. Medit., SW Asia; (Australia). BM, LANC, NMW, OXF, RNG. 1, 20*, 22*, 24*, 33*, 174, 220, 354(12:274).

stipulaceum Thunb.
 • A wool casual. S Africa. RNG. 17, 747*.

subterraneum L. **alien variants**
Var. *oxaloides* (Bunge ex Nyman) Rouy (subsp. *oxaloides* Bunge ex Nyman) from SE Europe, Turkey and Caucasia has occurred frequently as a wool casual. BM, K. 370(1:118-119), 747.

tenuifolium Ten.
 Pre-1930 only. E Medit. SLBI. 7, 21, 39*, 354(4:11; 9:686), 683*, 747*.
 Possibly better treated as *T. bocconei* Savi var. *tenuifolium* (Ten.) Griseb.

thalii Villars
 Pre-1930 only. Europe, Morocco. 7, 21, 27*, 267, 683*, 747*.

tomentosum L.   Woolly Clover
 ••• A wool, esparto and tan-bark casual. Medit., Atlantic Is; (Australasia). BM, CGE, E, NMW, OXF, RNG. 1, 14, 20*, 24*, 31*, 44*, 212, 251.

vesiculosum Savi
 Pre-1930 only. S Europe, Turkey, Russia. 9, 21, 39*, 683*, 737*, 747*.

wormskioldii Lehm.   Springbank Clover
   Pre-1930 only. Western N America, Mexico. BM, OXF. 2, 41*, 42,
   300(1907:39), 747*, 1221*, 1270. *T. involucratum* Ortega, non Lam.;
   T. spinulosum Douglas ex Hook.

LOTONONIS (DC.) Ecklon & C.Zeyher
cf. **lupinifolia** (Boiss.) Benth.
   • A wool casual. S Spain, NW Africa. RNG, HbCGH. 21, 30*, 46.
   Amphinomia lupinifolia (Boiss.) Pau; *Leobordea lupinifolia* Boiss.

THERMOPSIS R.Br.
montana Nutt. ex Torrey & A.Gray   False Lupin
   • A persistent or established garden escape; naturalised for at least 25 years at
   the gravel pits at Oundle (Northants), now gone; also on the riverside at
   Canonbie (Dumfriess) and as a garden relic at Houbie, Fetlar (Shetland).
   Western N America. OXF, ZCM. 7, 20, 41*, 54*, 243, 253, 370(2:195),
   371(343:28), 1110*. *T. fabacea* auct., non DC.

BAPTISIA Vent.
sp.
   • Reported from an old tip at Bannerdown, Bath (N Somerset). HbEJC. 477.
tinctoria (L.) R.Br.   Wild-indigo
   • A garden relic long persistent on the site of Warley Place gardens (S Essex).
   Eastern N America. 38*, 40, 51, 313(31:370), 699*, 1110*.

LUPINUS L.
albus L.   White Lupin
   • A casual, probably imported for crop evaluation trials. E Medit. OXF.
   2, 20, 29*, 44*, 300(1906:102), 303(37:16; 48:35), 354(8:392), 683*, 737*.
angustifolius L.   Narrow-leaved Lupin
   •• A relic or escape from cultivation, and a grain and wool casual. Medit.;
   (Australasia). K, NMW, OXF, RNG. 1, 14, 20, 24*, 29*, 32*, 44*, 235,
   683*, 1242*. Incl. *L. linifolius* Roth; *L. reticulatus* Desv.
[*annuus* sphalm.   An unpublished name, perhaps in use as an aggregate
   commercial name for annual species of lupin. A wool casual. 251.]
arboreus Sims   Tree Lupin
   •••• An established escape from cultivation; introduced to stabilise dunes and
   for game cover, now naturalised on sandy ground by the coast and on railway
   banks, mainly in southern England and East Anglia; increasing. California;
   (Australasia). 12, 20, 22*, 34*, 41*.
luteus L.   Annual Yellow-lupin
   •• (but confused with *L. arboreus*). An escape or relic of cultivation persistent
   on sandy soils; also a wool casual. W Medit.; (Australasia). BM, RNG. 7,
   21, 22*, 29*, 30*, 44*, 145, 235, 683*.
micranthus Guss.   Bitter Blue-lupin
   • A casual escape or relic of cultivation. No modern records. Medit. NMW.
   7, 21, 32*, 44*, 235, 683*, 737*. *L. hirsutus* L.(1763), non L.(1753).

nootkatensis Donn ex Sims   Nootka Lupin
●●● An established escape from cultivation; abundantly naturalised on river
shingle in various places along the rivers Tay, Dee, Spey and Glass in
Scotland, and on gravelly heaths in Orkney and Shetland; known since 1862 by
the river Dee, having escaped from the grounds of Balmoral Castle; very rarely
recorded in England and NW Ireland.   NE Asia, W Canada.   BM, CGE, K,
LANC, OXF, RNG.   1, 20, 22*, 33*, 35*, 267, 319(11:347), 1242*.
L. perennis auct., non L.
[*perennis* L.   In error for *L. nootkatensis*. 2, 280.]

polyphyllus group   Garden Lupins
●●●●● Established garden escapes; locally abundant on railway banks and river
shingle in England and Scotland.   Western N America; the hybrids originating
in cultivation. 7, 20, 22*, 41*, 56*, 191. Many modern records probably
refer to Russell Lupins, *L. polyphyllus* Lindley × *L. arboreus* (*L.* × *regalis*
Bergmans, RNG), or backcrosses with *L. arboreus*.   *L. polyphyllus* × *L.
nootkatensis* (*L.* × *pseudopolyphyllus* C.P.Smith) occurs on river shingle at
Stanley and at Caputh (both Mid Perth) and *L.* × *regalis* × *L. nootkatensis*
may also occur there. 20, 370(14:113).

LABURNUM Fabr.

alpinum (Miller) Bercht. & J.S.Presl   Scottish Laburnum
●●● Introduced or a garden escape on roadsides and woodland edges, especially
in Scotland and Wales; reproducing in a few places, as in natural woodland
between Chislehurst and Sidcup (W Kent).   In use as a hedging plant in
Co Londonderry and Tyrone, but seedlings are rare.   Alps, Apennines,
Balkans. BM, E, LANC, NMW. 2, 20, 59*, 60*, 194, 305*(52:1,4-5),
319*(23:109-110), 331(64:115), 458, 606*, 631*.   *Cytisus alpinus* Miller;
L. scoticum sphalm.

anagyroides Medikus   Laburnum
●●●●● An established garden escape; widespread throughout the British Is,
sometimes naturalised and freely reproducing in quarries and on bushy waste
ground. C & S Europe. 2, 20, 34*, 59*, 631*.   *L. laburnum* (L.) Doerfler;
L. vulgare Bercht. & J.S.Presl; *Cytisus laburnum* L.   A hybrid with
L. alpinum (*L.* × *watereri* (Wettst.) Dippel; *L.*'Vossii') has been recorded.
335(113:78), 432.

CYTISUS Desf.

battandieri Maire   Pineapple Broom
● Recorded as self-seeding at Cheltenham (E Gloucs). Morocco. 51, 54*,
61*, 62*, 63*, 303(55:33).
[*cantabricus* (Willk.) Reichb.f.   In error for *C. striatus*. LIV. 371(374:8;
379:25).]

multiflorus (L'Hér. ex Aiton) Sweet   White Broom
●● An established garden escape; naturalised on a few banks and commons in
Wales and SE England; known since 1957 on a railway bank at Dorking
(Surrey). SW Europe. OXF, RNG. 13, 20*, 30*, 61, 371(352:28), 684*,

826*(22:14-17).   *C. albus* (Lam.) Link, non Hacq.   A hybrid with native
C. scoparius (*C.* × *dallimorei* Rolfe) has been recorded.   RNG.   370(15:169).

nigricans L.   Black Broom
    •• A garden escape established on waste ground at the disused marshalling
    yard at Feltham (Middlesex), in old gravel pits at Aylesford and on waste
    ground at Forstal (both E Kent), and on a roadside at Inverpolly National
    Nature Reserve (W Ross).   Europe.   20*, 22*, 26*, 29*, 236, 303*(26:20-21),
    371(340:28; 364:29), 480.   *Lembotropis nigricans* (L.) Griseb.

striatus (Hill) Rothm.   Hairy-fruited Broom
    ••• Introduced on roadside banks, and now well established and spreading in
    a few widely scattered localities; abundant between Aberystwyth and Penparcau
    (Cards).   SW Europe.   BM, E, NMW, RNG.   20*, 30*, 61, 303(20*:1,12;
    27:20),   325*(106:432-433),   370(12:49,170,191),   826*(22:14-17).
    C. grandiflorus hort., non DC.; *Genista striata* Hill; *Sarothamnus striatus* (Hill)
    Samp.; *S. welwitschii* Boiss. & Reuter.
[*villosus* Pourret (*C. triflorus* L'Hér.)   In error for *Genista monspessulana*.
    358(1875:115,163).]

CHAMAECYTISUS Link
purpureus (Scop.) Link   Purple Broom
    Pre-1930 only.   Europe.   21, 23*, 29*, 55*, 63*, 220.   *Cytisus purpureus*
    Scop.

supinus (L.) Link   Clustered Broom
    • Introduced on a roadside near Avon (S Hants) and on the banks of the Birtley
    by-pass (Co Durham).   Europe, SW Asia.   21, 22*, 29*, 30*, 31*, 167,
    371(340:28), 546*.   *Cytisus capitatus* Scop.; *C. supinus* L.

SPARTIUM L.
junceum L.   Spanish Broom
    ••• An established garden escape and introduction; often planted in quantity
    and reproducing freely on banks by railways and new roads in southern
    England, and naturalised on coastal cliffs between Jaywick and Frinton
    (N Essex); increasing.   Medit.   BM, LANC, NMW, OXF, RNG.   1, 20*, 22*,
    29*, 31*, 126, 683*, 1330.

GENISTA L.
aetnensis (Raf. ex Biv.) DC.   Mount Etna Broom
    • A persistent or established garden escape; naturalised on Chobham Common
    (Surrey) and on waste ground at Rosherville, near Northfleet (W Kent).
    Sardinia, Sicily.   HbKWP.   20*, 51*, 58*, 199, 303(64:47), 371(397:35), 458,
    683*.

florida L.
    • Introduced on the banks of the M1 motorway (Beds).   Portugal, Spain,
    Morocco.   21, 30*, 61, 63*, 303(27:21).

hispanica L.   Spanish Gorse
    ••• An established garden escape or relic in a few places in the south;
    naturalised since 1927 on Constitution Hill, Aberystwyth (Cards).   SW Europe.

BM, NMW, OXF. 7, 20*, 27*, 29*, 30*, 370(12:170,191). All records appear to be referable to subsp. *occidentalis* Rouy (*G. occidentalis* (Rouy) Coste).

linifolia L. Needle-leaved Broom
• A persistent garden relic. No modern records. W Medit., Canary Is. 21, 30*, 63*, 403, 737*. *Teline linifolia* (L.) Webb & Berth.; *Cytisus linifolius* (L.) Lam.

monspessulana (L.) L.Johnson Montpellier Broom
••• An established garden escape; naturalised by railways, on roadside banks, waste ground and in old gravel pits, mainly in SE England and the Channel Is. Medit., Azores. BM, CGE, LIV, OXF, RNG. 7, 20*, 30*, 63*, 191, 199, 303(20*:13; 21:19). *Cytisus candicans* (L.) DC.; *C. monspessulanus* L.; *C. monspeliensis* sphalm.; *Teline monspessulana* (L.) K.Koch.

radiata (L.) Scop. Southern Greenweed
• A garden escape recorded from Poole (Dorset) and from Durham. C & SE Europe. 21, 23*, 27*, 29*, 63*, 164, 349(3:259).

sagittalis L. Winged Greenweed
• Introduced. No modern records. C & SE Europe. BM. 21, 22*, 23*, 27*, 312*(n.s. t.332), 349(1:320), 546*, 683*. *Chamaespartium sagittale* (L.) P.Gibbs; *Genistella sagittalis* (L.) Gams.

tenera (Jacq.) Kuntze Madeira Broom
• A casual or persistent garden escape. Madeira, Canary Is. RNG. 51, 61, 63*, 199, 312*(t.2265), 546*, 683*. *G. cinerea* hort., non DC.; *G. virgata* (Aiton) Link, non Lam.; *Cytisus tener* Jacq.

tinctoria L. **alien variants**
Var. *elatior* (Koch) F.Schultz (*G. elatior* Koch), OXF; var. *ovata* (Waldst. & Kit.) F.Schultz (*G. ovata* Waldst. & Kit.); and var. *virgata* Koch (*G. virgata* Willd., non Lam., non (Aiton) Link) have been recorded. 354(3:17; 9:554), 458.

ADENOCARPUS DC.
complicatus (L.) Gay
• Introduced. No modern records. Medit. SUN. 21, 29*, 30*, 63*, 166, 737*. *Spartium complicatum* L.; incl. *A. commutatus* Guss.

ELAEAGNACEAE
ELAEAGNUS L.
[*angustifolia* L. Unconfirmed; probably always in error for *E. umbellata*. 303(36:28; 37:16), 370(19:173).]

commutata Bernh. Silver-berry
• A garden escape, or perhaps introduced, in hedges at Milborne St Andrew and Stourpaine (both Dorset). N America. RNG. 51*, 61*, 63*, 312*(t.8369), 1284*. *E. argentea* Pursh, non Moench. *E.*'Quicksilver' (probably of *E. angustifolia* × *E. commutata* origin) may have been overlooked; this taxon appears to have partly superseded true *E. commutata* in gardens. 325*(118:76-77).

glabra Thunb.
- A relic of cultivation in the Channel Is. China, Japan. 43, 51, 61, 220, 617*.

macrophylla Thunb.
- A garden relic on the site of Warley Place gardens (S Essex) and introduced in hedges in Dorset; seedlings reported from waste ground in Middlesex and W Kent. Korea, Japan. K. 43, 51, 52*, 61, 371(407:41), 403, 458.

multiflora Thunb.   Cherry Elaeagnus
- Introduced at Poole and on the sea-wall at Hamworthy (both Dorset). China, Japan. RNG. 43, 51*, 61*, 63*, 403, 617*.

pungens Thunb.
- Introduced or a garden relic, long persistent in Guernsey; also reported from Chislehurst Common (W Kent) and Chobham Common (Surrey). Japan. RNG. 43, 52*, 61, 63*, 371(355:27), 458. Incl. *E. simonii* hort. ex Carrière.

umbellata Thunb.   Elaeagnus
- • A persistent garden escape; in hedges at Amroth (Pembs) and Kings Mills in Guernsey (Channel Is), and in an old Bargate sandstone pit at Godalming (Surrey); seedlings have been reported from Littlebrook Marshes and Dartford Heath (both W Kent). Himalayas, China, Japan. LANC, NMW, RNG. 20, 43, 54*, 61, 156, 199, 220, 303(24:16), 617*, 685*, 1128*. Incl. *E. parvifolia* Royle.

HALORAGACEAE
Haloragidaceae

HALORAGIS Forster & Forster f.

micrantha (Thunb.) R.Br. ex Siebold & Zucc.   Creeping Raspwort
- Established in wet peaty heath near Lough Bola (W Galway), associated with *Juncus planifolius*. SE Asia, Australasia. BM, TCD. 20, 36*, 303(51:48), 544*, 649*, 1065*, 1170*, 1263*. *Gonocarpus micranthus* Thunb.

MYRIOPHYLLUM L.

aquaticum (Vell.Conc.) Verdc.   Parrot's-feather
- ••• Introduced and persistent in a few ponds and streams, mainly in the south, as at Duddleswell in Ashdown Forest (E Sussex) and at Horrabridge (S Devon); increasing. S America; (widespread). BM, JSR, LTR, NMW, RNG. 20, 52*, 87*, 174, 189, 303(36:29), 370(11:375-376; 12:259), 564*, 736*, 825(12:14-15), 847*(36:307-323), 1110*, 1347*. *M. brasiliense* Cambess.; *M. proserpinacoides* Gillies ex Hook. & Arn.

heterophyllum Michaux   Various-leaved Water-milfoil
- Introduced or an escape from aquaria, established for many years in a canal near Halifax (SW Yorks), probably now gone. Eastern N America; (Europe). BM, LANC, LIV, RNG. 16, 20, 21, 38*, 40*, 354(12:677; 13:160), 826(19:98-101), 1110*.

verrucosum Lindley   Red Water-milfoil
- A wool casual which persisted for a few years in gravel pits at Eaton Socon (Beds). No modern records. Australia. BM, LIV, LTN, OXF, RNG. 14, 20, 37*, 370*(1:63), 520*, 533*.

GUNNERACEAE

GUNNERA L.

manicata Linden ex André   Brazilian Giant-rhubarb
- - (but confused with *G. tinctoria*). Introduced and a long-persistent garden escape by lakes and rivers and in damp woodland in widely scattered localities; not spreading by seed. Colombia, S Brazil. NMW. 16, 20, 52*, 54*, 137, 220, 252, 303(9:17), 347(10:160-166), 354(12:278). *G. brasiliensis* Schindler.

peltata Philippi
- A persistent garden escape. No modern records. Polynesia. 354(11:253), 624*, 626*, 832*(9:13).

tinctoria (Molina) Mirbel   Giant-rhubarb
- - - (but confused with *G. manicata*). An established garden escape; widely naturalised by lakes and streams, in wet meadows, in woodland and on sea cliffs; abundant and spreading freely by seed in parts of the Channel Is and western Ireland. Chile. BM, CGE, OXF, RNG. 2, 20, 51a, 117*, 220, 275*, 303(9:17), 347(10:160-166), 349(3:403), 650*, 1110*. *G. chilensis* Lam.; *G. scabra* ?auct., non Ruíz Lopez & Pavón.

LYTHRACEAE

LYTHRUM L.

alatum Pursh   Angled Purple-loosestrife
- A casual garden escape. Eastern N America. 38*, 40, 51, 197, 223, 1319*.

junceum Banks & Sol.   False Grass-poly
- - - - A bird-seed, grain and wool casual, and garden escape. Medit., SW Asia; (New Zealand). 1, 3*, 14, 19, 20, 30*, 31*, 197, 303*(22:15). *L. acutangulum* auct., non Lag.; *L. flexuosum* auct., non Lag.; *L. graefferi* Ten.; *L. monanthum* Link ex Steudel; *L. griffithii* sphalm.

virgatum L.   Slender Purple-loosestrife
Pre-1930 only. C & E Europe. 21, 29*, 52*, 210, 300(1907:56-57). A 1904 record from near Cowley (Middlesex) appears to be in error for *L. alatum* × *L. salicaria* L. BM.

CUPHEA P.Browne

× **purpurea** hort. ex Lemaire
- A casual greenhouse escape. Originated in cultivation. 51, 54, 199, 593*, 1011*. *C. llavea* Lex. × *C. procumbens* Cav.; *C. miniata* hort., ?non Brongn.

THYMELAEACEAE

THYMELAEA Miller

passerina (L.) Cosson & Germ.   Annual Thymelaea
  Pre-1930 only. Eurasia, N Africa. 7, 21, 22*, 23*, 25*, 354(6:43).

MYRTACEAE

LEPTOSPERMUM Forster & Forster f.

lanigerum (Aiton) Smith   Woolly Tea-tree
  • A garden escape established in Abbey Wood on Tresco (Scilly). Australia,
  Tasmania. 20, 37*, 52*, 61, 63*, 207, 569*. Incl. *L. pubescens* Lam.

liversidgei R.Baker & Henry Smith   Lemon-scented Tea-tree
  • A garden escape; seedlings reported from Abbey Wood on Tresco (Scilly).
  Australia.   51,   61,   458,   590*,   725*,   821*(18:t.3),   829*(39:t.2),
  840*(n.s. 5:t.12).

scoparium Forster & Forster f.   Broom Tea-tree
  • A garden escape well established in Abbey Wood on Tresco (Scilly);
  seedlings reported from walls in Penzance (W Cornwall). Australasia. 20, 36,
  52*, 61*, 207, 458, 514*, 680*.

sericeum Labill.   Swamp Tea-tree
  • A garden escape; seedlings reported from Abbey Wood on Tresco (Scilly).
  Australia. 51, 458, 514, 533*, 569*, 590*, 743, 1278*. *L. myrtifolium* Sieber
  ex DC.

EUCALYPTUS L'Hér.

coccifera Hook.f.   Tasmanian Snow-gum
  • Introduced. Tasmania. LANC. 51, 60*, 166, 549*.

dalrympleana Maiden   Broad-leaved Kindlingbark
  • Introduced. E Australia, Tasmania. 20, 58*, 166, 549*, 569*.

globulus Labill.   Southern Blue-gum
  • Introduced or a garden escape on Herm (Channel Is); introduced for forestry
  in Ireland and occasionally self-sown. Tasmania. 20, 32*, 36c*, 51b*, 58*,
  61, 349(3:324).

gunnii Hook.f.   Cider Gum
  • Introduced; abundant in woods and by roadsides at Brightlingsea (N Essex).
  Tasmania. CGE, RNG. 20, 36c*, 51b*, 58*, 59*, 60*, 61, 191, 166.

parvifolia Cambage   Small-leaved Gum
  • Introduced. SE Australia. 166, 531*, 547*, 576*.

pulchella Desf.   White Peppermint-gum
  • A garden escape established from self-sown seed in Abbey Wood on Tresco
  (Scilly). Tasmania. BM. 20, 36c*, 51a, 61, 207. *E. linearis* Dehnh.

LUMA A.Gray

apiculata (DC.) Burret   Chilean Myrtle
  •• An established garden escape; in hedges and on roadsides at Glengarriff
  (W Cork); abundantly self-sown in woodland gardens on the Lizard
  (W Cornwall), Tresco (Scilly) and Valentia Is (S Kerry); increasing. Chile,
  Argentina.   20, 54*, 58*, 60*, 61, 63*, 319(19:369), 325*(118:42-44),

334(2:8), 370(9:342), 371(403:40). *Eugenia apiculata* DC.; *Myrceugenia
apiculata* (DC.) Niedenzu; *Myrtus luma* auct., non Molina. Much confused
nomenclaturally, being incorrect in 20 and 1177; this is the prolific seeder that
usually escapes, although one or two other species may also occur, viz.
Amomyrtus luma (Molina) Legrand & Kausel; *Myrtus lechleriana* (Miq.) Sealy;
M. luma Molina; *M. luma* Barnéoud, non Molina, non hort.; *M. luma* hort.,
non Molina. 436.

UGNI Turcz.
molinae Turcz.   Chilean Guava
• A garden escape; seedlings reported from woodland on Tresco (Scilly).
Bolivia, Chile. 36c*, 51, 61, 63*, 458. *Myrtus ugni* Molina.

PUNICACEAE
PUNICA L.
granatum L.   Pomegranate
• A casual. No modern records. SW Asia; (widespread). 21, 29*, 32*, 68*,
80*, 248, 354(11:479).

ONAGRACEAE
EPILOBIUM L.
alsinoides Cunn.
Pre-1930 only. New Zealand. 3*, 36, 36c, 126, 370(9:142), 651*. Post-
1930 records are probably in error for *E. pictum*.
brunnescens (Cockayne) Raven & Engelhorn   New Zealand Willowherb
••••• An established garden escape; well naturalised on damp, stony ground,
mainly in the north and west; an aggressive weed in gardens; increasing.
New Zealand. 2, 20*, 22*, 25*, 33*, 370(9:140-142). *E. nerteroides* auct.,
non Cunn.; *E. nummulariifolium* auct., non Cunn.; *E. pedunculare* auct., non
Cunn. Hybrids with *E. ciliatum* and native *E. montanum* have been recorded
from Co Antrim. 20, 1292.
ciliatum Raf.   American Willowherb
••••• Possibly originally a timber alien, established and locally abundant as
a weed throughout the British Is; naturalised in woods, on commons and
railway banks, especially in SE England; increasing. N America. 12, 20*,
22*, 33*, 35*, 370(17:279-288). *E. adenocaulon* Hausskn. Hybrids with the
following native species have been recorded: *E. hirsutum* (*E.* × *novae-civitatis*
Smejkal); *E. lanceolatum*; *E. montanum*; *E. obscurum*; *E. palustre*
(*E.* × *fossicola* Smejkal, nom. nud.); *E. parviflorum*; *E. roseum*
(*E.* × *nutantiflorum* Smejkal, nom. nud.); *E. tetragonum* (incl. *E. lamyi*)
(*E.* × *iglaviense* Smejkal, nom. nud.). BM, E, LANC, LSR, NMW. 20.
[*collinum* C.Gmelin In error. 354(5:763).]

komarovianum A.Léveillé   Bronzy Willowherb
•• A persistent garden escape and lawn weed. New Zealand. DBY, K, RNG.
20*, 22*, 36c, 166, 212, 370(9:274), 371(355:26), 1228*, 1362*.
E. inornatum Melville.

pedunculare Cunn.   Rockery Willowherb
•• (but confused with *E. brunnescens*). An established garden escape;
abundantly naturalised on roadsides near Leenane (W Galway). New Zealand.
ABD, BM, CGE, DBN, E, RNG. 20, 22*, 36, 36c, 349(1:37), 370(9:140-
142), 1363*. *E. linnaeoides* Hook.f.

pictum Petrie
• A garden weed. New Zealand. RNG. 36, 36c, 126, 370(9:142), 1362*.

ZAUSCHNERIA C.Presl
californica C.Presl   Californian Fuchsia
• A garden relic persistent for many years at Sand Point, Kewstoke
(N Somerset). California, Mexico. HbEJC. 42, 51*, 54*, 61, 312(n.s. t.19),
370(15:62). *Z. cana* E.Greene; *Z. mexicana* C.Presl; *Epilobium canum*
(E.Greene) Raven.

[*CHAMERION* (Raf.) Raf.
dodonaei (Villars) Holub Dubious records. LIV. 280. *Chamaenerion dodonaei*
(Villars) Schur; *C. rosmarinifolium* Moench; *Epilobium dodonaei* Villars;
E. rosmarinifolium Haenke.]

OENOTHERA L.
affinis Cambess.
• A wool casual. S America; (Australia). BM, CGE, E, RNG. 21, 36c, 37*,
805*(64:425-626). *O. mollissima* auct. pro parte, non L.
[*ammophila* Focke In error for *O. cambrica*. 370(14:1-34).]

biennis L.   Common Evening-primrose
••••• An established garden escape; locally naturalised in many places in
England, Wales and southern Scotland; known since 1805 on the coastal dunes
of S Lancs. Europe, E Asia. 1, 20*, 22*, 370*(14:17). *Onagra biennis* (L.)
Scop. Hybrids with *O. cambrica* (WAR, RNG); male *O. glazioviana*
(*O. × albivelutina* Renner, nom. nud.; BM, CGE, LTR, MANCH, OXF) and
multiple hybrids have been recorded. 20, 1218. Contrary to much literature,
this species is not known as native in N America.

cambrica Rostański   Small-flowered Evening-primrose
••••• An established garden escape; long naturalised on dunes and by
railways in many places in England, Wales and the Channel Is. Probably
N America. 7, 20*, 22*, 156, 201, 370*(14:13), 878*(23:285-293).
O. ammophila auct., non Focke; *O. novae-scotiae* auct., non Gates;
O. parviflora auct., non L. Hybrids with *O. biennis, O. fallax* and
O. glazioviana have been recorded. 20, 303(61:19-33).

canovirens Steele
• Recently recorded at Tillingdown Hill (Surrey); also casual in a garden,
possibly from bird-seed. N America; (Europe). K, LIV, MANCH. 21,

342(Feb91), 370*(14:26-27), 371(394:36).   *O. renneri* H.Scholz; *O. strigosa*
auct., non (Rydb.) Mackenzie & Bush.   The record from Borthwick
(Midlothian) was in error for *O. salicifolia*. According to 805(63:382-383), the
correct name is *O. villosa* Thunb., but this is contradicted by other authors.

[*chicaginensis* De Vries ex Renner & Cleland (*O. chicagoensis* Renner ex Cleland
& Blakeslee)   In error for *O. cambrica*. 370(14:7).]

[*coronifera* Renner   In error for *O. glazioviana*. 370(14:7).]

drummondii Hook.
   • A wool casual.   Texas, Mexico; (Australia).   RNG.   14, 44*, 51,
   312*(t.3361), 520*.

fallax Renner
   ••• Established, occurring both as a spontaneous hybrid and as a garden
   escape in a few widely scattered localities, especially on the coastal dunes of
   S Lancs. BM, CGE, E, NMW, OXF, RNG. 20*, 370*(14:21-22). A stable
   derivative of *O. glazioviana* (female) × *O. biennis* (male) origin.
   O. cantabrigiana B.Davis; *O. velutirubata* Renner. Hybrids with *O. cambrica*
   and backcrosses with both parents have been recorded. LANC. 20.

[*fruticosa* L. A dubious record. 300(1896:236,245).]

glazioviana Micheli ex C.Martius   **Large-flowered Evening-primrose**
   ••••• An established garden escape; locally naturalised on dunes and by
   railways in many places throughout the British Is. N America; (widespead).
   2, 20*, 22*, 370*(14:17). *O. erythrosepala* Borbás; *O. grandiflora* auct., non
   L'Hér.; *O. lamarckiana* auct., non Ser.; *O. vrieseana* Léveillé. Hybrids with
   O. cambrica (*O.* × *britannica* Rostański; ABS, WAR; 20*, 370*(14:20)),
   female *O. biennis* and multiple hybrids have been recorded.

[*grandiflora* L'Hér. In error for *O. glazioviana*. LANC. 370(14:7).]

humifusa Nutt.
   Pre-1930 only. Eastern N America. 38*, 40, 51, 251. *Raimannia humifusa*
   (Nutt.) Rose.

indecora Cambess.
   Pre-1930 only. S America. OXF. 7, 9, 16, 354(7:436; 8:210), 370(14:8),
   598, 805*(64:425-626), 896*(24:179), 904*(1921:99). *O. argentinae* Léveillé
   & Thell.; incl. *O. catharinensis* Cambess.

laciniata Hill   **Cut-leaved Evening-primrose**
   • A wool and oil-seed casual. USA, C & S America; (S Africa). BM, CGE,
   K, LTN, NMW, RNG. 2, 9, 14, 21, 22*, 38*, 134, 145, 370(14:30,453),
   577a, 1268*, 1294*. *O. sinuata* L.

[*lamarckiana* Ser. In error for *O. glazioviana*. 370(1-34).]

aff. **lipsiensis** Rostański & Gutte
   • Recorded from Walton Common (Surrey), vector unknown. Probably
   N America. K. 126, 199, 370(14:8), 859*(9:63-88).

[*longiflora* L. Unconfirmed. The BM, CGE, E & RNG vouchers are probably
all in error for *O. affinis*. 3.]

missouriensis Sims   Missouri Evening-primrose
- A garden weed at Capel Hendre, Ammanford (Carms). Southern USA.
NMW. 38*, 51, 305(55:20), 312*(t.1592), 566*, 594*, 1255*. *Megapterium missouriense* (Sims) Spach.

[*nuda* Renner ex Rostański   In error for an unidentified plant near to *O. victorinii*. CGE. 370(14:7).]

[*odorata* Jacq.   In error for *O. stricta*. 370(14:1-34).]

parviflora L.
- ●● (but over-recorded for *O. cambrica*). Locally plentiful in a chalk pit at Stone (W Kent), vector unknown. Eastern N America. BM, OXF. 22*, 370*(14:27-28). *O. pachycarpa* Renner ex Rudloff.

perangusta Gates
- ●● (but confused with *O. parviflora)*. A casual, vector unknown. Canada. BM, K, NMW, RNG, UCSW. 20, 339(16:238-249), 370*(14:24-25).

perennis L.   Small Sundrops
- A casual garden escape. Eastern N America. RNG. 1, 16, 38*, 51, 354(12:490), 371(376:25), 805*(64:381-424). *O. pumila* L.; *Kneiffia pumila* (L.) Spach.

rosea L'Hér. ex Aiton   Rosy Evening-primrose
- A persistent grain alien. USA, C & S America; (widespread). BM, BRISTM, CGE, LIV, NMW, SLBI. 1, 10, 20, 21, 22*, 29*, 303(18:13), 349(1:170), 370(14:29), 371(391:34), 546*, 598*, 707*. Easily mistaken for an *Epilobium* species.

rubricaulis Kleb.
- ●● (but confused with *O. parviflora*). Established in grassland over a large area between Swanscombe and Greenhithe (W Kent), elsewhere at docks and by railways, vector unknown. C & E Europe. BM, CGE, K, LIV, RNG. 2, 20, 21, 22*, 339(16:238-249), 370*(14:23-24), 458. *O. muricata* L. pro parte.

rubricuspis Renner ex Rostański
Pre-1930 only. N America. NMW. 21, 22*, 370*(14:28-29).

salicifolia Desf. ex G.Don
- (but confused with *O. canovirens*). Persisted for several years on a railway tip at Borthwick Bank, Edinburgh (Midlothian), vector unknown. N America; (Europe). BM, CGE, E, NMW, OXF, RNG. 370(14*:25-26; 15:131). *O. bauri* Boedijn; *O. depressa* E.Greene; *O. hungarica* Borbás; *O. multiflora* Gates; *O. strigosa* var. *depressa* (E.Greene) Gates; *O. villosa* sensu Raven & Dietrich, non Thunb.

stricta Ledeb. ex Link   Fragrant Evening-primrose
- ●●● An established wool alien and garden escape; naturalised, sometimes abundantly, on dunes in southern England and Wales and in the Channel Is. S America; (widespread). BM, CGE, E, NMW, OXF, RNG. 1, 14, 20*, 22*, 201*, 617*. *O. agari* Gates; *O. odorata* auct., non Jacq.; *O. striata* Ledeb. ex Link; *O. suaveolens* Desf.

[*strigosa* (Rydb.) Mackenzie & Bush   Unconfirmed. All records appear to be referable to the segregates *O. canovirens* or *O. salicifolia*. 339(16:245).]

tetraptera Cav.
- A casual at the John Innes Institute at Merton (Surrey), vector unknown. Mexico, C America; (widespread). CGE. 199, 370(14:30), 577a, 598*, 649*.

aff. **victorinii** Gates & Catcheside   Victorin's Evening-primrose
Pre-1930 only. Eastern N America. BM. 327*(49:t.26), 370(14:8), 1210*.
O. biennis auct., non L.

CLARKIA Pursh
amoena (Lehm.) Nelson & Macbr.   Godetia
- • A casual garden escape. Western N America. E, RNG. 3, 20, 41*, 54*, 56*, 66*, 199, 236, 500*. *C. grandiflora* hort.; *Godetia amoena* (Lehm.) G.Don; incl. *G. grandiflora* Lindley; *Oenothera amoena* Lehm.; *O. whitneyi* A.Gray.

bottae (Spach) F. & M.Lewis
Pre-1930 only. Western N America. 7, 42, 354(3:20), 500*, 670*. *Godetia bottae* Spach; *Oenothera bottae* (Spach) Torrey & A.Gray.

concinna (Fischer & C.Meyer) E.Greene   Red-ribbons
Pre-1930 only. California. 42, 52*, 237, 500*.

pulchella Pursh
- A casual garden escape. Western N America. NMW, OXF. 1, 41*, 51, 52*, 66, 263, 354(5:380), 370(15:169), 458, 500*.

purpurea (Curtis) Nelson & Macbr.
- A casual garden escape. Western N America. NMW. 1, 41*, 42, 51, 192, 300(1907:39), 354(12:41), 500*. *Godetia purpurea* (Curtis) G.Don; *Oenothera purpurea* Curtis; incl. *C. viminea* (Spach) Nelson & Macbr.; *G. viminea* Spach.

quadrivulnera (Douglas) Nelson & Macbr.
- A casual, possibly from bird-seed. Western N America. BM, NMW, OXF, SLBI. 1, 4, 41*, 42, 51, 354(7:185,572), 423, 500*. *Godetia quadrivulnera* Douglas; incl. *C. tenella* (Cav.) F.& M.Lewis; *G. tenella* (Cav.) Spach ex Steudel; *Oenothera tenella* Cav.

unguiculata Lindley   Clarkia
- •• A casual garden escape. Western N America. NMW. 7, 20, 42, 500*, 506*, 650*, 680*, 1011*. *C. elegans* Douglas, non Poiret.

FUCHSIA L.
'Corallina' hort. ex Lynch
- A garden escape established on Lundy Island (N Devon) and on Lleyn Peninsula (Caerns). Originated in cultivation. 51b, 436, 902*(13:26), 1101. *F. cordifolia* Benth. × *F. globosa* hort.; *F.*'Exoniensis' hort. ex Paxton.

excortica (Forster & Forster f.) L.f.   Tree Fuchsia
- Reported as seedlings in Cornwall. New Zealand. 36c, 51, 61, 63*, 301*(t.857), 458, 1209*, 1286*. *Skinnera excortica* Forster & Forster f.

× **hybrida** hort. ex Vilm.   Cultivated Fuchsia Hybrids
- Casual garden escapes. Originated in cultivation. OXF. 54*, 63*, 67*, 347(1:181-186), 458, 475, 1011*. A complex of hybrids involving several species.

magellanica Lam. sensu lato   Fuchsia
●●●● An established garden escape; naturalised in SW England, Wales,
Ireland, the Isle of Man and Scotland. S America. 2, 20, 54*, 63*, 65*,
303*(34:15). Incl. *F. coccinea* Curtis, non Sol.; *F.*'Corallina'; *F. gracilis*
Lindley; *F. macrostema* Ruíz Lopez & Pavón; *F.*'Riccartonii' hort. ex Tillery.

GAURA L.
coccinea Pursh   Scarlet Gaura
● A casual in 1935 at Pembery Burrows (Carms), vector unknown. No modern
records.   N America.   BM, NMW, OXF, SLBI.   38*, 41*, 51,
354(11:175,256).

LOPEZIA Cav.
coronata Andrews   Crown-jewels
● A casual garden escape. No modern records. Mexico, C America. OXF.
51, 52*, 354(12:41), 600*, 1011*, 1110*.

NYSSACEAE
Davidiaceae
DAVIDIA Baillon
involucrata Baillon   Dove-tree
● Introduced. SW China. 58*, 59*, 60*, 166.

CORNACEAE
CORNUS L.
alba L.   White Dogwood
●●● Introduced in woods and for game cover, perhaps established in a few
places. NE Europe, Siberia, China. BM, CGE, OXF, SLBI. 7, 20, 63*, 185,
220, 613*, 684*. *C. tatarica* Miller; *Swida alba* (L.) Holub; *Thelycrania alba*
(L.) Pojark. Confused with, and possibly conspecific with, *C. sericea*.
alternifolia L.f.   Pagoda Dogwood
● A garden relic on the site of Warley Place gardens (S Essex). Eastern
N America. 38*, 40, 61, 313(31:370).
canadensis L.   Creeping Dogwood
● A persistent nursery weed recorded from Seal (W Kent). NE Asia,
N America. 38*, 41*, 51, 54*, 62*, 371(346:27), 482. *Chamaepericlymenum
canadense* (L.) Asch. & Graebner.
capitata Wallich   Bentham's Dogwood
Pre-1930 only. Himalayas, China. 7, 58*, 61, 63*, 65*, 347*(1:85-105).
Benthamia fragifera Lindley; *Benthamidia capitata* (Wallich) Hara;
Dendrobenthamia capitata (Wallich) Hutch.
florida L.   Flowering Dogwood
● A garden relic at Thorndon Park (E or W Essex). Eastern N America. 51*,
56*, 59*, 61*, 61a, 191.
kousa Hance
● A casual garden escape. E Asia. 43, 61*, 61a, 63*, 458.
[*macrophylla* Wallich   In error for *C. sericea*.   199.]

mas L.   Cornelian-cherry

●●● An established garden escape; bird-sown, or introduced and naturalised, in hedges, woodland and scrub in a few widely scattered localities; abundant at Hargham (W Norfolk). Europe, W Asia. BM, CGE, LANC, RNG. 7, 20, 22*, 26*, 51*, 59*, 370(9:274). *C. mascula* L.

sericea L.   Red-osier Dogwood

●●●● An established garden escape; widely naturalised throughout the British Is in woods and by rivers, sometimes forming extensive dense thickets by suckering. N America. 1, 20, 22*, 38*, 41*, 156, 185, 370(18:33-36). *C. stolonifera* Michaux; *Swida sericea* (L.) Holub; *Thelycrania sericea* (L.) Dandy; *T. stolonifera* (Michaux) Pojark.

AUCUBA Thunb.

japonica Thunb.   Spotted-laurel

●●● An established garden escape; seedlings have been recorded from a few places in the south; naturalised over an area of several acres in woodland near Trevarrack (W Cornwall). E Asia. 20, 56*, 63*, 66*, 119, 236, 305(44:23), 458.

GRISELINIA Forster f.

littoralis (Raoul) Raoul   New Zealand Broadleaf

●● A garden escape; seedlings have been recorded from Penzance (W Cornwall), Rhu (Dunbarton), on sea cliffs near Kilmelford (Main Argyll), at the head of Killary harbour (W Galway), and unlocalised (Co Down). New Zealand. RNG. 20, 36, 52*, 61, 62*, 66*, 213, 371(349:27; 370:30), 437, 467, 1118*.

GARRYACEAE

GARRYA Douglas ex Lindley

elliptica Douglas ex Lindley   Silk-tassel

● A garden relic on the site of Warley Place gardens (S Essex), and introduced on rocks near the Observatory in Avon Gorge (W Gloucs). California, Oregon. 42, 51*, 52*, 61*, 65*, 312*(n.s. t.220), 313(31:370), 350(47:37), 371(349:27).

laurifolia Hartweg   Large-leaved Silk-tassel

● Introduced or a garden relic in Guernsey (Channel Is). Mexico. 51, 51b, 61, 371(355:27), 710*. Incl. *G. macrophylla* Hartweg; *G. oblonga* Benth.

SANTALACEAE

THESIUM L.

cf. **arvense** Horvátovszky

Pre-1930 only. Eurasia. 9, 21, 39, 683*. *T. ramosum* Hayne.

[*humile* Vahl   A dubious record. 262, 354(5:784).]

CELASTRACEAE
EUONYMUS L.
fortunei (Turcz.) Hand.-Mazz.
- A garden escape persistent for many years in Darenth Woods (W Kent). China. 61, 61a, 63*, 312*(n.s. t.181), 347(3:133-166), 371(414:13), 458. Incl. *E. radicans* Siebold ex Miq. forma *carrierei* (Vauvel) Rehder.

hamiltonianus Wallich sensu lato   Himalayan Spindle
- A garden escape persistent on a roadside on Thursley Common (Surrey). Asia. HbACL. 20, 61, 61a, 63*, 199, 312*(n.s. t.548), 347(3:133-166), 685*, 712*. Incl. *E. lanceifolius* Loes.; *E. sieboldianus* Blume; *E. yedoensis* Koehne.

japonicus L.f.   Evergreen Spindle
- ●●● An established garden escape; naturalised on cliffs and in woods in a few places in the south; increasing. Japan. BM, NMW, RNG. 12, 20, 61, 63*, 236, 347(3:133-166), 684*.

latifolius (L.) Miller   Large-leaved Spindle
- ●● Introduced, self-sown and now established in a few woods and hedges, mainly in southern England; known since 1904 at Hopetoun (W Lothian); increasing. SE Europe, SW Asia. BM, CGE, E, RNG, SLBI. 2, 20, 29*, 61*, 126, 300(1904:176), 347(3:133-166), 475, 684*.

planipes (Koehne) Koehne
- A garden escape, probably bird-sown in Maitland Wood at Edge (E Gloucs). E Asia. 43, 61, 62*, 63*, 323(27:162), 347(3:133-166). *E. latifolius* var. *planipes* Koehne; *E. sachalinensis* (F.Schmidt) Maxim. pro parte, non *E. latifolius* var. *sachalinensis* F.Schmidt.

CELASTRUS L.
orbiculatus Thunb.   Oriental Bittersweet
- A garden escape established on wooded roadside at Shottermill (Surrey) and on the edge of an abandoned garden at West Porlock (N Somerset). E Asia. BM. 51*, 54*, 61, 61a, 199.

AQUIFOLIACEAE
ILEX L.
× **altaclerensis** (hort. ex Loudon) Dallimore   Highclere Holly
- ●● (but overlooked as native *I. aquifolium*). An introduced or persistent garden escape well represented in Dunbarton. Originated in cultivation. NMW. 20, 58*, 59*, 60*, 61, 61a, 63*, 199, 331(69:142), 347*(5:65-81), 467. *I. aquifolium* L. × *I. perado* Aiton; incl. *I.*'Balearica'; *I.*'Camelliifolia'; *I.*'Mundyi'; *I.*'Wilsonii'.

cornuta Lindley & Paxton   Horned Holly
- A garden relic on the site of Warley Place gardens (S Essex). China, Korea. 51*, 61, 63*, 313(31:370), 543*.

BUXACEAE

SARCOCOCCA Lindley
confusa Sealy
- Established on the site of a former garden as a well-naturalised colony at Hextable (W Kent). Origin obscure. 51, 52*, 61, 63*, 327*(92:117-159), 458. *S. humilis* hort. & Stapf pro parte.

hookeriana Baillon
- A garden relic long persistent on the site of Warley Place gardens (S Essex) and at Hextable (W Kent). China. 51*, 61, 63*, 327*(92:117-159), 371(352:28), 458. Incl. var. *humilis* Rehder & Wilson; *S. humilis* Stapf pro max. parte.

BUXUS L.
balearica Lam.   Balearic Box
Pre-1930 only. W Medit. RNG. 21, 46*, 59*, 60*, 281.

EUPHORBIACEAE

SEIDELIA Baillon
cf. **mercurialis** Baillon
A casual in 1930 on a cinder heap at Dover (E Kent), vector unknown. S Africa. OXF. 354(9:571), 564*, 592, 1012*, 1098*. *S. firmula* (Prain) Pax & O.Hoffm.

CHROZOPHORA Adr.Juss.
tinctoria (L.) Adr.Juss.   Turn-sole
- A tan-bark casual. Medit., SW Asia. HbALG, HbEJC. 21, 24*, 29*, 44*, 303*(35:10-11). Possibly conspecific with *C. obliqua* (Vahl) Adr.Juss.

MERCURIALIS L.
annua L.   Annual Mercury
●●●●● An established grain alien, locally abundant in southern England, showing a preference for rich, garden soils; probably increasing. First recorded in 1548 as 'beginneth now to be known in gentleman's gardens'. An early colonist, which has been claimed as a native. Eurasia, N Africa. 1, 20, 22*, 33*, 35*, 1339.

RICINUS L.
communis L.   Castor-oil-plant
●●● An oil-seed and coir casual; also a short-lived garden escape. Africa; (widespread). BM, BON, MNE, NMW, OXF, RNG. 2, 20, 29*, 30*, 44*, 303*(24:17), 371(343:30), 825*(7:87-91).

EUPHORBIA L.
amygdaloides L. subsp. **robbiae** (Turrill) Stace   Leathery Wood-spurge
●● An established garden escape; naturalised on Bookham Common (Surrey) and Bromley Common (W Kent). Turkey. OXF. 20, 119, 199, 303*(40:26-27), 312*(n.s. t.208). *E. amygdaloides* var. *robbiae* (Turrill) R.-Smith; *E. robbiae* Turrill.

cf. **arguta** Banks & Sol.
- Collected from Beckenham Place Park (W Kent). No modern records. E Medit. RNG. 21, 44*.

"*carthusiania*"
- Reported from waste ground at Kew (Surrey). 342(Feb89). The correct identity of this plant is not known; no such name has been found. 463.

ceratocarpa Ten.   Horned Spurge
- Persisted for some years at Barry docks (Glam), vector unknown. No modern records. Italy, Sicily. LANC, NMW, OXF, RNG, SLBI. 7, 10, 20, 21, 51a, 354(7:998; 8:588), 683*.

characias L.   Mediterranean Spurge
- •• An established garden escape and relic known since 1953 on the site of a former garden at Sand Point, Kewstoke (N Somerset) and persistent on waste ground at Wisley (Surrey); seedlings have been reported from New Ash Green (W Kent) and Lambeth (Surrey) and elsewhere; increasing. Medit. OXF, HbEJC. 2, 20, 32*, 54*, 55*, 199, 312*(n.s. t.482), 347*(5:129-156), 370(15:62), 458. Incl. *E. veneta* Willd.; *E. wulfenii* Hoppe ex Koch. The two subspecies recognised in 20 and 21 are doubtfully distinct. 1338*.

corallioides L.   Coral Spurge
- •• A garden escape established for over 180 years at Slinfold (W Sussex); persistent elsewhere. Italy, Sicily. BM, LANC, NMW, OXF, RNG. 1, 20*, 22*, 174, 288, 347*(5:129-156).

cuneifolia Guss.
- A casual at Hull docks (SE Yorks), vector unknown. No modern records. C Medit. 21, 46*, 285.

cyparissias L.   Cypress Spurge
- •••• An established garden escape, or possibly native in SE England; naturalised in grassland in scattered localities, mainly in England; possibly formerly brought in from the continent with horses, now long established on a few racecourses. Europe. 1, 20, 22*, 33*, 35*, 209, 258*.

dulcis L.   Sweet Spurge
- ••• An established garden escape; naturalised on roadsides and riverbanks, especially in Scotland and Wales; known since 1894 by the river Conon (E Ross). C & S Europe. BM, CGE, E, NMW, OXF, RNG. 1, 20*, 22*, 26*, 277, 349(5:404). *E. purpurata* Thuill.

esula group   Twiggy Spurges
- ••••• Probably originally brought in with grain or timber, now long established in grassland, hedgerows and on waste ground, especially in SE England. Eurasia. 1, 20*, 22*, 33*. Incl. *E. esula* L.; *E. intercedens* Podp., non Pax; *E. mosana* Lej.; *E. pinifolia* Lam.; *E. podperae* Croizet; *E. pseudocyparissias* Jordan; *E.* × *pseudovirgata* (Schur) Soó; *E. tommasiniana* Bertol.; *E. uralensis* Fischer ex Link; *E. virgata* Waldst. & Kit., nom. illeg.; *E. waldsteinii* (Soják) Czerep. Most records are referable to *E.* × *pseudovirgata* (*E. esula* × *E. waldsteinii*) which occurs in several forms; *E. esula* occurs mostly in Scotland. *E. waldsteinii*, according to 316(22:90), has been known since 1989 from one site at Govan (Lanarks).

A hybrid with *E. cyparissias* (*E.* × *pseudoesula* Schur) has also been recorded.
IPS, LANC, NMW. 20, 156, 369(27:25). *E. boissieriana* (Woronow) Prokh.
(*E. virgata* var. *orientalis* Boiss.) has not been reported, but *E. boissieriana* ×
E. esula was found established in 1989 on railway sidings at Adderley Park,
Birmingham (Warks). K, HbJWP. 459.

falcata L.   Sickle Spurge
- A casual in a garden at Ashbourne (Derbys), perhaps from bird-seed.
Europe, N Africa, SW Asia. DBY, OXF. 2, 21, 22*, 29*, 44*, 46*,
300(1907:41), 354(7:210), 370(14:425). *E. acuminata* Lam.

griffithii Hook.f.   Griffith's Spurge
- A garden escape persistent or established on a roadside verge at
?Duncombletts (?Lancs). Himalayas. 51a, 54*, 55*, 56*, 347(5:129-156),
371(349:28), 1255*.

hirta L.   Asthma-plant
- A bird-seed casual recorded from the grounds of Buckingham Palace
(Middlesex). Tropics. HbEJC. 19, 36c, 44*, 48*, 331(58:67), 617*, 1247*.
Chamaesyce hirta (L.) Millspaugh.

maculata L.   Spotted Spurge
- - A persistent weed in botanic gardens and nurseries, possibly brought in with
imported cacti. N America; (S Europe). BIRA, BM, K, LTR, NMW, RNG.
20, 38*, 156, 303(13*:1,21; 15:12; 16:18; 42:21), 683*, 737*.
Chamaesyce maculata (L.) Small.

mellifera Aiton   Canary Spurge
- A garden escape on cliffs at Mousehole (W Cornwall) and seedlings on walls
near Tresco Abbey gardens (Scilly). Madeira, Canary Is. 51a, 66*,
303(36:23), 325*(104:433), 371(403:39), 530*, 634*.

myrsinites L.   Broad-leaved Glaucous-spurge
Reported in 1992 as self-sown on the retaining wall of a garden at Godalming
(Surrey). S Europe, Turkey, Caucasus. 21, 24*, 32*, 39, 342(Feb93), 615*,
630*, 683*. *E. pontica* Prokh.

oblongata Griseb.
- A casual in a meadow at Newport (Wight), and reported in 1992 from both
a pathside at Godalming (Surrey) and a trackside at Podimore (N Somerset),
probably a garden escape. SE Europe. K, HbKWP. 21, 39, 51a, 303(22:16;
23:26), 423, 523*. The record for Kings Lynn (W Norfolk) was of a cultivated
(not wild) plant. 303(23:26).

[*palustris* L. In error. 156, 193(p.216).]

peplus L. **alien variants**
Var. *minima* DC. (*E. peploides* Gouan) has been recorded as a casual garden
weed brought in with garden plants from Italy. 21, 354(10:481), 683, 737.

[*pilosa* L. In error for *E. villosa* and *E. corallioides*.]

[*prostrata* Aiton (*Chamaesyce prostrata* (Aiton) Small) In error for *E. maculata*.
303(42:21-22).]

[*salicifolia* Host   In error for *E. esula* group. 1, 187, 300(1906:180).]
segetalis L.   Corn Spurge
  • A bird-seed casual. W Medit., Atlantic Is. RNG. 21, 36c*, 683*, 737*.
serpens Kunth
  • A bird-seed and wool casual. Tropical America; (widespread). E, RNG.
  19, 21, 38*, 40*, 303(13:21), 350(30:19), 371(416:7), 390(123:247), 683,
  737*. *Chamaesyce serpens* (Kunth) Small.
spinosa L.
  Pre-1930 only. Medit. 7, 21, 29, 32*, 166, 683*.
taurinensis All.
  • Reported from waste ground at Kew (Surrey). S Europe, SW Asia. 2, 21,
  39*, 300(1907:41), 342(Feb89).
valerianifolia Lam.
  • A casual at Southampton (S Hants), vector unknown. No modern records.
  Crete, Cyprus, SW Asia. HbJPMB. 39, 44*, 354(13:111). Incl. *E. cybirensis*
  Boiss. This species was incorrectly equated to *E. akenocarpa* Guss. in 21.
  463.
villosa Waldst. & Kit. ex Willd.   Hairy Spurge
  • A casual at Cardiff (Glam), vector unknown. Formerly established for about
  350 years in woodland and hedgebanks near Bath (N Somerset). Eurasia. LIV,
  NMW, OXF, RNG. 2, 12, 20, 22*, 29*, 46*, 156, 1279*. *E. pilosa* auct.,
  non L. A hybrid with native *E. amygdaloides* (*E.* × *turneri* Druce) has been
  recorded in error. 1311.

PEDILANTHUS Necker ex Poit.
tithymaloides (L.) Poit.   Devil's-backbone
  Pre-1930 only. Tropical America. 7, 49, 51, 166, 503*, 681*. *Euphorbia
  tithymaloides* L.

RHAMNACEAE
RHAMNUS L.
alaternus L.   Mediterranean Buckthorn
  •• An established garden escape; plentifully naturalised in Marl Woods,
  Llandudno Junction and on Great Ormes Head (both Caerns). Medit. BM, K,
  NMW, OXF, SLBI. 7, 20*, 24*, 31*, 44*, 156.
purshiana DC.   Cascara Sagrada
  • Introduced. Western N America. 36c, 41*, 42, 61, 119. *Frangula
  purshiana* (DC.) Cooper.

VITACEAE
VITIS L.
coignetiae Pull. ex Planchon   Crimson-glory-vine
  • A persistent garden escape or relic abundant on a bombed site at Ludgate Hill
  (Middlesex); also recorded from the derelict gardens of Easton Lodge (N Essex)
  and on roadsides near Hextable (W Kent). E Asia. 20, 21, 54*, 63*, 191,
  303(59:46), 371(407:41).

vinifera L. Grape-vine
●●●● An established escape from cultivation; naturalised in hedges, on river banks and waste ground; known since 1935 on the banks of the river Thames at Kew (Surrey); also a food refuse casual on tips. Origin obscure; (widespread). 2, 20, 54*, 63*, 192, 209, 220, 553*, 613*.

PARTHENOCISSUS Planchon

henryana (Hemsley) Graebner ex Diels & Gilg Chinese Virginia-creeper
● Reported from waste ground near London Bridge Station (Surrey). China. 51, 52*, 54*, 55*, 342(Feb90), 458. *Ampelopsis henryana* (Hemsley) Rehder; *Cissus henryana* hort.; *Vitis henryana* Hemsley.

inserta (A.Kerner) Fritsch False Virginia-creeper
●●● A garden escape established on waste ground, tips and dunes in the south. N America. NMW. 20, 21, 52*, 61, 191, 192, 236, 826*(13:75-77). *P. quinquefolia* auct., non (L.) Planchon; *P. vitacea* (Knerr) A.Hitchc.; *Ampelopsis inserta* A. Kerner. Possibly conspecific with *P. quinquefolia*; an intermediate has been recorded from old mine-waste below St Agnes Beacon (W Cornwall) and may be overlooked elsewhere. 213.

quinquefolia (L.) Planchon Virginia-creeper
●●●● An established garden escape; naturalised, especially on railway banks and roadsides, in widely scattered localities. Eastern N America. 2, 20, 22*, 55*, 56*, 209. *Ampelopsis quinquefolia* (L.) Michaux; *Cissus quinquefolia* (L.) Sol. ex Sims; *Hedera quinquefolia* L.; *Vitis hederacea* Ehrh.; *V. quinquefolia* (L.) Lam.

tricuspidata (Siebold & Zucc.) Planchon Boston-ivy
●● A persistent garden escape on old walls and river banks. China, Japan. LANC, LTR, NMW, OXF. 7, 20, 51*, 52*, 54*, 56*, 156, 354(12:712). *Ampelopsis tricuspidata* Siebold & Zucc.; *A. veitchii* hort.; *Vitis inconstans* Miq.; *V. thunbergii* Druce, non Siebold & Zucc.

LINACEAE

LINUM L.

austriacum L.
● A casual or persistent garden escape; possibly also introduced with grass and clover seed mixtures. Eurasia. CGE, HbEJC, HbPM. 21, 22*, 234, 445, 595b*, 623*.

grandiflorum Desf. Crimson Flax
● A bird-seed casual and garden escape. N Africa. BM. 1, 51, 52*, 55*, 258, 331(37:203), 354(9:685), 600*.

narbonense L.
● Recorded as a garden weed. W & C Medit. 21, 30*, 52*, 54*, 55*, 236, 354(9:259), 737*. *L. reflexum* Aiton.

perenne L. subsp. **montanum** (Schleicher ex DC.) Ock.
● Recorded in 1976 from chalk grassland at the edge of Darenth Wood (W Kent). Alps, Juras. 21, 236, 331(56:85), 525*, 623a*, 645*(t.674).

L. montanum Schleicher ex DC.   A 1979 record from Reading (Berks) may
also be referable to this subsp. RNG.

suffruticosum L.   White Flax
   Pre-1930 only. SW Europe. 7, 21, 22*, 29*, 30*, 354(4:191), 737*. Incl.
   L. jimenezii Pau.

tenue Desf.
   ● A bird-seed casual. W Medit. 19, 21, 30*, 45*, 258, 303(20:12), 737*.

tenuifolium L.
   ● Collected in 1951 from Greenham Common (Berks), vector unknown.
   No modern records. C & S Europe, SW Asia. CGE. 21, 22*, 31*, 546*,
   595b*, 623*.

usitatissimum L.   Flax, incl. Linseed
   ●●●●● A persistent bird-seed and oil-seed alien and an escape from or relic of
   cultivation.   White-flowered forms are frequent from bird-seed mixtures.
   Origin obscure; (widespread). 1, 19, 20, 22*, 26*, 27*, 33*, 264. Incl.
   L. crepitans (Boenn.) Dumort.; *L. humile* Miller. The short, oil-producing
   crop plant, linseed, is now widely grown in southern England.

POLYGALACEAE
POLYGALA L.
chamaebuxus L.   Shrubby Milkwort
   ● A casual garden escape. No modern records. Europe. 21, 22*, 27*, 28*,
   52*, 245.

STAPHYLEACEAE
STAPHYLEA L.
colchica Steven   Caucasian Bladder-nut
   ● A garden relic at Headington, Oxford (Oxon). Caucasus. 51*, 61, 62*, 63*,
   371(349:29).

pinnata L.   Bladder-nut
   ●●● A garden escape established in a few places, as in a copse near Otford
   (W Kent) and in an old woodland hedge at Devizes (N Wilts); introduced and
   sometimes self-sown elsewhere; decreasing. Europe, SW Asia. BM, CGE,
   GLR, NMW, OXF, RNG. 1, 20, 26*, 31*, 63*, 126, 477, 1110*, 1141*.

SAPINDACEAE
KOELREUTERIA Laxm.
paniculata Laxm.   Pride-of-India
   ●● Introduced and a garden escape; self-sown saplings occur on waste ground
   in several localities in SE England. China. NMW. 20, 58*, 59*, 60*,
   303(49:30), 370(2:106).

CARDIOSPERMUM L.
halicacabum L.   Balloon-vine
- A garden escape collected in 1992 from behind a hotel at Sandwich (E Kent).
Tropical America; (widespread). BM. 21, 38*, 48, 51, 312*(t.1049), 510*,
564*, 1208*, 1347*.

HIPPOCASTANACEAE
AESCULUS L.
carnea Zeyher   Red Horse-chestnut
- •••• Introduced; self-sown at Aldershot (N Hants), and abundantly so at
Lane End, Darenth (W Kent). Originated in cultivation. 16, 20, 58*, 59*,
60*, 119, 191, 371(361:27), 458.   Probably of *A. hippocastanum* ×
A. pavia L. origin.
flava Sol.   Yellow Buckeye
- Introduced or a garden escape in a hedgerow near Farthinghoe (Northants);
also reported from Yealand Conyers (W Lancs) and Weston, near Otley
(MW Yorks).   Eastern N America. LIV. 58*, 59*, 60*, 61, 191,
371(355:27), 432. *A. octandra* Marshall.
hippocastanum L.   Horse-chestnut
- ••••• Introduced and reproducing in widely scattered localities. SE Europe.
1, 20, 58*, 59*, 60*, 191.
indica (Wallich ex Cambess.) Hook.   Indian Horse-chestnut
- Introduced and freely reproducing in Queen's Wood, Highgate (Middlesex);
also recorded from Angmering (W Sussex). Himalayas. 58*, 59*, 60*, 61,
122, 331(69:142).
parviflora Walter   Bottlebrush Buckeye
- A garden relic reported as suckering extensively at Stone and Hextable (both
W Kent). Southern USA. 51, 58*, 61, 63*, 325*(114:78-80), 458, 543*.

ACERACEAE
ACER L.
cappadocicum Gled.   Cappadocian Maple
- •• Introduced and reproducing from seeds or suckers in Darley Park (Derbys),
at Grayswood (Surrey), Farnborough Common (W Kent), Trent Park and
Hadley Wood (both Middlesex). Asia. BM. 20*, 58*, 59*, 60*, 199, 234,
331(69:142), 458. *A. laetum* C.Meyer; *A. pictum* auct., non Thunb.
heldreichii Orph. ex Boiss.   Heldreich's Maple
- Introduced. Balkans. 21, 59*, 60*, 61, 684*, 1295*, 1353.
japonicum Thunb.   Downy Japanese-maple
- A garden relic on the site of Warley Place gardens (S Essex) and introduced
on the calcareous downs at Avon Gorge (W Gloucs). Japan. 59*, 60*, 61,
313(31:370), 350(47:35), 1110*.
macrophyllum Pursh   Oregon Maple
- Seedlings reported from the grounds of Trinity College in the city of Dublin.
Western N America. 41*, 59*, 60*, 61*, 1366.

monspessulanum L.   Montpellier Maple
• Introduced.   Medit.   BM, OXF.   7, 12, 21, 58*, 59*, 60*, 61, 151, 313(22:96).

negundo L.   Ashleaf Maple
•• Introduced, and a persistent garden escape, in a few places in the south; seedlings have been reported on walls at Hampton Court (Middlesex) and Lambeth (Surrey).   N America.   BM, OXF.   7, 20*, 58*, 59*, 60*, 126, 236. *Negundo aceroides* Moench; *N. negundo* (L.) Karsten.

palmatum Thunb.   Smooth Japanese-maple
• Introduced; self-sown seedlings reported from central London (Middlesex). E Asia.   RNG.   58*, 59*, 60*, 61, 61a, 63*, 191, 166, 303(64:41), 1110*.

platanoides L.   Norway Maple
••••• An established garden escape; naturalised in widely scattered localities throughout the British Is.   Europe, SW Asia.   1, 20*, 58*, 59*, 60*.

pseudoplatanus L.   Sycamore
••••• An established garden escape; long naturalised throughout the British Is. Europe, SW Asia.   1, 20*, 35*, 58*, 59*, 60*.

rubrum L.   Red Maple
• Introduced; reported as self-seeding in Brookwood Cemetery (Surrey). Eastern N America.   38*, 58*, 59*, 60*, 61, 342(Feb89), 1301.

rufinerve Siebold & Zucc.   Grey-budded Snake-bark-maple
• Reported as self-seeding in Brookwood Cemetery (Surrey).   Japan.   HbKWP. 58*, 59, 60*, 61, 312*(t.5793), 342(Feb89).

saccharinum L.   Silver Maple
• Introduced; apparently self-seeding on Wimbledon Common (Surrey). Eastern N America.   16, 20*, 58*, 59*, 60*, 119, 331(60:89), 350(47:35). *A. dasycarpum* Ehrh.

saccharum Marshall   Sugar Maple
• Self-sown near Bedgebury and reported from a hedgerow at Penshurst (both W Kent).   Eastern N America.   OXF.   38*, 58*, 59*, 60*, 61, 236, 403. *A. saccharinum* Wangenh., non L.

tataricum L.   Tartar Maple
• Self-sown in woodland at New Ash Green (W Kent).   SE Europe, Turkey. 20, 58*, 60*, 61, 458.

truncatum Bunge   Shantung Maple
• Introduced.   E Asia.   20, 61, 191, 563*, 712*.   *A. cappadocicum* subsp. *truncatum* (Bunge) E.Murray; incl. *A. mono* Maxim.; *A. pictum* Thunb. pro parte.

ANACARDIACEAE
RHUS L.
coriaria L.   Tanners' Sumach
• Reported from a railway bank at Chearsley (Bucks), possibly a garden escape.   Medit.   21, 24*, 30*, 32*, 51, 307(8:2).

hirta (L.) Sudw. Stag's-horn Sumach
- ••• An established garden escape; naturalised by extensive suckering on railway banks and roadsides in a few places, mainly in the south. Eastern N America. BM, CGE, NMW, OXF. 20, 38*, 59*, 119, 236, 258, 370(12:286; 13:84), 505*, 606*. *R. typhina* L.; *Datisca hirta* L.

COTINUS Miller
coggygria Scop. Smoke-tree
- • Introduced and reproducing in a wooded part of Dulwich Park (Surrey) and recorded from a hedge at Rush Court (Berks); also a purple-leaved form reported as self-sown by the river at Shoreham (W Kent). S Europe. OXF. 20, 31*, 32*, 51*, 61*, 61a, 119, 199, 458. *Rhus cotinus* L.

MANGIFERA L.
indica L. Mango
- • A food refuse casual. N India, Burma; (tropics as a crop). 51, 68*, 69*, 80*, 126.

SIMAROUBACEAE

AILANTHUS Desf.
altissima (Miller) Swingle Tree-of-heaven
- •••• Introduced and a garden escape, reproducing from seed and suckers, sometimes extensively, especially in SE England; increasing. China. 12, 20, 21, 22*, 58*, 59*, 60*, 126, 209, 236, 370(11:273). *A. glandulosa* Desf.; *Toxicodendron altissima* Miller.

RUTACEAE

RUTA L.
graveolens L. Rue
- •• A persistent garden escape. SE Europe. BM, LIV, OXF. 7, 20, 21, 29*, 56*, 228, 236, 600*.
montana (L.) L.
Pre-1930 only. Medit. 21, 45*, 46*, 288.

HAPLOPHYLLUM Adr.Juss.
tuberculatum (Forsskål) Adr.Juss.
Pre-1930 only. N Africa, SW Asia. NMW, OXF. 7, 39, 44*, 317*(t.3931), 354(3:156), 622*. *Ruta tuberculata* Forsskål.

CITRUS L.
auriantium L. Bitter Orange, incl. Sweet Orange
- •• A food refuse casual. SE Asia; (widespread as a crop). 2, 21, 68*, 71*, 80*, 192, 684*.
limon (L.) Burm.f. Lemon
- •• A food refuse casual. SE Asia; (widespread as a crop). 21, 68*, 80*, 126, 371(367:29), 606*, 684*. *C. limonum* Risso. *C.* × *limonia* Osbeck, a name recorded in error for the lemon, is the hybrid Lemandarin.

paradisi Macfad.   Grapefruit
- A food refuse casual. Origin obscure; (widespread as a crop). RNG. 21, 68*, 80*, 126, 606*, 684*. Possibly of *C. maxima* (Burm.) Merrill × *C. sinensis* (L.) Osbeck origin.

reticulata Blanco   Mandarin Orange, incl. Tangerine, Satsuma, Clementine
- A food refuse casual. SE Asia; (widespread as a crop). BM. 21, 67*, 68*, 606*, 684*. Incl. *C. deliciosa* Ten.

PONCIRUS Raf.
trifoliata (L.) Raf.   Japanese Bitter-orange
- A garden relic. China. 51*, 52*, 62*, 369(22:48). *Aegle sepiaria* DC.; *Citrus trifoliata* L.

CHOISYA Kunth
ternata Kunth   Mexican Orange
- Introduced and now established as a huge spreading patch in Darenth Wood (W Kent); also established in a wood near Hartley Green (W Kent). Mexico. 20, 55*, 61*, 65*, 66*, 303(58:41), 312*(n.s. t.318), 371(391:33).

SKIMMIA Thunb.
japonica Thunb.
- A garden escape, collected from the south side of Brading Down (Wight). Japan, Philippines. RNG. 51*, 61*, 63*, 312*(t.8038), 649*.

CORREA Andrews
backhouseana Hook.
- An established garden escape abundantly naturalised in woods on Tresco (Scilly). Tasmania. HbEJC. 51a, 55*, 61, 63*, 312*(n.s. t.289), 458, 566*. *C. speciosa* Aiton var. *backhouseana* (Hook.) Benth.; *C. virens* hort., ?non Hook., vix Smith.

ZYGOPHYLLACEAE
ZYGOPHYLLUM L.
fabago L.   Syrian Bean-caper
- Established for some years, probably as a garden escape, at Mitcham (Surrey), now gone. SE Europe, SW Asia. BM, RNG. 13, 21, 24*, 30*, 44*, 126, 209.

TRIBULUS L.
terrestris L.   Small Caltrops
- A wool, grain and animal-feed casual. Medit.; (widespread). NMW, OXF, RNG, SLBI. 1, 9, 17, 21, 24*, 29*, 32*, 44*, 303(50:30), 354(11:470).

OXALIDACEAE
OXALIS L.
articulata Savigny   Pink-sorrel
- ••••• An established garden escape; locally naturalised on dunes, cliffs and roadsides, mainly in the Channel Is, southern England, Wales and Ireland; a

pest in cultivated land, particularly in the bulbfields of Scilly; increasing. Eastern S America. 1 (as *O. violacea*), 7, 20*, 22*, 21, 25*, 207, 220, 546a*, 1216*. *O. floribunda* Lehm.; *O. semiloba* auct., non Sonder; *O. rosea* auct., non Jacq. Incl. *O. rubra* A.St.Hil. (*O. articulata* subsp. *rubra* (A.St.Hil.) Lourt.). According to 443 and 546a*, *O. floribunda* is a distinct species and is rare in the British Is.

[aff. *bulbifera* Knuth   Unconfirmed; probably always in error for *O. tetraphylla*. 443.]

conorrhiza Jacq.

- A wool casual. S America. RNG. 535*, 557, 653*, 696*, 1050*, 1196*. *O. andicola* Gillies ex Hook. & Arn.; *O. commersonii* Pers.; *O. corniculata* var. *serpens* Knuth; *O. sexenata* Savigny; *O. sternbergii* Zucc.

corniculata L.   Procumbent Yellow-sorrel

●●●●● (but confused with *O. exilis*). Established as a garden escape and as a persistent weed in churchyards and gardens, mainly in southern England and Wales; increasing. Origin obscure; (widespread). 1, 20, 22*, 23*, 34*, 44*. *O. repens* Thunb.; *Xanthoxalis corniculata* (L.) Small. Purple-tinged forms have been recorded as var. *atropurpurea* Van Houtte ex Planchon.

debilis Kunth   Large-flowered Pink-sorrel

●●●● Established as a garden escape and a pestilent weed in gardens, especially in SE England. Tropical America; (widespread). 2, 20*, 22*, 126*, 546a*, 649*, 737*, 856*(7:11-12), 1285*. Incl. *O. corymbosa* DC.; *O. martiana* Zucc.

decaphylla Kunth

- Long established as a weed on the promenade at Douglas (Man). Arizona, New Mexico, Mexico. HbEJC. 20, 103(as *O. lasiandra*), 443, 608*, 627, 897*(4:t.11), 910*(4:455-615). *O. grayii* (Rose) Knuth; *Ionoxalis decaphylla* (Kunth) Rose; *I. grayii* Rose. The record from Christchurch (S Hants) was an error for *O. lasiandra*.

dillenii Jacq.   Sussex Yellow-sorrel

- (but confused with *O. stricta*). A long-persistent weed of arable fields near Pulborough (W Sussex), last seen 1984; also recorded from Sark (Channel Is), Leics and Cumberland. Eastern N America. LANC, LSR, RNG. 13, 20, 22*, 38*, 40*, 175, 349(3:293), 370(4:58), 546a*(as *O.stricta*). *O. corniculata* auct., non L.; *O. navierei* Jordan; *O. stricta* auct., non L.

[aff. *drummondii* A Gray   Unconfirmed; probably in error for *O. tetraphylla*. 371(374:8), 443.]

exilis Cunn.   Least Yellow-sorrel

●●●● (but confused with *O. corniculata*). Established as a garden escape and a persistent weed in gardens, especially in the south. Australasia. 16, 20, 21, 33*(as *O. corniculata*), 192, 207, 209, 370(10:290). *O. corniculata* var. *microphylla* Hook.f.; *O. repens* auct., non Thunb.

incarnata L.   Pale Pink-sorrel
●●●● An established garden escape; naturalised by roadsides and on hedgebanks, mainly in SW England, Scilly and the Channel Is; also a persistent weed in gardens. S Africa. 16, 20*, 22*, 220*, 370(4:66). *O. carinata* sphalm.

lasiandra Zucc.
● A relic of cultivation persistent in turf in an abandoned nursery at Christchurch (S Hants), now gone. Mexico. BM, RNG. 21, 51b, 371(364:28), 511, 856*(7:11-22), 910*(4:455-615). *Ionoxalis lasiandra* (Zucc.) Rose. The record from Douglas (Man) is in error for *O. decaphylla*. 103, 443. Widely confused in the literature with *O. lasiandra* R. Graham, which appears to be a synonym of *O. articulata*, or is closely related to it; references 312*(t.3896), 379*(80:332) and 842*(1916:78) belong here.

latifolia Kunth   Garden Pink-sorrel
●●●● An established garden escape and pestilent weed in gardens, market gardens and nurseries, especially in SW England and the Channel Is. Mexico, C & S America, W Indies. 7, 20*, 22*, 117*, 126*, 201, 220, 370(4:63-65), 383(62:75-79), 546a*, 856*(7:11-22), 1285*. *O. vespertilionis* A.Gray, non Zucc.

megalorrhiza Jacq.   Fleshy Yellow-sorrel
● An established garden escape naturalised on rocks, walls and hedgebanks in Scilly and on a wall at St Just (W Cornwall). Chile, Peru. BM, OXF, RNG. 13, 20, 207*, 208a*, 349(3:174), 371(415:27), 1117*, 1291*(1:t.1). *O. carnosa* auct., non Molina.

perennans Haw.   Woody-rooted Yellow-sorrel
● A wool casual. Australia; (New Zealand). RNG. 36c, 37c*, 590*. In S Australia this species apparently merges into *O. exilis*.

pes-caprae L.   Bermuda-buttercup
●●●● An established garden escape; abundantly naturalised in Scilly, and a persistent weed in Devon and the Channel Is; increasing. S Africa; (widespread). 2, 20, 22*, 30*, 44*, 207*, 546a*, 737*. *O. cernua* Thunb.; *O. libica* Viv.

purpurata Jacq.
● A casual garden or greenhouse escape. S Africa. 51*, 199, 371(361:26), 511*, 590*. *O. bowiei* Herbert ex Lindley

rosea Jacq.   Annual Pink-sorrel
●● Persistent as a weed in a few gardens and nurseries in the south, and reported as established on a high wall at Gulval, near Penzance (W Cornwall). Chile. HbEJC. 20*, 51, 201, 301*(1827:t.1123), 371(374:8), 458, 680*, 856*(6:22-32), 1216*. *O. floribunda* Lindley, non Lehm.; *O. rubra* hort., non Jacq. Early records were in error for *O. articulata*. 370(4:54).

[*semiloba* Sonder In error for *O. articulata*. 370(4:66).]

stricta L.   Upright Yellow-sorrel
●●●● (but confused with *O. dillenii*). Established as a garden escape and a persistent weed in gardens, churchyards and on arable land, mainly in the south. E Asia, N America. 2, 20, 22*, 26*, 34*, 209, 546a*(as *O. fontana*).

O. dillenii auct., non Jacq.; *O. europaea* Jordan; *O. fontana* Bunge; incl.
O. rufa Small.

tetraphylla Cav.   Four-leaved Pink-sorrel
• Introduced and persistent in abundance for more than 50 years in fields at
La Haule, Jersey (Channel Is); recorded from a few other localities in Jersey
and from Tresco (Scilly). Mexico. BM, JSY, OXF, RNG, SLBI. 13, 20*,
201, 207, 370(4:65), 532*, 649*. Incl. *O. deppei* Lodd. ex Sweet. Pink-
flowered bulbous plants with only three leaflets, found at Greenisland, near
Belfast (Co Antrim) and Logan (Wigtowns), originally identified as *O. bulbifera*
or *O. drummondii*, are referable to *O.* aff. *tetraphylla*. BM, RNG, HbEJC.
20, 371(374:8), 443.

valdiviensis Barnéoud   Chilean Yellow-sorrel
•• A garden escape and a persistent weed in gardens in widely scattered
localities. Chile. BM, RNG, SLBI. 13, 20, 51b, 349(3:399; 4:283),
370(4:54), 573*, 669*, 824*(t.626), 856*(6:22-32). *O. valdiviana* hort.
ex Vilm.

[*violacea* L. In error for *O. articulata* and *O. corymbosa*. 370(4:63).]

GERANIACEAE

GERANIUM L.
bohemicum L.
Pre-1930 only. Europe, Turkey. 7, 21, 22*, 25*, 99*, 354(3:16), 595b*,
623*.

brutium Gasp.
• A garden escape persistent since 1990 at Chaffcombe (S Somerset).
SE Europe, Turkey. 21, 39, 51b, 99*, 423, 623*, 683*. *G. molle* L. subsp.
brutium (Gasp.) Davis; *G. villosum* auct., non Ten., nec Miller.

canariense Reuter
• A garden escape recorded from Guernsey (Channel Is). Canary Is. 99*,
303(51:45), 530*.

carolinianum L.   Carolina Crane's-bill
Pre-1930 only. N America. BM, OXF. 7, 38*, 40*, 99*, 354(4:406), 1128*.

collinum Stephan ex Willd.
Pre-1930 only. Eurasia. 7, 21, 39*, 99*, 172, 1024*.

divaricatum Ehrh.   Spreading Crane's-bill
Pre-1930 only. Eurasia. OXF. 7, 21, 22*, 23*, 27*, 354(5:373; 4:10),
826*(10(3)suppl.:1-8).

endressii Gay   French Crane's-bill
•••• An established garden escape; naturalised by roadsides and on railway
banks, in churchyards and woodlands, in widely scattered localities. Pyrenees.
3*, 20, 22*, 34*, 99*.

eriostemon DC.
• A casual or persistent garden escape at St Bernard's Wier, below
Upper Helensburgh (Dunbarton). E Asia. 99*, 303(50:27).

himalayense Klotzsch   Himalayan Crane's-bill
  • (but perhaps overlooked as native *G. pratense*). Established in a churchyard
  at Carshalton (Surrey) and on the sites of former gardens at Warley (S Essex)
  and Farnborough (W Kent); also recorded from Helensburgh and Renton
  (Dunbarton).   Himalayas, Pamirs.   20, 51, 51a, 99*, 126, 223, 454.
  G. grandiflorum Edgew. non L.; *G. meeboldii* Briq.; *G. pratense* auct., non L.
  Some records may be referable to the hybrid with native *G. pratense*
  (*G.* 'Johnson's Blue').

ibericum Cav.   Caucasian Crane's-bill
  •• (but confused with *G.* × *magnificum*).   An established garden escape;
  naturalised for many years in an old churchyard at Llangwyryfon (Cards) and
  on a wooded area of sand dunes near Longniddry (E Lothian).   SW Asia.
  BM, E, SLBI.   7, 20, 22*, 99*, 370(9:46; 14:442).

macrorrhizum L.   Rock Crane's-bill
  ••• An established garden escape; naturalised on roadsides in a few widely
  scattered localities; known for 100 years at Postbridge, Dartmoor (S Devon).
  S Europe.   BM, CGE, E, LANC, OXF, RNG.   1, 20, 22*, 27*, 34*, 99*,
  189, 325*(118:340-342), 354(1:438).

[*maculatum* L. In error for *G. endressii*. 354(2:499).]

maderense Yeo   Giant Herb-Robert
  • An established garden escape naturalised on cliffs in Scilly, and covering an
  old roof and self-sown on nearby ground at Câtel, Guernsey (Channel Is).
  Madeira.   20, 36c*, 99*, 371(385:34), 1069*.

× **magnificum** N.Hylander   Purple Crane's-bill
  •••• A long-persistent garden escape; naturalised on roadsides and railway
  banks in widely scattered localities. Originated in cultivation. 20, 99*, 126*,
  192, 680*. *G. ibericum* × *G. platypetalum*.

[*microphyllum* Hook.f. (*G. potentilloides* Hook.f., non L'Hér.; *G. retrorsum*
  A.Cunn., non L'Hér.).   In error for *G. potentilloides* and *G. submolle*.
  349(4:414; 5:224-226). A non-weedy species which is unlikely to occur.]

× **monacense** Harz   Munich Crane's-bill
  • An established garden escape; long naturalised near Edinburgh and on a
  roadside at Hurst Green (E Sussex). Originated in cultivation. RNG, SLBI.
  20, 99*, 303(13:24), 371(346:26 as *G. reflexum*), 465.   *G. phaeum* ×
  G. reflexum; *G. punctatum* hort.; *G. reflexum* auct., non L. The plant from
  Hurst Green has been named as *G.* 'Eric Clement'. 99.

nodosum L.   Knotted Crane's-bill
  ••• An established garden escape; naturalised in woodland in scattered
  localities in England and Wales. S Europe. BM, LANC, NMW, OXF, SLBI.
  1, 20, 22*, 34*, 99*. Early records were confused with *G. endressii*.

× **oxonianum** Yeo   Druce's Crane's-bill

●●● An established garden escape; naturalised on roadsides, railway banks and in woodland in a few widely scattered localities; long naturalised in High Wood, Durham. Originated in cultivation. BM, E, LANC, NMW, OXF, RNG. 20, 99*, 167, 209. *G. endressii* × *G. versicolor*; incl. *G. endressii* var. *thurstonianum* Turrill.

phaeum L.   Dusky Crane's-bill

●●●● An established garden escape; in many places throughout the British Is, sometimes long naturalised in woodlands and on hedgebanks. S & C Europe. 1, 20, 22*, 27*, 35*, 99*. Incl. *G. fuscum* L.; *G. lividum* L'Hér.

[*platypetalum* Fischer & C.Meyer   Unconfirmed; probably always in error for *G.* × *magnificum*. 349(7:389).]

potentilloides L'Hér. ex DC.

● A wool alien persistent by the railway at Bordon Station (N Hants), probably now gone. No modern records. Australasia. BM, CGE, E, K. 36c, 349(5:224-226), 514, 533, 569*, 743, 839*(1965:89). *G. microphyllum* auct., non Hook.f.

psilostemon Ledeb.   Armenian Crane's-bill

●● A persistent garden escape. Turkey, Caucasus. E. 20, 54*, 99*, 126, 153, 187, 303(16:18), 335(108:33). *G. armenum* Boiss.

[*pyrenaicum* Burman f.   Accepted, with reservations, as native.]

reflexum L.   Reflexed Crane's-bill

● A garden escape recorded from Morland (Westmorland) and Little Ouse (W Norfolk). SE Europe. 2, 21, 51, 99*, 135, 248. The records for Edinburgh and Hurst Green were in error for *G.* × *monacense*.

[*retrorsum* L'Hér. ex DC. (*G. pilosum* Willd., non Cav.)   In error for *G. potentilloides* and *G. submolle* 349(5:224-226).]

rubescens Yeo   Greater Herb-Robert

● A persistent garden escape or relic recorded from Guernsey (Channel Is) and Man. Madeira. 20*, 36c*, 99*, 220, 371(382:24). *G. anemonifolium* L'Hér., non Aiton.

sibiricum L.

Pre-1930 only. Eurasia. OXF. 6, 7, 21, 99*, 354(8:302,611), 826*(10(3)suppl.:1-8). *G. ruthenicum* Uechtr.

submolle Steudel   Alderney Crane's-bill

●● Established; recorded only from a few places in Guernsey, Jersey and Alderney (Channel Is). S America. BM, JSY, K. 20*, 201, 220, 370(8:47). *G. core-core* auct., non Steudel; *G. microphyllum* auct., non Hook.f.

subulatostipulatum Knuth

● A persistent garden escape collected in 1987 from the edge of a car park outside the RHS Gardens at Wisley (Surrey). Mexico. BM. 710*.

tuberosum L.   Tuberous Crane's-bill

Pre-1930 only. Medit. LNHS. 7, 21, 44*, 99*, 192, 354(2:499), 615*, 700*.

versicolor L.   Pencilled Crane's-bill
●●●● An established garden escape; widespread on hedgebanks, railway banks, in grassy places and woods, especially in the south. SE Europe. 1, 20, 22*, 35*, 99*. *G. discolor* sphalm.; *G. striatum* L.

MONSONIA L.
angustifolia E.Meyer ex A.Rich.
● A wool casual. E & S Africa. BM, OXF, RNG. 7, 48*, 143, 146, 162, 354(5:18), 568*, 739*, 1047*, 1102, 1191. *M. biflora* auct., non DC.
[*biflora* DC. In error for *M. angustifolia*. 1102*.]
brevirostrata Knuth   Short-fruited Dysentery-herb
●● A wool casual. S Africa. ABD, BM, CGE, E, LTN, RNG. 3, 14, 20, 133, 143, 739*, 1191, 1294*.

ERODIUM L'Hér.
acaule (L.) Bech. & Thell.
● A casual on a roadside at Dorking (Surrey), vector unknown. Formerly a wool casual. No modern records. Medit. BM, E, NMW. 2, 3, 21, 44*, 156, 370(2:195), 600*. *E. romanum* (Burman f.) L'Hér.
aureum Carolin
● A wool casual. Probably SW Asia; (Australia). E, HbEJC. 37c*, 590*, 819(1967:102).
botrys (Cav.) Bertol.   Mediterranean Stork's-bill
●●●● A wool casual. Medit.; (Australia). 3*, 14, 20*, 44*, 303*(31:13-14), 500*, 815*(42:417-425).
brachycarpum (Godron) Thell.   Hairy-pitted Stork's-bill
●●● A wool casual. N Africa; (Australia, California, Chile). BM, CGE, EXR, LTN, OXF, RNG. 7, 14, 20*, 37c*, 143, 236, 590*, 809*(20:78), 813*(53:79,81), 815*(42:417-425), 1143*. *E. botrys* var. *brachycarpum* Godron; *E. obtusiplicatum* (Maire, Weiller & Wilczek) J.Howell.
chium (L.) Willd.   Three-lobed Stork's-bill
●● A wool casual. Medit. BM, E, NMW, RNG, SLBI. 2, 20*, 45*, 46*, 145, 303(1:16), 374(1950:58). Incl. *E. littoreum* Léman.
ciconium (L.) L'Hér.
● A wool casual. Medit. BM, LTN, RNG. 1, 21, 44*, 45*, 46*, 145, 371(388:37), 825*(2:49-55).
crinitum Carolin   Eastern Stork's-bill
●●● A wool casual. Australia. BM, CGE, E, LTN, NMW, RNG. 14, 20*, 303*(31:13-14), 349(3:284), 514*, 533*, 569*, 825*(2:49-55). *E. cygnorum* auct., non Nees.
cygnorum Nees   Western Stork's-bill
●● (but confused with *E. crinitum*). A wool casual. Australia. BM, CGE, E, NMW, OXF, RNG. 1, 3*, 20*, 37*, 349(3:285). Some records are referable to subsp. *glandulosum* Carolin.
geoides A.St.Hil.
● A wool casual. Uruguay, Brazil. BM, NMW, OXF, RNG. 7, 146, 354(5:553), 599*, 653, 708*.

glaucophyllum (L.) L'Hér.
- An esparto casual. No modern records. N Africa, SW Asia. RNG. 44*, 45*, 46*, 332(11:24), 349(3:441).

gruinum (L.) L'Hér.   Long-beaked Stork's-bill
- A wool casual. E Medit. LTN, HbCGH. 21, 29*, 32*, 44*, 370(16:167).

hymenodes L'Hér.
Pre-1930 only. NW Africa. 2, 46*, 51, 51a.

laciniatum (Cav.) Willd.
- (but confused with *E. geoides*). A wool and esparto casual. Medit. BM, CGE, E, LIV, RNG, TCD. 2, 15, 21, 30*, 44*, 145, 348(20:30), 599*. *E. triangulare* auct., vix (Forsskål) Muschler.

malacoides (L.) L'Hér.   Soft Stork's-bill
- • A wool casual. Medit., SW Asia; (Australia). BM, CGE, E, NMW, OXF, RNG. 1, 14, 20*, 22*, 24*, 44*, 590*.

manescavii Cosson   Manescav's Stork's-bill
- A garden escape established on a roadside verge NE of Wookey (N Somerset); now gone from Caterham (Surrey). French Pyrenees. BM, RNG. 21, 23*, 27*, 28*, 57*, 350(31:29).

petraeum (Gouan) Willd.   Rock Stork's-bill
- A casual garden escape. No modern records. W Medit. RNG. 21, 23*, 27*, 28*, 30*.

[*reichardii* (Murray) DC. (*E. chamaedryoides* L'Hér.) Unconfirmed; probably always in error for *E.* × *variabile* Leslie. 442.]

ruthenicum M.Bieb.
Pre-1930 only. E Europe, Caucasus. 7, 21, 354(5:281), 728*. *E. serotinum* Steven.

salzmannii Del.
- A wool casual. W Medit., Egypt. RNG. 21, 683, 825*(2:49-55). *E. aethiopicum* auct., non *Geranium aethiopicum* Lam.; *E. cicutarium* (L.) L'Hér. subsp. *jacquinianum* (Fischer, C.Meyer & Avé-Lall.) Briq.; *E. jacquinianum* Fischer, C.Meyer & Avé-Lall. Other segregates from native *E. cicutarium* may have been overlooked; plants from wool waste are especially variable.

stephanianum Willd.
- • A wool casual. C Asia, China. BM, E, LTN, OXF, RNG. 7, 14, 147, 303(45:22), 364(27:323), 617*, 636*, 638*, 825*(2:49-55), 1115*, 1135*.

× **variabile** Leslie
- A casual or persistent garden escape. Originated in cultivation. 53, 371(361:26; 428:7 both as *E. reichardii*), 543*. *E. corsicum* Léman × *E. reichardii*.

PELARGONIUM L'Hér.

capitatum (L.) Aiton   Rose Geranium
- A casual in a mangold field, vector unknown. No modern records. S Africa. 36c, 51, 97, 354(13:259), 562*.

× **hortorum** L.Bailey   Zonal Geranium
- A casual or persistent garden escape; reported as spreading by seed at Burnham-on-Sea (N Somerset). Originated in cultivation. 20, 36c, 54*, 66*, 67*, 126, 236, 371(470:41), 423, 617*. Incl. *P.* × *hybridum* Aiton; *P. salmoneum* R.A.Dyer.

inodorum Willd.   Kopata Geranium
- A wool casual. No modern records. Australasia. K, RNG. 14, 36c, 37, 514, 520*.

tomentosum Jacq.   Peppermint-scented Geranium
- A garden escape established near Tresco Abbey Gardens (Scilly). S Africa. 20, 51*, 97*, 207, 599*.

LIMNANTHACEAE
LIMNANTHES R.Br.
douglasii R.Br.   Meadow-foam
- ••• A persistent garden escape, perhaps established on the seashore at Brodick, Isle of Arran (Clyde Is) and on a cliff near the Devil's Frying Pan, Cadgwith (W Cornwall). Western N America. BM, CGE, E, LANC, NMW, OXF. 1, 20, 41*, 54*, 303*(19:14-15), 415. Incl. *L. sulphurea* Loudon.

TROPAEOLACEAE
TROPAEOLUM L.
majus L.   Nasturtium
- •••• A persistent garden escape. Andes. 2, 20, 36c*, 52*, 54*, 191, 590*, 617*. Most records may be referable to *T. majus* × *T. peltophorum* Benth. 51b, 303(63:31).

minus L.   Dwarf Nasturtium
- •• A persistent garden escape. Andes. 2, 51, 156, 184, 191, 312*(t.98).

peregrinum L. sensu lato   Canary-creeper
- •• A persistent garden escape. Peru, Ecuador. BM, LIV, OXF. 2, 20, 51, 52*, 54*, 65*, 192, 353(24:45). Incl. *T. aduncum* Smith; *T. canariensis* hort.

speciosum Poeppig & Endl.   Flame Nasturtium
- •• A persistent or established garden escape, mainly in Scotland and Ireland; reported from hedges well away from houses near Knockdee (Caithness) and by the Lagan Canal at Aghalee (Co Antrim). Chile. ZCM. 20, 36c*, 52*, 54*, 65*, 223, 371(370:28), 417, 425.

BALSAMINACEAE
IMPATIENS L.
balfourii Hook.f.   Kashmir Balsam
- Introduced by the river Thames at Ham (Surrey), now gone. Himalayas. 20, 21, 22*, 199, 347*(5:86-102), 506*, 546a*, 1011*(fig. as *I. glandulifera*).

balsamina L.   Balsam
- A casual garden escape. SE Asia. 21, 55*, 126, 347(5:86-102), 353(24:45), 371(370:30), 593*, 600*.

capensis Meerb. Orange Balsam

●●●●● An established garden escape; naturalised, sometimes abundantly, on river banks, especially in the south; increasing. Eastern N America. 1, 20, 22*, 33*, 35*, 192, 209. *I. biflora* Walter; *I. fulva* Nutt.

glandulifera Royle Indian Balsam

●●●●● An established garden escape; naturalised, sometimes abundantly, on river banks and waste ground throughout the British Is, especially in C & N England and Wales; increasing. Himalayas. 1, 20, 22*, 26*, 33*, 35*. *I. roylei* Walp., nom. illeg. Not to be confused with *I. glandulifera* Arn., an unrelated endemic of Ceylon.

parviflora DC. Small Balsam

●●●●● An established alien; possibly originally brought in with Russian timber, now naturalised in shady places in scattered localities in Britain, especially in SE England; increasing. C Asia; (Europe). 1, 20, 22*, 26*, 33*, 35*, 235.

sultanii Hook.f. Busy-Lizzie

● A casual garden escape. Originated in cultivation. 51, 66*, 303(37:31), 347*(5:86-102), 593*, 600*. Incl. *I. holstii* Engl. Probably a derivative from hybrids between the tropical African *I. walleriana* Hook.f. and other species.

ARALIACEAE

HEDERA L.

colchica (K.Koch) K.Koch Persian Ivy

●●● An established garden escape; locally abundantly naturalised in woodland, as on the wooded common at Weybridge (Surrey) and at Ardrishaig (Kintyre). Balkans, Turkey, Caucasus. LANC, NMW, RNG. 20, 65*, 92*, 126, 137, 191, 303*(36:12-14), 1110*. *H. helix* var. *colchica* K.Koch; incl. *H.* 'Dentata'.

helix L. subsp. **canariensis** (Willd.) Cout. Canary Ivy

●● An established garden escape; well naturalised in Glen Maye (Man). W Medit., Atlantic Is. NMW. 20, 92*, 103, 156, 303*(36:12-14), 371(337:29), 1110*. *H. algeriensis* Hibberd; *H. canariensis* Willd.

helix L. subsp. **poetarum** (Bertol.) Nyman Yellow-berried Ivy

Pre-1930 only. SE Europe. CGE. 20, 21, 631*, 733*. *H. chrysocarpa* Walsh; *H. poetarum* Bertol.

['Hibernica' Accepted, with reservations, as native. Probably a selection from native *H. helix* subsp. *hibernica*.]

ELEUTHEROCOCCUS Maxim.

henryi Oliver

● Introduced or a garden escape at Box Hill (Surrey). China. 51, 61, 61a, 312*(t.8316), 370(367:30), 617*. *Acanthopanax henryi* (Oliver) Harms.

ARALIA L.

chinensis L. Chinese Angelica-tree

● A garden escape established by suckering on a field edge near Lasswade (Midlothian); also recorded from Wateringbury (W Kent). China. 20, 371(391:33), 617*, 631*.

elata (Miq.) Seemann   Japanese Angelica-tree
- Introduced and long persistent by a muddy pond at Dannemarche, Jersey (Channel Is); also recorded in 1992 from Carrington in Greater Manchester. E Asia. 20, 61*, 201, 371(428:7), 505*, 617*, 631*. *Dimorphanthus elatus* Miq.

racemosa L.   American-spikenard
- An established introduction or garden escape well naturalised in Badger Dingle (Salop). N America. 7, 20, 38*, 259, 354(8:27), 1110*.

spinosa L.   Hercules'-club
- Reported as self-seeding on a bank of the A20 road at Mottingham (W Kent). Eastern N America. 38*, 51, 458, 1110*, 1319*.

FATSIA Decne. & Planchon
japonica (Thunb.) Decne. & Planchon
- A garden escape reported in 1989 from Harpford Wood (S Devon); also as seedlings in Cornwall. Japan. 51, 53*, 61*, 63*, 312*(t.8638), 390(122:?), 458, 1207*. *Aralia japonica* Thunb.

APIACEAE
Umbelliferae. Incl. *Hydrocotylaceae*
HYDROCOTYLE L.
moschata Forster f.   Hairy Pennywort
- An established garden escape; naturalised in abundance on grassy roadside banks on Valentia Is (S Kerry). A small-leaved variant, of uncertain identity, persisted for a time as a lawn weed at Broadwater Down, Tunbridge Wells (W Kent) and on a golf course at Largs (Ayrs); recently recorded, also as a lawn weed, at Tregothnan (E Cornwall). New Zealand. BM, CGE, E, RNG. 20*, 22*, 36c*, 213, 303(11:11), 349(3:288,403), 370(18:93-95). *H. microphylla* Cunn.; *H. sibthorpioides* Colenso, non Lam.

novae-zelandiae DC.   New Zealand Pennywort
- A persistent weed in lawns at Parkhill, Arbroath (Angus) and in Cornwall, probably garden escapes. New Zealand. BM, CGE, RNG. 20*, 36c*, 187, 371(355:28). *H. microphylla* auct., non Cunn. All records are referable to var. *montana* Kirk.

ranunculoides L.f.   Floating Marsh-pennywort
- An established nursery garden escape; reported in 1991 as abundant along the river Chelmer and the Chelmer-Blackwater canal at Chemsford; also in a gravel pit at North Shoebury (all S Essex). N America; (S Europe). 21, 38*, 41*, 410, 1357*.

sibthorpioides Lam.
- Persistent as a weed in paving at Highfields School, Wolverhampton (Staffs). Tropics. 20, 21, 44*, 51a, 303(48:36). The identity of this plant needs confirmation.

TRACHYMENE Rudge
cyanopetala (F.Muell.) Benth.   Purple Trachymene
- A wool casual. Australia. K, RNG. 17, 37*, 370(10:283), 520*. *Didiscus cyanopetalus* F. Muell.

pilosa Smith   Dwarf Trachymene
- A wool casual. Australia. RNG, HbTBR. 17, 37*, 370(10:283), 520*. *Didiscus pilosus* (Smith) Domin, non Benth.; *D. pusillus* (DC.) F.Muell.

BOWLESIA Ruíz Lopez & Pavón
incana Ruíz Lopez & Pavón
- A wool and bird-seed casual. S USA, Mexico, S America. CGE, E, K, RNG.   1, 12, 17, 181(pp.81,365), 371(357:24), 1165*, 1233*, 1357*. *B. tenera* Sprengel.

HACQUETIA Necker ex DC.
epipactis (Scop.) DC.   Hacquetia
- A long-persistent garden relic near Landford (S Wilts). Europe. 21, 23*, 27*, 29*, 57*, 371(346:27).

ASTRANTIA L.
carniolica Jacq.   Carnic Masterwort
- A casual garden escape. Alps. 21, 51a, 223, 325*(107:361), 596*.

major L.   Astrantia
•••• An established garden escape; widely naturalised in woods, quarries, on railway banks, roadsides and by rivers; known since about 1840 in Stoke Wood, near Craven Arms (Salop). Europe. 1, 20, 22*, 26*, 27*, 95*. Both subsp. *major* and subsp. *elatior* (Friv.) Maly (*A. elatior* Friv.) occur (see BM). Many recent records are referable to subsp. *carinthiaca* (Jacq.) Arcang. 434.

maxima Pallas
- (but perhaps overlooked as *A. major*). A garden escape. SW Asia. BM, E, RNG.   7, 20, 39*, 51, 52*, 126, 300(1909:41), 354(5:655; 9:652). Incl. *A. helleborifolia* Salisb.

minor L.   Lesser Masterwort
Pre-1930 only. Europe. 1, 21, 23*, 27*, 28*, 193(p.165).

ERYNGIUM L.
amethystinum L.
- A long-persistent or established garden escape known since 1963 on dunes at Llandudno (Caerns); also a garden relic at Penarth (Glam). C & SE Europe. NMW, RNG. 20, 24*, 29*, 51*, 52*, 370(13:334; 14:192).

[*bourgatii* Gouan   In error for *E. giganteum*. 335(113:78), 442.]

[*dichotomum* Desf.   In error for *E. tripartitum*. 371(376:26).]

giganteum M.Bieb.   Tall Eryngo
- A garden escape established since 1986 on the side of a gravel pit at Ben Rhydding, near Otley (MW Yorks); casual elsewhere. Turkey, Caucasia. BM. 20, 335(114:54), 354(11:480), 594*, 634*, 680*, 730*. *E. haussknechtii* Bornm.

× **oliverianum** Delaroche
   Pre-1930 only. OXF. 51, 53*, 54*, 354(9:22). Originated in cultivation.
   Probably *E. giganteum* × *E. planum*.

planum L.   Blue Eryngo
   •• A garden escape established since before 1965 on sandy ground at
   Littlestone-on-Sea (E Kent) and recorded from a few other places in the south.
   Eurasia. EXR, OXF, RNG. 7, 20*, 22*, 236, 612*.

tripartitum Desf.
   • A garden escape persistent on a roadside at Headley Mill (S Hants),
   probably now gone; also recorded from waste ground at Sandwich Bay
   (E Kent). Origin obscure, probably a hybrid between *E. planum* and an
   unknown species. E, MNE. 51, 53*, 54*, 55*, 236, 371(376:26).

ECHINOPHORA L.
   [*capitata* sphalm. In error for *Echinaria capitata* (L.) Desf. 1, 447.]

spinosa L.
   Pre-1930 only; possibly a native plant now extinct. Medit. OXF. 1, 21, 24*,
   29*, 163, 176, 370(8:397-399).

MYRRHOIDES Heister ex Fabr.

nodosa (L.) Cannon
   Pre-1930 only. Europe, N Africa, SW Asia. LIV, OXF, RNG, SLBI. 2, 21,
   44*, 46*, 300(1907:39). *Physocaulis nodosus* (L.) Koch; *Chaerophyllum
   nodosum* (L.) Crantz.

CHAEROPHYLLUM L.

aromaticum L.   Broad-leaved Chervil
   • Collected from Milton Road, Cambridge, probably a garden escape.
   No modern records. Europe. CGE. 1, 21, 159, 354(5:764), 683*.

aureum L.   Golden Chervil
   •• An established garden escape; known since 1909 in abundance in water-
   meadows near Callander (W Perth), locally frequent at Calton Hill (Midlothian)
   and well naturalised in the grounds of Buckingham Palace (Middlesex), now
   gone; casual elsewhere. Europe, SW Asia. BM, CGE, LANC, NMW, OXF,
   RNG. 1, 20*, 22*, 35*, 95*, 192, 300(1911:168), 349(3:232), 451,
   826*(17:158-160). *Myrrhis aurea* (L.) All.

hirsutum L.   Hairy Chervil
   • Established since 1979 on a roadside in Westmorland, possibly as a garden
   escape; also collected from Dungannon (Tyrone). Europe. LANC, RNG.
   20*, 22*, 26*, 27*.

ANTHRISCUS Pers.

cerefolium (L.) Hoffm.   Garden Chervil
   ••• A garden escape established for over 100 years on rocks near Ross
   (Herefords); casual or persistent elsewhere; decreasing. Europe, N Africa,
   SW Asia. BM, E, LANC, LSR, OXF, RNG. 1, 20*, 22*, 80*, 88*, 283,
   475. *Cerefolium cerefolium* (L.) Schinz & Thell.; *C. sativum* Besser;

Chaerefolium sylvestre (L.) Schinz & Thell.; *Chaerophyllum cerefolium* (L.) Crantz; *C. sativum* Lam.

SCANDIX L.
australis L.
> Pre-1930 only. Medit., SW Asia. BM, LCN, LIV, LSR, SLBI. 1, 9, 21, 44*, 45*, 258, 303*(53:32), 354(1:415). Incl. *S. falcata* Loudon; *S. grandiflora* L.

iberica M.Bieb.
> Pre-1930 only. SW Asia. E. 2, 9, 21, 44*, 300(1906:103), 354(5:108,656), 1271*. *Scandicium ibericum* (M.Bieb.) Thell.

[*pecten-veneris* L. Accepted, with reservations, as native.]

stellata Banks & Sol.
> • A bird-seed casual. Medit., SW & C Asia. OXF. 1, 9, 19, 21, 44*, 46*, 151, 258, 303(26:18), 729*, 1271*. *S. pinnatifida* Vent.; *Scandicium pinnatifidum* (Vent.) Thell.; *S. stellatum* (Banks & Sol.) Thell.

OSMORHIZA Raf.
chilensis Hook. & Arn.
> • A wool casual. S America. HbTBR. 370(10:282), 665*, 669*. According to 864(45:181-182) the correct name is *O. berteroi* DC.

[*MYRRHIS* Miller
odorata (L.) Scop. Accepted, with reservations, as native.]

CHAETOSCIADIUM Boiss.
trichospermum (L.) Boiss.
> Pre-1930 only. E Medit. RNG. 44*. *Scandix trichosperma* L.

CORIANDRUM L.
sativum L.   Coriander
> ••••• A bird-seed and spice casual. Formerly more often an escape from cultivation and a grain casual. Origin obscure. 1, 19, 20*, 22*, 95*, 191, 303(21:24).

tordylium (Fenzl) Bornm.
> Pre-1930 only. E Medit. OXF. 8, 39*, 185, 354(9:22). *C. tordylioides* Boiss.

BIFORA Hoffm.
radians M.Bieb.
> •• A grain casual. C & S Europe, SW Asia. BM, E, SLBI. 1, 9, 21, 22*, 24*, 119, 182, 475, 1141*. *Anidrum radians* (M.Bieb.) Kuntze.

testiculata (L.) Roth
> • A grain casual. No modern records. Medit., SW Asia. OXF. 1, 21, 44*, 45*, 46*, 199, 248, 371(331:21). *Anidrum testiculatum* (L.) Kuntze; *Coriandrum testiculatum* L.

SMYRNIUM L.
olusatrum L.   Alexanders
●●●●● An established escape from former cultivation; long naturalised in many
places throughout the British Is, especially near the sea. W Europe, Medit.
1, 20*, 22*, 35*, 95*.

perfoliatum L.   Perfoliate Alexanders
●●● An escape from botanic gardens, long persistent as a weed in parks and
gardens in widely scattered localities in England; casual elsewhere.
C & S Europe, SW Asia. BM, LANC, LNHS, RNG, SUN. 12, 20*, 22*,
29*, 95*, 192, 209.

PIMPINELLA L.
affinis Ledeb.
● An established alien, probably from grass seed, found in 1990 in quantity at
Nelson Hill, Southampton (S Hants). Turkey, Caucasus, Iran. HbRPB. 39,
404, 630, 668*.

anisum L.   Anise
● A casual, perhaps a spice or bird-seed alien. No modern records. Origin
obscure. BM, GLR. 7, 9, 21, 22*, 44*, 192, 245, 1156*.

cf. **hirtella** (Hochst.) A.Rich.
● A casual garden weed, vector unknown. Tropical Africa. HbEJC. 253,
327(110:327-372), 736. *Tragium hirtellum* Hochst. Early works, like 676,
wrongly considered this plant to be conspecific with Mediterranean
P. peregrina L. Our plant is possibly in error for *P. affinis*.

AEGOPODIUM L.
podagraria L.   Goutweed
●●●●● An established garden escape, abundantly naturalised throughout most
of the British Is, yet rarely far from habitation; an ancient introduction
as a medicinal and pot herb, which has been claimed as a native. Eurasia.
1, 20, 22*, 33*, 35*.

AETHUSA L.
[*cynapium* L. subsp. *agrestis* (Wallr.) Dostál   Accepted, with reservations, as
native.]
cynapium L. subsp. **cynapioides** (M.Bieb.) Nyman
● Recorded from a suburban footpath near Bromley (W Kent) and from Jersey
(Channel Is). C Europe. 20, 21, 303(59:49), 1146*. *A. cynapioides* M.Bieb.

FOENICULUM Miller
vulgare Miller subsp. **piperitum** (Ucria) Cout.   Bitter Fennel
Pre-1930 only. Medit. 7, 21, 24*, 354(4:415; 5:765). *F. piperitum* (Ucria)
Sweet.
[*vulgare* Miller subsp. *vulgare*   Accepted, with reservations, as native.]

ANETHUM L.
graveolens L.   **Dill**
  •••• A persistent wool, spice, bird-seed and grain alien.   SW Asia;
(widespread). 1, 9, 14, 19, 20*, 44*, 95*, 126. *Peucedanum graveolens* (L.)
Hiern, non S.Watson.

CACHRYS L.
uechtritzii (Boiss. & Hausskn.) Herrnst. & Heyn
  • A casual at Barry docks (Glam), vector unknown.   No modern records.
Turkey. OXF, SLBI.   10, 39, 354(9:829; 10:22). *Prangos uechtritzii* Boiss.
& Hausskn. According to 472, the RNG specimen is probably referable to
C. sicula L.

BUPLEURUM L.
croceum Fenzl
  Pre-1930 only.   Turkey, SW Asia.   OXF, SLBI.   2, 39, 300(1905:100),
354(5:108), 1233*, 1271*.
fruticosum L.   **Shrubby Hare's-ear**
  •• An established garden escape; naturalised in a few widely scattered localities
in the south, as on a railway bank at South Darenth (W Kent). Medit. BM,
LANC, OXF, RNG, SLBI. 1, 20*, 95*, 126, 191, 312*(n.s. t.408), 684*.
[*glumaceum* Sibth. & Smith (*B. aristatum* Bartling)   Unconfirmed; probably
always in error for *B. odontites* or native *B. baldense*. LANC. 9, 121, 247,
281, 300(1909:41).]
lancifolium Hornem. sensu lato   **False Thorow-wax**
  •••• A bird-seed casual. Medit., SW Asia. 1, 19, 20*, 22*, 95*, 126*,
303(15:13). *B. heterophyllum* Link; *B. protractum* var. *heterophyllum* (Link)
Boiss.; *B. subovatum* var. *heterophyllum* (Link) H.Wolff. Incl. *B. subovatum*
Link ex Sprengel (*B. protractum* Hoffsgg. & Link; *B. intermedium* (Urv.)
Steudel; *B. rotundifolium* var. *intermedium* Urv.). Modern records are all of
the segregate *B. subovatum*.
nodiflorum Sibth. & Smith
  Pre-1930 only. E Medit. 7, 39, 44*, 300(1908:102), 555*, 668*.   Incl.
B. nanum Poiret (*B. proliferum* Del.)
odontites L.
  •• A grain and bird-seed casual. Medit.; (widespread). BM, CGE, LSR,
OXF. 1, 19, 20, 21, 44*, 46*, 160, 354(6:283). *B. fontanesii* Guss.
ex Caruel.
petraeum L.   **Rock Hare's-ear**
  • A casual on a tip at Welwyn (Herts), vector unknown. No modern records.
Alps. 21, 23*, 27*, 354(10:827).
[*rotundifolium* L.   Accepted, with reservations, as native.]
semicompositum L.
  • A wool casual. Medit., SW Asia. E, RNG. 3, 14, 21, 32*, 44*, 45*,
546*. *B. glaucum* Ledeb.

CUMINUM L.
cyminum L.   Cumin
•• A bird-seed and spice casual. Probably Egypt; (Africa, SW Asia). BM, E, LTN, RNG. 7, 19, 20*, 46*, 80*, 126, 145, 191, 248, 1233*.

CICLOSPERMUM Lagasca
leptophyllum (Pers.) Britton & E.Wilson   Slender Celery
•• A wool and grain casual. Tropical & subtropical America; (widespread). E, LTN, NMW, RNG. 3*, 14, 21, 182, 320(61:129-133), 520*, 533*, 546a*. *Apium ammi* Urban, nom. illeg., non (L.) Crantz; *A. leptophyllum* (Pers.) F.Muell. ex Benth.; *A. tenuifolium* (Moench) Thell., nom. illeg.; *Pimpinella leptophylla* Pers.; *Sison ammi* sensu Jacq., non L.

TRACHYSPERMUM Link
ammi (L.) Sprague ex Turrill   Ajowan
••• A spice casual; perhaps also a bird-seed and oil-seed alien. Origin obscure. BM, CGE, E, LTN, OXF, RNG. 7, 19, 20*, 95*, 126, 371(340:27). *T. copticum* (L.) Link; *Ammi copticum* L.; *Carum aromaticum* (L.) Druce, non Salisb.; *C. copticum* Benth. & Hook.; *Sison ammi* L. Much confused nomenclaturally with *Ciclospermum leptophyllum*.

SPERMOLEPIS Raf.
divaricatus (Walter) Britton
• A wool casual. Southern USA, Mexico. HbEJC. 38*, 370(10:283), 1205*, 1331*.
echinatus (Nutt.) Heller
• A wool casual. Southern USA, Mexico. HbEJC. 38*, 40, 370(10:283), 1205*, 1357*.

PETROSELINUM Hill
crispum (Miller) Nyman ex A.W.Hill   Garden Parsley
•••• An established escape from cultivation; naturalised in quarries, on rocks, walls and roadsides, mainly in the south and east. Probably W Asia; (widespread). 1, 20*, 35*, 95*. *P. hortense* Hoffm.; *P. petroselinum* (L.) Karsten; *P. sativum* Hoffm.; *Apium crispum* Miller; *A. petroselinum* L.; *Carum petroselinum* (L.) Benth.

RIDOLFIA Moris
segetum (L.) Moris   False Fennel
•• A wool and bird-seed casual. Medit. BM, CGE, LTN, RNG. 1, 14, 19, 20*, 44*, 45*, 370(9:189), 737*. *Carum ridolfia* Benth. & Hook.f.; *C. segetum* (L.) Benth. & Hook.f. ex Arcang.

AMMI L.
majus L.   Bullwort
•••• A wool, grain, agricultural seed and bird-seed alien sometimes persisting a few years. Medit.; (widespread). 1, 11, 14, 19, 20*, 22*, 95*, 235, 737*. Incl. *A. glaucifolium* L.

visnaga (L.) Lam.   Toothpick-plant
••• A bird-seed, grain and wool alien sometimes persisting a few years.
Medit. BM, E, LIVU, LSR, OXF, RNG. 1, 11, 14, 19, 20*, 22*, 95*, 737*.
Daucus visnaga L.; *Visnaga daucoides* P.Gaertner.

FALCARIA Fabr.
vulgaris Bernh.   Longleaf
•••• An established grain and agricultural seed alien; locally naturalised in
fields, chalk pits, gravel workings and on roadsides, especially in southern
England and the Channel Is. Eurasia. 1, 11, 20*, 22*, 26*, 95*, 144.
F. falcaria (L.) Karsten; *F. rivinii* Host; *Prionitis falcaria* (L.) Dumort.;
Sium falcaria L.

[*CARUM* L.
carvi L. Accepted, with reservations, as native.]

ANGELICA L.
archangelica L.   Garden Angelica
•••• An established escape from former cultivation; long naturalised by canals
and rivers, especially by the river Thames in the London area and by the
Manchester Ship Canal. N Europe, W Asia. 1, 20*, 22*, 34*, 95*, 126, 230,
1083*. *Archangelica archangelica* (L.) Karsten; incl. *A. decurrens* Ledeb.;
A. officinalis Hoffm.
pachycarpa Lange
Collected in 1993 from St Peter-in-the-Wood, Guernsey (Channel Is), vector
unknown. Spain, Portugal; (New Zealand). STP. 21, 36c, 565*, 895*(3:137-
161), 1189*.

LEVISTICUM Hill
officinale Koch   Lovage
••• A relic and escape from cultivation established or long persistent in
widely scattered localities. SW Asia. BM, CGE, LANC, OXF, RNG, ZCM.
1, 20*, 22*, 25*, 95*, 156, 199, 1141*. *L. levisticum* (L.) Karsten;
L. paludapifolium Asch.

CAPNOPHYLLUM Gaertner
peregrinum (L.) Lange
Pre-1930 only. W Medit. OXF. 2, 12, 21, 44*, 46*, 300(1906:103),
354(5:31). *C. dichotomum* Lagasca; *Tordylium peregrinum* L.

FERULA L.
communis L.   Giant Fennel
• An established alien, possibly brought in with building materials, or a garden
escape, persistent for over 30 years at Oundle (Northants), now gone; also
recently recorded from a park at South Shields (Co Durham) and by the A11
trunk road near Mildenhall roundabout (W Suffolk). Medit. BM. 13, 20, 21,
32*, 44*, 303(50*:1,30-32; 51:31-33).

PEUCEDANUM L.
austriacum (Jacq.) Koch
- Recorded in 1986 from Broadstone (Dorset), possibly a garden escape. C & SE Europe. BM. 51b, 351(108:212), 546*, 683*, 1190*.

ostruthium (L.) Koch   Masterwort
- •••• An established escape or relic of former cultivation; locally naturalised by rivers and on roadsides, especially in Scotland. C & S Europe. 1, 20*, 22*, 26*, 95*. *Imperatoria ostruthium* L.

PASTINACA L.
sativa L. subsp. **urens** (Req. ex Godron) Čelak.
- Recorded without locality from Suffolk. Europe, SW Asia. 21, 39, 13, 546*, 695*. *P. opaca* Bernh.; *P. teretiuscula* Boiss.; *P. umbrosa* Steven ex DC.; *P. urens* Req. ex Godron.

HERACLEUM L.
canescens Lindley
Pre-1930 only. NW Himalayas. 181, 604, 1063*. *H. hirsutum* Edgew.

mantegazzianum group   Giant Hogweeds
- ••••• Established garden escapes; naturalised and dominant in many places throughout the British Is, especially by rivers; increasing. SW Asia. 1, 20*, 22*, 34*, 95*. Incl. *H. giganteum* Fischer; *H. mantegazzianum* Sommier & Levier; *H. persicum* Desf. ex Fischer; *H. sibiricum* sphalm.; *H. stevenii* Manden.; and *H. villosum* Fischer. The name *H. lehmannianum* Bunge has also been suggested. Most records are referable to *H. mantegazzianum*. *H. persicum* has been reported from Cheltenham (E Gloucs), Aldeburgh (E Suffolk), Usk (Mons), Cambridge and Edinburgh (CGE, RNG); and plants thought to be *H. stevenii* from Wateringbury (W Kent). *H. lanatum* Michaux and *H. maximum* Bartram may also occur. 365(3:19). Hybrids with native *H. sphondylium* have also been recorded. E, OXF. 20, 1311.

TORDYLIUM L.
aegyptiacum (L.) Lam.
- A grain casual. No modern records. E Medit., SW Asia. 1, 39, 44*, 354(12:42).

[*maximum* L. Accepted, with reservations, as native.]
[*officinale* L. In error for *T. maximum*. 354(5:765).]
cf. **syriacum** L.
- Collected in 1931 from Stoneybrough, Thirsk (NW Yorks), vector unknown. Formerly a ballast alien. E Medit. BM. 7, 39, 44*, 286.

LASER Borkh. ex P.Gaertner, Meyer & Scherb.
trilobum (L.) Borkh.   Laser
- A casual on a quayside at King's Lynn (W Norfolk), vector unknown. No modern records. Formerly persistent for many years at Cherry Hinton (Cambs). Europe, SW Asia. BM, LIV, OXF. 1, 20, 21, 22*, 39*, 235, 1099. *Laserpitium trilobum* L.; *Siler trilobum* (L.) Crantz.

LASERPITIUM L.
latifolium L.   Broad-leaved Sermountain
- A casual at Hull (SE Yorks), vector unknown. No modern records. Europe. HbFEC. 21, 22*, 23*, 26*, 134.

TORILIS Adans.
[*arvensis* (Hudson) Link subsp. *arvensis*   Accepted, with reservations, as native.]
arvensis subsp. **neglecta** (Schultes) Thell.
- A wool and bird-seed casual.   S Europe, SW Asia.   14, 19, 251, 370(10:281), 737*.   *T. infesta* (L.) Sprengel; *T. neglecta* Schultes; incl. *T. africana* Sprengel.

leptophylla (L.) Reichb.f.
- • A wool, grain and bird-seed casual. Medit.   BM, CGE, E, LTN, OXF, RNG. 1, 11, 14, 19, 21, 39*, 44*, 45*, 95.   *Caucalis leptophylla* L.

stocksiana (Boiss.) Grossh.
- A wool casual. Iraq, Iran, Afghanistan. CGE, K, RNG. 524, 690, 1271*.

tenella (Del.) Reichb.f.
Pre-1930 only. E Medit., SW Asia. 2, 21, 39, 44*, 300(1905:100), 1271*. *Caucalis tenella* Del.

ucranica Sprengel
Pre-1930 only.   C & S Europe, SW Asia.   7, 21, 39*, 354(7:438). *T. macrocarpa* sphalm.; *T. microcarpa* Besser; *T. microcarpa* var. *aculeata* Boiss.; *Caucalis grandiflora* (Boiss.) Druce, non L. var. *aculeata* (Boiss.) Druce.

CAUCALIS L.
bischoffii Kozo-Polj.
- A grain casual. No modern records. Europe, SW Asia. BRIST. 7, 9, 21, 354(7:772-774; 12:42), 623a, 715*, 1320*.   *C. daucoides* L. var. *muricata* (Bischoff) Gren. & Godron; *C. muricata* Bischoff, non Crantz.   Possibly conspecific with *C. platycarpos*.

platycarpos L.   Small Bur-parsley
- ••• A grain, bird-seed and wool casual; decreasing and now very rarely found. Formerly a long-persistent weed on arable land. Medit., SW Asia. BM, CGE, E, LANC, NMW, OXF. 1, 8, 19, 20, 22*, 26*, 33*, 191, 303*(15:14-15). *C. daucoides* L.(1767), non L.(1753); *C. lappula* Grande; *C. royeni* (L.) Crantz.

TURGENIA Hoffm.
latifolia (L.) Hoffm.   Greater Bur-parsley
- ••• A grain and bird-seed casual; decreasing, and now very rarely found. Europe, N Africa, SW Asia. BM, CGE, E, NMW, OXF, RNG. 1, 19, 20, 22*, 25*, 44*, 303*(15:14-15), 354(12:490). *Caucalis latifolia* L.

ORLAYA Hoffm.
daucoides (L.) Greuter   Flat-fruited Orlaya
- A casual on a building site at Cambridge, vector unknown. Medit.,
SW Asia. CGE. 21, 39*, 44*, 303*(15:14-15). *O. kochii* Heyw.;
O. platycarpos auct.; *O. topaliana* Beauv.; *Caucalis daucoides* L.(1753),
non L.(1767); *C. platycarpos* L.(1767), non L.(1753).
grandiflora (L.) Hoffm.   Large-flowered Orlaya
- A casual at Bristol docks (W Gloucs), vector unknown. No modern records.
Europe, N Africa, SW Asia. SLBI. 1, 21, 22*, 26*, 32*, 175, 206, 245,
1141*. *Caucalis grandiflora* L.; *Daucus grandiflorus* (L.) Scop.

DAUCUS L.
carota L. subsp. **maritimus** (Lam.) Battand.   Mediterranean Carrot
- A casual, vector unknown. Medit. 21, 556*, 683*. *D. maritimus* Lam.; *D.
parviflorus* Desf.
carota L. subsp. **sativus** (Hoffm.) Arcang.   Carrot
- •• A casual escape from cultivation. Origin obscure. 7, 20, 21, 683, 1238*,
1280*. *D. carota* var. *sativus* Hoffm.; *D. sativus* (Hoffm.) Roehl. Many
authors do not separate this taxon from native subsp. *carota*.
crinitus Desf.
- A wool casual. No modern records. W Medit. LTN. 14, 21, 45*, 46*,
145.
cf. **durieua** Lange
Pre-1930 only. SW Europe, N Africa, SW Asia. E, OXF. 3, 21, 44*, 46*,
354(5:31), 737*, 914*(12:228-229). *D. hispanicus* (Lam.) Druce, non Gouan;
Durieua hispanica (Lam.) Boiss. & Reuter. Incl. *D. subsessilis* Boiss.
glochidiatus (Labill.) Fischer, C.Meyer & Avé-Lall.   Australian Carrot
- •• A wool casual. Australasia. ABD, CGE, E, LTN, OXF, RNG. 3*, 14,
20*, 37*, 143, 277, 303*(31:13-14), 370(10:282). *D. brachiatus* Sieb.
ex DC.; *D. platyacanthus* Thell.; *Daycus glochidiatus* sphalm.; *Scandix
glochidiata* Labill. Possibly conspecific with *D. montanus*.
guttatus Sibth. & Smith
- A wool casual. Medit. K, RNG. 17, 21, 44*, 683*, 722*. *D. bicolor*
Sibth. & Smith; *D. setulosus* Guss. ex DC.
littoralis Sibth. & Smith
Pre-1930 only. E Medit. 2, 39*, 44*, 300(1905:100), 658, 722*. *D. glaber*
(Forsskål) Thell., non Opiz ex Čelak.
montanus Humb. & Bonpl. ex Schultes & Schultes f.
- A wool casual. Mexico, S America. HbEJC. 370(10:282), 665*.
muricatus (L.) L.
- A bird-seed casual. Medit. 9, 19, 21, 45*, 46*, 191, 683*.
pusillus Michaux   Rattle-snake-weed
- A wool casual. America. BM, E, K, RNG. 17, 41*, 42, 370(10:282).

AGROCHARIS Hochst.
gracilis Hook.f.
- A wool casual. Tropical W Africa. RNG. 145(as *Caucalis melanantha*), 509*.

melanantha Hochst.
- A wool casual. Tropical E Africa. RNG. 14, 48*, 1336*. *Caucalis melanantha* (Hochst.) Hiern.

PSEUDORLAYA (Murb.) Murb.
pumila (L.) Grande
Pre-1930 only. Medit., SW Asia. OXF. 2, 21, 39*, 44*, 46*, 149. *P. maritima* (L.) Murb.; *Caucalis maritima* Gouan; *C. pumila* L.; *Daucus pumilus* (L.) Hoffsgg. & Link; *Orlaya maritima* (L.) Koch.

LOGANIACEAE
DESFONTAINEA Ruíz Lopez & Pavón
spinosa Ruíz Lopez & Pavón
Pre-1930 only. Andes. OXF. 7, 51, 52*, 61*, 62*, 63*, 354(5:38).

GENTIANACEAE
[*GENTIANELLA* Moench
aurea (L.) K.Smith In error. 253.
ciliata (L.) Borkh. Accepted, with reservations, as native.
tenella (Rottb.) Börner In error. 253.]

GENTIANA L.
acaulis group   **Trumpet Gentians**
- Introduced and established for over 30 years in chalk grassland near Betchworth (Surrey). Formerly introduced in several places in the north, and established on the headland of Knoydart (Westerness). C & S Europe. 20, 22*, 23*, 26*, 27*, 199, 331(40:17,19), 354(4:421; 5:667). Incl. *G. acaulis* L.; *G. clusii* Perrier & Song.; *G. excisa* C.Presl; *G. kochiana* Perrier & Song.

asclepiadea L.   **Willow Gentian**
- An established garden escape recorded from Twyford Lodge (W Sussex), Chelwood Gate (E Sussex), Holden Clough (MW Yorks) and Rhu (Dunbarton). C & S Europe, SW Asia. OXF. 20, 22*, 23*, 25*, 27*, 303(37:15), 312*(n.s. t.191), 370(1:120), 371(343:28), 1181*.

cruciata L.   **Cross Gentian**
Pre-1930 only. Eurasia. MNE. 21, 22*, 23*, 25*, 27*, 236.

lagodechiana (Kusn.) Grossh. ex Möller
- Introduced in an old bomb crater at Brockham Hill (Surrey), now gone. Caucasus. RNG. 51, 55*, 199, 331(30:7), 594). *G. septemfida* Pallas var. *lagodechiana* Kusn.

[*SWERTIA* L.
perennis L. A dubious record. BM, LIV. 354(5:778).]

APOCYNACEAE
VINCA L.

difformis Pourret   Intermediate Periwinkle
- A persistent or established garden escape recorded from Chevening Park (W Kent. SW Europe. MNE. 20, 24*, 32*, 51*, 52*, 371(403:39). *V. acutiflora* Bertol.

[*herbacea* Waldst. & Kit. In error for *V. major* var. *oxyloba*.]

major L.   Greater Periwinkle
- ••••• An established garden escape; naturalised in woods and hedgerows in many places throughout the British Is, especially in the south and east. Medit., SW Asia. 1, 20, 22*, 26*, 52*, 56*, 349(5:341). *V. herbacea* auct., non Waldst. & Kit.; *V. pubescens* Urv.; incl. subsp. *hirsuta* auct., non (Boiss.) Stearn; var. *oxyloba* Stearn; *V.*'Oxyloba'.

[*minor* L. Accepted, with reservations, as native.]

ASCLEPIADACEAE
VINCETOXICUM Wolf

aff. **nigrum** (L.) Moench   Black Swallow-wort
- A casual, perhaps a seed impurity, in a garden at Biddenden (E Kent). SW Europe. 21, 29, 30, 312*(t.2390), 371(394:35), 737*, 825(2:64). *Cynanchum nigrum* (L.) Pers.

NOLANACEAE
NOLANA L.f.

acuminata (Miers) Miers ex Dunal
- A casual garden escape. Chile. 51b, 342(Feb89), 511, 594*, 728a*(t.305), 1013*. *N. atriplicifolia* D.Don ex Sweet; *N. grandiflora* Lehm. ex G.Don; *N. paradoxa* subsp. *atriplicifolia* (D.Don ex Sweet) Mesa.

humifusa (Gouan) I.M.Johnston   Trailing Chilean-bellflower
Pre-1930 only. Peru. 2, 51, 267, 312*(t.731), 650*. *N. prostrata* L.

paradoxa Lindley   Chilean-bellflower
- A casual garden escape. Chile. 51*, 312*(t.2604), 456, 510*, 610*, 674*, 893*(6:43-47).

SOLANACEAE
NICANDRA Adans.

physalodes (L.) Gaertner   Apple-of-Peru
- •••• A grain, bird-seed, wool and oil-seed alien and garden escape established in mild areas; known for about 50 years as a weed of cultivation on Tresco (Scilly) and in Jersey (Channel Is). Peru; (Australasia). 1, 3*, 14, 18, 19, 20*, 22*, 29*, 201, 207, 1247*.

LYCIUM L.
barbarum L. sensu lato   Chinese Boxthorn
●●●●● An established garden escape; widely naturalised in hedges, by railways
and on dunes.   China.   7, 20, 22*, 33*, 52*.   *L. chinense* Miller;
L. europaeum hort., non L.; *L. halimifolium* Miller; *L. ovatum* Veillard;
L. rhombifolium Dippel.

EXODECONUS Raf.
glutinosa (Schldl.) ined.
Pre-1930 only.   Mexico.   7, 313(22:42), 554, 843(69:82-84).   *Cacabus
mexicanus* S.Watson; *Physalis eximia* Standley; *P. glutinosa* Schldl.   Possibly
an error for *Physalis peruviana*.   6.

SCOPOLIA Jacq.
carniolica Jacq.   Scopolia
● A persistent garden escape or relic on waste ground at Wisley (Surrey), on
the site of Warley Place gardens (S Essex) and in a shrubbery at Brinton
(E Norfolk).   C Europe.   RNG.   20, 23*, 29*, 53*, 57*, 199, 235,
303(32:19), 371(352:28).

PHYSOCHLAINA Don
orientalis (M.Bieb.) Don
Pre-1930 only.   SW Asia.   OXF.   7, 39, 51, 312*(t.2414), 354(5:570).

HYOSCYAMUS L.
albus L.   White Henbane
●● A casual from Egyptian woollen rags; possibly also a bird-seed casual.
Medit.   BM, E, LIV, NMW, OXF, RNG.   1, 21, 24*, 32*, 44a*, 191, 251.
aureus L.   Golden Henbane
Pre-1930 only.   E Medit.   1, 21, 29*, 32*, 44a*, 166.
muticus L.
● A casual at the Crown Wallpaper tip, Darwen (S Lancs) and on a tip at
Dagenham (S Essex), vector unknown.   N Africa.   BON, RNG.   7, 46*, 251,
354(8:33; 10:628).
reticulatus L.
● A grain casual.   No modern records.   SW Asia, NE Africa.   LCN.   21,
44a*, 160.

SALPICHROA Miers
origanifolia (Lam.) Thell.   Cock's-eggs
●● An established garden or greenhouse escape; naturalised in several
places on the coasts of southern England and the Channel Is; known since 1927
from the Undercliff at St Lawrence (Wight).   Temperate S America;
(widespread).   BFT, BM, CGE, E, OXF, RNG.   13, 20*, 22*, 51*, 117*,
158, 349(1:157; 2:136), 576*.   *S. rhomboidea* (Gillies & Hook.) Miers;
Physalis origanifolia Lam.

PHYSALIS L.
acutifolia (Miers) Sandwith   Wright's Ground-cherry
  • A wool casual. SW USA, Mexico. BM, K, RNG. 7, 42, 236, 328(14:231-233), 500*, 699*, 1248*. *P. wrightii* A.Gray; *Saracha acutifolia* Miers.
alkekengi L. sensu lato   Japanese-lantern
  •••• A persistent garden escape, perhaps established in a few places on railway banks and in old quarries in the south. Eurasia. 1, 20*, 22*, 26*, 54*, 617*. Incl. *P. franchetii* Masters; *P.* × *bunyardii* hort.
angulata L.   Cut-leaved Ground-cherry
  •• A wool and oil-seed casual. Tropical N America; (tropics, subtropics). DGS, E, RNG. 7, 12, 21, 38*, 201, 303(21:18; 35:14), 369(26:58), 582*.
fendleri A.Gray
  Pre-1930 only. SW USA. BM. 7, 42, 354(7:446), 500*, 652, 699, 1303. *P. hederifolia* A.Gray var. *cordifolia* (A.Gray) Waterfall.
foetens Poiret
  Pre-1930 only. Mexico, Tropical America. OXF. 7, 9, 354(4:423), 707, 710*.
ixocarpa Brot. ex Hornem.   Tomatillo
  ••• A wool and bird-seed casual. Mexico; (widespread). BM, CGE, E, LIV, OXF, RNG. 3*, 14, 20*, 36c, 38*, 303(21:18), 354*(4:203), 1280*.
minima L.
  Pre-1930 only. Tropics. OXF. 7, 48, 354(5:39; 7:1055), 576*, 676, 1006*. *P. lagascae* Roemer & Schultes; *P. micrantha* Link.
cf. **pendula** Rydb.   Lance-leaved Ground-cherry
  • A wool casual and a casual greenhouse weed. N America. RNG, HbCGH. 38*, 40, 699, 1218.
peruviana L.   Cape-gooseberry
  ••• A grain, bird-seed, wool, oil-seed and food refuse casual. S America; (widespread). BM, E, LTN, NMW, OXF, RNG. 7, 11, 14, 19, 20*, 52*, 80*, 903(46:109-112), 1006*. *P. edulis* Sims. Some old records may be referable to the similar *P. pruinosa* L. 38*, 915*(85:413-416).
philadelphica Lam.   Large-flowered Tomatillo
  •• A food refuse and grain casual. C America. E. 20, 36c, 236, 303*(21:18), 331(64:120), 350(37:29). *P. ixocarpa* auct., non Brot. ex Hornem. Much confused in the literature with Tomatillo; the vernacular name has been applied to both taxa, but was probably intended for this species (e.g. as in 576*). Its occurrence in sewage sludge and on refuse tips suggests that this is the major crop plant. 1280 claims that horticultural hybrids also exist.
pubescens L.   Low Ground-cherry
  •• (but over-recorded for *P. peruviana*). A casual, vector unknown. Formerly a wool alien. No confirmed modern records. N & S America; (widespread). E, LIV, OXF. 3*, 9, 21, 36c, 38*, 41*, 191, 192, 617*.
viscosa L.   Stellate Ground-cherry
  • A casual, vector unknown. USA, Mexico, S America; (Australia). RNG. 38*, 197, 545, 576*, 699, 825(1:65-71). *P. pensylvanica* L.

CAPSICUM L.
annuum L.   Sweet Pepper
••• A bird-seed and food refuse casual. Tropical America; (widespread).
K, LIV, RNG. 9, 19, 20, 51*, 67*, 68*, 80*.

LYCOPERSICON Miller
esculentum Miller   Tomato
••••• A persistent food refuse alien sometimes abundant on sewage farms and
sea shores. S America; (widespread as a crop). 1, 20, 22*, 66*, 68*, 80*.
L. lycopersicum (L.) Karsten; *Solanum lycopersicum* L.
pimpinellifolium Miller   Currant Tomato
• A casual garden escape. Peru, Ecuador. HbACL. 51, 66, 199, 510*.
Probably better treated as *L. esculentum* var. *pimpinellifolium* (Jusl.) Miller.

SOLANUM L.
acaule Bitter
Pre-1930 only. Bolivia, Argentina. OXF. 354(9:31), 822(11:391-394;
12:453), 1051*. Recorded as var. *subexinterruptum* Bitter.
americanum Miller   Small-flowered Nightshade
•• A wool and oil-seed casual. Tropics, subtropics. E, HTN, K, RNG.
14, 21, 36c*, 328(27:95-114), 331(62:106), 500*, 620*, 1109*. Two distinct
varieties have occurred, var. *americanum* (incl. *S. adventitium* Polgár) and
var. *nodiflorum* (Jacq.) Edmonds (*S. nodiflorum* Jacq.). Other taxa may also
have been recorded under this name.
[*angustifolium* Miller (*S. cornutum* Lam.)   In error for *S. rostratum*.
303(17:18), 882*(11:359-426). This plant, confined to Mexico, is not a weedy
species and is unlikely to occur; it has been wrongly equated to *S. rostratum* in
some recent works. 413, 882*(2:359-426).]
[*aviculare* Forster f.   In error for *S. laciniatum*. 207.]
capsicoides All.   Cockroach-berry
• A casual at Cardiff and Porthcawl (both Glam) and in N Essex, vector
unknown. Central America; (tropics). NMW, OXF. 7, 48*, 49, 51*, 156,
303*(35:12,15), 354(11:491), 1006*, 1330, 1352*. Incl. *S. aculeatissimum*
Jacq.; *S. ciliatum* Lam.
carolinense L.   Horse-nettle
• An oil-seed casual. Eastern N America. RNG. 18, 20, 36c, 38*, 40, 236,
731*. All plants from Australia under this name are referable to *S. dimidiatum*
Raf., from SE USA. 576.
chenopodioides Lam.   Tall Nightshade
•• Established extensively as a weed in the Bermondsey area (Surrey) and in
small numbers since 1958 in Guernsey and more recently in Jersey (both
Channel Is); elsewhere short lived. Argentina; (widespread). BM, K, NMW,
OXF, RNG. 20*, 36c*, 220*, 303(28:8; 45:24; 54:35), 354(9:276, 11:34;
12:684), 869*(16:1-78), 887*(4:1-367). *S. gracile* Moric. ex Dunal, non
Sendtner; *S. gracilius* Herter; *S. ottonis* N.Hylander; *S. sublobatum* Willd.
ex Roemer & Schultes.

commersonii Dunal
Pre-1930 only.   S America.   E.   3, 327*(20:t.44; 38:t.39-43), 554, 633, 842*(1902:338-339); 1906:304), 1067*, 1227*.   *S. tenue* Sendtner.

curtipes Bitter
• A wool casual. Paraguay. RNG. 14, 348(20:32), 822(11:228). Probably synonymous with *S. americanum* var. *americanum*. 328(27:95-114).

eleagnifolium Cav.   Silvery-leaved Nightshade
• A wool casual. No modern records. S America; (widespread). RNG. 14, 21, 31*, 37*, 44a*, 66*, 191.

fontanesianum Dunal
Pre-1930 only. Brazil. 7, 301*(t.177), 354(5:39), 554, 1200*.

aff. **jamesii** Torrey   Wild Potato
• A wool casual. USA. BM, RNG. 40, 51, 312*(t.6766), 652*.

laciniatum Aiton f.   Kangaroo-apple
•• A garden escape established since the 1920s in Scilly and the Channel Is; recently reported as a weed in several places in southern England; increasing. Australasia.   BM, JSY, LTN, RNG.   20, 36c, 117*, 201, 207, 220, 312*(t.349), 519*, 583*, 686*.   Possibly only a tetraploid variant of *S. aviculare*.

linnaeanum Hepper & P.-M.L.Jaeger   Apple-of Sodom
Pre-1930 only. N Africa; (S Europe, S Africa, Australasia). LIV. 21, 36c, 37c*, 328(42:433-435), 576, 1074*, 1263*, 1369. *S. hermannii* auct., non Dunal; *S. sodomeum* auct., non L. The plant in S France known as *S. juvenale* Thell. (546*, 822(5:161), 1332) appears to be a synonym with date priority. The true *S. sodomeum* L. (*S. hermannii* Dunal, nom. illeg.; *S. indicum* auct., non L.) is a native of India and Sri Lanka and is best referred to unambiguously as *S. violaceum* Ortega, non R.Br.; the closely allied *S. anguivi* Lam. from tropical Africa and Madagascar has been wrongly equated to it. 327*(76:287-295), 1370.

maritimum Meyen ex Nees
Pre-1930 only. Chile. OXF. 7, 354(6:616; 7:886).

melongena L.   Aubergine
• A food refuse casual. India; (widespread as a crop). 21, 54*, 68*, 71*, 80*, 192, 197, 617*. *S. esculentum* Dunal.

nigrum L. subsp. **schultesii** (Opiz) Wessely
••• (but perhaps overlooked). Casual or persistent, vector unknown; frequent on the eastern side of London. Europe. BM, K, RNG. 20, 192, 303*(24:21-22), 1352*. *S. schultesii* Opiz.

physalifolium Rusby   Green Nightshade
•••• A wool, spice, agricultural seed, bird-seed and oil-seed alien sometimes established as an abundant weed on cultivated ground; increasing. S America; (Australasia).   7, 14, 19, 20*, 22*, 207, 303*(35:12-13), 327*(92:1-38), 1352*.   *S. nitidibaccatum* Bitter; *S. sarachoides* auct., non Sendtner. All records are referable to var. *nitidibaccatum* (Bitter) Edmonds. A hybrid with native *S. nigrum* (*S.* × *procurrens* Leslie) has been recorded. CGE, LTN, RNG. 303*(35:13).

pseudocapsicum group   Winter-cherries
●●● Escapes from cultivation and bird-seed casuals. Eastern S America; (tropics, subtropics). BM, E, LANC, LIV, NMW, RNG. 7, 19, 20, 36c, 54*, 67*, 199, 593*, 617*, 719*, 1347*. Incl. *S. diflorum* Vell.Conc. (*S. capsicastrum* Link ex Schauer); *S. pseudocapsicum* L. and hybrids. Modern records of garden or greenhouse escapes are probably all referable to *S. diflorum*.

pygmaeum Cav.
● Established for some years on a canal path between Hanwell and Southall (Middlesex), last seen c.1970, vector unknown. Chile, Argentina. BM, E, K, LANC, OXF, RNG. 20, 192, 538*, 1352. *S. pseudocapsicastrum* auct., non L.

pyracanthum Jacq.
● A casual greenhouse escape. Tropical Africa, Madagascar. HbEJC. 51b, 199, 312*(t.2547), 554, 1162*, 1306*. *S. pyracanthos* Poiret.

rostratum Dunal   Buffalo-bur
●●● A wool, grain, bird-seed and oil-seed casual. USA, Mexico; (Australasia). BM, CGE, E, NMW, OXF, RNG. 1, 11, 14, 19, 20*, 303*(14:12,15), 887*(4:1-367), 1352*. *S. cornutum* auct., non Lam.; *S. heterandrum* Pursh.

sarachoides Sendtner   Leafy-fruited Nightshade
●● (but over-recorded for *S. physalifolium*). Persistent on tips and waste ground, vector unknown. S America. LANC, RNG. 20, 303(24:21), 327*(92:1-38), 826*(8:98-105), 1352*. *S. chenopodioides* auct., non Lam.; *S. nitidibaccatum* auct., non Bitter; *S. physalifolium* auct., non Rusby.

scabrum Miller   Garden Huckleberry
● A casual garden escape; also in 1990 on fields spread with sludge near Crockenhill (W Kent). Origin unknown, probably Africa. CGE, HbCGH. 21, 148, 303(28:8; 32:16; 59:46), 370(12:280), 458, 856*(7:33-35), 869*(16:1-78), 1280*. *S. guineense* Lam., non L.; *S. intrusum* Soria; *S. melanocerasum* Willd.; *S. nigrum* var. *guineense* L.; *S. nigrum* var. *melanocerasum* (Willd.) Dunal; *S. pterocaulon* Reichb., non Dunal.

sisymbriifolium Lam.   Red Buffalo-bur
●●● A wool, oil-seed, bird-seed and agricultural seed casual. S America; (widespread). BM, CGE, E, NMW, OXF, RNG. 7, 14, 18, 19, 20, 38*, 303*(36:1,10), 312*(t.2568, t.2828, t.3954), 319(18:144-145), 583*, 1352*. *S. balbisii* Dunal.

[*sodomeum* L. In error for *S. linnaeanum*.]

torvum Swartz   Devil's-fig
● A wool casual. Tropics. RNG. 14, 49, 604, 676, 724*, 1006*, 1109*.

triflorum Nutt.   Small Nightshade
●●● A wool and agricultural seed alien established at Wolferton (W Norfolk), Icklingham (W Suffolk) and on Holy Is (Cheviot); casual elsewhere. Western N America, Chile, Argentina; (Australia). BM, E, LANC, NMW, OXF, RNG. 1, 14, 20*, 22*, 38*, 303*(19:15), 825(3*:90-91; 4:108-109), 1352*. Var. *ponticum* (Prodán) Borza (var. *dentatum* Ooststr.; *S. ponticum* Prodán), has been recorded. 354(10:346), 825*(3:90-91; 4:108-109).

tuberosum L.   Potato
●●●●● A persistent relic or escape from cultivation throughout the British Is.
S America; (widespread). 2, 20, 22*, 36c*, 68*, 80*.

vernei Bitter & Wittm.
● A weed established since 1985 in shrubbery beds in Whiteknights Park at
Reading (Berks). Argentina. RNG, HbEJC. 303(41:14), 377*(1944:t.28-29),
833*(26:472), 849*(2:1-13), 1067*. *S. ballsii* Hawkes.

villosum Miller
●● A wool, bird-seed and oil-seed casual. Eurasia, N Africa; (N America).
BM, MNE, NMW, OXF, RNG, SLBI. 4, 17, 18, 19, 20, 44a*. *S. luteum*
Miller; incl. *S. alatum* Moench; *S. miniatum* Bernh. ex Willd.; *S. nigrum* var.
rubrum Miller; *S. puniceum* C.Gmelin.

CYPHOMANDRA C.Martius ex Sendtner
crassicaulis (Ortega) Kuntze   Tree Tomato
● A food refuse casual at Beddington Sewage Farm (Surrey), first found in
flower in 1989, although seedlings on tips may have been overlooked in the
past. Peru, S Brazil. 51b, 303(64:40), 312*(t.7682), 510*, 1160*, 1238*,
1280*. *C. betacea* (Cav.) Miers; *Solanum betaceum* Cav.

DATURA L.
ferox L.   Angel's-trumpets
●●● A wool casual. China; (widespread). BM, CGE, E, LANC, NMW,
RNG. 12, 14, 21, 36c, 51a, 583*, 826*(11:14-15).

innoxia Miller   Recurved Thorn-apple
● A wool casual. Tropical America; (widespread). RNG. 13, 14, 21, 31*,
44a*, 51a, 201, 325*(115:318), 371(334:21), 583*, 617*, 1109*. *D. metel*
auct., non L.; *D. meteloides* DC. ex Dunal.

metel L.
● Recently reported from Toynton All Saints (N Lincs), perhaps a garden
escape. Origin obscure; (tropics, subtropics). 7, 21, 29*, 32*, 38*, 51a, 55*,
161, 354(12:497). *D. fastuosa* L. The Colchester (N Essex) record is in error
for *D. innoxia*. RNG.

stramonium L.   Thorn-apple
●●●●● A wool, hides, grain, bird-seed and oil-seed alien and escape from
cultivation, sporadic in occurrence, sometimes persisting many years.
N America; (widespread). 1, 14, 18, 19, 20, 22*, 33*, 34*, 44*, 51a, 191,
209. Incl. *D. inermis* Juss. ex Jacq.; *D. laevis* L.f.; *D. stramonium* var. *tatula*
(L.) Torrey; *D. tatula* L.

NICOTIANA L.
acuminata (Graham) Hook.
● Reported from waste ground at South Millfields, Clapton (Middlesex).
S America. SLBI. 7, 51, 75*, 312*(t.2919), 458.

alata Link & Otto   Sweet Tobacco
●●● A garden escape sometimes persisting a few years. S America.
BM, LIV, NMW, OXF, SLBI. 7, 20, 52*, 53*, 54*, 75*, 856*(7:37-41),
1352*. *N. affinis* hort. ex T.Moore; incl. *N.* 'Grandiflora'. Hybrids with
N. forgetiana (*N.* × *sanderae* hort. ex Will.Watson) have been recorded as
garden escapes. OXF, RNG. 20, 55*, 1352*.

forgetiana Hemsley   Red Tobacco
●● A casual garden escape. S America. 20, 312*(t.8006), 804*(t.227).
Hybrids with *N. alata*, q.v., have been recorded.

glauca Graham   Tree Tobacco
● A wool casual. S America; (widespread). BM, E, RNG. 17, 21, 32*,
36c*, 37*, 44*, 75*, 312*(t.2837).

goodspeedii Wheeler   Small-flowered Tobacco
●● A wool casual. Australia. BM, RNG. 14, 37, 75*, 335(912:33), 576*.

ingulba J.Black
● A wool casual. Australia. RNG. 14, 37*, 75*, 576, 920*(57:156), 1025*.
N. rosulata (S.Moore) Domin subsp. *ingulba* (J.Black) P.Horton.

langsdorfii J.A.Weinm.
● A casual on sewage sludge at Stone (W Kent), perhaps a garden escape.
Brazil. 51, 312*(t.2221, t.2555), 325*(103:164), 371(428:8), 595a*,
805*(18:109).

longiflora Cav.   Long-flowered Tobacco
● A casual at Bristol (W Gloucs), Cardiff (Glam) and in Fife (unlocalised),
vector unknown. S America. E, NMW. 38*, 51, 75*, 156, 354(11:35).

occidentalis Wheeler   Native Tobacco
● A wool casual. No modern records. Australia. LTN, RNG. 14, 37, 75*,
145, 576, 583*.

rustica L.   Wild Tobacco
●● A wool casual and garden escape. Formerly a persistent or recurrent escape
from extensive cultivation for the tobacco trade and for insecticide. S America;
(N America). BM, CGE, E, K, OXF, RNG. 1, 14, 20, 22*, 25*, 29*, 75*,
220, 245. 303(17:29). *N. rusticana* sphalm.

suaveolens Lehm.   Australian Tobacco
● A wool casual. Australia. E, K, LTN, RNG. 3, 14, 37, 51, 75*, 143, 199,
540*, 576*.

sylvestris Speg. & Comes
● A persistent relic of cultivation in the Channel Is. Argentina. 20, 51*, 52*,
53*, 75*, 201, 220, 1352*.

tabacum L.   Tobacco
●● (but confused with *N. alata*). A wool casual and relic or escape from
cultivation. Tropical America; (widespread). BM, E, K, OXF, RNG, SLBI.
7, 12, 20, 22*, 29*, 52*, 68*, 75*, 192.

velutina Wheeler   Velvet Tobacco
● A wool casual. No modern records. Australia. RNG. 14, 37*, 75*, 540*,
576*. *N. suaveolens* var. *parviflora* Benth.

PETUNIA Juss.

× **hybrida** (Hook.) Vilm.   Petunia
- ●●● Casual or persistent garden escapes. Originated in cultivation. BM, LANC, NMW, OXF, RNG, SLBI. 20, 53*, 54*, 55*, 56*, 192, 617*, 1352. *P.* × *punctata* Paxton. A complex of hybrids involving *P. axillaris* (Lam.) Britton, Sterns & Pogg., *P. integrifolia* (Hook.) Schinz & Thell., *P. violacea* Lindley and other species. Incl. *P.* × *atkinsiana* D.Don (*P. axillaris* × *P. violacea*).

SALPIGLOSSIS Ruíz Lopez & Pavón
sinuata Ruíz Lopez & Pavón   Painted-tongue
- • A casual garden escape. Chile. OXF. 21, 36c*, 52*, 53*, 66*, 199, 342(Feb89), 354(9:276).

SCHIZANTHUS Ruíz Lopez & Pavón
pinnatus Ruíz Lopez & Pavón   Poor-man's-orchid
- • A casual garden escape. Chile. OXF, RNG. 7, 21, 52*, 53*, 55*, 135, 354(4:204), 1207*, 1352.

× **wisetonensis** hort.
- • A casual garden escape. Originated in cultivation. 51b, 54, 342(Feb90), 1072*, 1155*, 1179*. Possibly *S. pinnatus* × *S. grahamii* Gillies (*S. retusus* Hook.'Grahamii'). Incl. *S.*'Compactus'; *S.*'Cherry Shades'.

BROWALLIA L.
americana L.
- • A casual in a garden at Lymington (S Hants), vector unknown. Tropical America. HbEJC. 371(385:35), 511, 575*, 576, 1009, 1358*. Incl. *B. demissa* L.

speciosa Hook.   Bush-violet
- • A weed since 1979 in the botanical garden in Whiteknights Park, Reading (Berks). Colombia. RNG. 51*, 53*, 321*(t.4339), 353(40:32), 594*.

CONVOLVULACEAE

DICHONDRA Forster & Forster f.
micrantha Urban   Kidney-weed
- • Established since 1955 on dunes near Hayle (W Cornwall), vector unknown; also a wool casual. E Asia; (widespread). BM, CGE, E, K, LANC, RNG. 16, 17, 20, 212, 302(2:22), 349(2:21). *D. repens* auct., non Forster & Forster f.

[*repens* Forster & Forster f.   In error for *D. micrantha*. 302(2:22). *D. sericea* Sw.]

CONVOLVULUS L.
althaeoides L.   Mallow-leaved Bindweed
- • A grain and bird-seed casual. Medit. BM, RNG. 2, 11, 19, 21, 24*, 30*, 31*, 44*, 300(1907:41). Incl. *C. elegantissimus* Miller; *C. tenuissimus* Sibth. & Smith.

betonicifolius Miller
- A casual on the bank of the river Wye (Hereford), vector unknown. No modern records. E Medit. 21, 39*, 44*, 354(12:286). *C. hirsutus* M.Bieb.

erubescens Sims   Australian Bindweed
- A wool casual. Australia. LTN. 14, 147, 348(20:32), 520*, 520a*, 533*, 569*, 666*.

pentapetaloides L.
- A grain casual. No modern records. Medit. OXF. 11, 21, 39*, 44*, 354(12:49).

sabatius Viv.   Ground Blue-convolvulus
- A bird-seed casual; also a garden escape on a few walls on Tresco and St Mary's (Scilly). Medit. 21, 46*, 52*, 53*, 371(357:24), 436, 1117*. Incl. *C. mauritanicus* Boiss.

siculus L.   Small Blue-convolvulus
- A casual at Sharpness docks (W Gloucs), vector unknown, and at the seed trial grounds at Mistley (N Essex). Medit. RNG. 7, 21, 24*, 44*, 46*, 182, 191, 354(6:616).

stachydifolius Choisy
- A casual at Bristol, vector unknown. No modern records. SW Asia. 39*, 44*, 350(28:22), 354(11:491).

tricolor L.   Tricolour Convolvulus
- • A bird-seed casual and garden escape. Formerly a grain and ballast alien. Medit. BM, NMW, OXF, RNG. 1, 19, 21, 24*, 29*, 30*, 44*, 135.

CALYSTEGIA R.Br.

[*dahurica* (Herbert) Don (*Convolvulus dahuricus* (Herbert) Druce; *Volvulus dahurica* (Herbert) Junger) In error for *C. pulchra*. This N Asian plant is rare or unknown in gardens and so unlikely to escape.]

pulchra Brummitt & Heyw.   Hairy Bindweed
- •••• (but confused with native *C. sepium* subsp. *roseata*). An established garden escape; naturalised on banks and in hedgerows in widely scattered localities throughout the British Is; increasing.   Probably NE Asia; (N & C Europe). 7, 20*, 22*. *C. sepium* subsp. *pulchra* (Brummitt & Heyw.) Tutin, nom. inval.; *C. silvatica* subsp. *pulchra* (Brummitt & Heyw.) Rothm., nom. inval.; *C. silvatica* var. *pulchra* (Brummitt & Heyw.) Scholz, nom. inval.; *C. sylvestris* var. *pulchra* (Brummitt & Heyw.) Scholz, nom. inval.; *Convolvulus dubius* J.Gilbert. Hybrids with native *C. sepium* (*C.* × *scanica* Brummitt) and *C. silvatica* (*C.* × *howittiorum* Brummitt) have been recorded. LIVU. 20, 328(35:333-334).

sepium (L.) R.Br. subsp. **americana** (Sims) Brummitt
- (but perhaps overlooked as native subsp. *roseata*). Atlantic coasts of N & S America, Azores. 12, 21, 38a*, 312*(t.732), 354(5:80). *Convolvulus americanus* (Sims) Greene; *C. sepium* L. var. *americanus* Sims.

sepium (L.) R.Br. subsp. **spectabilis** Brummitt
- Established on waste ground near Arthog (Merioneth), now gone. Probably Siberia; (Scandinavia). 20, 51b, 156, 327(64:73).

silvatica (Kit.) Griseb. subsp. **fraterniflora** (Mackenzie & Bush) Brummitt
• An oil-seed casual. USA. RNG. 38*, 192, 328(35:332). *C. fraterniflora*
(Mackenzie & Bush) Brummitt.

silvatica (Kit.) Griseb. subsp. **silvatica**   Large Bindweed
••••• An established garden escape; naturalised in woodlands and hedgerows
in many places throughout the British Is, especially in England and Wales.
Medit., SW Asia. 7, 20*, 22*, 25*, 33*, 354(13:265-268). *C. inflata* auct.;
C. sepium subsp. *silvatica* (Kit.) Battand.; *C. sylvestris* (Waldst. & Kit. ex
Willd.) Roemer & Schultes; *Convolvulus silvaticus* Kit.; *Volvulus inflatus* Druce
pro parte. A hybrid with native *C. sepium* (*C.* × *lucana* (Ten.) G.Don has
been recorded, and about London is more common than either parent. OXF,
RNG. 20, 371(427:38-39).

MERREMIA Dennst. ex Endl.
sibirica (L.) Haller f.
• An oil-seed casual. No modern records. N Asia. 354(11:34), 604, 630,
1135*, 1198*. *Convolvulus sibiricus* L.; *Ipomoea sibirica* (L.) Pers.

IPOMOEA L.
batatas (L.) Lam.   Sweet-potato
•• A food refuse casual. Tropical America; (tropics, subtropics). HTN,
LTN, RNG. 21, 51*, 68*, 69*, 126, 144, 145, 146, 192, 199, 1156*.
Batatas edulis Choisy.

hederacea Jacq.   Ivy-leaved Morning-glory
•• An oil-seed and grain casual. Tropical & subtropical America.
BM, LANC, MNE, OXF, RNG, SLBI. 20, 38*, 53*, 126, 192, 331(58:63),
354(11:34), 371(343:29). *Pharbitis hederacea* Choisy. Var. *integriuscula*
Gray has been recorded. 192. The author citation is often wrongly quoted as
(L.) Jacq. 436. Contrary to 20, the non-weedy tropical *I. nil* (L.) Roth is
unlikely to occur.

lacunosa L.   White Morning-glory
• An oil-seed and grain casual. Eastern N America. BM, RNG. 18, 20, 38*,
40, 126, 303*(35:28-29), 331(27:33).

muricata (L.) Jacq.
• A casual, vector unknown. Tropical America; (tropics). BM. 64*,
371(379:26), 604, 676.

purpurea Roth   Common Morning-glory
•• A grain, bird-seed and oil-seed casual. Tropical America; (Asia). BM, E,
LANC, MNE, OXF, RNG. 7, 11, 18, 19, 20, 29*, 38*, 64*, 303*(35:28-29),
710*, 1247*. *Convolvulus hederaceus* L.; *C. purpureus* L., nom illeg.;
Pharbitis purpurea (Roth) Voigt. The author citation is usually wrongly
quoted as (L.) Roth. 436.

cf. **trichocarpa** Elliott
• An oil-seed casual. USA. HbEJC. 38*, 49, 126, 236, 646. *I. coccinea* L.
may have been overlooked; it has been grown from soya-bean waste. BM.

tricolor Cav.   Morning-glory
- An oil-seed casual and garden escape. Tropical America. 51, 54*, 64*, 126, 325*(113:226), 371(370:30). *I. rubrocaerulea* Hook.

CUSCUTACEAE

CUSCUTA L.

americana L.
   Pre-1930 only. Tropical America. BM. 49, 156, 1068*, 1109*, 1367*.

approximata Bab.
   Pre-1930 only. Eurasia, N Africa. CGE. 2, 21, 44*, 350(30:112), 715*.
   C. planiflora var. *approximata* (Bab.) Engelm. pro parte.

australis R.Br.   Australian Dodder
   •• (but over-recorded for *C. campestris*). A bird-seed casual. No modern records. Eurasia, Africa, Australia. BM, NMW, OXF, RNG. 7, 19, 21, 46*, 129, 191, 354(9:31), 520*, 569*, 666*, 715*, 1352*. Incl. *C. breviflora* Vis.; *C. cesatiana* Bertol.; *C. tinei* Inz.

campestris Yuncker   Yellow Dodder
   ••• A locally abundant wool, agricultural and horticultural seed, and bird-seed casual found on crops of carrot, tomato, beetroot, lucerne and various ornamental plants, rarely persistent for a few years; increasing. N & tropical America; (widespread). BM, CGE, E, NMW, OXF, RNG. 12, 14, 19, 20, 22*, 44*, 303(14:13; 17*:1,16), 715*, 847*(29:13-24), 885*(13:10), 1205*, 1247*, 1352*. *C. arvensis* auct., non Beyr. ex Engelm.; *C. arvensis* Beyr. ex Engelm. var. *calycina* (Engelm.) Engelm.; *C. pentagona* var. *calycina* Engelm.; *C. suaveolens* auct., non Ser.

epilinum Weihe   Flax Dodder
   • A serious weed of flax fields in the 19th century, deliberately exterminated; last recorded in 1968. Origin obscure; (Europe). BM, LIV, NMW, OXF, RNG, SLBI. 1, 20, 22*, 38*, 44*, 156, 258, 288, 303(32:10), 354(10:534), 715*, 1352*, 1367*. *C. densiflora* Soyer-Will., non Hook.f.; *C. vulgaris* J.S.Presl & C.Presl.

[*pentagona* Englm. (*C. arvensis* Beyr. ex Engelm.) In error for *C. campestris*. 354(12:694; 13:118), 715*.]

planiflora Ten.
   Pre-1930 only. Medit., N America. 1, 21, 39, 44*, 45*, 46, 48*, 300(1905:101,103), 715*.

suaveolens Ser.   Lucerne Dodder
   •• A casual found on various native and cultivated plants, particularly lucerne. No modern records. S America; (widespread). BM, CGE, LIV, NMW, OXF, RNG. 1, 21, 320*(1908:241-244), 354(11:407), 500*, 545, 715*, 1205*, 1352*. *C. racemosa* var. *chiliana* Engelm.; incl. *C. corymbosa* Choisy; *C. hassiaca* Pfeiffer; *C. racemosa* Martius.

[*trifolii* Bab. Now considered to be conspecific with native *C. epithymum*.]

POLEMONIACEAE

POLEMONIUM L.

caeruleum L. subsp. **himalayanum** (Baker) H.Hara
- A casual or persistent garden escape on Wisley Common (Surrey). Himalayas. 458, 685*.

foliosissimum (A.Gray) A.Gray
A casual garden escape reported in 1993. Western USA. 51, 312*(n.s. t.310), 458, 1255*, 1257*. *P. caeruleum* var. *foliosissimum* A.Gray; incl. *P. archibaldae* Nelson; *P. robustum* Rydberg.

pauciflorum S.Watson
- A casual weed in gardens and in the car park at the RHS Gardens, Wisley (Surrey), vector unknown. USA. 51, 134, 315*(6:97), 331(54:65), 371(340:28), 511.

GILIA Ruíz Lopez & Pavón
achilleifolia Benth.
Pre-1930 only. California. BM, CLE, LIV. 1, 42, 51, 156, 220, 312*(t.3440, t.5939), 500*, 600*. Incl. *G. multicaulis* Benth. Some records are in error for *G. capitata*.

capitata Sims   Blue-thimble-flower
- A bird-seed casual and garden escape. Formerly a grain and ballast alien. Western N America. BM, E, LIV, NMW, OXF, RNG. 1, 41*, 51, 52*, 54*, 66*, 371(394:36), 500*.

cf. **inconspicua** (Smith) Sweet
Pre-1930 only. Western N America. OXF. 42, 699, 1165, 1306*. *G. millefoliata* Fischer & C.Meyer; *Ipomopsis inconspicua* Smith. Not to be confused with *Gilia inconspicua* Douglas ex Hook., a synonym of *G. tenuiflora* Benth.

laciniata Ruíz Lopez & Pavón
Pre-1930 only. S America. 1, 51, 511, 705a*, 833*(8:216), 1050*.

tricolor Benth.   Bird's-eyes
- A casual or persistent garden escape. California. BM, E. 1, 42, 51, 54*, 55*, 135, 369(26:57), 370(14:426), 500*.

IPOMOPSIS Michaux
congesta (Hook.) V.Grant
Pre-1930 only. Western N America. 41*, 42, 354(9:361), 699*. *Gilia congesta* Hook.

NAVARRETIA Ruíz Lopez & Pavón
[**cotulaefolia** (Benth.) Hook. & Arn. (*Gilia cotulaefolia* (Benth.) Steudel) In error for *Ipomopsis congesta*. 195, 354(9:361).]
intertexta (Benth.) Hook.
Pre-1930 only. Western N America. CLE. 1, 9, 41*, 42, 195. *Gilia intertexta* Steudel.

squarrosa (Eschsch.) Hook. & Arn.   Skunkweed
- A casual in gardens, probably from bird-seed; persistent in quantity in a tree nursery in Windsor Great Park (Berks). Formerly a wool and grain alien. Western N America; (Australasia). BM, E, OXF, RAMM, RNG. 3, 7, 37*, 41*, 42, 51*, 179, 303(23:12), 312*(t.2977), 354(7:46; 10:533), 370*(19:259-263), 420, 519*. *Gilia pungens* Douglas ex Hook.; *G. squarrosa* (Eschsch.) Hook. & Arn.; *Hoitzia squarrosa* Eschsch.   Not to be confused with *G. pungens* (Torrey) Benth., which is a synonym of the shrubby *Leptodactylon pungens* (Torrey) Rydb., also from western N America.

LINANTHUS Benth.
androsaceus (Benth.) E.Greene
- A casual impurity in horticultural seed. Western N America. BM. 2, 42, 51*, 52*, 139, 303(34:22), 371(331:20), 600*. *Gilia androsacea* (Benth.) Steudel; incl. *G. lutea* Steudel; *G. micrantha* Steudel; *Leptosiphon hybridus* hort. ex Vilm.

grandiflorus (Benth.) E.Greene
- A casual garden escape. California. 42, 51, 303(48:36), 312*(t.3578). *Gilia densiflora* (Benth.) Benth.

COLLOMIA Nutt.
biflora (Ruíz Lopez & Pavón) Brand
- A casual, probably an impurity in horticultural seed. S America. BM. 36c, 51*, 303(50:27), 371(391:34), 546a*, 1231*. *C. cavanillesii* Hook. & Arn.; *C. coccinea* Lehm.; *Phlox biflora* Ruíz Lopez & Pavón.

grandiflora Dougl. ex Hook.
- A grain casual, and probably a garden escape. Western N America. BM, LCN, OXF. 1, 21, 41*, 160, 258, 354(5:667; 7:199), 546a*, 650*. *C. coccinea* auct., non Lehm.

linearis Nutt.
Pre-1930 only. N America. BM, E, LSR. 1, 9, 21, 38*, 41*, 51, 112, 354(5:667), 546a*, 666*.

PHLOX L.
drummondii Hook.   Annual Phlox
- A casual garden escape. Southern USA, Mexico. RNG. 36c, 40, 52*, 54*, 55*, 370(19:262), 458, 1268*.

paniculata group   Border Phlox
- ••• Persistent garden escapes. Eastern N America. LIV. 7, 20, 36c, 38*, 52*, 54*, 55*, 56*, 354(4:203).   Incl. *P. decussata* Lyon ex Pursh; *P. × hortorum* Bergm.; *P. maculata* L.; *P. paniculata* L.

subulata L.   Moss Phlox
- A casual garden escape. Eastern USA. 38*, 51b, 53*, 370(19:262), 564*, 565*, 594*, 1207*, 1355*.

HYDROPHYLLACEAE

NEMOPHILA Nutt.

maculata Benth. ex Lindley   Fivespot
   • A casual garden escape. California. 42, 51, 53*, 66*, 371(352:26), 500*.

menziesii Hook. & Arn.   Baby-blue-eyes
   •• A grain casual and garden escape. Western N America. OXF, RNG. 7,
   8, 21, 22*, 51*, 52*, 53*, 264, 500*. *N. insignis* Douglas ex Benth.; incl.
   N. liniflora Fischer & C.Meyer.

PHACELIA A.L.Juss.

campanularia A.Gray   California-bluebell
   • A casual garden escape. Western N America. 7, 42, 51, 52*, 54*, 55*,
   198, 303(60:35), 354(2:505), 416, 474, 1352*.

ciliata Benth.
   •• A grain casual. Western N America. BM, CLE, LIV, NMW, OXF, RNG.
   1, 21, 42, 191, 500*, 699*, 1166*.

circinata Jacq.f.
   Pre-1930 only. S America. E. 3, 1161*.

congesta Hook.   Blue-curls
   Pre-1930 only. Texas, New Mexico. 7, 51, 312*(t.3452), 354(4:422), 699*.

crenulata Torrey ex S.Watson
   Pre-1930 only. Western USA. BM. 42, 500*, 511.

linearis (Pursh) Holzinger   Linear-leaved Phacelia
   • A casual on shingle of the river Exe (S Devon), vector unknown. Western
   USA. BM. 42, 500*, 511. *Hydrophyllum lineare* Pursh.

parviflora Pursh   Small-flowered Phacelia
   • Probably a grain casual. No modern records. Eastern N America. OXF.
   1, 38*, 40, 285, 354(1:40; 5:667; 7:584), 699*. *P. dubia* (L.) Trel.

ramosissima Douglas ex Lehm.   Branching Phacelia
   Pre-1930 only. Western N America. 41*, 42, 179, 500*, 699*.

tanacetifolia Benth.   Phacelia
   ••• A casual or persistent grain, grass and agricultural seed alien and garden
   escape; cultivated in recent years as an aid in the control of insect pests and as
   a food plant for bees, and probably escaping from this source; increasing.
   California. BM, LTN, NMW, OXF, RNG, ZCM. 1, 8, 20*, 22*, 42,
   303(37:15; 39:9; 43*:1,17; 48:35; 63:31), 325*(118:32-35), 500*, 546a*,
   699*, 1166*.

cf. **viscida** (Benth. ex Lindley) Torrey   Sticky Phacelia
   • A casual on a refuse tip at Oxford, vector unknown. No modern records.
   Western N America. OXF, HbJPMB. 42, 51, 312*(t.3572), 349(2:110),
   500*, 699*, 826*(18:27-28). *Eutoca viscida* Benth. ex Lindley.

BORAGINACEAE
HELIOTROPIUM L.
amplexicaule Vahl   Summer Heliotrope
  Pre-1930 only. Argentina.   7, 21, 51, 312*(t.3096, t.8480), 354(4:422).
  H. anchusifolium Poiret; *Tournefortia heliotropioides* Hook.
arborescens L.   Cherry-pie
  • A casual garden escape. S America. 7, 51, 52*, 66*, 192, 354(3:25), 458.
  H. peruvianum L.
curassavicum L.   Seaside Heliotrope
  Pre-1930 only. N & S America; (widespread). OXF. 21, 38*, 39*, 40*, 41*,
  354(5:295; 8:125).
europaeum L.   European Turnsole
  •• A wool and oil-seed casual. Europe, N Africa, SW Asia; (Australia). BM,
  LANC, LTR, NMW, OXF, RNG. 1, 14, 21, 22*, 24*, 25*, 44*, 285, 520a*.

ARNEBIA Forsskål
pulchra (Roemer & Schultes) Edmondson   Prophet-flower
  Pre-1930 only. SW Asia. 7, 39*, 51*, 54*, 136. *A. echioides* (L.) A.DC.;
  Echioides longiflorum (K.Koch) I.M.Johnston; *Macrotomia echioides* (L.)
  Boiss.

NEATOSTEMA I.M.Johnston
apulum (L.) I.M.Johnston
  Pre-1930 only. Medit., Canary Is. NMW. 7, 21, 24*, 39*, 156, 166,
  354(3:25). *Lithospermum apulum* (L.) Vahl. This plant is easily confused with
  Amsinckia spp.

LITHOSPERMUM L.
arvense L. **alien variants**
  Var. *caeruleum* Coss. & Germ. (var. *medium* (Chev.) Druce) has been recorded
  pre-1930 only. 150, 300(1907:40), 354(12:286), 1281.

LITHODORA Griseb.
diffusa (Lagasca) I.M.Johnston
  • A casual or persistent garden escape. SW Europe. 21, 30*, 32*, 53*, 57*,
  371(391:34). *Lithospermum diffusum* Lagasca.

CERINTHE L.
glabra Miller   Smooth Honeywort
  Pre-1930 only. Europe, SW Asia. 1, 21, 22*, 23*, 27*, 151, 300(1906:103).
  C. aspera Roth.
major L.   Greater Honeywort
  • A casual in a garden, vector unknown. No modern records. Medit. BM.
  21, 24*, 36c, 45*, 144. Incl. *C. gymnandra* Gasp.

minor L.   **Lesser Honeywort**
  • A grain casual. No modern records. Europe, SW Asia. BM, CLE, E,
OXF, SLBI. 2, 8, 21, 22*, 23*, 26*, 27*, 349(1:39), 1352*. Incl. subsp.
auriculata (Ten.) Domac; *C. minor* var. *hispida* Turrill.

ALKANNA Tausch
lutea DC.
  Pre-1930 only. W Medit. BM. 1, 12, 21, 163, 546*, 683*. *Nonea lutea*
  DC., non (Desr.) DC.
tinctoria (L.) Tausch   **Dyer's Alkanet**
  Pre-1930 only. E Medit. 24*, 39*, 282. Incl. *Anchusa tinctoria* (L.) L.

ECHIUM L.
arenarium Guss.
  A casual in 1930. Medit. 21, 354(9:253,276), 546*, 737*, 894*(1:76). The
  record for Barry docks (Glam) was in error for *E. rosulatum*. 156, 354(8:916).
[*humile* Desf. (incl. *E. pycnanthum* Pomel) In error for *E. rosulatum*. 10, 156,
  354(8:633).]
italicum L.   **Pale Bugloss**
  Pre-1930 only. Europe, SW Asia. BM, CGE, K, NMW, OXF, RNG. 1, 21,
  24*, 29*, 31*, 39*, 166, 354(5:39; 8:748).   *E. pyramidale* Lapeyr.;
  E. pyramidatum DC.
lusitanicum L.
  Pre-1930 only. SW Europe. BM, LCN. 21, 30*, 160.
parviflorum Moench
  Pre-1930 only. Medit. OXF. 21, 46*, 354(9:30), 546*. *E. calycinum* Viv.
pininana Webb & Berth.   **Giant Viper's-bugloss**
  ••• A garden escape self-sown in several places in the Channel Is, Scilly,
  W and E Cornwall and at Llandudno (Caerns). Canary Is. HbEJC. 20, 36c*,
  51b, 201, 207, 213, 220, 303(46:31), 312*(n.s. t.269), 530*, 894*(2:37-115).
[*pustulatum* Sibth. & Smith   Now considered to be conspecific with native
  E. vulgare. 21, 318(15:147), 354(9:254).]
rosulatum Lange   **Lax Viper's-bugloss**
  • Established since 1927 at Barry docks (Glam), vector unknown, still there in
  1984, but perhaps now gone. SW Europe. CGE, NMW, RNG, SLBI. 20, 21,
  156, 1189*. *E. humile* auct., non Desf.
× **scilloniensis** hort.
  • A garden escape self-sown in several places on Tresco and at Hugh Town on
  St Mary's (both Scilly).   Originated in cultivation.   207, 458, 1180.
  E. pininana × *E. webbii* Coincy.
[*strictum* L.f. Unconfirmed; probably always in error for *E. pininana*. RNG.]
tuberculatum Hoffsgg. & Link
  Pre-1930 only. SW Europe, NW Africa. OXF. 7, 21, 354(5:39), 737*.

PULMONARIA L.

cf. **affinis** Jordan
- • Recorded from Glenfarg (Mid Perth), vector unknown. SW Europe. CGE. 7, 21, 22*, 51, 347(4:100-111), 354(5:570), 371(397:36), 1290*.

angustifolia L.
- • (but confused with native *P. longifolia* and with *P.*'Mawson's Blue'). A persistent garden escape. E Europe. BM. 20, 21, 22*, 23*, 27*, 199, 347*(4:100-111), 449. *P. azurea* Besser; *P.*'Bowles' Variety'. Not to be confused with *P. angustifolia* auct. angl., a synonym of *P. longifolia*.

'Mawson's Blue'   Mawson's Lungwort
- •• An established garden escape; recently recorded from Surrey, S Essex, Midlothian and Dunbarton; well naturalised by Urr Water south of Corsock (Kirkcudbrights). Possibly a hybrid of garden origin. E, HbACL, HbEJC. 20*, 199, 303*(25:1,16), 316(20:478), 350(42:106). *P. longifolia* auct., non (Bast.) Boreau.

cf. **montana** Lej.   Mountain Lungwort
- • A garden relic persistent at Jarn Mound, Boars Hill (Berks). W & C Europe. RNG. 21, 22*, 23*, 27*, 546*, 1290*. *P. angustifolia* subsp. *tuberosa* Gams pro parte; *P. tuberosa* auct., non Schrank; *P. vulgaris* Mérat pro parte.

officinalis L. sensu lato   Lungwort
- •••• An established garden escape; widely naturalised in woods and by roadsides throughout Britain. Europe. 1, 20*, 22*, 27*, 347*(4:100-111). Some records are referable to subsp. *obscura* (Dumort.) Murb. (*P. officinalis* var. *immaculata* Opiz) which is perhaps native in a few localities in the south. A hybrid with native *P. longifolia* has been recorded. BM, K. 1311.

rubra Schott   Red Lungwort
- •• An established garden escape; in hedges and woodland in a few scattered localities, mainly in the north; well naturalised in woodland near Blacklunans, Glen Shee (E Perth). SE Europe. BM, E, K, LANC. 20, 31*, 119, 303(26:16), 370(14:194), 371(385:34), 543*, 623a*, 1352*.

saccharata Miller   Bethlehem-sage
Recorded in 1930 only. France, Italy. SLBI. 21, 53*, 54*, 55*, 347*(4:100-111), 354(9:362,690). *P. picta* Rouy; *P. saccharoides* sphalm.

NONEA Medikus

caspica (Willd.) Don
Pre-1930 only. Russia, SW Asia. 2, 21, 39*, 320(44:35). *N. picta* (M.Bieb.) Sweet.

lutea (Desr.) DC.   Yellow Nonea
- •• A casual garden escape and persistent garden weed. E Europe, SW Asia. BM, NMW, RNG. 20, 21, 39*, 234, 303(16*:1,18; 17:28), 370(14:426), 371(343:30), 1152*.

pulla (L.) DC. sensu lato   Brown Nonea
- • A casual at Cheeklaw (Berwicks), vector unknown. No recent records. SW Asia; (E Europe). BM, OXF. 7, 21, 22*, 25*, 26*, 29*, 39*, 288,

354(5:38), 623a*, 683*, 1041, 1352*. *Lycopsis pulla* L.; incl. subsp. *monticola* Rech.f.

rosea (M.Bieb.) Link   Pink Nonea
- A casual or persistent garden escape. Formerly a grain alien. SW Asia. BM, OXF. 7, 20, 21, 22*, 39*, 150, 234, 349(6:60), 354(4:422). Incl. *N. versicolor* (Steven) Sweet; *Anchusa latifolia* hort.

ventricosa (Sibth. & Smith) Griseb.
- A casual at Felixstowe docks (E Suffolk), vector unknown. No modern records. S Europe, SW Asia. OXF, RNG. 2, 21, 31*, 39*, 44*, 300(1907:40), 354(12:496; 5:570). *N. alba* DC.

vesicaria (L.) Reichb.
  Pre-1930 only. W Medit. OXF. 7, 21, 30*, 45*, 46*. *N. decumbens* Moench.

SYMPHYTUM L.

asperum Lepechin   Rough Comfrey
- ●●● (but mostly in error for *S.* × *uplandicum*). An established escape from cultivation; known for over 30 years by the old station-yard at Fortrose (E Ross). SW Asia. BM, HIN, LANC, NMW, RNG, TCD. 1, 20, 22*, 153, 319(21:498), 370(10:297). *S. asperrimum* Donn ex Sims; incl. *S. armeniacum* Buckn.

brachycalyx Boiss.
  Pre-1930 only. SW Asia. 39, 354(8:325).

bulbosum C.Schimper   Bulbous Comfrey
- ●● A garden escape established in a few places, mostly in southern England and Wales; naturalised in abundance, perhaps for 100 years, by a stream at Abbotsbury (Dorset). SE Europe, Turkey. BM, DOR, EXR, K, NMW, RNG. 20*, 31*, 349(4:43), 525*, 546*, 595b*.

caucasicum M.Bieb.   Caucasian Comfrey
- ●● A garden escape persistent in widely scattered localities. SW Asia. BM, NMW, OXF, RNG. 2, 20, 52*, 53*, 258.

grandiflorum DC.   Creeping Comfrey
- ●●●● An established garden escape; well naturalised by roadsides and on river banks, especially in the south and east. Caucasus, Turkey. 12, 20*, 22*, 55*. Incl. *S. ibericum* Steven.

'Hidcote Blue'   Hidcote Comfrey
- ●● An established garden escape; in woods and on roadsides, mainly in the south; plentifully naturalised in Roundhill Wood (Salop). Originated in cultivation. OXF. 20*, 303*(30:16-17). Probably *S. grandiflorum* × *S. asperum* × *S. officinale*; *S. tauricum* auct. pro parte, non Willd.

officinale L. subsp. **uliginosum** (A.Kerner) Nyman
  Pre-1930 only. C Europe. MANCH. 21, 251, 623a*. *S. uliginosum* A.Kerner.

orientale L.   White Comfrey
●●●● An established garden escape; abundant in hedgebanks and on waste ground, especially in the south; increasing. SW Russia, Turkey. 1, 20, 22*, 33*, 39*, 258*, 546*, 715*. *S. tauricum* auct. pro parte, non Willd.

tauricum Willd.   Crimean Comfrey
● (but confused with *S. orientale*). A garden escape established in a hedgerow at Swaffham Prior (Cambs). E Europe, SW Asia. BM, LTR, OXF, MANCH. 1, 13, 20, 31*, 336(17:37), 368(32:1-173), 370(12:286), 715*. *S. orientale* Pallas, non L.

tuberosum L. subsp. **nodosum** (Schur) Soó
● Recorded without locality from NE Ireland. C Europe. RNG. 21, 239, 319(16:21), 352(45B:243), 683, 715*. *S. leonhardtianum* Pugsley; *S. nodosum* Schur. Possibly not worthy of subspecific rank.

× **uplandicum** Nyman   Russian Comfrey
●●●●● Formerly cultivated for fodder and more recently as a green manure, now widely established by roadsides and on river banks throughout most of the British Is. Originated in cultivation. 2, 20, 22*, 25*, 33*, 51*, 191, 370(3:280; 4:117). *S. asperum* × *S. officinale*; *S. asperrimum* auct., non Sims; *S. caeruleum* auct.; *S. densiflorum* auct.; *S. lilacinum* auct.; *S. × peregrinum* auct., non Ledeb. Backcrosses with *S. officinale* (incl. *S. × bicolor* Buckn.) and hybrids with *S. tuberosum* have been recorded. 20.

BRUNNERA Steven
macrophylla (Adams) I.M.Johnston   Great Forget-me-not
●●● (but under-recorded for *Omphalodes verna*). An established or persistent garden escape in scattered localities. Caucasus. ABD, BM, E, LANC, OXF, RNG. 7, 20*, 54*, 55*. *B. myosotidiflora* (Lehm.) Steven; *Anchusa myosotidiflora* Lehm.; *Myosotis macrophylla* Adams.

ANCHUSA L.
arvensis (L.) M.Bieb. subsp. **orientalis** (L.) Nordh.
Pre-1930 only. SE Europe, NE Africa, Asia. 7, 21, 39*, 44*, 160, 354(7:47), 1352*. *A. orientalis* (L.) Reichb.f., non L.; *Lycopsis orientalis* L.

azurea Miller   Garden Anchusa
●●● An established fodder alien naturalised since the 1914-18 war on dunes at Upton Towans (W Cornwall); also a grain and bird-seed casual and garden escape. Eurasia, N Africa. BM, LANC, LTN, NMW, OXF, RNG. 1, 11, 19, 20*, 22*, 30*, 44*, 207, 354(9:542), 370(11:181). *A. italica* Retz.

capensis Thunb.   Cape Alkanet
● A wool casual. S Africa (Australia). RNG. 14, 520a*, 554, 578a*, 822(22:286-322), 1064*, 1294*. *A. riparia* A.DC.

ochroleuca M.Bieb.   Yellow Alkanet
● Long established on the site of an army camp at Upton Towans (W Cornwall), possibly a fodder alien; casual elsewhere. E Europe, SW Asia. BM, CGE, E, LANC, OXF, RNG. 7, 13, 20*, 22*, 25*, 354(9:660), 370(11:181), 1352*. Hybrids with *A. azurea* and *A. officinalis*

(*A.* × *baumgartenii* (Nyman) Guşul.; *A. ochroleuca* var. *baumgartenii* Nyman) have been recorded at Upton Towans. 370(11:181), 623a*.

officinalis L.   Alkanet
●●● An established fodder alien naturalised since the 1914-18 war on dunes at Upton Towans (W Cornwall), possibly now gone; also a persistent garden escape and bird-seed alien; decreasing. Europe, Turkey. BM, CGE, E, NMW, OXF, RNG. 1, 20, 22*, 23*, 26*, 34*, 144, 370(11:181). *A. angustifolia* L.; incl. *A. procera* Besser ex Link. A hybrid with *A. ochroleuca*, q.v., has been recorded.

strigosa Labill.
Pre-1930 only. E Medit., SW Asia. BM. 1, 39, 44*, 197, 528*, 1186*. The 1908 record from Mirfield (SW Yorks) may be in error for *A. stylosa*.

stylosa M.Bieb.
● A casual at Bristol (W Gloucs), vector unknown. No modern records. SE Europe, SW Asia. 2, 9, 21, 39, 350(27:157), 715*, 822*(26:t.86), 1172*.

undulata L.
● A grain casual. Medit. BM, LCN, OXF, SLBI. 1, 21, 30*, 32*, 44*, 119, 354(13:303). Incl. *A. amplexicaulis* Sibth. & Smith; *A. hybrida* Ten.

variegata (L.) Lehm.
● A casual, probably from soil associated with *Cyclamen* introduced from Rhodes. S Aegean. HbALG. 21, 39, 303(34:22), 815*(20:192-210). *Lycopsis variegata* L. Many illustrations labelled as *L. variegata* (e.g. 623a*) are of the closely allied S European *A. cretica* Miller.

CYNOGLOTTIS (Guşul.) Vural & Kit Tan
barrelieri (All.) Vural & Kit Tan   False Alkanet
●● A persistent garden escape in a disused quarry at Kilkenny, Andoversford (E Gloucs), in a gravel pit at Beckford (Worcs) and at Stutton (E Suffolk); probably casual elsewhere. SE Europe, SW Asia. BM, RNG. 20, 27*, 39*, 137, 182, 144, 303(45:20), 370(19:288), 371(397:35), 623a*, 1352*. *Anchusa barrelieri* (All.) Vitman; *Buglossum barrelieri* All.

[*PENTAGLOTTIS* Tausch
sempervirens (L.) Tausch ex L.Bailey   Accepted, with reservations, as native.]

BORAGO L.
officinalis L.   Borage
●●●● A persistent or established garden escape and bird-seed alien on roadsides and waste ground, mainly in the south. Medit. 1, 20, 22*, 23*, 30*, 33*.

pygmaea (DC.) Chater & Greuter   Slender Borage
●● An established garden escape; naturalised in a few places in the south and west; known since 1932 on heathy ground on Jethou (Channel Is). Corsica, Sardinia. BM, CGE, E, LANC, NMW, RNG. 13, 20, 57*, 156, 212, 220, 680*, 1352*. *B. laxiflora* (DC.) Fischer, non Poiret; *Campanula pygmaea* DC.

TRACHYSTEMON D.Don
orientalis (L.) Don   Abraham-Isaac-Jacob
●●●● An established garden escape; naturalised in shady places, mainly
in England, sometimes in abundance. Bulgaria, Turkey, Caucasus. 7, 20,
22*, 25*, 57*. *Borago orientalis* L.; *Psilostemon orientale* (L.) DC.
ex DC. & A.DC.

MERTENSIA Roth
ciliata G.Don
● A garden escape persistent on a roadside at Pett, near Rye (E Sussex).
Eastern N America. BM. 41*, 51, 371(388:36).
virginica (L.) Pers.   Virginia-bluebells
●● (but confused with *M. ciliata*). A garden escape. Eastern N America.
1, 38*, 52*, 54*, 55*, 353(37:31-36), 354(5:388), 371(388:36).
M. pulmonarioides Roth; *Pneumaria pulmonarioides* (L.) Roth; *P. virginica*
(L.) Hill; *Pulmonaria virginica* L.

AMSINCKIA Lehm.
calycina (Moris) Chater   Hairy Fiddleneck
●●● (but greatly over-recorded for its congeners). A grain and bird-seed
casual. S America; (Australasia). BM, DBY, E, NMW, OXF, RNG. 1, 21,
22*, 36c*, 37*, 303(9:12-14), 665*. Incl. *A. angustifolia* Lehm.; *A. hispida*
I.M.Johnston, nom. illeg.; *Benthamia angustifolia* (Lehm.) Druce; *B. hispida*
(Ruíz Lopez & Pavón) Druce. The true identity of these records remains
unresolved; according to 20, most of them probably belong to *A. micrantha*.
lycopsoides (Lehm.) Lehm.   Scarce Fiddleneck
●●● (but over-recorded for *A. micrantha*). An established grain and seed alien;
known on the Farne Is (Cheviot) since 1922, now naturalised in abundance on
Inner Farne; elsewhere, on sandy soils in England. Western N America;
(Australia). CGE, E, LCN, NMW, OXF, RNG. 1, 8, 20, 41*, 303(9:12-14),
500*, 915*(82:141-150). *Benthamia lycopsoides* (Lehm.) Lindley ex Druce;
Lithospermum lycopsoides Lehm.
micrantha Suksd. group   Common Fiddlenecks
●●●● An established wool, grain, bird-seed, carrot seed, grass seed and linseed
alien; locally abundant on sandy soils, especially in East Anglia; increasing.
Western N America; (Australia). 7, 8, 14, 19, 20*, 25*, 41*, 264,
303*(9:1,12-14), 500*. Incl. *A. eastwoodae* J.F.Macbr.; *A. intermedia* Fischer
& C.Meyer; *A. intermedia* var. *eastwoodae* (J.F.Macbr.) Jepson & Hoover;
A. menziesii (Lehm.) Nelson & J.F.Macbr.; *A. parviflora* Heller; *Benthamia
intermedia* (Fischer & C.Meyer) Druce; *B. menziesii* (Lehm.) Druce;
B. parviflora (Heller) Druce. In addition, *A. retrorsa* Suksd. may have been
overlooked. 825(2:66-67). Most records are referable to *A. micrantha*
sensu stricto, but some of the segregates undoubtedly occur; others may be
errors of identity, or synonyms. The literature on this genus abounds in errors
and contradictions.

spectabilis Fischer & C.Meyer   Seaside Fiddleneck
Pre-1930 only. Western N America. 1, 41*, 42, 251. *Benthamia spectabilis*
(Fischer & C.Meyer) Druce.

tessellata A.Gray   Tessellate Fiddleneck
• A casual at ports, vector unknown. Western N America. NMW. 41*, 42,
156, 303(9:12-14), 915*(82:141-150), 1352*.

PLAGIOBOTHRYS Fischer & C.Meyer
australasicus (A.DC.) I.M.Johnston
Pre-1930 only. Western Australia. E, OXF. 3*, 37, 300(1910:45), 520a,
554, 1352*. *Eritrichium australasicum* A.DC. pro parte, non Benth.

cf. **campestris** E.Greene   Fulvous Popcorn-flower
• A casual in 1968 in an orchard at East Malling research station (W Kent), in
large quantity, possibly a grass seed alien. California. BM, RNG. 42,
371(355:26), 500*. *P. fulvus* Hook. & Arn. var. *campestris* (E.Greene)
I.M.Johnston.

canescens Benth.   Valley Popcorn-flower
Pre-1930 only. Western N America. 9, 42, 500*, 743.

elachanthus (F.Muell.) I.M.Johnston   Hairy Forget-me-not
• A wool casual. Australia. E. 37*, 133, 520*, 743, 1074*. *Eritrichium
australasicum* Benth. pro parte, non A.DC.; *Heliotropium elachanthum*
F.Muell.

procumbens (Colla) A.Gray
• A casual at Bristol (W Gloucs), vector unknown. No modern records.
Chile. 354(11:490). *Myosotis procumbens* Colla.

scouleri (Hook. & Arn.) I.M.Johnston   White Forget-me-not
•• A persistent grass seed alien; locally abundant on silt and gravel in shallow
flood hollows in levelled tip areas at Setley and at nearby Beaulieu airfield
(both S Hants); also in new-sown grass at Thurso (Caithness), on a road re-
building site at Salisbury (S Wilts) and collected from Shetland. Western
N America. BM, E, K, OXF, RNG. 20, 41*, 303(14:11), 370(17:191),
371(400:36), 500*, 1057*. *Allocarya scouleri* (Hook. & Arn.) E.Greene;
Myosotis scouleri Hook. & Arn.

stipitatus (E.Greene) I.M.Johnston   Stipitate Forget-me-not
• A casual at Avonmouth (W Gloucs), vector unknown. No modern records.
Western N America. OXF. 7, 42, 354(5:38; 7:777; 10:534), 500*, 1352*.
Allocarya stipitata E.Greene; *Echinospermum stipitatum* (E.Greene) Druce;
Lappula stipitata (E.Greene) Druce.

[*tenellus* (Nutt.) A.Gray (*Allocarya tenella* (A.Gray) Druce)   In error for
P. canescens. 9, 354(5:295).]

ASPERUGO L.
procumbens L.   Madwort
••• (but confused nomenclaturally with *Asperula arvensis*). An established
alien known since 1848 and now abundantly naturalised on the shore below
cliffs at Auchmithie (Angus); elsewhere a casual or persistent grain and wool

alien. Eurasia, N Africa. BM, E, LANC, NMW, OXF, RNG. 1, 8, 14, 20, 22*, 25*, 33*, 44*, 159, 187, 825*(13:176-185).

MYOSOTIS L.
alpina Lapeyr.
>   Pre-1930 only. Pyrenees. 21, 151, 546*, 1090*. *M. pyrenaica* Pourret.

[*balbisiana* Jordan   In error for native *M. discolor*. 220.]

dissitiflora Baker
*   (but perhaps overlooked as native *M. sylvatica*). A casual garden escape. Origin obscure, possibly Switzerland. BM, NMW, OXF. 2, 51, 148, 223, 312*(t.7589), 354(6:848).

[*stricta* Link ex Roemer & Schultes   In error. BM.]

verna Nutt.
>   Pre-1930 only. N America. 38*, 41*, 278, 1143*. *M. virginica* (L.) Britton, Sterns & Pogg., non *M. virginiana* L.; *Lycopsis virginica* L.

[*welwitschii* Boiss. & Reuter   In error for native *M. secunda*. The voucher specimen in NMW for the record in 21 has been redetermined.]

ERITRICHIUM Schrader ex Gaudin
nanum (L.) Schrader ex Gaudin   King-of-the-Alps
>   Pre-1930 only. Northern hemisphere. NMW. 7, 21, 23*, 26*, 27*, 29*, 170. *E. terglouense* (Hacq.) A.Kerner; *E. tergloviense* Endl. ex Putterl.

LAPPULA Gilib.
deflexa (Wahlenb.) Garke   Nodding Stickseed
*   A grain casual. No modern records. Eurasia, Canada. CGE. 1, 21, 22*, 25*, 38*, 182, 683*. *Echinospermum deflexum* (Wahlenb.) Lehm.

echinophora (Pallas) Kuntze
>   Pre-1930 only. SE Russia. SLBI. 2, 21, 300(1906:103), 630, 1105*, 1244*. *L. minima* (Lehm.) Druce; *Echinospermum echinophorum* (Pallas) Bornm.; *E. minimum* Lehm.; *Heterocaryum echinophorum* (Pallas) Brand; *Myosotis echinophora* Pallas.

marginata (M.Bieb.) Gürke
*   A wool casual. Europe. CGE, E. 7, 21, 133, 335(108:33), 683*. *L. patula* (Lehm.) Menyh.; *Echinospermum patulum* Lehm.

squarrosa (Retz.) Dumort.   Bristly Stickseed
*** A grain, grass seed, bird-seed, oil-seed and wool casual. Eurasia; (N America). BM, CGE, E, NMW, OXF, RNG. 1, 8, 13, 14, 19, 20*, 22*, 26*, 27*, 303(36:28; 48*:1), 683*. *L. echinata* Gilib.; *L. lappula* (L.) Karsten; *L. myosotis* Moench; *Echinospermum lappula* (L.) Lehm.; *Rochelia lappula* (L.) Roemer & Schultes.

szovitsiana (Fischer & C.Meyer) Druce
>   Pre-1930 only. Asia. 1, 39*, 630*, 1262*(as *L. echinophora*). *Echinospermum szovitsianum* Fischer & C.Meyer; *Heterocaryum szovitsianum* (Fischer & C.Meyer) A.DC. Dubiously conspecific with *L. echinophora*, as treated by 564.

HACKELIA Opiz
floribunda (Lehm.) I.M.Johnston **Large-flowered Stickseed**
- A casual garden weed, vector unknown. N America. RNG. 38*, 40*, 41*, 898(26:1-121). *Lappula floribunda* (Lehm.) E.Greene.

OMPHALODES Miller
linifolia (L.) Moench
Pre-1930 only. W Medit. 2, 21, 29*, 300(1906:103), 354(8:748), 737*, 1352*.

verna Moench **Blue-eyed-Mary**
•••• (but over-recorded for *Brunnera macrophylla*). An established garden escape and former introduction; naturalised in woods in widely scattered localities, sometimes in abundance for many years, as in Cwm Woods, Aberystwyth (Cards) and Treiorwerth woods (Anglesey); probably decreasing. Europe. 1, 20, 22*, 23*, 25*, 33*, 246, 305(42:40), 354(9:127). *O. omphalodes* (L.) Voss; *Cynoglossum omphalodes* L.; *Picotia verna* (Moench) Roemer & Schultes.

CYNOGLOSSUM L.
amabile Stapf & Drumm. **Chinese Hound's-tongue**
•• A casual garden escape. Tibet, China; (widespread). BM, K, RNG, SLBI. 36c*, 51*, 199, 312*(t.9334; n.s. t.82), 371(340:28), 893*(17:145-148), 1347*.

australe R.Br. **Australian Hound's-tongue**
- A wool casual. Australia. BM, CGE, E, RNG. 14, 37*, 514, 533, 549, 590*, 1078*.

montanum L.
Pre-1930 only. SW Asia. 39*, 263.

nervosum Benth. ex Hook.f. **Himalayan Hound's-tongue**
- A garden escape in the newly made verge of the Gloucester-Ledbury road near Playley Green (E Gloucs). Himalayas. RNG. 51, 57*, 349(4:289), 604.

VERBENACEAE
VERBENA L.
bonariensis L. **Argentinian Vervain**
•• (but confused with *V. litoralis*). A casual garden escape. Eastern S America; (widespread). CGE, RNG. 20, 36c, 51*, 53*, 191, 192, 328*(45:101-120). *V. patagonica* hort.

bracteata Lagasca & Rodriguez **Bracted Vervain**
- A casual in a quarry at Barham Crossing (E Suffolk), vector unknown. No modern records. N America. 7, 38*, 40*, 41*, 51, 354(6:617), 369(8:192), 500*. *V. bracteosa* Michaux.

elegans Kunth **Elegant Vervain**
- A casual garden escape. No modern records. Texas, Mexico. HbJPMB. 51, 349(2:110), 707. *V. moranensis* Willd. ex Sprengel.

hastata L.   American Vervain
- A casual garden escape. N America. 1, 38*, 40*, 41*, 51, 151, 199, 354(7:205). *V. paniculata* Lam.

hispida Ruíz Lopez & Pavón   Hairy Vervain
- A casual, perhaps a garden escape. No modern records. Southern S America. OXF. 51, 354(12:51), 705*, 849*(88:116).

× **hybrida** Voss   Hybrid Verbena
- • A casual garden escape. Originated in cultivation. 20, 51, 51b, 52*, 54*, 55*, 192, 199, 371(334:21). A complex of hybrids involving *V. peruviana* (L.) Druce and other S American species.

litoralis Kunth   Brazilian Vervain
- • • (but confused with *V. bonariensis*). A wool casual. S America; (widespread). ABD, BFT, CGE, E, LTR, RNG. 7 & 14 (as *V. bonariensis*), 36c, 328*(45:101-120). *V. bonariensis* auct., non L.; *V. brasiliensis* Vell.

menthifolia Benth.   Mint-leaved Vervain
- A wool casual. Western N America, Mexico. K. 14, 42, 145, 349(4:334), 500*, 1357.

platensis Sprengel
Pre-1930 only. Southern S America. 6, 51, 312*(t.3694). *V. teucrioides* Gillies & Hook.

rigida Sprengel   Slender Vervain
- • A persistent garden escape. Argentina, Brazil. RNG, HbEJC. 20, 52*, 53*, 54*, 245, 363*(25:50), 371(397:34). *V. venosa* Gillies & Hook.

supina L.   Trailing Vervain
- A wool and bird-seed casual. Medit., Canary Is; (Australia). E, K, RNG. 1, 14, 19, 21, 37*, 44*, 145, 737*.

tenuisecta Briq.
- A grain casual. S America. OXF. 7, 9, 42, 51, 52*, 55*, 354(5:298). *V. tenera* Sprengel. The similar *V. erinoides* Lam. from Brazil and Peru may have been overlooked. 826*(25:53).

urticifolia L.   Nettle-leaved Vervain
- A wool casual. No modern records. Eastern N America. 14, 38*, 40*, 51.

PHYLA Lour.
nodiflora (L.) E.Greene   Frogfruit
Pre-1930 only. Widespread. OXF. 7, 21, 29*, 36c*, 44*, 51*, 354(7:779; 9:130). *Lippia nodiflora* (L.) Michaux; *Verbena nodiflora* L.

CARYOPTERIS Bunge
× **clandonensis** A.Simmonds ex Rehder   Bluebeard
- A casual garden escape. Originated in cultivation. 51*, 54*, 61*, 312*(n.s. t.75), 347(11:16), 371(346:28). *C. incana* (Thunb.) Miq. × *C. mongholica* Bunge.

CLERODENDRUM L.
trichotomum Thunb.
- A garden escape persistent on a railway bank near Dorking Town station (Surrey), now gone; and by a footpath in Winsley, Bradford-on-Avon (N Wilts). E Asia. 51, 54*, 61*, 62*, 63*, 199, 462.

LAMIACEAE
Labiatae
STACHYS L.
annua (L.) L.   Annual Yellow-woundwort
- ●●● A grain and oil-seed alien sometimes persisting several years; decreasing. Formerly established in abundance on the downs in W Kent, now gone. Eurasia. BM, LIV, LTN, NMW, OXF. 1, 20, 22*, 46*, 134, 176, 303*(12:1,12), 354(10:480), 358(1873:209). *Betonica annua* L.; incl. *S. longibracteata* Bréb.

byzantina K.Koch   Lamb's-ear
- ●●● A garden escape persistent or established on roadsides in scattered localities, mainly in the south. SE Europe, SW Asia. NMW, OXF, RNG. 2, 20, 39*, 52*, 54*, 55*, 222, 325*(118:240-242). *S. lanata* Jacq., non Crantz; *S. olympica* auct., vix Poiret.

coccinea Ortega
Pre-1930 only. Texas, Mexico. LIV. 51, 312*(t.666), 354(9:279).

cretica L.
Pre-1930 only. Medit., SW Asia. NMW, OXF. 1, 9, 21, 24*, 31*, 44a*, 263. *S. italica* Miller; incl. *S. salicifolia* sphalm.; *S. salviifolia* Ten.

heraclea All.
Pre-1930 only. W Medit. OXF. 7, 21, 151, 288, 320(44:50), 546*, 683*. *S. intermedia* Ten.

macrantha (K.Koch) Stearn
- A garden escape persistent on the bank of a stream at Hartley, near Kirkby Stephen (Westmorland). Turkey, Iran, Caucasus. 39, 51, 52*, 52*, 53*, 55*, 303(52*:1,30; 54:33). *S. grandiflora* (Willd.) Benth.; *Betonica grandiflora* Willd.; *B. macrantha* K.Koch.

ocymastrum (L.) Briq.
- A casual at Bristol (W Gloucs), vector unknown. No modern records. W Medit. 21, 30*, 45*, 46*, 354(11:39). *S. hirta* L.

recta L.   Perennial Yellow-woundwort
- Established since 1923 at Barry docks (Glam), vector unknown; a casual elsewhere. Europe, SW Asia. CGE, LANC, NMW, OXF, RNG, SLBI. 7, 10, 20, 22*, 23*, 29*, 31*, 119, 156, 166.

BALLOTA L.
hispanica (L.) Benth.
- A casual introduced in a bomb crater on Brockham Hill (Surrey). No modern records. S Europe. RNG. 12, 21, 683*. *B. acuta* Briq.; *B. rupestris* (Biv.) Vis.

nigra L. subsp. **nigra**
- A casual at docks, vector unknown. No modern records. Europe, SW Asia.
BM, BRIST, CGE, GLR, RNG. 1, 9, 20, 22*, 39*, 212, 245, 826*(6:3-7).
B. nigra subsp. *ruderalis* (Sw.) Briq.; *B. ruderalis* Sw. Confused with native
subsp. *meridionalis* (Bég.) Bég. (subsp. *foetida* (Vis.) Hayek).
pseudodictamnus (L.) Benth.
- A garden escape recorded without locality. SE Europe. 21, 39*, 57*, 62*,
371(346:29).

MOLUCCELLA L.
laevis L.   Bells-of-Ireland
- A casual garden escape or horticultural seed impurity. SW Asia. 2, 21.
44a*, 54*, 55*, 261, 371(349:29).

LEONURUS L.
cardiaca L.   Motherwort
•••• An established grain alien and garden escape; naturalised in widely
scattered localities; known for over 50 years at Storeton (Cheshire); decreasing.
Eurasia. 1, 9, 20, 22*, 25*, 26*, 33*, 230.
marrubiastrum L.   False Motherwort
Pre-1930 only. Eurasia. 1, 21, 22*, 247.

LAMIASTRUM Heister ex Fabr.
galeobdolon (L.) Ehrend. & Polatschek subsp. **argentatum** (Smejkal) Stace
•••• An established garden escape; sometimes naturalised in abundance,
particularly in S Devon where it is beginning to oust native species. Origin
obscure. 20, 55*, 303*(46:9-11), 370(17:477), 908(47:247-248), 1097*,
1275*. *Lamium galeobdolon* (L.) L. subsp. *argentatum* (Smejkal) J.Duvign.;
L. galeobdolon forma *argentatum* (Smejkal) Mennema; *Galeobdolon argentatum*
Smejkal. Much confused with variegated forms of native subsp. *montanum*.

LAMIUM L.
bifidum Cirillo
Pre-1930 only. Medit. 1, 21, 195, 362(1901:40), 683*.
garganicum L.
- A casual at a farm near Haweswater (Westmorland), vector unknown.
No modern records. Medit. OXF. 1, 21, 31*, 44a*, 51, 51a, 354(2:418;
11:498), 683*. Incl. *L. bithynicum* Benth.; *L. laevigatum* DC., non L.
maculatum (L.) L.   Spotted Dead-nettle
••••• A persistent or established garden escape in widely scattered localities.
Eurasia. 1, 20, 22*, 26*, 27*, 33*. *Lamium album* L. var. *maculatum* L.;
incl. *L. laevigatum* L., non DC.; *L. rubrum* Wallr., non Hudson; *L. rugosum*
Aiton. *L.*'Album' has been reported. 458. A hybrid with native *L. album*
(*L.* × *holsaticum* E.H.Krause) may have been overlooked. 826*(23:17-23).

orvala L.
  ● A long-persistent garden relic at Warley Place gardens (S Essex). Europe.
  RNG. 21, 23*, 29*, 55*, 126, 313(31:370).

WIEDEMANNIA Fischer & C.Meyer
orientalis Fischer & C.Meyer
  Pre-1930 only. SW Asia. LCN, LIV, OXF, SLBI. 1, 8, 9, 21, 44a*, 160.
  W. erythrotricha (Boiss.) Benth.

GALEOPSIS L.
ladanum L.   Broad-leaved Hemp-nettle
  ●● (but over-recorded for native *G. angustifolia*). A casual, vector unknown.
  Eurasia. BM, CLE, DUE, E, LIV, OXF. 1, 20, 22*, 23*, 27*, 370*(5:143-
  149), 825*(5:248-255). *G. calcarea* Schönheit ex Steudel; *G. intermedia*
  Villars.

PHLOMIS L.
fruticosa L.   Jerusalem Sage
  ●● A long-persistent garden escape on dunes and walls; freely seeding and
  completely naturalised by the cliff walk at Daddy-hole Plain, Torquay
  (S Devon); known for over 60 years at Ball Hill, Stowell (S Somerset), now
  gone. Medit. BM, LANC, OXF, RNG, SLBI. 1, 20*, 29*, 31*, 32*, 52*,
  55*, 227, 248, 390(123:247). *P. reticosa* sphalm.
russelliana (Sims) Benth.   Turkish Sage
  ●● An established garden escape; known for over 30 years on a roadside at
  Mells (N Somerset), on a railway embankment at Langwathby (Cumberland),
  and on a roadside bank near Coultra (Fife). Turkey. CLE, K, LANC, RNG.
  13, 20*, 39*, 52*, 55*, 248, 370(19:150), 378(14:74,80). *P. lunariifolia*
  Sibth. & Smith var. *russelliana* Sims; *P. viscosa* hort., non Poiret.
samia L.   Greek Sage
  ● (but confused with *P. russelliana*). A persistent garden escape. SE Europe,
  Turkey. NMW. 7, 20, 24*, 31*, 39*, 156, 312*(t.1891), 354(8:316).
  Prasium samia sphalm.

SIDERITIS L.
hyssopifolia L.
  Pre-1930 only. SW Europe. 21, 30, 239, 525*, 546*, 595b*, 633. *Stachys
  scordioides* Poiret; incl. *Sideritis guillonii* Timb.-Lagr.
lanata L.
  Pre-1930 only. E Medit. 7, 21, 39, 354(4:206), 1172*.
montana L.   Mountain Ironwort
  ●● A grain casual. No modern records. Medit. BM, CGE, E, NMW, OXF,
  RNG. 1, 9, 21, 29*, 45*, 46*, 160.
romana L.
  ● A casual, probably a grain alien. No modern records. Medit. 7, 12, 21,
  24*, 30*, 32*, 46*, 289, 354(10:539).

MARRUBIUM L.
alysson L.
Pre-1930 only. Medit. K, LCN, OXF. 1, 9, 21, 44a*, 45*, 46*, 160, 258.
peregrinum L.
Pre-1930 only.   Eurasia.   OXF.   7, 9, 21, 29*, 31*, 39*, 354(5:48),
826*(11:88-91).

PRASIUM L.
majus L.
Pre-1930 only. Medit. 7, 21, 24*, 29*, 32*, 44a*, 267.

SCUTELLARIA L.
altissima L.   Somerset Skullcap
   • An established garden escape, naturalised since 1929 in a wooded valley near
   Mells (N Somerset); more recently recorded from the margin of Pasture Wood
   at Holmsbury St Mary, near Guildford (Surrey) and at an unpublished location
   in Roxburghshire. SE Europe, SW Asia. BM, OXF, SLBI. 20, 22*, 39*,
   248, 303(29:12), 349(3:47), 354(9:34). *S. columnae* auct., non All.
   [*columnae* All.   In error for *S. altissima.*   349(3:47), 354(9:34).]
hastifolia L.   Norfolk Skullcap
   • An established alien introduced and now well-naturalised by vegetative spread
   in woodland near Brandon (W Norfolk). Europe, SW Asia. BM, OXF, RNG.
   20, 22*, 25*, 235, 269, 370*(2:18-21). Old records from Ickleford Common
   (Herts) and Tintagel (E Cornwall) remain unconfirmed. 7, 241, 354(2:506).

TEUCRIUM L.
chamaedrys L.   Wall Germander
   ••• A garden escape long naturalised on downland between Beachy Head and
   Cuckmere Haven (E Sussex), but not seen recently; also established on walls
   and rocks in widely scattered localities; known since 1714 on the walls of
   Camber Castle (E Sussex) and since 1848 on limestone crags at South Cornelly
   (Glam). Medit. BM, DOR, LANC, NMW, OXF, RNG. 1, 20, 22*, 25*,
   26*, 33*, 122, 156, 212, 303(49:11). Some records may be referable to a
   hybrid with *T. lucidum* L. 20.
flavum L.
Pre-1930 only. Medit. NMW. 21, 31*, 39*, 270.
fruticans L.   Shrubby Germander
   • A casual garden escape. W Medit. 21, 24*, 30*, 32*, 303(36:23).
[*regium* Schreber   In error, for an unspecified reason. 354(5:783), 1149.]
resupinatum Desf.
   • A bird-seed casual. SW Spain, NW Africa. K, RNG. 7, 19, 21, 45*, 46*,
   303(22:14), 371(367:29).
[*scordium* L. subsp. *scordioides* (Schreber) Maire & Petitm. (*T. scordioides*
   Schreber) In error for native subsp. *scordium.* 354(5:783).]
spinosum L.
Pre-1930 only. Medit. 7, 9, 21, 44a*, 46*.

AMETHYSTEA L.
caerulea L.
- An oil-seed casual. N Asia. 9, 43, 51, 134, 312*(t.2448), 354(9:279), 650*, 1241*.

AJUGA L.
[*chamaedrys* sphalm. In error for *Teucrium chamaedrys*. 108.]
genevensis L.   Cornish Bugle
- A fodder alien which persisted for many years on dunes near Hayle (W Cornwall) and on chalk grassland near Churn (Berks); casual elsewhere. Europe, SW Asia. BM, LANC, LIV, NMW, OXF, RNG. 2, 20, 22*, 26*, 27*, 33*, 119, 156, 212, 354(5:299-305; 9:542), 825*(4:208-210).
iva (L.) Schreber
Pre-1930 only. Medit. 1, 21, 24*, 44a*, 45*, 195.

CEDRONELLA Moench
canariensis (L.) Webb & Berth.   Canary Islands Balm
- A casual horticultural seed impurity. Madeira, Canary Is. RNG. 21, 36c*, 51, 371(367:29), 530*, 650*, 844*(6:169-171). *C. triphylla* Moench.

NEPETA L.
[*cyanea* Steven In error for *N.* × *faassenii*. K.]
× **faassenii** Bergmans ex Stearn   Garden Cat-mint
- ●●● A persistent garden escape. Originated in cultivation. BM, E, LANC, NMW, OXF, RNG. 20, 34*, 54*, 370(2:108). *N. nepetella* × *N. racemosa*; *N.* × *mussinii* auct., non Sprengel ex Henckel. Contrary to most descriptions, at least some variants are fertile and merge into the variable *N. racemosa*.
grandiflora M.Bieb.
- A casual garden escape. Caucasus. OXF. 7, 12, 21, 264, 265, 354(5:573), 1257*.
nepetella L.   Lesser Cat-mint
- Recorded in 1990 from Granton (E Lothian). W Medit. 2, 21, 23*, 27*, 46*, 300(1904:178), 371(422:9).
nuda L.   Hairless Cat-mint
Pre-1930 only. Eurasia. OXF, SLBI. 7, 21, 22*, 25*, 29*, 300(1909:43).
racemosa Lam.
- ●● A casual garden escape. Caucasus. NMW, OXF, SLBI. 7, 20, 21, 303(60:35), 325*(75:403), 370(2:108). *N. mussinii* Sprengel ex Henckel.
[*tatarica* sphalm. In error for native *N. cataria*. 154.]

DRACOCEPHALUM L.
moldavica L.   Moldavian Dragon-head
Pre-1930 only. Asia; (E Europe, N America). 2, 21, 38*, 40, 197, 300(1905:101), 826*(11:50-53), 1110*.
parviflorum Nutt.   American Dragon-head
- ●●● A grain, bird-seed and wool casual, rarely persistent for a few years. N America. BM, CGE, E, NMW, OXF, RNG. 2, 19, 38*, 40*, 41*, 145, 182, 303*(17:14-15), 652*.

thymiflorum L.
> Pre-1930 only. Russia. SLBI. 2, 21, 25*, 300(1910:46), 320(1871:245), 1138*, 1241*.

[*GLECHOMA* L.
hirsuta Waldst. & Kit. In error for native *G. hederacea.* 354(5:699).]

LALLEMANTIA Fischer & C.Meyer
iberica (M.Bieb.) Fischer & C.Meyer
> • A grain casual. No modern records. Caucasus; (E Europe). LCN, OXF, RNG, SLBI. 2, 8, 21, 44a*, 354(13:110), 667*, 825(7:89-90), 826*(11:50-53; 13:15-17). *Dracocephalum ibericum* M.Bieb.

peltata (L.) Fischer & C.Meyer
> Pre-1930 only. SW Asia. 2, 39, 300(1904:178; 1905:101), 565*, 826*(13:15-17), 1110*. *Dracocephalum ocymifolium* Miller; *D. peltatum* L.

PRUNELLA L.
grandiflora (L.) Scholler   Large Selfheal
> • A persistent garden escape on railway embankments near Beaulieu Road station (S Hants) and between Harpenden and Wheathampstead (Herts); also on waste ground at Wisley (Surrey). Europe. RNG. 21, 22*, 29*, 30*, 303(41:20), 342(Feb89).

laciniata (L.) L.   Cut-leaved Selfheal
> ••• (but confused with native *P. vulgaris* and hybrids). Thoroughly established, vector unknown, in grassland in southern England. Europe, N Africa, SW Asia. BM, CGE, LANC, OXF, RNG. 2, 20, 22*, 23*, 25*, 33*, 191, 209, 245, 303(41:20). *P. vulgaris* var. *laciniata* L. The hybrid with *P. vulgaris* (*P. × intermedia* Link; *P. × hybrida* Knaf) has been recorded frequently, almost invariably superseding its parents wherever the two species meet. LANC, RNG. 20, 191, 209, 236.

vulgaris L. **alien variants**
> Var. *lanceolata* (Barton) Fern. (*P. pennsylvanica* Willd.) from N America has been recorded pre-1930 only. OXF. 2, 40, 41*, 197.

PHYSOSTEGIA Benth.
virginiana (L.) Benth.   Obedient-plant
> • A casual garden escape. Eastern N America. OXF, RNG. 38*, 51, 52*, 54*, 55*, 199, 354(9:759), 371(357:24). *Dracocephalum virginianum* L.

MELISSA L.
officinalis L.   Balm
> •••• An established garden escape, especially in the south. Medit, SW Asia. 1, 20, 22*, 25*, 33*, 34*.

ZIZIPHORA L.
capitata L.
> • A casual at South Darenth (W Kent), vector unknown. SE Europe, SW Asia. 21, 39, 44*, 303(28:25), 528*, 630*, 1282*.

clinopodioides Lam.
  Pre-1930 only. W Asia. 1, 39, 51, 139, 528*, 630*.
taurica M.Bieb.
  Pre-1930 only. W Asia. LCN. 7, 21, 160, 354(4:207), 630*, 1282*.

SATUREJA L.
hortensis L.   Summer Savory
  ● A casual garden escape. Medit. CGE, NMW, OXF, RNG. 1, 21, 26*, 80*, 88*, 135, 156.
montana L.   Winter Savory
  ● A garden escape long established on old walls at Beaulieu Abbey (S Hants) and Mells Manor (N Somerset). S Europe, N Africa, Caucasus. BM, CGE, LANC, OXF, RNG. 1, 13, 20*, 22*, 24*, 29*.

CLINOPODIUM L.
grandiflorum (L.) Stace   Greater Calamint
  ● A persistent garden escape. No modern records. S Europe, SW Asia. LANC, OXF, RNG. 7, 20, 23*, 27*, 29*, 39*, 167, 354(11:411; 12:52). *Calamintha grandiflora* (L.) Moench; *Satureja grandiflora* (L.) Scheele.
rotundifolium (Pers.) ined.   Round-leaved Calamint
  Pre-1930 only. Medit., SW Asia. LCN, OXF. 1, 21, 39*, 44a*, 160, 258, 354(8:211). *Acinos rotundifolius* Pers.; *Calamintha rotundifolia* (Pers.) Benth.; *Satureja rotundifolia* (Pers.) Briq.; incl. *A. graveolens* (M.Bieb.) Link; *C. graveolens* (M.Bieb.) Link, non (Benth.) Kuntze; *Thymus graveolens* M.Bieb.

HYSSOPUS L.
officinalis L.   Hyssop
  ●● A persistent garden escape, in quantity on oolitic limestone quarry waste at Edge (E Gloucs); probably casual elsewhere. Formerly established on old castle and abbey walls. Medit. BM, LCN, NMW, RNG, SLBI. 1, 20*, 22*, 23*, 27*, 324(24:276).

ORIGANUM L.
majorana L.   Pot Marjoram
  ● A casual garden escape. Medit. BM, NMW. 21, 29*, 39*, 71*, 156, 192, 370(12:287), 371(409:41). *Majorana hortensis* Moench.
[*majoricum* Cambess. In error for native *O. vulgare*. 371(376:25; 382:25).]
onites L.
  Pre-1930 only. Medit. 1, 6, 21, 39*, 88*, 313(22:42).
virens Hoffsgg. & Link
  ● Persistent on waste ground at Langley (Bucks), vector unknown. W Medit., Atlantic Is. RNG, HbEJC. 21, 303(9:18), 737, 1027, 1030*. *O. vulgare* subsp. *virens* (Hoffsgg. & Link) Bonnier & Layens. Pre-1930 records were in error for native *O. vulgare*. 354(1:229; 5:781).

THYMUS L.
vulgaris L. Culinary Thyme

•• An established garden escape; on walls in a few scattered localities in the south and in grassland near Darenth Wood (W Kent). Medit. RNG. 7, 20*, 24*, 29*, 88*, 199. A hybrid with native *T. pulegioides* (*T.* × *citriodorus* (Pers.) Schreber) was recorded pre-1930. NMW. 1311.

MENTHA L.
[*crispa* auct., non L. An ambiguous name which has been used for several species and hybrids. Most records probably refer to *M.* × *piperita*.]

[× *dalmatica* Tausch (*M. arvensis* L. × *M. longifolia*)   In error for *M.* × *gracilis*.]

[× *dumetorum* Schultes (*M. aquatica* L. × *M. longifolia*)   In error for *M.* × *piperata*. 20.]

× **gracilis** Sole Bushy Mint

••• An established garden escape in scattered localities throughout most of the British Is. BM, DHM, DTN, LSR, NMW, SUN. 20*, 22*, 33*, 88*. *M. arvensis* × *M. spicata*; *M.* × *cardiaca* (Gray) Baker; *M.* × *gentilis* auct., non L.; incl. *M. hackenbruchii* Briq.

longifolia (L.) Hudson Horse Mint

Pre-1930 only. Europe. 20, 21, 22*, 29*, 1311. Other records are in error for *M. spicata* or its hybrids.

× **piperita** L. Peppermint, incl. Eau-de-Cologne Mint

••• An established garden escape in scattered localities throughout the British Is. BM, DHM, OXF, SLBI. 20*, 22*, 33*, 88*. *M. aquatica* × *M. spicata* ; *M.* × *dumetorum* auct., non Schultes; incl. *M.* × *citrata* Ehrh.; *M. ciliata* sphalm.

requienii Benth. Corsican Mint

••• An established garden escape; naturalised in damp, grassy places in a few woodlands in the south, on peat banks at the head of Killary Harbour (W Galway) and on the north slope of Slieve Gullion (Co Armagh). C Medit. BM, CGE, E, NMW, OXF, RNG. 1, 20, 22*, 319(1:141; 18:83), 376*, 683*.

× **rotundifolia** (L.) Hudson False Apple-mint

•• An established garden escape in scattered localities, mainly in the north. Europe. OXF. 20*, 22*. *M. longifolia* × *M. suaveolens* Ehrh.; *M.* × *amaurophylla* Timb.-Lagr.; *M.* × *niliaca* Juss. ex Jacq.; *M. nouletiana* Timb.-Lagr.; *M. spicata* L. var. *rotundifolia* L.

× **smithiana** R.A.Graham Tall Mint

•• An established garden escape in scattered localities throughout the British Is, mainly in the south. BM, DHM, LSR. 20*, 22*, 33*. *M. aquatica* × *M. arvensis* × *M. spicata*; *M. rubra* Smith, non Miller.

spicata L.   Spear Mint
  ●●●●● An established garden escape; well naturalised in many places
  throughout the British Is. Origin obscure. 1, 20, 22*, 25*, 33*, 35*.
  M. crispa L.; *M. longifolia* auct., non Hudson; *M. niliaca* auct., non Juss.
  ex Jacq.; *M. scotica* R.A.Graham; *M. viridis* (L.) L.

× **villosa** Hudson   Apple-mint, incl. Bowles' Mint
  ●●●● An established garden escape in scattered localities throughout the
  British Is.   20*, 22*, 33*, 88*.   *M. spicata* × *M. suaveolens* Ehrh.;
  M. × *alopecuroides* Hull; *M.* × *cordifolia* auct., ?Opiz; *M.* × *nemorosa*
  Willd.; *M.* × *niliaca* auct., non Juss. ex Jacq.; *M.* × *scotica* auct.,
  non R.A.Graham; *M. velutina* Lej.

× **villosonervata** Opiz   Sharp-toothed Mint
  ●● An established garden escape in scattered localities in Britain. DHM,
  LANC, OXF, SUN, ZCM. 20*. *M. longifolia* × *M. spicata*; *M. longifolia*
  var. *horridula* auct., non Briq.; *M. nouletiana* auct., non Timb.-Lagr.;
  M. × *villosa* auct., non Hudson. A recorded hybrid with native *M. suaveolens*
  was probably in error for *M.* × *villosa*.

PERILLA L.
frutescens (L.) Britton
  ● A casual garden escape. India, China. 21, 51, 52*, 54*, 458. Incl.
  P. nankinensis (Lour.) Decne.

LAVANDULA L.
angustifolia group   Lavenders
  ●●● Persistent garden escapes; bird-sown on walls and in chalk pits in the
  south. Medit.; hybrids originating in cultivation. 20, 23*, 26*, 27*, 63*, 199,
  331(62:109), 393(29:187). Incl. *L. angustifolia* Miller; *L. latifolia* Medikus;
  L. spicata auct.; *L.* × *intermedia* Lois. (*L. angustifolia* × *L. latifolia*).

ROSMARINUS L.
officinalis L.   Rosemary
  ●● A garden escape, perhaps established on limestone cliffs west of Great
  Ormes Head (Caerns) and on walls in a few places in SE England and the
  Channel Is. Medit. LANC, RNG. 20*, 24*, 63*, 88*, 199, 236, 303(36:23).

SALVIA L.
aethiopis L.   African Sage
  Pre-1930 only. Eurasia. NMW, OXF. 2, 8, 21, 29*, 30*, 39, 51b*,
  354(11:274), 546*.
× **andrzejowskii** Błocki
  Pre-1930 only. SW Asia. 7, 354(2:506).
argentea L.   Silver Sage
  Pre-1930 only. Medit. LIV. 2, 21, 29*, 45*, 52*, 53*, 244, 370(18:205).
barrelieri Etlinger
  Pre-1930 only. SW Spain, NW Africa. 1, 21, 30*, 51b, 737*. *S. bicolor*
  Lam.; *S. inamoena* Vahl.

[*bertolonii* Vis.   Now considered to be conspecific with native *S. pratensis*. 354(5:573; 10:538).]

bornmuelleri Hausskn.
Pre-1930 only. W Asia. OXF. 2.

[*canescens* C.Meyer   A dubious record. 220.]

[*ceratophylloides* Ard.   Now considered to be conspecific with native *S. pratensis*. 197.]

[*controversa* Ten.   Now considered to be conspecific with native *S. verbenaca*. LCN. 160, 247.]

glutinosa L.   Sticky Clary
•• An established garden escape; naturalised in woods and on roadsides in a few widely scattered localities; long naturalised in abundance in woodland near the river Tay at Caputh (Mid Perth). Eurasia. BM, CGE, K, RNG. 1, 20, 22*, 23*, 26*, 27*, 349(5:343), 370(14:113), 546*.

nemorosa L.   Balkan Clary
• (but confused with its hybrids *S.* × *digenea* and *S.* × *sylvestris*). A grain alien long naturalised on dunes at Phillack Towans (W Cornwall); casual elsewhere. Eurasia. BIRM, BM, E, NMW, OXF, SLBI. 1, 9, 21, 22*, 23*, 31*, 39*, 51b*, 212, 354(10:837), 715*, 825*(4:72-74). *S. sylvestris* auct., non L.(1753).   Hybrids with native *S. pratensis* (*S.* × *sylvestris* L.) and backcrosses with *S. villicaulis* (*S.* × *digenea* Borbás; *S.* × *superba* Stapf; *S. virgata* hort., non Jacq.) have been recorded as casual garden escapes. BM. 371(334:21), 715*, 825*(4:72-74), 1311.

[*nutans* L.   In error for native *S. pratensis*. 371(337:29; 340:26).]

officinalis L.   Sage
••• A persistent garden escape. S Europe. ABS, BM, LCN, LIV, OXF, SLBI. 1, 21, 26*, 56*, 71*, 88*, 126, 222, 546*.

reflexa Hornem.   Mintweed
••• A wool, grass seed, grain and bird-seed casual. N America; (Australasia). BM, CGE, E, LIV, OXF, RNG. 14, 19, 20*, 51b, 303*(31:1,17), 370(1:318). *S. lanceolata* auct., non Willd.

[*rubella* Jordan & Fourr.   Now considered to be conspecific with native *S. verbenaca*. 320(1879:344).]

sclarea L.   Clary
••• A garden escape established on old walls in a few places in the south. Medit., SW Asia. BM, NMW, OXF, RNG, SLBI. 7, 20, 29*, 31*, 44*, 51b*, 54*, 349(3:324,410), 371(385:34), 546*. *Stachys sclarea* sphalm.

splendens Ker Gawler   Scarlet Sage
• A casual garden escape. Brazil. 51b*, 52*, 53*, 56*, 126, 135, 301*(t.687), 715*.

[*triloba* L.f.   In error for *S. reflexa*. 258, 433.]

verticillata L.   Whorled Clary
•••• An established grain alien and garden escape; naturalised by roads and railways, mainly in C & S England. Europe, SW Asia; (N America). 1, 20*, 22*, 23*, 26*, 51b, 209, 546*.

villicaulis Borbás
- A casual on grassy waste ground, Belvedere marshes (W Kent), vector unknown. No modern records. SE Europe, NW Turkey. BM. 12, 21, 39, 144, 623a*. *S. amplexicaulis* Reichb., non Lam.; *S. sylvestris* sensu Boiss. pro parte, non L. According to 715*, the correct name is *S. villicaulis*, since the type illustration in 1187* of *S. amplexicaulis* Lam., non Reichb., nec Heuff., nec Neilr., is of the N American *S. urticifolia* L.

virgata Jacq.
- A grain alien persistent at Silloth docks, near Carlisle (Cumberland). No modern records. SE Europe, SW & C Asia. CGE, LIV, OXF. 21, 39, 51b, 354(3:332; 4:73); 374(1951:45), 528*, 667*, 683*, 722*. Incl. *S. garganica* Ten.; *S. sibthorpii* Sibth. & Smith; *S. similata* Hausskn.

viridis L.   Annual Clary
- ●●● A bird-seed casual and garden escape. Medit., SW Asia. BM, CGE, E, LTN, OXF, RNG. 1, 20*, 29*, 54*, 55*, 147, 199, 247, 371(403:38), 546*. *S. horminum* L.

MONARDA L.
citriodora Cerv. ex Lagasca   Lemon Beebalm
- A wool casual. Southern USA, Mexico. CGE, E, RNG. 17, 38, 545, 699*, 1240*. *M. aristata* Nutt., non Hook.; *M. tenuiaristata* (A.Gray) Small.

didyma L.   Oswego-tea
Pre-1930 only. Eastern N America. OXF. 3, 38*, 40, 51*, 52*, 54*, 354(4:206).

ELSHOLTZIA Willd.
ciliata (Thunb.) N.Hylander
- A casual garden escape. Asia. BM, CGE. 1, 9, 21, 22*, 38*, 192, 197, 371(340:26; 416:8), 617*. *E. cristata* Willd.; *E. patrinii* (Lepechin) Garcke.

OCIMUM L.
basilicum L.   Sweet Basil
- A casual on a tip at Greenhithe (W Kent), vector unknown. Tropics. 26*, 39, 51, 54*, 56*, 80*, 458.

micranthum Willd.   Wild Basil
Pre-1930 only. Tropical America. 6, 646, 1358*.

COLEUS Lour.
blumei Benth.   Painted-nettle
- A casual garden escape. SE Asia. 51, 53*, 312*(t.4754), 331(72:119), 594*, 649*, 1255*. *Ocimum scutellarioides* L.; *Plectranthus scutellarioides* (L.) R.Br.; *Solenostemon blumei* (Benth.) Launert. According to 51b and other recent literature *Solenostemon scutellarioides* (L.) Codd is the preferred name.

PLANTAGINACEAE

PLANTAGO L.

afra L. Glandular Plantain

●● (but under-recorded as *P. arenaria*). A grain and bird-seed casual. Medit., SW Asia. BM, CGE, E, NMW, RNG. 1, 19, 20, 24*, 44a*, 277, 825*(2:129-135). *P. cynops* L.; *P. indica* auct., non L.; *P. psyllium* L.(1762), non L.(1753).

albicans L.

Pre-1930 only. Medit. 2, 21, 44a*, 46*, 300(1906:104).

[*alpina* L. In error for native *P. maritima*. 358(1865:283; 1866:39).]

arenaria Waldst. & Kit. Branched Plantain

●●●● (but confused with *P. afra*). A bird-seed, wool and probably agricultural seed alien, naturalised in Breckland. Eurasia, N Africa; (Australia). 1, 14, 19, 20, 22*, 25*, 29*, 44a*, 371(397:35). *P. indica* L., nom. illeg.; *P. psyllium* L. pro parte; *P. ramosa* Asch., nom. illeg.; *P. scabra* Moench, nom. illeg.; *Psyllium scabrum* Holub, nom. illeg.

[*argentea* Chaix. In error for *P. monosperma*. 354(8:35).]

aristata Michaux Bracted Plantain

● A casual at Avonmouth Docks (W Gloucs), probably an impurity in American clover or grass seed. N America. BM, E, K, OXF, RNG. 7, 36c, 38*, 41*, 42, 350(30:20), 354(6:620), 825*(2:129-135), 1285, 1319*. *P. patagonica* var. *aristata* A.Gray.

exigua Juss. ex Murray

Pre-1930 only. Iran, Afghanistan, Pakistan; (Egypt). 328(30:669-673), 354(8:634), 585*, 672, 729, 817*(1:94). *P. pumila* L.f. Some modern records of *P. afra* may be referable to this species. Grown in 1977 from seed impurity in cumin. K.

lagopus L. Hare's-foot Plantain

● A bird-seed, grain and esparto casual. Medit., SW Asia. BM, K, NMW, OXF, RNG, SLBI. 1, 9, 15, 19, 21, 24*, 32*, 44a*, 160, 182. *P. lusitanica* L.

loeflingii L.

● An esparto casual. W Medit., SW Asia. OXF. 7, 15, 21, 46*, 354(5:574), 825*(2:129-135).

[*macrorhiza* Poiret Unconfirmed. All records appear to be in error for native *P. coronopus*. BM. 354(2:583; 3:27-28).]

[*monosperma* Pourret A dubious record. 354(8:35).]

notata Lagasca

Pre-1930 only. Medit. SLBI. 7, 9, 21, 44a*, 46*, 300(1911:102), 729*.

ovata Forsskål Blond Plantain

● An esparto casual, possibly also from seed imported for sale in Asian-owned grocery shops. Medit. BM, RNG. 2, 21, 44a*, 46*, 199, 354(13:67), 418.

patagonica Jacq.

Pre-1930 only. N America, Argentina, Chile. 7, 38*, 41*, 300(1905:102-103); 1906:178), 391(1903-1904:7). *P. gnaphalioides* Nutt.; *P. purshii* Roemer & Schultes.

rugelii Decne.   Blackseed Plantain
  Pre-1930 only. Eastern N America. 38*, 40*, 300(1905:102; 1906:178),
  349(2:276), 1285, 1319*. Still occurs as a contaminent of grass and clover
  seed but no records of casuals, as claimed in 1285, have been found. 418.
sempervirens Crantz   Shrubby Plantain
  Pre-1930 only. SW Europe. BM, OXF, RNG. 7, 20, 21, 29*, 61, 123,
  320(6:271, 294), 354(6:36), 546*, 825*(2:129-135). *P. cynops* L.(1762),
  non L.(1753).
[*serraria* L. In error for native *P. coronopus*. 1215.]
tenuiflora Waldst. & Kit.
  Pre-1930 only. E Europe. 7, 21, 136, 715*.
varia R.Br.   Variable Plantain
  ● A wool casual. Australia, Tasmania; (New Zealand). BM, E, RNG. 3, 14,
  37, 133, 549*, 583*.
virginica L.   Virginian Plantain
  ● A wool casual. N America. BM, E, OXF, RNG. 1, 3, 17, 38*, 40, 42,
  300(1911:102), 354(5:674), 1319*. The S American *P. myosuros* Lam. may
  have been overlooked as this species. 825*(2:129-135).

BUDDLEJACEAE
BUDDLEJA L.
alternifolia Maxim.   Alternate-leaved Butterfly-bush
  ● A long-persistent garden escape in Darenth Wood (W Kent) and at Otley
  (MW Yorks); also recorded from Bedminster Down (N Somerset), now gone.
  China. 20, 54*, 61*, 62*, 63*, 370(15:429), 371(391:33; 414:13).
davidii Franchet   Butterfly-bush
  ●●●●● An established garden escape; on walls, cliffs, waste ground and in
  scrub, often well naturalised, especially in towns and cities in southern
  England; increasing. China. 7, 20, 22*, 62*, 63*. *B. magnifica* hort.;
  B. nanhoensis hort.; *B. spectabilis* hort., non Kunth & Bouché; *B. variabilis*
  Hemsley.
globosa Hope   Orange-ball-tree
  ●● A persistent garden escape or relic. Chile, Peru. E, RNG. 20, 54*, 62*,
  63*, 370(343:28).
lindleyana Fortune ex Lindley
  Pre-1930 only. China. 7, 51, 61, 63*, 354(7:776).
[*spectabilis* Kunth & Bouché In error for *B. davidii*. 199.]
× **weyeriana** Weyer
  ● A persistent garden escape or relic near Truro (E or W Cornwall), at
  Highbridge (N Somerset), by the river Wey at Guildford (Surrey) and at
  Hextable (W Kent). Originated in cultivation. 20, 62*, 63*, 212, 371(379:25).
  B. davidii × *B. globosa*.

OLEACEAE

JASMINUM L.
humile L.   Italian Jasmine
Pre-1930 only.   C Asia; (SE Europe).   21, 63*, 301*(t.178,t.350),
312*(t.1731), 594*, 597*, 617*, 1304. Recorded as var. *revolutum* (Sims)
Stokes (*J. revolutum* Sims; *J. humile* 'Revolutum').

nudiflorum Lindley   Winter Jasmine
• A garden escape or relic established in Tugley Wood (Surrey), on earth banks
at Stone and on waste ground at Tunbridge Wells and Green Street Green (all
W Kent); also recorded from a wall in Cambridge city (Cambs). China. 20,
21, 54*, 56*, 65*, 199, 236, 303(60:35), 322(36:136-148), 347*(10:148), 458.

officinale L.   Summer Jasmine
•• An established garden escape or relic in a few scattered localities in the
south, with self-sown seedlings reported from South Tottenham (Middlesex);
abundant at Hextable (W Kent). C Asia. 20, 21, 29*, 54*, 56*, 236, 258,
303(41:14; 64:41), 347*(10:148), 458.

FORSYTHIA Vahl
[*europaea* Degen & Baldacci   Unconfirmed records. 204, 258, 369(21:44).]

giraldiana Lingelsh.   Early Forsythia
• (but confused with *F.* × *intermedia* and *F. suspensa*). A persistent garden
escape on Dartford Heath and at Biggin Hill (both W Kent); now gone from
Hyndland, Glasgow (Lanarks).   Two Surrey records lack confirmation of
identity. China. HbPM. 51, 61, 63*, 199, 236, 371(361:27), 617*.

× intermedia hort. ex Zabel   Forsythia
•• A persistent garden escape at Banstead Heath (Surrey); introduced or casual
elsewhere.   Originated in cultivation. 20, 54*, 56*, 61, 63*, 156, 199.
F. suspensa × *F. viridissima* Lindley; *F.* × *suspensa* auct., non (Thunb.) Vahl;
F. 'Spectabilis'.

ovata Nakai   Korean Forsythia
• A garden relic by a ruined cottage at East End (Oxon). Korea. RNG. 51,
63*, 312*(t.9437), 638*.

suspensa (Thunb.) Vahl   Golden-bell
• A persistent garden escape on old walls on Stone Marshes and at Orpington
(both W Kent); also recorded near Fornham (W Suffolk), now gone. China;
(Japan). 20, 61, 63*, 65*, 369(21:44), 370(15:429), 458, 617*.   Incl.
F. sieboldii Dippel.

FRAXINUS L.
americana L.   White Ash
• Introduced. Eastern N America. 21, 58*, 59*, 166.

latifolia Benth.   Oregon Ash
• Introduced. Western N America. OXF. 41*, 58*, 59*, 61, 354(10:973).
F. oregona Nutt.

ornus L.   Manna Ash
•• A garden escape by Barnes station (Surrey); probably introduced elsewhere.
S Europe, SW Asia. BM, LANC, LIV, OXF. 7, 20, 58*, 59*, 60*, 119,
204, 303(45:22).

pennsylvanica Marshall   Red Ash, incl. Green Ash
• Introduced. Eastern N America. 21, 38*, 58*, 59*, 60*, 119, 166.

SYRINGA L.
× **persica** L.   Persian Lilac
• Introduced. Asia. 51, 62*, 63*, 278, 313(31:370).

vulgaris L.   Lilac
••••• An established garden escape; naturalised by suckering in hedgerows
and thickets in widely scattered localities throughout the British Is. SE Europe.
1, 20, 21, 22*, 62*, 63*.

LIGUSTRUM L.
lucidum Aiton f.   Tree Privet
• A garden relic and escape; seedlings on waste ground at New Ash Green;
also reported from Chislehurst Common and Dartford Marshes (all W Kent).
China. RNG. 21, 36c*, 59*, 60*, 236, 248, 331(37:198), 458. *L. japonicum*
auct., non Thunb.

ovalifolium Hassk.   Garden Privet
••••• An established garden escape; occasionally bird-sown and naturalised
in woods, hedges and thickets, and recorded throughout most of the British Is.
Japan. 12, 20, 22*, 36c*, 52*, 236.

sinensis Lour.   Chinese Privet
Pre-1930 only. China. 36c*, 51, 61, 62*, 354(9:30).

strongylophyllum Hemsley
• A casual in a garden, probably bird-sown. C China. 51, 61, 312*(t.8069),
370(15:169).

PHILLYREA L.
latifolia L.
• Introduced and persistent at Abbotsbury (Dorset). Medit. RNG. 21, 24*,
60*, 63*. Incl. *P. media* L.

OSMANTHUS Lour.
× **burkwoodii** (Burkwood & Skipwith) P.Green
• Introduced and persistent at Abbotsbury (Dorset). Originated in cultivation.
51, 52*, 61*, 63*, 458. *O. decorus* (Boiss. & Balansa) Kasapl. × *O. delavayi*
Franchet; × *Osmarea burkwoodii* Burkwood & Skipwith.

heterophyllus (G.Don) P.Green
• A garden escape, recorded in 1992, on a roadside bank at Farningham
(W Kent). Japan, Taiwan. 53*, 61*, 63*, 458, 597*, 649*. *O. ilicifolius*
(Hassk.) Carrière; *Ilex heterophyllus* G.Don; *Olea ilicifolia* Hassk.

SCROPHULARIACEAE
VERBASCUM L.
blattaria L. **Moth Mullein**

•••• An established garden escape found mainly in C and S England, though seldom in the same locality for many years. Eurasia, N Africa. 1, 20, 22*, 33*, 35*. The Asian *V. macrocarpum* Boiss. may have been overlooked as this species. 825*(3:52).

boerhavii L.

Pre-1930 only. W Medit. 7, 9, 21, 546*, 683*. Incl. *V. lydium* Boiss.

bombyciferum Boiss. **Broussa Mullein**

•• A persistent garden escape. Turkey. RNG. 20, 52*, 220, 303*(34:20), 312*(n.s. t.422), 371(374:9), 1218. *V.* 'Broussa'. Hybrids with *V. phlomoides* and native *V. nigrum* have been recorded. 135.

chaixii Villars **Nettle-leaved Mullein**

•• A grain alien and garden escape, sometimes persistent. Eurasia. BM, NMW, OXF, RNG, SLBI. 2, 20, 30*, 54*, 160, 325*(113:327). Incl. *V. austriacum* Schott ex Roemer & Schultes. A white-flowered form is established at Pagham (W Sussex). 371(425:8).

creticum (L.) Cav. **Cretan Mullein**

• A grain casual and garden escape. W Medit. LANC, OXF. 1, 21, 29*, 174, 350(28:248), 354(13:304), 650*. *Celsia cretica* L.

densiflorum Bertol. **Dense-flowered Mullein**

•• (but confused with *V. phlomoides* and hybrids). A casual or persistent alien in a few places in SE England, perhaps a garden escape. Europe, SW Asia. BM, OXF, RNG. 1, 20, 22*, 26*, 126, 325*(117:542). *V. thapsiforme* Schrader. Putative hybrids with native *V. nigrum* and *V. pulverulentum* have been recorded. 371(379:26), 1311.

olympicum Boiss.

• A garden escape. Turkey. CGE, OXF. 7, 39, 51, 53*, 325*(113:326), 336(14:28), 371(346:29).

ovalifolium Donn ex Sims

• A casual at Portishead docks (N Somerset) and Avonmouth docks (W Gloucs), vector unknown. No modern records. SE Europe, SW Asia. OXF. 4, 9, 21, 312*(t.1037), 354(7:48; 11:35). Incl. *V. pulchrum* Velen.

phlomoides L. **Orange Mullein**

•••• A persistent garden escape, mainly in southern England and Wales. Europe, SW Asia. 1, 20*, 22*, 23*, 56*. Incl. var. *albiflorum* (Rouy) Wilm. Hybrids with native *V. lychnitis*, *V. nigrum* (*V.* × *brockmuelleri* Ruhmer; and *V. thapsus* (*V.* × *kerneri* Fritsch) have been recorded. OXF, RNG.

phoeniceum L. **Purple Mullein**

•• A bird-seed casual and garden escape. S Europe, SW Asia. BM, OXF, RNG, SLBI. 1, 19, 20, 22*, 25*, 26*, 52*, 325*(117:540-542). A hybrid with native *V. nigrum* (*V.* × *ustulatum* Čelak.) has been recorded. 336(14:27 28).

pyramidatum M.Bieb.   Caucasian Mullein
  •• (but confused with *V. chaixii*). Probably a garden escape; well established
  along the disused railway between Swaffham Prior and Fordham (Cambs) and
  on a railway bank at Melton Mowbray (Leics). Caucasus. BM, CGE, OXF,
  RNG. 7, 20, 303*(13:16; 19:1,21; 34:21), 370(12:286). Hybrids with native
  V. nigrum and *V. thapsus* have been recorded. 303*(19:1,21; 34:21).
sinuatum L.
  • A wool casual. Medit., SW Asia. BM, LIV, NMW, OXF, RNG. 7, 16,
  17, 21, 24*, 44a*, 45*, 235.
speciosum Schrader   Hungarian Mullein
  •• An established garden escape; naturalised since 1930 on a railway bank at
  Gt Blakenham (E Suffolk). SE Europe, SW Asia. BM, JSY, OXF, RNG.
  7, 20, 31*, 258, 1279*. *V. longifolium* DC., non Ten. A hybrid with native
  V. thapsus (*V.* × *duersteinensis* Teyber) has been recorded. 371(382:23).
thapsus L. subsp. **giganteum** (Willk.) Nyman
  • A garden escape. S Spain. 21, 371(346:29), 737*. *V. giganteum* Willk.

SCROPHULARIA L.
canina L.   French Figwort
  • A former ballast alien established for many years at Cardiff and Barry docks
  (both Glam) and at Newport (Mons), perhaps now gone. Medit., SW Asia.
  BM, NMW, OXF, RNG. 1, 10, 20, 22*, 23*, 25*, 27*, 156. Incl. *S. hoppei*
  Koch; *S. ramosissima* Lois.
peregrina L.   Nettle-leaved Figwort
  • A persistent garden weed at Shapwick (Dorset), vector unknown; also
  recorded from Lewes and Mark Cross (both E Sussex). Medit. RNG. 7, 21,
  31*, 32*, 39*, 288, 349(7:506), 354(10:106).
scopolii Hoppe ex Pers.   Italian Figwort
  • A casual at Oxford (Oxon), vector unknown. No modern records. C & S
  Europe, SW Asia. OXF. 4, 21, 23*, 27*, 39*, 354(7:48; 9:363; 12:287).
vernalis L.   Yellow Figwort
  •••• An established garden escape; naturalised in shady places, especially in
  the south; formerly introduced in plantations and old gardens, and still
  surviving in several localities more than 150 years later. Eurasia. 1, 20, 22*,
  23*, 27*, 33*.

COLLINSIA Nutt.
heterophylla Buist ex Graham   Chinese-houses
  • A casual garden escape or horticultural seed impurity. California. BM, K,
  RNG, SLBI. 1, 42, 51*, 52*, 55*, 300(1908:102). *C. bicolor* Benth.,
  non Raf.

PHYGELIUS E.Meyer ex Benth.
capensis E.Meyer ex Benth.   Cape Figwort
  •• An established garden escape; spreading by both seed and suckers near Bray
  (Co Wicklow) and spreading on the site of a long-abandoned nursery at Hythe

(E Kent); also recorded from Connemara (W Galway). S Africa. 20, 36c*, 51*, 57*, 347*(9:233), 371(331:20), 458, 1294*.

MIMULUS L.
× **burnetii** S.Arn.   Coppery Monkeyflower
  ●●● An established garden escape, mainly in the north and west. Originated in cultivation. LANC, OXF, RNG. 20, 370(14:426; 15:135), 594*. *M. cupreus × M. guttatus*; *M. × cupreus* auct., non Dombrain.

cardinalis Douglas ex Benth.   Scarlet Monkeyflower
  ● A casual in a garden, vector unknown. Western N America. 42, 51, 54*, 55*, 371(376:25), 596*, 1257*.

[*cupreus* Dombrain   Unconfirmed records. Probably always in error for its hybrids. 370(6:371-376).]

guttatus DC.   Monkeyflower
  ●●●●● An established garden escape; naturalised by lowland streams and lakes throughout most of the British Is. Western N America. 1, 20, 22*, 33*, 34*, 35*. *M. langsdorffii* Donn ex E.Greene, nom. illeg.; *M. luteus* auct., non L. A hybrid with *M. tigrinus* hort. (*M. × hybridus* hort. ex Siebert & Voss) has been recorded. 199.

luteus L. sensu lato   Blood-drop-emlets
  ●●●● (but over-recorded for *M. guttatus* and hybrids). An established garden escape; locally naturalised by streams and lakes, mainly in Scotland. S America. 2, 20, 22*, 25*, 33*, 1257*. Incl. *M. nummularis* C.Gay; *M. rivularis* Nutt.; *M. smithii* auct., non Lindley, vix Paxton; *M. variegatus* Lodd., nom. nud.; *M. youngeana* Hook.

× **maculosus** T.Moore   Scottish Monkeyflower
  ●● An established garden escape; recently recorded from Bicknoller (S Somerset) and Deuchrie (E Lothian). Originated in cultivation. NMW. 20, 371(422:9), 423. *M. cupreus × M. luteus*. A hybrid with *M. guttatus* has been recorded. LANC. 20, 370(14:426).

moschatus Douglas ex Lindley   Musk
  ●●●● An established garden escape; naturalised in damp and shaded places in widely scattered localities. Western N America. 1, 20, 22*, 25*, 29*, 33*.

ringens L.   Allegheny Monkeyflower
  ● A casual garden escape. Eastern N America. 38*, 40, 51, 199, 1257*.

× **robertsii** Silverside   Hybrid Monkeyflower
  ●●●● An established garden escape; in scattered localities over most of the British Is, mainly in the north and west. Originated in cultivation. BM, LANC. 20, 370(18:210-212), 594*, 1246. *M. guttatus × M. luteus* var. *rivularis* Lindley; *M. × luteus* auct., non L.; incl. *M.*'A.T.Johnson'; *M. × caledonicus* Silverside, nom. nud.

LIMOSELLA L.
capensis Thunb.
  Pre-1930 only. S Africa. OXF. 7, 354(6:34), 568*.

GRATIOLA L.
officinalis L.   Gratiole
Pre-1930 only. Eurasia. 2, 21, 22*, 25*, 29*, 278.

CALCEOLARIA L.
biflora Lam.
• A garden escape at Northiam (W Sussex). Chile, Argentina. 51, 53*, 54, 122, 650*, 665*, 669*.
× herbeohybrida Voss   Florists' Slipperwort
• A casual greenhouse escape. Originated in cultivation. 54*, 55*, 371(416:8). C. hybrida hort. ex Vilm.
integrifolia Murray   Bush Slipperwort
• A garden escape self-sown on walls in the Channel Is. Chile. 20, 54*, 55*, 66*, 220.
[mexicana Benth. In error for C. tripartita; this tropical American species is closely related, but appears to be rare in cultivation and to have no tendency to weediness.]
tripartita Ruíz Lopez & Pavón sensu lato   Annual Slipperwort
••• A persistent bird-seed and wool alien, and an escape from botanic and other gardens; a weed of cultivated ground in scattered localities. Mexico, C & S America; (widespread). BM, E, K, NMW, OXF, RNG. 3, 14, 19, 20, 36c, 51, 312*(t.2405), 349(5:338-341), 511*, 596*, 805*(66:202-208). Incl. C. chelidonioides Kunth; C. glutinosa Heer & Regel; C. gracilis Kunth; C. mexicana auct., non Benth.; C. pinnata L.; C. scabiosifolia Roemer & Schultes.

JOVELLANA Ruíz Lopez & Pavón
sinclairii (Hook.) Kränzlin
Pre-1930 only. New Zealand. 36, 51*, 349(5:340), 312*(t.6597), 709*, 1228*. Calceolaria sinclairii Hook.

ALONSOA Ruíz Lopez & Pavón
caulialata Ruíz Lopez & Pavón
• A casual in a garden, probably a seed impurity. Mexico, Peru. OXF, RNG. 7, 51, 354(6:33), 804*(t.203), 1013*. A. meridionalis (L.f.) Kuntze.
incisifolia Ruíz Lopez & Pavón   Cut-leaved Mask-flower
• A casual garden or greenhouse escape. Peru, Bolivia. NMW, OXF. 7, 51, 156, 288, 312*(t.471), 354(5:296; 7:586), 510*, 511, 565*, 1216*, 1258*. Celsia urticifolia Sims.
peduncularis (Kuntze) Wettst.
Pre-1930 only (possibly in error for A. warscewiczii). S America. BM. 7, 354(4:18), 511.
warscewiczii Regel   Mask-flower
• A casual seed impurity. Peru. E. 51, 54*, 55*, 277, 593*, 594*, 650*.

ANARRHINUM Desf.
bellidifolium (L.) Willd.   **Daisy-leaved Toadflax**
Pre-1930 only. SW Europe. RNG. 7, 21, 22*, 25*, 29*, 354(4:204), 737*.
A. anarrhinum (L.) Druce; *Antirrhinum bellidifolium* L.

ANTIRRHINUM L.
majus L.   **Snapdragon**
• • • • • An established garden escape; naturalised on cliffs, in quarries and on
old walls, especially in the south and east. W Medit. 1, 20, 21, 22*, 23*,
33*, 53*.
siculum Miller
• A casual garden weed, vector unknown. No modern records. C Medit. 21,
44a*, 235.

CHAENORHINUM (DC. ex Duby) Reichb.
origanifolium (L.) Kostel.   **Malling Toadflax**
• An established garden escape, known for over 100 years on old walls at West
Malling (W Kent); also a shortly persistent introduction in a bomb crater at
Brockham (Surrey). W Medit. BM, LANC, OXF, RNG. 12, 20, 23*, 27*,
30*, 199, 236, 349(2:144), 354(13:32). *Linaria origanifolia* (L.) DC.
rubrifolium (Robill. & Castagne ex DC.) Fourr.
Pre-1930 only. Medit. 7, 21, 368(1914:234), 683*, 737*. *Linaria rubrifolia*
Robill. & Castagne ex DC. Early records from West Malling were in error for
C. origanifolium. 354(13:32).

MISOPATES Raf.
calycinum (Vent.) Rothm.   **Pale Weasel's-snout**
• • A bird-seed casual. W Medit. RNG, HbEJC. 2, 19, 20, 21, 303(23:12),
683*, 737. *M. orontium* var. *grandiflorum* (Chav.) Valdés; *Antirrhinum
calycinum* Vent.; *A. elegans* Ten.; *A. orontium* L. var. *grandiflorum* Chav.

ASARINA Miller
procumbens Miller   **Trailing Snapdragon**
• • • An established garden escape; naturalised on sandstone cliff above
Broadmarsh, Nottingham (Notts); elsewhere well established on old walls in a
few scattered localities. Pyrenees, Cevennes. ABD, BM, CGE, E, LANC,
RNG. 13, 20, 27*, 30*, 55*, 303*(43:20-21). *Antirrhinum asarina* L.

CYMBALARIA Hill
hepaticifolia (Poiret) Wettst.   **Corsican Toadflax**
• (but perhaps overlooked). A garden plant persistent as a weed in a nursery
at Seal (W Kent) and in Edinburgh Botanic Gardens (Midlothian); accidentally
brought in with nursery plants elsewhere. Corsica. 20, 51, 303(39:9),
371(346:27; 352:27), 650*, 683*. *Linaria hepaticifolia* (Poiret) Steudel.

longipes (Boiss. & Heldr.) A.Chev.
- A garden escape established on a wall at Strete (S Devon). Greece, Crete, Turkey. RNG. 21, 31*, 39, 658, 853*(12:58,242). *Linaria longipes* Boiss. & Heldr.; *L. cymbalaria* subsp. *longipes* (Boiss. & Heldr.) Hayek.

muralis P.Gaertner, Meyer & Scherb. subsp. **muralis**   Ivy-leaved Toadflax
●●●●● A garden escape established throughout the British Is; mainly on old walls, but also in quarries and on rocks, shingle and limestone pavement. S Europe. 1, 20, 22*, 29*, 33*, 35*. *C. cymbalaria* (L.) Druce, nom. illeg.; *Linaria cymbalaria* (L.) Miller. Forma *toutonii* (A.Chev) Cuf. (*C. toutonii* A.Chev.) has been recorded from Cambridge University Botanic Garden (Cambs). 303(60:62), 546a*.

muralis P.Gaertner, Meyer & Scherb. subsp. **visianii** (Kumm. ex Jáv.) D.Webb
- (but possibly overlooked). Recorded from Wisley (Surrey), probably as a garden escape. Italy, Yugoslavia. HbACL. 21, 303(60:62), 683. *C. muralis* var. *pilosa* (Vis.) Degen; *Linaria cymbalaria* var. *pilosa* Vis.; *L. pilosa* auct., non (Jacq.) DC.

pallida (Ten.) Wettst.   Italian Toadflax
●●● An established garden escape; on walls, mainly in the north and west. Formerly naturalised in quantity on shingle at Bardsea (Furness). Italy. BM, CGE, E, LANC, NMW, OXF, RNG. 7, 20, 27*, 66*, 354(7:420), 683*. *Antirrhinum pallidum* Ten.; *Linaria pallida* (Ten.) Ten. ex Guss.

KICKXIA Dumort.
elatine (L.) Dumort. subsp. **crinita** (Mabille) Greuter
- A casual at Bristol (W Gloucs), vector unknown. No modern records. Medit. 21, 39*, 44a*, 354(10:471). *Linaria sieberi* Reichb.

LINARIA Miller
aeruginea (Gouan) Cav.
- A garden escape persistent on waste ground at Wisley (Surrey). SW Europe. NMW. 7(as *L. pinifolia*), 21, 30*, 156, 342(Feb89), 354(1920:33), 458, 737*. *L. pinifolia* auct., non (Poiret) Thell.; *L. reticulata* auct. eur., non (Smith) Desf.

alpina (L.) Miller
- A garden escape found in 1986 at Braye, Alderney (Channel Is). C & S Europe. 21, 22*, 23*, 29*, 51, 546*, 1031.

angustissima (Lois.) Borbás   Narrow-leaved Toadflax
- A casual in a cornfield near Headley (Surrey), vector unknown. No modern records. Europe. LANC, OXF. 7, 9, 12, 21, 27*, 683*. *L. italica* Trevir.

anticaria Boiss. & Reuter
Probably pre-1930 only. S Spain. 21, 30, 51, 197, 737*.

arenaria DC.   Sand Toadflax
- Introduced on dunes at Braunton Burrows (N Devon) last century, now well established; also recorded in the 1930s between Bloxworth and Bere (Dorset). SW Europe. BM, CGE, LANC, OXF, RNG, TOR. 7, 20, 22*, 25*, 33*, 165, 217.

arvensis (L.) Desf.   Corn Toadflax
Pre-1930 only. Europe, N Africa; (Australia). BM. 7, 21, 22*, 25*, 36c,
156. The record from Bristol was in error for *L. chalepensis*. 9, 354(4:514).
bipartita (Vent.) Willd.
  ● A casual garden escape or seed impurity. W Morocco. OXF, RNG.
  1, 45*, 51, 119, 371(334:21), 650*. *L. tripartita* sphalm. A hybrid with
  L. maroccana has been collected. OXF.
caesia (Pers.) Dietr.
Pre-1930 only. C & S Spain. 2, 21, 263, 737*, 1323*.
chalepensis (L.) Miller
  ● A casual at Southampton (S Hants) and Ipswich(E Suffolk), vector unknown.
  No modern records. Medit., SW Asia. LCN, OXF. 1, 21, 29, 32*, 39,
  44a*, 258, 354(13:110).
dalmatica (L.) Miller   Balkan Toadflax
  ●●● An established garden escape; known since 1916 on railway banks near
  Langport (S & N Somerset). SE Europe, SW Asia. BM, CGE, E, NMW,
  OXF, RNG. 3, 20, 29*, 31*, 36c*, 52*, 248. *L. genistifolia* (L.) Miller;
  L. macedonica Griseb.; *Antirrhinum dalmaticum* L.
× **dominii** Druce
  ●●● A persistent alien, mainly at ports and by railways. LANC, LSR, NMW,
  OXF, RNG. 7, 20, 156, 370(14:53-57). *L. purpurea* × *L. repens* (L.) Miller.
  Hybrid swarms have also been recorded.
[*glauca* (L.) Chaz.   In error for *L. supina*. LANC. 484.]
incarnata (Vent.) Sprengel
  ● A casual, vector unknown. SW Spain, Portugal, Morocco. NMW. 21, 156,
  192, 737*. *L. bipartita* auct., non (Vent.) Willd.
maroccana Hook.f.   Annual Toadflax
  ●●● A bird-seed and grass seed casual and garden escape. Morocco.
  BM, CGE, E, LSR, OXF, RNG. 7, 19, 20, 36c*, 51, 52*, 53*, 66*, 213,
  312*(t.5983).
micrantha (Cav.) Hoffsgg. & Link
Pre-1930 only.   Medit., SW Asia.   1, 7, 21, 44a*, 354(4:18).   Incl.
  L. parviflora Desf., non (Jacq.) Hal.
pelisseriana (L.) Miller   Jersey Toadflax
  ● A casual at Chesterfield (Derbys) and Combs (E Suffolk), vector unknown.
  Possibly once native in Jersey (Channel Is), but now extinct. W Europe,
  Medit., SW Asia. DBY. 1, 20, 22*, 25*, 31*, 201, 258, 370(15:400).
[*pinifolia* (Poiret) Thell. (*L. reticulata* (Smith) Desf.) In error for *L. aeruginea*.
  354(6:33).]
purpurea (L.) Miller   Purple Toadflax
  ●●●●● An established garden escape; naturalised in many localities throughout
  the British Is; increasing. C Medit. 1, 20*, 22*, 25*, 33*, 34*, 36c*, 264.
  Incl. *L.* 'Canon J.Went'. A hybrid with native *L. vulgaris* has been recorded.
  RNG. 265.
reflexa (L.) Desf.
Pre-1930 only. C Medit. OXF. 4, 21, 45*, 46*, 354(7:48).

saxatilis (L.) Chaz.
   • A casual on granite chippings at Gloucester docks (E Gloucs). SW Europe.
21, 323(25:415), 737*.
simplex (Willd.) DC.
   Pre-1930 only. Medit., SW Asia. BM. 21, 44a*, 45*, 199. *L. parviflora*
(Jacq.) Hal., non Desf.
spartea (L.) Hoffsgg. & Link
   Pre-1930 only. SW Europe. BM, OXF, SLBI. 1, 21, 249, 737*.
supina (L.) Chaz.   Prostrate Toadflax
   ••• A ballast alien long established at Par (E Cornwall), Plymouth (S Devon)
and Burry Port (Carms); elsewhere casual or persistent by railways, sometimes
in abundance. W Medit. BM, LANC, NMW, OXF, RAMM, RNG. 1, 20,
22*, 23*, 30*, 35*, 156, 189, 212, 217. A hybrid with native *L. repens*
(*L.* × *cornubiensis* Druce) has been recorded at Par harbour. OXF.
354(10:556).
triphylla (L.) Miller
   Pre-1930 only. Medit. LIV, NMW, OXF. 1, 5, 21, 29*, 32*, 44a*, 258.
viscosa (L.) Dum.-Cours.
   Pre-1930 only. C & W Medit. NMW, OXF. 4, 21, 45*, 46*, 354(6:617),
737*. Incl. *L. heterophylla* Desf.; *L. stricta* (Sibth. & Smith) Guss.,
non Hornem.

NUTTALLANTHUS D.A.Sutton
canadensis (L.) D.A.Sutton
   Pre-1930 only. N America. 1, 21, 38*, 354(1:40), 1323*. *Linaria canadensis*
(L.) Dum.-Cours.

NEMESIA Vent.
melissifolia Benth.
   • A wool casual. S Africa. RNG. 17, 735*.
strumosa Benth.   Cape-jewels
   •• A casual garden escape. S Africa. E. 7, 20, 51, 52*, 53*, 66*, 126, 264,
371(391:34).
versicolor E.Meyer ex Benth.
   • A garden casual. S Africa. RNG. 51, 561*, 628*, 655*, 1274*.

DIGITALIS L.
ferruginea L.   Rusty Foxglove
   • Introduced and persistent for some years in a bomb crater and chalk pits at
Brockham (Surrey). No modern records. S Europe, SW Asia. BM, LANC,
RNG. 13, 20, 21, 29*, 31*, 57*, 209, 331(30:6).
grandiflora Miller   Yellow Foxglove
   • Introduced and persistent for a few years in a bomb crater at Brockham Hill
(Surrey). No modern records. Europe, SW Asia. BM, LANC, RNG. 7, 12,
20, 21, 22*, 26*, 29*, 52*, 280, 354(9:128). *D. ambigua* Murray.

lanata Ehrh.   Grecian Foxglove
  • A relic of war-time cultivation for the pharmaceutical industry persistent for
several years at Dartford (W Kent); also introduced and persistent in a bomb
crater and chalk pits at Brockham (Surrey). No modern records. SE Europe,
Turkey. BM, LANC, OXF, RNG, SLBI. 12, 13, 20, 21, 31*, 209,
331(27:39), 826*(19:92-95).

lutea L.   Straw Foxglove
  •• An established garden escape; naturalised and spreading near Betchworth
from its original introduction in a bomb crater at Brockham in the 1940s, and
on the chalk banks of the Leatherhead by-pass (both Surrey); also naturalised
on the sides of an old quarry north of Seal (W Kent), on a road verge at
Codford St Mary (Wilts) and a roadside chalk bank at Portchester (S Hants).
W Europe. BM, K, LANC, OXF, RNG. 7, 20, 22*, 29*, 52*, 126, 209,
303(39:9), 371(346:27). A hybrid with native *D. purpurea* (*D.* × *fucata*
Ehrh.; *D.* × *purpurascens* Roth) has been recorded as spontaneous, with its
parents, in a garden. 371(382:23).

ERINUS L.
alpinus L.   Fairy Foxglove
  •••• An established garden escape; naturalised in widely scattered localities
throughout the British Is, usually on old walls. W & C Europe. 1, 20, 22*,
27*, 33*, 34*.

VERONICA L.
acinifolia L.   French Speedwell
  •• A garden weed accidentally brought in with plants from nursery gardens in
southern England; established for 40 years or more at Beaminster (Dorset).
Formerly abundant in cornfields at Chiddingfold (Surrey). Europe, Turkey.
BM, OXF, RNG, SLBI. 2, 20, 22*, 165, 209, 303(19:16), 623a*, 683*.
V. alsinefolia sphalm.

austriaca L.   Large Speedwell
  •• A persistent garden escape; naturalised in scattered localities, especially on
dunes. Europe, SW Asia. BM, LANC, OXF, RNG. 7, 20, 22*, 27*, 29*,
30*, 303(45:24), 371(352:28), 683*. Incl. *V. jacquinii* Baumg.;
V. pseudochamaedrys Jacq.; *V. teucrium* L. Most records are referable to
subsp. *teucrium* (L.) D.Webb.

bachofenii Heuffel
  • A persistent garden escape on a disused tip at Bedminster Down
(N Somerset). Eastern Europe. HbEJC. 21, 371(385:35), 623a*.

calycina R.Br.   Cup Speedwell
  • A wool casual. Australia. BM, RNG. 17, 37, 549, 569*, 583*, 743.

campylopoda Boiss.
  Pre-1930 only. Asia. 1, 39*, 44a*.

chamaepithyoides Lam.
  Pre-1930 only. Spain. BM. 7, 21, 136, 195, 354(9:276). *V. digitata* Vahl,
non Lam.

crista-galli Steven   Crested Field-speedwell
•• An established garden escape; long naturalised on roadsides in the Batheaston-Bathford area (N Somerset), at Henfield (W Sussex), now gone, and at Cork (Mid & E Cork). Caucasus. BM, K, NMW, OXF, RNG, SLBI. 1, 20, 370*(12:75,129-132). The similar *V. argute-serrata* Regel & Schmalh. (*V. bornmuelleri* Hausskn.) from SW & C Asia may have been overlooked. 826*(25:3-5).

cymbalaria Bod.   Pale Speedwell
• Recorded in 1985 from a road verge at Goldsithney, near Penzance (W Cornwall). Medit. BM. 20, 21, 31*, 32*, 44a*, 303*(40:1,14).

filiformis Smith   Slender Speedwell
••••• An established garden escape; a weed of mown grass throughout the British Is, also on tips and naturalised in quarries; increasing. Turkey, Caucasus. 12, 20, 22*, 25*, 26*, 33*, 349(2:197-217; 4:384).

[*fruticulosa* L. Dubious records. 1, 111, 231, 354(5:778).]

gentianoides Vahl
• A casual garden escape. Crimea, SW Asia. BM, OXF, RNG. 7, 21, 51*, 53*, 56*.

glauca Sibth. & Smith
Pre-1930 only. S Balkans. 1, 21, 312*(t.7759), 327*(49:454), 511.

grandis Fischer & Sprengel
Pre-1930 only. N Asia. OXF. 51, 354(9:31), 511, 1296*. *V. koenitzeri* hort.

grisebachii Walters
Pre-1930 only. Bulgaria, Turkey. LCN, LSR, SLBI. 1, 21, 39*, 160, 184, 258, 825(7:89). *V. chamaepitys* Griseb., non Pers.

longifolia L.   Long-leaved Speedwell
••• (but confused with *V. paniculata* and native *V. spicata*). A persistent or established garden escape; in scattered localities, mainly in the south. Eurasia. BM, E, LANC, NMW, OXF, RNG. 7, 13, 20, 22*, 23*, 52*, 683*. *Pseudolysimachion longifolium* (L.) Opiz.

opaca Fries
Pre-1930 only. Europe. 21, 22*, 260, 623a*, 683*.

orientalis Miller
Pre-1930 only. Crimea, SW Asia. OXF. 7, 21, 354(5:571), 645*.

paniculata L.
•• (but mostly in error for *V. longifolia*). A persistent garden escape. SE Europe. BM, OXF. 7, 16, 21, 119, 258, 623a*, 1089*. *V. spuria* auct., non L.

peregrina L.   American Speedwell
•••• An established weed of cultivated ground, especially in Ireland; formerly accidentally brought in to nursery gardens with imported plants, now widely distributed, mainly in gardens. Var. *xalapensis* (Kunth) St John & Warren has occurred as a wool casual. N & S America; (Australia). 1, 20, 22*, 38*, 349(6:215-220), 1279*.

persica Poiret   Common Field-speedwell

●●●●● A well-established weed of cultivated ground throughout the British Is;
also a bird-seed alien. Formerly brought in with agricultural seed, especially
clover. SW Asia; (widespread). 1, 19, 20, 22*, 26*, 35*. *V. buxbaumii*
Ten., non F.W.Schmidt; *V. tournefortii* C.Gmelin pro parte, non Vill.
[*praecox* All. Accepted, with reservations, as native.]

prostrata L.   Prostrate Speedwell

• A garden escape persistent on an old wall at Beaulieu Abbey (S Hants). No
  modern records. Eurasia. BM, OXF. 7, 21, 22*, 23*, 53*, 245, 354(6:34),
  623a*, 683*. *V. rupestris* hort., non Salisb.

repens Clarion ex DC.   Corsican Speedwell

• A garden escape established as a weed of mown grass in northern England
  and Scotland. Spain, Corsica. BM, E. 1, 20, 130, 251, 277, 370(13:337),
  683*. *V. reptans* Kent, nom. illeg.

spicata L. subsp. **orchidea** (Crantz) Hayek
  Pre-1930 only. SE Europe. BM. 21, 623a*. *V. orchidea* Crantz.

urticifolia Jacq.   Nettle-leaved Speedwell
  Pre-1930 only. C & S Europe. BM. 21, 22*, 26*, 623a*.

VERONICASTRUM Moench
virginicum (L.) Farw.   Culver's-root

• A garden escape persistent for a few years. Eastern N America. 38*, 51,
  52*, 53*, 54*, 223. *Leptandra virginica* (L.) Nutt.; *Veronica virginica* L.
  The genus is not to be confused with *Veronicastrum* Fabr., which is a synonym
  of *Veronica*.

HEBE Comm. ex A.L.Juss.
× **amabilis** (Cheeseman) Cockayne & Allan

• A garden escape established on waste ground at Helensburgh (Dunbarton).
  New Zealand. 36, 51, 61, 370(17:192), 881*(8:375-388). Possibly *H. elliptica*
  × *H. salicifolia*, and therefore synonymous with *H.* × *lewisii*; *H.* × *blanda*
  (Cheeseman) Pennell; *H. salicifolia* var. *gracilis* T.Kirk; *Veronica* × *amabilis*
  Cheeseman.

'Autumn Glory'

• A persistent garden escape at Llanaber (Merioneth). Originated in
  cultivation. 61, 62*, 63*, 70*, 303(41:18), 325*(112:37). Possibly
  H. × *franciscana* × *H. pimeleoides* (Hook.f.) Cockayne & Allan.

barkeri (Cockayne) Wall   Barker's Hebe

• An established garden escape in Devon and Cornwall, localities not
  published. Chatham Is. 20*, 36, 61, 70, 1275*. *Veronica barkeri* Cockayne.

brachysiphon Summerh.   Hooker's Hebe

• A garden escape persistent or established in Scilly; seedlings reported from
  a railway track at Winfrith and as pavement weeds in East Wareham (both
  Dorset). New Zealand. 20*, 36, 52*, 61, 70, 403, 1275*. *Veronica
  brachysiphon* (Summerh.) Bean; *V. traversii* Masters, non Hook.f.

cupressoides (Hook.f.) Cockayne & Allan
- Collected from a conifer plantation at Powerstock (Dorset). New Zealand. RNG. 36*, 51, 61, 70*, 403, 609*. *Veronica cupressoides* Hook.f.

dieffenbachii (Benth.) Cockayne & Allan   Dieffenbach's Hebe
- An established garden escape in Devon and Cornwall, localities not published. Chatham Is. 20*, 36, 61, 70, 1275*. *Veronica dieffenbachii* Benth.

[*elliptica* (Forster f.) Pennell (*Veronica decussata* Aiton, non hort.; *V. elliptica* Forster f.) Recorded from S Kerry, Co Clare and W Galway, but the records require confirmation. Other records are in error for *H.* × *franciscana*.]

× **franciscana** (Eastw.) Souster   Hedge Veronica
- ●●● An established garden escape; long naturalised by self seeding on cliffs in SW England and the Channel Is where it is used as a hedging plant. Originated in cultivation. BM, E, K, NMW, OXF, RNG. 13, 20*, 61, 70*, 207*, 212, 220, 303(41:18; 43:17), 1275*. *H. elliptica* × *H. speciosa*; *H.* 'Blue Gem'; *H.* × *lewisii* auct. angl., non (J.Armstr.) Wall; *V.* 'Blue Gem'; *Veronica decussata* hort., non Aiton; *V.* × *franciscana* Eastw.; *V. lobelioides* hort.

aff. **leiophylla** (Cheeseman) Cockayne & Allan
- A garden escape; self-sown seedlings reported from Mytchett (Surrey). New Zealand. 36, 61, 70*, 342(Feb89), 609*. An uncertain and variable taxon, possibly of hybrid origin.

× **lewisii** (J.Armstr.) Wall   Lewis's Hebe
- ●● (but confused with *H.* × *franciscana*). An established garden escape in SW England and the Channel Is. Originated in cultivation; (New Zealand). BM, LANC. 20*, 36, 61, 70*, 303(43:18), 1275*. *H. elliptica* × *H. salicifolia*; *Veronica* × *lewisii* J.Armstr.

parviflora (Vahl) Cockayne & Allan
Pre-1930 only. New Zealand. 7, 36*, 61, 61a, 70, 320(45:60), 609*, 686*. *Veronica parviflora* Vahl; incl. *H. angustifolia* (A.Rich.) Cockayne & Allan; *V. angustifolia* A.Rich., non Fischer ex Link.

salicifolia (Forster f.) Pennell   Koromiko
- ●● (but confused with hybrids). An established garden escape; naturalised abundantly by seed in a few widely scattered localities, especially near the sea. New Zealand, Chile. BM, OXF, RNG. 7, 20*, 36*, 61, 70*, 1275*. *Veronica salicifolia* Forster f. A hybrid with *H. speciosa* (*H.* × *andersonii* (Lindley & Paxton) Cockayne) has been recorded. 122.

[*speciosa* (R.Cunn. ex Cunn.) Cockayne & Allan (*Veronica speciosa* R.Cunn. ex Cunn.) Recorded from Mid Cork, E Cork, Co Clare and Tyrone, but the records require confirmation. Some records may be referable to *H.* × *franciscana*. 303:41:18.]

stricta (Banks & Sol. ex Benth.) L.Moore
- A garden escape recently recorded as self-sown on a wall in Norwich (E Norfolk). New Zealand. 393(28:353), 686*, 709*. *Veronica stricta* Banks & Sol. ex Benth.

CASTILLEJA Mutis ex L.f.
cf. **coccinea** (L.) Sprengel **Scarlet Painted-cup**
- A casual at Felixstowe docks (E Suffolk), vector unknown. No modern records. Eastern N America. RNG. 38*, 40, 51.

ORTHOCARPUS Nutt.
erianthus Benth. **Butter-and-eggs**
Pre-1930 only. California. OXF. 7, 41, 42, 51, 354(7:49), 500*.
purpurascens Benth. **Purple Owl-clover**
Pre-1930 only. Western N America. OXF. 7, 41, 42, 51, 354(8:409), 500*, 824*(t.1166).

MELAMPYRUM L.
[*arvense* L. Accepted, with reservations, as native.]
barbatum Waldst. & Kit. ex Willd.
- A casual at Dagenham (S Essex), vector unknown. No modern records. Europe. 21, 31*, 354(11:494).

[*EUPHRASIA* L.
stricta D.Wolff ex J.Lehm. Accepted, with reservations, as native.]

ODONTITES Ludwig
jaubertianus (Boreau) D.Dietr. ex Walp. subsp. **chrysanthus** (Boreau) P.Fourn.
- Long established on gravelly heath near Aldermaston (Berks), in quantity in the late 1960s but now much reduced, vector unknown; also recorded from Greenham Common (Berks). France. RNG. 20, 21, 22*, 353(31:24), 370(16:111), 546*. *O. chrysanthus* Boreau.
[*luteus* (L.) Clairv. In error for *O. jaubertianus*. 119, 353(31:24), 371(346:26).]

PARENTUCELLIA Viv.
latifolia (L.) Caruel **Southern Red-bartsia**
Pre-1930 only. Medit.; (Australasia). 7, 21, 22*, 24*, 29*, 44a*, 354(5:572). *Bartsia latifolia* Sibth. & Smith; incl. *P. flaviflora* (Boiss.) Nevski.

BELLARDIA All.
trixago (L.) All. **Trixago Bartsia**
Pre-1930 only. Medit. 2, 9, 21, 24*, 29*, 32*, 44a*, 300(1905:101). *Bartsia trixago* L.; *Trixago apula* Steven.

[*PENSTEMON* Schmidel
confertus Douglas ex Lindley In error for a mixture of *Hyssopus officinalis* and *Salvia reflexa*. LDSP. 381(4:69).]

RHINANTHUS L.
alectorolophus (Scop.) Pollich
Recorded in 1992-93 on a road verge at Wellesbourne (Warks), probably as a grass seed impurity. Europe. WARMS, HbDJH. 7, 21, 22*, 23*, 26*, 27*, 354(5:297). *R. hirsutus* Lam.; *R. major* L. pro parte.

PAULOWNIA Siebold & Zucc.
tomentosa (Thunb.) Steudel   Foxglove-tree
- Introduced; reported as reproducing in Middlesex. China. 20(p.703), 51*, 58*, 60*, 61, 166, 236, 436, 1128*. Saplings are readily confused with *Catalpa* spp. 826*(21:25-32).

GLOBULARIACEAE
Incl. *Hebenstretiaceae, Selaginaceae*
HEBENSTRETIA L.
dentata L.
- A wool casual. S Africa; (Australia). BM, E, RNG. 48*, 51, 590*, 735*.

SELAGO L.
cf. **corymbosa** L.
- A wool casual. S Africa; (Australia). RNG. 51, 514, 578a*, 590*, 628*.

OROBANCHACEAE
LATHRAEA L.
clandestina L.   Purple Toothwort
••• An established garden relic or escape; naturalised, often in damp places, mainly on roots of poplars and willows, in widely scattered localities. W Europe. BM, DBY, LANC, LCH, OXF, RNG. 7, 20, 22*, 25*, 29*, 308, 182*.

OROBANCHE L.
aegyptiaca Pers.
- A casual parasite on tomatoes in a nursery at St Leonards (E Sussex). No modern records. Medit. K, RNG. 21, 44a*, 349(1:345), 370(18:257-295), 724*, 1247*.
[*amethystea* Thuill. In error for native *O. minor*. 16, 354(9:289).]
[*arenaria* Borkh. In error for native *O. purpurea*. 354(5:779).]
[*cernua* Loefl. In error; no localised records found. This species may be overlooked. 20, 21, 825*(2:22-24).]
crenata Forsskål   Carnation-scented Broomrape
- (but confused with native *O. caryophyllacea*). Persisted for several years on native *Vicia tetrasperma* and *V. hirsuta* in a field at Cranham and recently reported from a field of peas in the same district; also recorded on beans in two gardens in nearby Upminster (all S Essex); a casual on *V. faba* of Italian origin at a seed testing site in Cambs. Medit. GLR, K, OXF. 2, 20*, 29*, 32*, 119, 303(43*:18-19; 59:10), 331(56:88), 370(15:161; 18*:257-295). *O. pruinosa* Lapeyr.; incl. *O. speciosa* DC.
flava C.Martius ex F.Schultz   Butterbur Broomrape
- Possibly accidentally brought in with imported plants, now established on *Petasites* and related genera in Oxford Botanic Garden (Oxon). C & SE Europe. RNG. 21, 23*, 27*, 370(15:170; 18*:257-295).
[*gracilis* Smith (*O. cruenta* Bertol.) A dubious record. 7, 300(1893:121).]

lucorum A.Braun   **Barberry Broomrape**
  • Possibly accidentally brought in with imported plants, now established on *Berberis* in Oxford Botanic Garden (Oxon); also recorded from St Andrews (Fife). Alps. E, RNG. 21, 370(15:170; 18*:257-295). The record from Epsom (Surrey) was in error for native *O. elatior*. 249, 354(5:779).
[*minor* Smith   Accepted, with reservations, as native.]
ramosa L.   **Hemp Broomrape**
  Pre-1930 only. Medit. BM, LIV, OXF. 1, 20, 22*, 24*, 30*, 258, 370*(18:257-295). *Phelypaea ramosa* (L.) C.Meyer.

GESNERIACEAE

[*HABERLEA* Friv.
rhodopensis Friv.   In error for native *Silene dioica*. 156, 303(1:10; 2:30).]

RAMONDA Rich.
myconi (L.) Reichb.   **Pyrenean-violet**
  • Introduced in 1921 and still surviving on a rock-face in Cwm Glas, Snowdon (Caerns). Pyrenees. 20, 23*, 27*, 28*, 30*, 371(374:8).

ACANTHACEAE

ACANTHUS L.
mollis L.   **Bear's-breech**
  •••• A garden escape or relic well established in many places, mainly in SW England; known since about 1800 on St Agnes (Scilly). W Medit. 1, 20, 29*, 32*, 207, 650*, 1279*. Incl. *A. latifolius* hort. ex Goeze, non Poiret.
spinosus L.   **Spiny Bear's-breech**
  •• An established garden escape or relic in a few places, mainly in southern England. E Medit. BM, RNG. 20, 24*, 29*, 56*, 370(13:84), 680*. Incl. *A. caroli-alexandri* Hausskn.; *A. spinosissimus* Pers. According to 407, modern cultivated plants are mostly hybrids involving *A. mollis*, *A. spinosus* and *A. hungaricus* (Borbás) Baenitz. These may have been overlooked.

PEDALIACEAE

SESAMUM L.
orientale L.   **Sesame**
  • A bird-seed and spice casual. Tropics. HbACL, HbCGH. 19, 21, 68*, 80*, 303(14:13), 331(63:145; 69:142), 532*. *S. indicum* L.

BIGNONIACEAE

ECCREMOCARPUS Ruíz Lopez & Pavón
scaber (D.Don) Ruíz Lopez & Pavón   **Chilean Glory-flower**
  • A garden escape and relic. Chile, Argentina. 61, 64*, 66*, 162, 303(25:10), 593*, 680*.

CAMPSIS Lour.
radicans (L.) Seemann ex Bureau   Trumpet-creeper
- Reported as self-sown seedlings at Marble Arch (Middlesex). Eastern N America. 51, 54*, 458, 613*, 680*. *Bignonia radicans* L.

CATALPA Scop.
bignonioides Walter   Indian-bean-tree
- Introduced; self-sown seedlings in the London area. Eastern N America. 51, 54*, 58*, 60*, 166, 199, 303(35:16), 371(407:41). Saplings are readily confused with *Paulownia tomentosa*.

CAMPANULACEAE
Incl. *Lobeliaceae*

CAMPANULA L.
alliariifolia Willd.   Cornish Bellflower
- ●●● An established garden escape; naturalised, mostly on railway banks in southern England, especially in Cornwall; increasing. SW Asia. BM, E, K, LANC, RNG. 12, 20, 53*, 54*, 55*, 212, 354(13:263), 826*(20:121-123).
bononiensis L.   Pale Bellflower
- A casual garden escape. No modern records. Eurasia. LTN. 21, 29*, 143, 683*.
carpatica Jacq.   Tussock Bellflower
- A garden escape established on walls and in paving at Yeoman's Row in London SW3 (Middlesex); casual elsewhere. Carpathians. 20, 21, 54*, 55*, 57*, 303*(38:20-23).
cochleariifolia Lam.   Fairy's-thimble
- A casual garden escape, possibly persistent on a wall in the Chew Valley (N Somerset). S Europe. RNG. 21, 22*, 23*, 27*, 28*, 371(370:30), 388(133:234). *C. pusilla* Haenke.
fragilis Cirillo
- A garden escape on a wall in St Peter Port in Guernsey (Channel Is), increasing there. S Italy. STP. 20, 21, 201, 303(57:52), 312*(t.6504), 683*.
garganica Ten.   Adriatic Bellflower
- A casual garden escape. Italy, Greece. 21, 55*, 56*, 122, 172, 303(48:35), 584*, 594*.
glomerata L. subsp. **elliptica** (Kit. ex Schultes) O.Schwarz
- A garden escape recorded without locality in W Kent. SE Europe. 21, 370(15:429).
lactiflora M.Bieb.   Milky Bellflower
- ●●● An established garden escape; naturalised by streams and on roadsides, mainly in Scotland. SW Asia. BM, E, LANC, OXF, RNG. 13, 20*, 55*, 56*, 57*, 277, 349(1:493), 1181*.
medium L.   Canterbury-bells
- ●●●● An established garden escape; locally abundant on chalk railway banks in SE England. France, Italy. 1, 12, 20, 22*, 36c*, 55*, 56*, 209. Incl. *C. calycanthema* hort.

persicifolia L.   Peach-leaved Bellflower
●●●● An established garden escape; naturalised in woods and on commons in
widely scattered localities.   Eurasia.   1, 20, 22*, 25*, 27*, 30*, 209.
C. latiloba A.DC.; incl. *C. subpyrenaica* Timb.-Lagr.

portenschlagiana Schultes   Adria Bellflower
●●●● (but confused with *C. poscharskyana*).   A garden escape established on
walls, mainly in the south.   Yugoslavia.   7, 20, 57*, 65*, 303*(33:10-11),
312*(n.s. t.256), 1275*.   *C. muralis* Portenschlag.   Some records may be
referable to *C. portenschlagiana* × *C. poscharskyana*.

poscharskyana Degen   Trailing Bellflower
●●● (but confused with *C. portenschlagiana*).   A garden escape established on
walls in the south; long naturalised on Whiteparish Common (S Wilts).
Yugoslavia.   BM, CGE, LANC, OXF, RNG.   13, 20, 21, 36c*, 57*, 261,
303*(33:10-11), 312*(n.s. t.334), 1275*.

pyramidalis L.   Chimney Bellflower
●● A garden escape self-sown on a few walls.in the south; long established on
an old wall at West Malling (W Kent).   Italy, Yugoslavia, Albania.   CGE,
LCN.   7, 20*, 24*, 236, 371(352:26).

rapunculoides L.   Creeping Bellflower
●●●●● An established garden escape; naturalised on roadsides and in grassy
places throughout the British Is.   Eurasia.   1, 20, 22*, 23*, 27*, 34*.   A
hybrid with native *C. trachelium* may have been overlooked. 826*(16:48-50).

rapunculus L.   Rampion Bellflower
●● An established garden escape; long naturalised at Pulborough (W Sussex)
and on railway banks near Kelvedon (N Essex) and Betchworth (Surrey);
decreasing. Formerly widely naturalised in England as a relic of cultivation for
its edible rootstock.   Eurasia, N Africa.   BM, E, LANC, OXF, RNG.   1, 20,
22*, 26*, 33*, 174, 191, 209.

rhomboidalis L.
● A garden escape established on the wooded bank of the river Esk near
Langholm (Dumfriess); also persistent on the bank of a sunken lane near Knock
(Westmorland).   Alps, Juras.   DFS, LANC.   20*, 22*, 27*, 29*, 303(20:12;
43:17), 370(12:357).
[*uniflora* L.   In error for native *C. rotundifolia*.   354(5:776).]

SYMPHYANDRA A.DC.
hofmannii Pant.
● A casual garden escape.   Yugoslavia.   21, 31*, 52*, 371(382:25).

LEGOUSIA Durande
falcata (Ten.) Fritsch
Pre-1930 only.   Medit.   7, 21, 44a*, 354(6:614), 546*, 737*.   *Specularia
falcata* (Ten.) A.DC.
pentagonia (L.) Druce
Pre-1930 only.   E Medit., SW Asia.   2, 21, 31*, 44a*, 300(1904:178;
1907:40).   *Specularia pentagonia* (L.) A.DC.

speculum-veneris (L.) Chaix   Large Venus's-looking-glass
•• A grain casual and garden escape; apparently persistent since the early
1940s in arable fields near Wooton St Lawrence (N Hants). Europe, SW Asia,
N Africa. BM, E, LSR, NMW, OXF, RNG. 1, 21, 22*, 32*, 44a*, 145,
199, 303(19:16; 41*:14-15), 401. *L. durandii* Delarbre; *Campanula speculum-veneris* L.; *Specularia speculum-veneris* (L.) A.DC.

TRIODANIS Raf.
biflora (Ruíz Lopez & Pavón) E.Greene
Pre-1930 only. USA, S America. OXF. 3, 38*, 40, 354(5:38), 534*, 1143*.
Legousia biflora (Ruíz Lopez & Pavón) Britton; *Specularia biflora* (Ruíz Lopez
& Pavón) Fischer & C.Meyer.
perfoliata (L.) Nieuwl.
Pre-1930 only. USA. 2, 38*, 40, 41*, 197. *Legousia perfoliata* (L.) Britton;
Specularia perfoliata (L.) A.DC.

WAHLENBERGIA Schrader ex Roth
[*gracilis* (Forster f.) Schrader (*Cervicina gracilis* (Forster f.) Druce). A dubious
record. OXF. 354(9:274).]
lobelioides (L.f.) Link
Pre-1930 only. N Africa, Madeira, Canary Is. OXF. 7, 45*, 354(6:32), 511.
W. pendula Schrader, nom. nud.; *Cervicina lobelioides* (L.f.) Druce;
C. pendula Schrader.
nutabunda (Guss.) A.DC.
• A casual on Rhum (N Ebudes). W Medit. BM. 21, 46*, 337(163:104).
trichogyna Stearn
• A wool casual. Australia; (New Zealand). RNG. 17, 37*, 51, 312*(n.s.
t.343). Incl. *W. consimilis* Loth.; *W. gracilis* hort., non (Forster f.) Schrader.

TRACHELIUM L.
caeruleum L.   Throatwort
•• A garden escape long established on walls in the Channel Is, known since
1892 at St Peter Port, Guernsey (Channel Is); more recently established on a
wall at Bexley (W Kent). W Medit. BM, E, OXF, RNG. 2, 20, 24*, 30*,
52*, 53*, 201, 220, 303(29:14), 312*(n.s. t.427), 320(30:346).

PHYTEUMA L.
nigrum F.W.Schmidt   Black Rampion
Pre-1930 only. Europe. 21, 22*, 23*, 105.
scheuchzeri All.   Oxford Rampion
• A garden escape established for 40 years on the wall of St John's College
gardens, Oxford (Oxon); also recorded from Holden Clough (MW Yorks).
Alps, N Apennines. OXF. 20, 23*, 27*, 53*, 57*, 370(11:254), 371(343:28).
spicatum L. subsp. **coeruleum** R.Schulz
Pre-1930 only. S Europe. NMW. 21, 156, 550*.

LOBELIA L.
erinus L. Garden Lobelia

●●●● A persistent garden escape and wool casual; often recorded as a pavement weed in towns. S Africa. 3, 17, 20, 52*, 53*, 54*, 55*.

filiformis Lam.

Pre-1930 only. S Africa. LIV, OXF. 7, 354(6:32; 9:125), 735*. Incl. *L. natalensis* A.DC.

gracilis Andrz.

• A wool casual and, as var. *major* Benth., perhaps a horticultural seed impurity. Australia. BM, LES. 3, 51, 133, 569*, 591*.

siphilitica L. Great Lobelia

• A persistent garden escape on waste ground at Wisley (Surrey). Eastern USA. 20, 38*, 40, 52*, 54*, 55*, 199.

MONOPSIS Salisb.
debilis (L.f.) C.Presl

• A wool casual. S Africa; (Australia). E, K, OXF. 3*, 7, 51, 133, 327*(93:323-328), 354(9:125). *Lobelia debilis* L.f.

PRATIA Gaudich.
angulata (Forster f.) Hook.f. Lawn Lobelia

●● A rock-garden plant now established as a lawn weed, mainly in parks and botanic gardens, in widely scattered localities; known since the 1930s in turf in the Royal Botanic Garden, Edinburgh (Midlothian). New Zealand. BM, CGE, E, RNG. 20, 36c, 52*, 55*, 57*, 199, 312*(n.s. t.171B), 325(85:276-280), 370(17:192). *Lobelia angulata* Forster f.; incl. *P.* 'Treadwellii'.

arenaria Hook.f. Chatham Island Pratia

• Established in a lawn at Branksome (Dorset). No modern records. New Zealand. RNG. 36, 36c, 51, 325(85:276-280), 709*. *P. angulata* var. *arenaria* (Hook.f.) Hook.f.

DOWNINGIA Torrey
elegans (Douglas ex Lindley) Torrey California Lobelia

●● A persistent agricultural and grass seed casual, usually in damp hollows or by newly-made ponds. Western N America. K, LTN. 20, 41*, 42, 303(23*:10-11; 24:24; 44:20; 59:50), 370(16:168), 826*(19:75-77).

RUBIACEAE
COPROSMA Forster & Forster f.
repens A.Rich. Tree Bedstraw

• An established garden escape; abundant near St Warna's Well, St Agnes, and bird-sown in woodland on Tresco and on walls at Porthcressa beach, St Mary's (all Scilly). New Zealand. RNG, HbEJC. 20, 36*, 55*, 207, 212, 458, 560*, 686*, 1117*. *C. baueri* auct., non Endl. Not to be confused with *C. repens* Hook.f., a synonym of *C. pumila* Hook.f., also from New Zealand.

NERTERA Banks & Sol. ex Gaertner
granadensis (Mutis ex L.f.) Druce   Beadplant
•• A garden or greenhouse escape established as a weed in turf in a few places
in the Helensburgh area (Dunbarton) but now mostly gone, and on a golf course
near Largs (Ayrs).   S America, Australasia.   13, 20, 51*, 52*, 53*,
316(18:585), 371(352:26), 669*. Incl. *N. depressa* Banks & Sol. ex Gaertner.

BORRERIA G.Meyer
verticillata (L.) G.Meyer
• An oil-seed casual. No modern records. Tropical America, Africa. CMM.
49, 285, 354(11:30), 534*, 710*, 1174*. *Spermacoce verticillata* L.

CRUCIANELLA L.
angustifolia L.
•• A bird-seed casual. Medit. E, RNG. 2, 21, 29*, 39*, 45*, 46*, 264,
277, 303(16:16; 22:14).
patula L.
Pre-1930 only. SW Europe, NW Africa. 7, 21, 48*, 354(3:163).

PHUOPSIS (Griseb.) Hook.f.
stylosa (Trin.) Benth. & Hook.f. ex B.D.Jackson   Caucasian Crosswort
••• An established garden escape in several scattered localities in the south and
west; naturalised since 1927 on a limestone bank at St Donat's (Glam).
Caucasus, Iran. BM, CGE, LANC, NMW, OXF, RNG. 1, 20, 52*, 53*,
57*, 156, 349(1:579), 1181*. *Asperula ciliata* auct, non Moench; *Crucianella
stylosa* Trin.

ASPERULA L.
arvensis L.   Blue Woodruff
•••• (but confused with *A. orientalis*). A grain and bird-seed alien, sometimes
persistent for a few years. Medit., SW Asia. 1, 8, 19, 20, 22*, 23*, 26*,
44a*, 371(394:36).
cf. **laevigata** L.
• Collected from Rochester (E Kent), vector unknown. Medit. SLBI. 21,
573*.
[*nitida* Sibth. & Smith   A dubious record. 354(4:14-15).]
orientalis Boiss. & Hohen   Annual Woodruff
• (but perhaps overlooked as *A. arvensis*). A casual garden escape. SW Asia.
NMW. 2, 20, 39*, 156, 303(33:10), 313(26:191), 350(37:30), 370(14:427),
650*. *A. azurea* Jaub. & Spach.
taurina L.   Pink Woodruff
•• An established garden escape or relic, mainly in Scotland; naturalised by
streams and in damp woods, abundantly so on the banks of Turret burn, Crieff
(Mid Perth). S Europe, SW Asia. BIRM, BM, CGE, LANC, OXF, RNG.
2, 20, 22*, 23*, 29*, 370(8:58).

tinctoria L. Dyer's Woodruff

- A shortly-persistent introduction in a bomb crater on Brockham Hill (Surrey); also a casual garden escape. Europe. LANC, RNG. 12, 21, 22*, 26*, 51, 165, 199. *Galium triandrum* N.Hylander.

GALIUM L.

album Miller subsp. *album* **alien variants**

Var. *aristatum* Druce (*G. aristatum* Smith, non L.) and var. *cinereum* (Smith) Druce (*G. cinereum* Smith; *G. diffusum* D.Don) have been recorded pre-1930 only. 2, 7. *G. capsiriense* Jeanb. ex Timb.-Lagr. possibly belongs here. CGE. 1281.

album Miller subsp. **pychnotrichum** (H.Braun) Krendl

- Recorded from Walthamstow reservoirs (Middlesex / S Essex). C & SE Europe, Turkey. 21, 39, 126.

aristatum L.

Pre-1930 only. C & SW Europe. 1, 21, 22*, 683*.

[*austriacum* Jacq. In error for native *G. pumilum*. 197.]

[*cinereum* All. Dubious records. 166, 187.]

divaricatum Pourret ex Lam.

- A wool casual. S Europe, SW Asia; (Australasia). RNG. 17, 21, 22*, 44a*, 683*. *G. floribundum* Sibth. & Smith from W Turkey may have been overlooked. 825*(87-88).

glaucum L.

Pre-1930 only. Europe. 7, 21, 51, 197, 354(4:269; 9:355), 683*, 695*. Incl. *G. biebersteinii* Ehrend.; *Asperula galioides* M.Bieb. pro parte.

murale (L.) All. Small Goosegrass

- A wool casual. Medit.; (Australasia). E. 3, 21, 44a*, 45*, 46*, 236, 683*, 825*(7:87-88).

setaceum Lam.

- A wool casual. Medit. LTN, RNG. 14, 21, 44a*, 45*, 46*, 145, 683*.

spurium L. False Cleavers

- ••• An established weed of cultivated land, mainly in southern England; known for more than 100 years at Saffron Walden (N Essex), possibly native there; also a wool and bird-seed casual; decreasing. Europe, W Asia, N Africa. LANC, NMW, RNG, SUN. 2, 14, 20, 22*, 33*, 191, 346(1:1123), 369(24:101), 683*. Incl. *G. vaillantii* DC.

tenuissimum M.Bieb.

Pre-1930 only. E Europe, Asia. 7, 21, 354(4:14), 623a*.

[*tricornutum* Dandy Accepted, with reservations, as native.]

verrucosum Hudson

- A bird-seed and spice casual. Medit., W Asia. CGE. 2, 19, 20, 21, 44a*, 236, 354(9:355), 371(388:36-37), 683*. *G. saccharatum* All.; *G. valantia* G.Weber.

virgatum Nutt. ex Torrey & A.Gray

- A wool casual. Eastern N America. E, RNG, HbEJC. 17, 38*, 40*.

CRUCIATA Miller
articulata (L.) Ehrend.
Pre-1930 only.   SW Asia, Egypt.   1, 39, 44a*, 729, 1259*.   *Galium articulatum* (L.) Roemer & Schultes; *Valantia articulata* L. Not to be confused with the perennial *Galium articulatum* Lam. from E Europe and the Caucasus.

VALANTIA L.
muralis L.
Pre-1930 only. Medit. 7, 21, 44a*, 46*, 354(5:32), 683*.

RUBIA L.
tinctorum L.   Madder
Pre-1930 only. E Medit., Asia. K. 12, 20, 22*, 44a*, 192, 234, 683*.

MARTYNIACEAE
PROBOSCIDEA Schmidel
louisianica (Miller) Thell.   Purple-flowered Devil's-claw
• A grain and wool casual. USA, Mexico; (Australia). BM. 11, 21, 38*, 350(30:19), 354(11:495), 371(334:22), 500*, 546a*, 593*, 666*, 710*, 1141*.
P. jussieui Steudel; *Martynia louisianica* Miller; *M. proboscidea* Gloxin.

IBICELLA Eselt.
lutea (Lindley) Eselt.   Yellow-flowered Devil's-claw
• A wool casual. S America; (S Africa, Australia, USA). BM, RNG. 14, 51, 147, 371(328:20; 334:22), 500*, 546a*, 600*, 666*, 903*(46:79-82).
Martynia lutea Lindley; *Proboscidea lutea* (Lindley) Stapf.

CAPRIFOLIACEAE
SAMBUCUS L.
canadensis L.   American Elder
••• An established garden escape and introduction; planted and naturalised along several stretches of railway bank in the north; cultivars have been recorded from hedges and waste ground elsewhere. Eastern N America. CGE, E, RNG, SUN. 20, 38*, 61, 166, 354(13:262), 371(334:21; 367:29), 606*, 613*. Incl. *S.*'Aurea'; *S.*'Chlorocarpa'.
[*ebulus* L. Accepted, with reservations, as native.]
racemosa L. sensu lato   Red-berried Elder
•••• An established garden escape; naturalised in woods and hedges, especially in the north. Eurasia, N America. 1, 20, 22*, 26*, 370(14:74-76), 684*, 1279*. Incl. *S. pubens* Michaux; *S. sieboldiana* (Miq.) Graebner.

VIBURNUM L.
× **bodnantense** Aberc.
• Recorded from near Bridgwater (S or N Somerset), probably a garden relic. Originated in cultivation. 51, 61*, 62*, 63*, 312*(n.s. t.113), 371(414:15).
V. farreri × *V. grandiflorum* Wallich.

farreri Stearn
- Introduced or a garden escape established in scrub near Parkgate House on Ham Common (Surrey). China. HbACL. 51*, 61, 63*, 199, 303(29:13). *V. fragrans* Bunge, non Lois.

lantanoides Michaux   Hobble-bush
Pre-1930 only. Eastern N America. 7, 38*, 40, 61, 62*, 63*. *V. alnifolium* Marshall.

× **rhytidophylloides** Valcken   Lantanaphyllum Viburnum
- A garden escape persistent in Darenth Wood and on a railway bank at Eynsford (both W Kent). Originated in cultivation. 20, 52*, 61, 303(50:35), 325*(113:304). *V. lantana* L. × *V. rhytidophyllum*.

rhytidophyllum Hemsley ex Forbes & Hemsley   Wrinkled Viburnum
- Introduced at Sutton Courtenay (Berks); also bird-sown seedlings at New Ash Green (W Kent). China. 20, 51, 52*, 61*, 62*, 119, 458. The record for Darenth Wood (W Kent) was in error for *V.* × *rhytidophylloides*. 236, 371(382:23), 458.

tinus L.   Laurustinus
••• An established garden escape; naturalised on chalk in a few scattered localities, especially on the sea cliffs between Dover and Deal (E Kent), and in the Avon Gorge (W Gloucs). Medit. BM, CGE, NMW, LANC, OXF, RNG. 7, 20, 24*, 52*, 63*, 236, 350(47:40).

SYMPHORICARPOS Duhamel

albus (L.) S.F.Blake   Snowberry
••••• An established garden escape; naturalised in many places throughout the British Is. Formerly planted for game cover. N America. 1, 20, 22*, 35*, 56*. *S. racemosus* Michaux; *S. rivularis* Suksd. Records are mostly referable to var. *laevigatus* (Fern.) S.F.Blake; var. *albus* has been reported from Borough Green (W Kent). 458.

× **chenaultii** Rehder   Pink Snowberry
•• A garden escape established in a few widely scattered localities. Originated in cultivation. CGE, NMW. 20, 61, 199, 303(32:20; 50:30; 58:36), 631*, 879*(16:36). *S. microphyllus* × *S. orbiculatus*.

Doorenbos Hybrids
- Persistent garden escapes reported from a meadow west of Borough Green, and in a hedge near Hextable Nature Reserve (both W Kent). Originated in cultivation. 61, 62*, 458, 631, 1139*. A complex of hybrids between *S. albus* var. *laevigatus, S.* × *chenaultii* and *S. orbiculatus*.

microphyllus Kunth
- (but confused with hybrids). Recorded from a disused railway yard at Woofferton station (Salop) and in a wood at Earlsmill, Darnaway (Moray). Mexico. E. 312*(t.4975), 370(15:400), 631*, 827*(10:324), 1085*. *S. montanus* Kunth. Other records, as garden escapes, have been redetermined as hybrids; the species is probably no longer in cultivation. 61, 303(32:20).

orbiculatus Moench   Coralberry
•• (but confused with hybrids).   A persistent garden escape.   Eastern
N America.   38*, 40, 51, 52*, 61, 126, 371(379:25), 586*, 631*.
S. symphoricarpos (L.) MacMillan.

ABELIA R.Br.
uniflora R.Br. ex Wallich
Pre-1930 only.   China.   7, 51, 61, 312*(t.4694), 354(2:415).

LEYCESTERIA Wallich
[*crocothyrsos* Airy Shaw   In error for *L. formosa*.   371(364:29; 388:37).]
formosa Wallich   Himalayan Honeysuckle
•••• An established garden escape; in scattered localities throughout the
British Is; sometimes planted for pheasant cover and naturalised; increasing.
Himalayas, Burma.   1, 20, 22*, 61*, 62*, 63*.

DIERVILLA Miller
lonicera Miller
Pre-1930 only.   Eastern N America.   BM, LIV.   1, 38*, 361(7:221),
380(p.531), 1349(suppl.).   *D. canadensis* Willd.; *D. diervilla* (L.) MacMillan,
nom. illeg.; *D. trifida* Moench; *Lonicera diervilla* L.

WEIGELA Thunb.
florida group   Weigela
•• Garden escapes persistent on an overgrown area by the railway at Redhill
(Surrey) and on a disused tip, Bedminster Down (N Somerset), casual
elsewhere.   E Asia.   OXF.   7, 20, 51*, 54*, 56*, 61, 62*, 354(8:210),
371(407:40-42), 424, 471, 631*, 649*, 1120*.   Incl. *W. floribunda* (Siebold &
Zucc.) K.Koch; *W. florida* (Bunge) A.DC.; *W. rosea* Lindley; *W. venusta*
L.Bailey; *Diervilla floribunda* Siebold & Zucc.; *D. florida* (Bunge) Siebold
& Zucc.; and putative hybrids; *D.*'Eva Rathke'; *D.*'Styriaca'.
japonica Thunb.
• Details not known to us.   Japan.   BM.   43, 51, 61, 882(2:23-24), 1085*.
Diervilla japonica (Thunb.) DC.

LONICERA L.
alpigena L.   Alpine Honeysuckle
Pre-1930 only.   C & S Europe.   LANC.   1, 21, 22*, 27*, 29*, 111, 195.
caerulea L.   Blue-berried Honeysuckle
• A garden escape or relic at Dalmahoy near Edinburgh (Midlothian).   Eurasia;
(N America).   BM, NMW, OXF, RNG.   7, 21, 22*, 23*, 27*, 63*,
354(4:198; 6:283), 371(346:28).   The record for Wye Dale (Derbys) was in
error for *L. involucrata*.   370(16:192).
caprifolium L.   Perfoliate Honeysuckle
•••• An established garden escape; naturalised in hedges and plantations,
mainly in southern England; claimed as possibly native in an ancient wood at
Acton (W Suffolk); known at Cherry Hinton (Cambs) since 1763, and at Bagley
Wood (Berks) since 1700; decreasing.   SE Europe, SW Asia.   1, 20, 22*, 63*,

65*, 234, 258. *L. perfoliatum* Edwards. A putative hybrid with native *L. periclymenum* has been recorded. 370(15:431).

cf. **chrysantha** Turcz.
- A casual on a refuse tip at New Years Green (Middlesex), probably a garden escape. Siberia, China, Japan. 51, 61, 461, 617*.

etrusca Santi   Etruscan Honeysuckle
Pre-1930 only. Medit. 1, 21, 29*, 44a*, 63*, 107. The record for Bristol was in error for *L.* × *italica*. Voucher specimen redetermined.

henryi Hemsley   Henry's Honeysuckle
- A garden escape well established and spreading, presumably bird-sown, on Holmwood Common (Surrey); also recorded from Martyr's Green and Coombe Hill (both Surrey) and from Sherrardspark Wood (Herts). China. BM, RNG. 20*, 63*, 199, 302(11:13; 72:118), 331(70:161), 349(4:416), 613*, 617*.

involucrata (Richardson) Banks ex Sprengel   Californian Honeysuckle
- •• A long-persistent or established garden escape and introduction; relics of old hedges of this species have survived for many years in Ireland. N America. BM, CGE, LANC, OXF, RNG. 13, 20*, 62*, 63*, 236, 259, 371(346:27). Incl. *L. ledebourii* Eschsch.

× **italica** Schmidt ex Tausch   Garden Honeysuckle
•• (but probably under-recorded for the parent species). A garden relic or persistent escape in a few places in the south. Origin obscure. BM, HbEJC. 20, 52*, 63*, 347(12:100-105), 350(41:89 as *L. etrusca*), 422, 458. *L. caprifolium* × *L. etrusca*; *L.* × *americana* auct., non (Miller) K.Koch. Some native forms of *L. periclymenum* closely resemble this hybrid in Dunbarton and elsewhere. 467.

japonica Thunb. ex Murray   Japanese Honeysuckle
•••• An established garden escape often forming extensive patches in the south; known since the 1930s at Bere Ferrers (S Devon). E Asia. 13, 20*, 22*, 52*, 63*, 65*, 212, 236, 534*. Incl. *L. halliana* hort. ex K.Koch.

morrowii A.Gray   Morrow's Honeysuckle
- A persistent garden escape or relic recorded in 1990 from Swanley (W Kent). Japan. 51, 53, 55, 303(59:46), 649*, 1128*.

nitida E.Wilson   Wilson's Honeysuckle
•••• An established garden escape; self-sown in woodland, hedges and scrub in scattered localities; also introduced and long persistent in derelict hedges. China. 13, 20*, 52*, 62*, 63*, 303(37:22).

pileata Oliver   Box-leaved Honeysuckle
•• A persistent garden escape; bird-sown at New Ash Green (W Kent) and at Ketton (Rutland). China. 20*, 52*, 62*, 63*, 222, 303(37:23), 342(Feb91).

× **purpusii** Rehder
- Introduced or a garden escape. Originated in cultivation. 51a, 61, 63*, 199, 312*(n.s. t.323). *L. fragrantissima* Lindley & Paxton × *L. standishii*.

standishii Jacq.
Pre-1930 only. China. BM. 51, 54*, 61, 63*, 312*(t.5709).

syringantha Maxim.   Lilac-scented Honeysuckle
- Probably a garden relic, reported from Hextable (W Kent). China, Tibet. 51*, 61, 62*, 63*, 458.

tatarica L.   Tartarian Honeysuckle
- A persistent garden escape or relic recorded from Hayling Is (S Hants), West Grinstead (W Sussex), Hextable (W Kent) and Oulton Broad (E Suffolk). W & C Asia. HbEJC. 21, 54*, 61, 63*, 122, 258, 369(21:41), 458.

trichosantha Bureau & Franchet
- Introduced and persistent in Clapham Woods (MW Yorks). China. HbEJC. 20, 51, 61, 335(114:106), 617, 631*, 805*(14:t.20). Incl. *L. ovalis* Batalin.

KOLKWITZIA Graebner
amabilis Graebner   Beauty-bush
- A persistent garden escape; self-sown on an old wall at Rochester (E Kent). China. 51*, 54*, 61, 63*, 236, 303(36:23), 617*.

VALERIANACEAE

VALERIANELLA Miller
coronata (L.) DC.
- A casual at Sharpness docks (W Gloucs), vector unknown. No modern records. Eurasia, N Africa. RNG, SLBI. 2, 21, 22*, 29*, 39*, 44a*, 182, 300(1904:177), 683*, 737*, 826*(10(1)suppl.:1-4).

discoidea (L.) Lois.
- A casual in Caerns, vector unknown. Medit. BM, E. 2, 9, 21, 39*, 45*, 156, 244, 683*, 737*, 826*(10(1)suppl.:1-4).

eriocarpa Desv.   Hairy-fruited Cornsalad
- ●●● (but confused with native *V. dentata* var. *mixta* Dufr.). A persistent weed of cultivated ground and dry open habitats; established in a few coastal areas of SW England; decreasing. Medit. BM, OXF, RNG. 1, 20*, 22*, 33*, 35*, 212, 213, 683*.

kotschyi Boiss.
- An esparto casual. SE Europe, S Asia. E, K. 21, 44a*, 1262*, 1351.

microcarpa Lois.
Pre-1930 only. S Europe, N Africa, Madeira. 21, 267, 546*, 683*, 737*.

FEDIA Gaertner
cornucopiae (L.) Gaertner   Horn-of-plenty
Pre-1930 only. Medit. OXF. 2, 21, 29*, 32*, 300(1907:40), 737*. *Mitrophora cornucopia* Raf.

VALERIANA L.
[*celtica* L. In error for native *V. dioica*. 195.]
phu L.
- An established garden escape, long naturalised on roadsides at Mydroilin (Cards). Turkey. 21, 51, 156, 250, 354(9:355), 695*, 1257*.

pyrenaica L.   Pyrenean Valerian
●●●● An established garden escape; naturalised by streams and in woods in the
north and west; known since 1782 at Blairadam (Fife). Pyrenees. 1, 20, 22*,
23*, 27*, 33*, 114, 1353*. *Valerianella pyrenaica* sphalm.

CENTRANTHUS DC.
calcitrapae (L.) Dufr.   Annual Valerian
● An escape from the Royal Botanic Gardens, established in Kew Green
churchyard (Surrey). Formerly established on walls at Enfield and Chelsea
(both Middlesex) and at Eltham (W Kent). Medit. BM, OXF, SLBI. 1, 20,
21, 46*, 126, 192, 303*(40:14-15), 331(62:107), 737*. *Kentranthus calcitrapa*
(L.) Druce.

macrosiphon Boiss.   Spanish Valerian
Pre-1930 only. S Spain. JSY, OXF. 7, 21, 201, 354(6:24), 594*, 650*,
737*. *Kentranthus macrosiphon* (Boiss.) Druce.

ruber (L.) DC.   Red Valerian
●●●●● An established garden escape; naturalised on railway banks, walls, cliffs
and in chalk pits, especially in the south; locally abundant. Medit. 1, 20, 22*,
25*, 26*, 35*. *Kentranthus ruber* (L.) Druce.

MORINACEAE
MORINA L.
longifolia Wallich ex DC.   Whorl-flower
Pre-1930 only. Himalayas. 7, 51, 52*, 53*, 57*, 300(1907:56; 1911:101).

DIPSACACEAE
DIPSACUS L.
laciniatus L.   Cut-leaved Teasel
●● A casual by the track at Charlbury station and on tips (all Oxon), vector
unknown; also reported in 1991-93 from roadsides in Middlesex, Surrey and
S Lancs, perhaps from wildflower/grass seed mixtures. Eurasia. BM, RNG,
SLBI.   1, 21, 22*, 29*, 44a*, 303(22:12; 60:35; 61*:1,39; 63:30-31),
354(12:791), 436, 683*. A hybrid with native *D. fullonum* has been recorded.
371(428:6).

sativus (L.) Honck.   Fuller's Teasel
●●●● A persistent bird-seed alien and escape from cultivation. Origin obscure.
1, 19, 20, 22*, 666*, 1275*. *D. fullonum* L. subsp. *sativus* (L.) Thell.
A hybrid with native *D. fullonum* has been recorded. 143.

strigosus Willd. ex Roemer & Schultes   Yellow-flowered Teasel
●● (but confused with native *D. pilosus*). An established escape from botanic
gardens; sporadic records over many years in the Cambridge area, possibly
naturalised for more than 150 years in a chalk pit at Cherry Hinton (Cambs);
also recorded from Oxford (Oxon). Russia, W Asia. BM, CGE, K, RNG. 7,
20, 22*, 349(5:123), 370(11:67; 12:286; 13*:126-128).

CEPHALARIA Schrader ex Roemer & Schultes

alpina (Lagasca) Roemer & Schultes   Alpine Scabious
- A casual, perhaps a garden escape. No modern records. Alps. 21, 23*, 27*, 354(10:22), 683*, 695*. *Lepicephalus alpinus* Lagasca; *Succisa alpina* (Lagasca) Moench.

gigantea (Ledeb.) Bobrov   Giant Scabious
- ●●● An established garden escape; on roadsides, river banks and by railways in scattered localities; increasing. Caucasus. E, LANC, LSR, OXF, RNG. 7, 13, 20, 39, 694*. *C. elata* (Hornem.) Schrader ex Roemer & Schultes; *C. tatarica* auct., non Schrader; *Lepicephalus tatarica* auct., non (Schrader) Lagasca; *Scabiosa gigantea* Ledeb.

syriaca (L.) Schrader
- ●● A casual, mainly at ports; perhaps from bird-seed or as a garden escape elsewhere. Medit, SW Asia. BM, LCN, OXF, RNG, SLBI. 1, 21, 44a*, 46*, 683*. *Lepicephalus syriacus* (L.) Lagasca.

transylvanica (L.) Roemer & Schultes
     Pre-1930 only. C & S Europe, SW Asia. LCN. 1, 21, 29*, 160, 197, 300(1909:42), 683*, 695*, 715*. *Lepicephalus transylvanicus* (L.) Lagasca; *Scabiosa transylvanica* L.; *Succisa transylvanica* (L.) Sprengel.

KNAUTIA L.

dipsacifolia (Schrank) Kreutzer
- Established in a hedgerow near Thames Lock, Weybridge (Surrey), now gone; vector unknown. C Europe. HbEJC. 21, 22*, 29*, 546*, 683*, 825(13:74-77). *K. sylvatica* Duby, non *Scabiosa sylvatica* L.; *Scabiosa dipsacifolia* Schrank. A variable species, perhaps divisible into six or more subspecies.

[**tatarica** (L.) Szabó (*Cephalaria tatarica* (L.) Schrader, non hort.; *Lepicephalus tataricus* (Schrader) Druce; *Scabiosa tatarica* L.) In error for *Cephalaria gigantea*. 51a, 144, 354(5:285).]

SUCCISELLA G.Beck

inflexa (Kluk) G.Beck
     Pre-1930 only. SE Europe. 21, 31, 197, 623a*, 683*, 715*. *Scabiosa australis* Wulfen; *S. inflexa* Kluk; *Succisa australis* (Wulfen) Reichb.

SCABIOSA L.

atropurpurea L.   Sweet Scabious
- ●●● An established garden escape; naturalised on chalk cliffs and railway banks in a few places on the south coast of England; known on cliffs at Folkestone (E Kent) since 1862; also an esparto casual. Medit. BM, JSY, LANC, OXF, RNG. 1, 15, 20*, 22*, 29*, 55*, 236. *Sixalis atropurpurea* (L.) Greuter & Burdet; incl. *Scabiosa maritima* L.

canescens Waldst. & Kit.
     Pre-1930 only. Europe. 7, 21, 22*, 25*, 346(1859), 354(9:688). *S. suaveolens* Desf. ex DC.; *S. virga-pastoris* Miller.

caucasica M.Bieb.   Pincushion-flower
• A garden relic or escape. Caucasus. 36c*, 51, 52*, 54*, 56*, 303(29:12).
ochroleuca L.
• A garden escape. C & SE Europe. 21, 22*, 52*, 313(31:29), 371(409:41).
S. columbaria L. var. *ochroleuca* (L.) Coulter.
prolifera L.
Pre-1930 only. E Medit. SLBI. 2, 32*, 39*, 44a*, 300(1904:238; 1905:100),
1259*.
tricuspidata sphalm.
• A garden escape at The Roughs in Sheffield (SW Yorks). The correct
identity of this plant is not known; no such name has been found. 1301.

ASTERACEAE
Compositae

ECHINOPS L.
bannaticus Rochel ex Schrader   Blue Globe-thistle
••• An established garden escape; naturalised on roadsides and railway banks
in widely scattered localities. SE Europe. E, RNG. 20, 54*, 236, 277.
E. ritro auct., non L.; incl. *E.* 'Taplow Blue'.

exaltatus Schrader   Globe-thistle
••• An established garden escape; known for many years on a roadside
between Ollerton and Worksop (Notts). C & E Europe. BM, E, K, OXF,
RNG. 20, 22*, 349(6:124). *E. commutatus* Juratzká; *E. strictus* Fischer
ex Sims.

humilis M.Bieb.
Pre-1930 only. C Asia. 51, 52*, 266. *E. lanatus* Stephan ex DC.
[*ritro* L. Unconfirmed; probably all records are referable to *E. bannaticus*.]
sphaerocephalus L.   Glandular Globe-thistle
••• An established garden escape; naturalised on railway banks and roadsides
in a few places, mainly in the south. Eurasia. BM, LANC, OXF, RNG.
7, 20, 22*, 26*, 39*, 129, 236.
[*spinosissimus* Turra (*E. viscosus* DC., non Schrader ex Reichb.)   In error for
E. sphaerocephalus. 371(370:30).]
strigosus L.   Rough-leaved Globe-thistle
• Recently recorded as a garden escape near Puckeridge Gate Road at the
Royal Aerospace Establishment at Farnborough (N Hants). Portugal, Spain,
NW Africa. 21, 29, 30*, 226, 737*.

XERANTHEMUM L.
annuum L.   Immortelle
Pre-1930 only. Europe, SW Asia. 7, 21, 29*, 31*, 39, 53*, 55*, 251,
300(1907:40).
cylindraceum Sibth. & Smith
Pre-1930 only. S Europe, SW Asia. 1, 21, 44a*, 138, 197.

inapertum (L.) Miller
Pre-1930 only. Medit. 7, 21, 30*, 31*(p.446), 45*, 368(1914:234), 737*.

ATRACTYLIS L.
cancellata L.
Pre-1930 only. Canary Is, Medit., SW Asia. 1, 21, 30*, 31*(p.446), 32*, 44a*, 195, 737*.
humilis L.
Pre-1930 only. W Medit. LCN. 21, 24*, 46*, 160.

CARLINA L.
acaulis L.   Stemless Carline-thistle
Pre-1930 only. Europe. 21, 22*, 23*, 25*, 29*, 219. *C. subacaulis* DC.; *C. subcaulescens* sphalm.
[*racemosa* L.   A dubious record. 354(5:772).]

ARCTIUM L.
tomentosum Miller   Woolly Burdock
•• A grain casual. Eurasia. BM, LIV, LTN. 2, 20, 21, 22*, 26*, 29*, 182, 187. Incl. *L. palladinii* Marcov. Easily confused with native species.

COUSINIA Cass.
tenella (Fischer & C.Meyer ex A.DC.) Fischer & C.Meyer
Pre-1930 only. Iran, Afghanistan. BM. 2, 39, 524, 554, 630, 1249*, 1271. *Lappa tenella* Fischer & C.Meyer ex A.DC.

SAUSSUREA DC.
discolor (Willd.) DC.   Heart-leaved Saussurea
Pre-1930 only. Europe. 21, 23*, 27*, 29*, 354(8:618).

CARDUUS L.
[*acanthoides* L.   Unconfirmed; probably all records are in error for native *C. crispus*. 20.]
argentatus L.
• A wool and grain casual. E Medit. BM, E, LTN, RNG. 2, 3, 14, 21, 39*, 44a*, 145, 277, 300(1905:100). *Cirsium argentatum* (L.) Druce; *Carbenia argentata* sphalm.
chrysacanthus Ten.
Pre-1930 only. Appennines. 21, 138, 683*.
hamulosus Ehrh.
Pre-1930 only. SE Europe, SW Asia. OXF. 7, 21, 354(3:327), 715*, 741*, 1089*.
leucophyllus Turcz.
Pre-1930 only. Mongolia. 2, 814(5:194).
macrocephalus Desf.   Giant Thistle
• A persistent grain alien or garden escape. No modern records. Medit. BM, RNG. 7, 9, 12, 13, 20, 21, 46*, 285, 324(8-11:17), 354(5:36). *C. nutans* L. subsp. *macrocephalus* (Desf.) Nyman.

nigrescens Villars
 Pre-1930 only. France. 7, 21, 266, 300(1907:40), 546*, 683. *C. hamulosus* auct., non Ehrh.

pycnocephalus L.   Plymouth Thistle
 ••• Established since 1868 on a cliff at Plymouth (S Devon); elsewhere a wool and bird-seed casual. Eurasia, N Africa; (Australasia). BM, CGE, E, NMW, OXF, RNG. 2, 14, 20, 22*, 39*, 44a*, 217, 354(13:109), 370(11:384). *C. tenuiflorus* auct., non Curtis; incl. *C. arabicus* Jacq. ex Murray.

thoermeri J.A.Weinm.
 •• A bird-seed and grain casual. E Europe. BM, K, LANC, STP. 21, 39*, 321(1-4:19), 354(11:32), 370(1:304), 371(397:36), 1218. *C. leiophyllus* Petrov.; *C. nutans* L. subsp. *macrolepis* (Peterm.) Kazmi.

vivariensis Jordan
 Pre-1930 only. SW Europe. 21, 281, 320(1896:478), 546*, 683*. *C. nigrescens* subsp. *vivariensis* (Jordan) Bonnier & Layens.

CIRSIUM Miller
arvense (L.) Scop. **alien variants**
 Var. *incanum* (Fischer) Ledeb., var. *mite* Wimmer & Grab. and var. *vestitum* Wimmer & Grab. (var. *argenteum* Payer) have been recorded, but are probably not worthy of recognition.

creticum (Lam.) Urv.
 Pre-1930 only. E Medit. OXF. 7, 21, 31*, 39*, 354(6:26). *C. polyanthemum* auct., non (L.) Sprengel.

edule Nutt.   Indian Thistle
 Pre-1930 only. Western N America. 7, 41*, 354(3:23), 500*.

erisithales (Jacq.) Scop.   Yellow Thistle
 • A persistent or established escape from a botanic garden, in two places in Nightingale Valley, Leigh Woods (N Somerset). Europe. 20*, 23*, 25*, 27*, 303(26:12), 350(41:89; 47*:43), 370(14:196).

italicum (Savi) DC.
 • A grain casual. No modern records. SE Europe, NW Turkey. 21, 39, 349(3:46), 683*. *Cnicus samniticus* Ten.

monspessulanum (L.) Hill
 Pre-1930 only. W Medit. BM. 2, 21, 30*, 46*, 320(44:396).

oleraceum (L.) Scop.   Cabbage Thistle
 •• An established garden escape; on damp soils in widely scattered localities; long naturalised on the Tay marshes below Perth (E Perth). Eurasia. BM, CGE, E, LANC, OXF, RNG. 2, 13, 20*, 22*, 26*, 34*, 303(49:11), 354(3:213). *Cnicus oleraceus* L.

rivulare (Jacq.) All.   Brook Thistle
 • A casual garden escape. C Europe. BM, LANC. 2, 21, 22*, 23*, 26*, 57*, 187, 300(1906:103), 1257*. *C. salisburgense* (Willd.) G.Don; *Cnicus rivularis* (Jacq.) Willd. Recorded from Angus as var. *atrosanguineum*, presumably in error for var. *atropurpureum* Blakelock. 312*(n.s. t.217).

NOTOBASIS (Cass.) Cass.
syriaca (L.) Cass.   Syrian Thistle
  • A grain casual. No modern records. Medit., SW Asia. BM, OXF, SLBI.
  2, 21, 24*, 29*, 32*, 44a*, 300(1906:103), 354(11:484). *Cirsium syriacum*
  (L.) Gaertner; *Cnicus syriacus* (L.) Roth.

GALACTITES Moench
tomentosa Moench
  •• A bird-seed casual. Medit. BM, NMW, SLBI. 2, 21, 24*, 29*,
  31*(p.446), 32*, 248, 303(45:22). *G. galactites* (L.) Druce; *G. pumila* Porta;
  Centaurea galactites L.; *Lupsia galactites* (L.) Kuntze.

ONOPORDUM L.
[*acanthium* L. Accepted, with reservations, as native.]
illyricum L.   Illyrian Thistle
  • A grain casual. No modern records. W Medit.; (Australia). OXF, SLBI.
  2, 8, 21, 29*, 354(12:493), 737*.
macracanthum Schousboe
  • A wool casual. W Medit. CGE, E, K, TCD. 17, 21, 30*, 45*, 737*.
nervosum Boiss.   Heraldic Thistle
  • A garden escape established in a sand pit at Crayford and reported from a
  chalk pit at Greenhithe (both W Kent); also recorded from Weston-super-Mare
  (N Somerset). SW Europe. 21, 30*, 371(343:29; 361:27), 458, 737*.
  O. arabicum hort., non L.
tauricum Willd.
  • A grain casual. No modern records. SE Europe, SW Asia. OXF, SLBI.
  2, 21, 31*, 251, 300(1907:40), 354(9:253).

ZOEGEA L.
leptaurea L.
  • A casual at Ryde (Wight), vector unknown. No modern records. SW Asia.
  OXF. 39*, 354(10:471), 528*(as *Crupina leptaurea*), 1262*. *Z. mianensis*
  Bunge ex Boiss.

CYNARA L.
cardunculus L.   Globe Artichoke, incl. Cardoon
  ••• A long persistent garden escape; known for more than 30 years at
  Bawdsey (E Suffolk). Medit. OXF. 2, 21, 29*, 30*, 55*, 80*, 258,
  327*(109:75-123), 349(1:54). Incl. *C. scolymus* L.

SILYBUM Adans.
marianum (L.) Gaertner   Milk Thistle
  ••••• An established garden escape and wool, bird-seed, grass seed and oil-
  seed alien; naturalised in meadows and on waste ground locally in the south,
  widespread as a casual. Medit.; (Australasia, N America). 1, 14, 19, 20, 22*,
  25*, 31*, 33*, 331(63:70). *Carduus marianus* L.; *Mariana lactea* Hill;
  M. mariana (L.) Hill.

AMBERBOA (Pers.) Less.
moschata (L.) DC.   Sweet-sultan
 • A casual garden escape. SW Asia. RNG. 7, 12, 21, 39*, 55*, 264,
 354(3:165), 594*. *Centaurea moschata* L.

VOLUTARIA Cass.
lippii (L.) Maire
 • A casual at Sharpness docks (W Gloucs), vector unknown. No modern
 records. N Africa, SW Asia; (S Europe). BM, RNG. 7, 21, 44a*, 45*, 46*,
 300(1910:45). *Amberboa lippii* (L.) DC.; *Centaurea lippii* L.; *Volutarella
 lippii* (L.) Cass.; incl. *A. tubuliflora* Murb.

CYANOPSIS Cass.
muricata (L.) Dostál
 • An esparto casual. No modern records. W Medit. 1, 15, 21, 45*, 46*.
 Amberboa muricata (L.) DC.; *Centaurea muricata* L.; *Volutarella muricata* (L.)
 Benth. & Hook.f.; *Volutaria muricata* (L.) Maire.

MANTISALCA Cass.
salmantica (L.) Briq. & Cavill.
 •• A grain and bird-seed casual. S Spain, NW Africa. BM, E, LCN, LIV,
 NMW, OXF. 1, 19, 21, 30*, 45*, 46*, 160, 277, 737*. *Centaurea clusii*
 (Spach) ined.; *C. duriaei* (Spach) Rouy; *C. microlonchus* Salzm.;
 C. salmantica L.; *Microlonchus clusii* Spach; *M. duriaei* Spach; *M. salmanticus*
 (L.) DC.

ACROPTILON Cass.
repens (L.) DC.   Russian Knapweed
 • A wool alien established for over 40 years on a railway bank at Hereford
 station (Herefs). E Europe, SW Asia; (N America, Australia). K, RNG. 14,
 20, 21, 41*, 283, 349(4:44), 500*, 1138*. *Centaurea picris* Pallas ex Willd.;
 C. repens L.

CENTAUREA L.
apiculata Ledeb.
 Pre-1930 only. SE Europe. GLR. 21, 245, 354(3:23), 623a*, 688*, 695*,
 715*. Incl. *C. spinulosa* Rochel ex Sprengel (*C. scabiosa* L. subsp. *spinulosa*
 (Rochel ex Sprengel) Arcang.).
 [*aspera* L.   Accepted, with reservations, as native.]
atropurpurea Waldst. & Kit.
 Pre-1930 only. Balkans. 2, 21, 181, 688*, 715*. *C. calocephala* Willd.
 [*aurea* Aiton   Dubious records. LCN. 160, 289.]
bruguieriana (DC.) Hand.-Mazz.
 • A casual at Cardiff (Glam), vector unknown.   No modern records.
 SW & C Asia. NMW. 10, 39, 354(11:485), 667*. *C. phyllocephala* Boiss.,
 nom. illeg.; *Tetramorphaea bruguieriana* DC.; incl. *C. belangeriana* Stapf.

calcitrapa L.   Red Star-thistle
•••• A grain, bird-seed, lucerne seed, wool and esparto alien established in
a few places in the south; decreasing, and now usually casual; known since
1839 on The Lines at Chatham (E Kent) and since 1765 on the Sussex coast.
Medit.; (Australasia). 1, 14, 19, 20*, 22*, 25*, 33*, 174, 236. A hybrid with
C. aspera has been recorded. 199.

[*calcitrapoides* L.   In error for *C. aspera* × *C. calcitrapa*. 199.]

centauroides L.
Pre-1930 only. S Italy. 1, 21, 354(1868:19), 683*.

[*cephalariifolia* Willk.   In error for native *C. scabiosa*. BM.]

cheiranthifolia Willd.
• A casual at Burstwick (SE Yorks), vector unknown. No modern records.
Turkey, Caucasus.   39, 285, 312*(t.1175), 899*(4:154-155).   Incl.
C. ochroleuca Mussin-Puschkin ex Willd.

cineraria L. group
• A garden escape perhaps established on sea-cliffs in SW England; also a bird-
seed casual. Italy, Sicily. OXF, HbEJC. 7, 21, 36c*, 54*, 220, 229,
371(403:38), 442, 683*. Incl. *C. candidissima* Lam.; *C. gymnocarpa* Moris
& de Notaris.

[*coronarium* sphalm.   Probably in error for *Chrysanthemum coronarium*. 121,
199.]

[*cyanus* L.   Accepted, with reservations, as native.]

dalmatica A.Kerner
Pre-1930 only. NW Yugoslavia. BM. 21, 870*(20:t.9).

dealbata Willd.
• A casual garden escape. SW Asia. 1, 21, 55*, 223, 506*, 594*, 657*.

depressa M.Bieb.
Pre-1930 only. SW & C Asia. RNG, SLBI. 1, 21, 39*, 258, 300(1905:100;
1906:48). *Carthamus depressus* sphalm.

diffusa Lam.
• A bird-seed and grain casual. SE Europe, Turkey; (N America). BM, E,
K, NMW, OXF, RNG. 1, 19, 21, 39*, 160, 354(10:830; 11:263; 13:110),
525*, 546*, 623*, 688*.   A hybrid with *C. rhenana* Boreau
(*C.* × *zimmermaniana* Lins.; *C.* × *psammogena* Gayer) was recorded in error
for *C. maculosa*. 354(5:37,288).

diluta Aiton   Lesser Star-thistle
•••• A bird-seed, grain and wool casual. SW Spain, NW Africa, Atlantic Is.
1, 14, 19, 20*, 45*, 46*, 150, 737*, 1285*. *C. raphanifolia* Salzm. ex DC.
Incl. *C. algeriensis* Cosson & Durieu. Some records, especially early ones,
were mis-named as *C. aspera*.

eriophora L.
• A bird-seed casual. SW Europe, NW Africa, Canary Is. K, RNG. 21, 45*,
46*, 150, 349(6:125), 737*.

hyalolepis Boiss.
•• A grain, wool and bird-seed casual. E Medit. BM, CGE, LCN, OXF, SLBI. 2, 11, 19, 21, 44a*, 160, 335(115:153), 448, 667*. *C. pallescens* Del. var. *hyalolepis* (Boiss.) Boiss.; *C. pallescens* auct. eur., non Del.

iberica Trevir. ex Sprengel   Iberian Star-thistle
• A grain casual. No modern records. SE Europe, SW Asia. BM, CGE, LCN, NMW, RNG, SLBI. 1, 9, 21, 44a*, 156, 350(30:307), 354(13:110), 500*, 623*, 685*, 688*, 724*, 1095*, 1368*. Incl. subsp. *holzmanniana* (Boiss.) Dostál; *Carthamus ibericus* sphalm.

intybacea Lam.
Pre-1930 only. SW Europe. BM, CGE, LCN. 1, 3, 21, 160, 187, 525*, 546*. *Cheirolophus intybaceus* (Lam.) Dostál.

jacea L.   Brown Knapweed
•• Formerly a persistent grain, lucerne and grass seed alien, now mostly replaced by hybrids with native *C. nigra*. Eurasia; (N America). BM, GLR, OXF, RNG. 1, 20*, 22*, 26*, 156, 209, 525*, 546*. *C. amara* L. pro parte; *C. angustifolia* Gugler, non Miller; incl. *C. subjacea* (G.Beck) Hayek. A complex of hybrids between *C. jacea* and native *C. nigra* sensu lato have been recorded as *C.* × *moncktonii* C.Britton (*C. surrejana* C.Britton) and *C.* × *drucei* C.Britton (*C. pratensis* auct.; *C. jungens* auct.). BM, OXF. 20*.

leucophaea Jordan   Whitish-leaved Knapweed
• A casual at Cardiff (Glam), vector unknown. SW Europe. NMW. 2, 21, 23*, 156, 201, 546*.

macrocephala Mussin-Puschkin ex Willd.   Yellow Hardhead
• A garden escape persistent in an abandoned orchard at Belstead and possibly bird-sown in a farmyard at Middleton (both E Suffolk). SW Asia. OXF. 7, 39, 51, 54*, 55*, 57*, 354(6:27), 369(22:44; 25:53).

maculosa Lam.   Spotted Knapweed
Pre-1930 only. W Europe. BM, OXF. 7, 9, 21, 36c*, 354(5:288; 7:578), 500*, 525*, 546*. *C. tenuisecta* Jordan.

melitensis L.   Maltese Star-thistle
••• A grain, wool, linseed, bird-seed and esparto casual. Medit.; (Australasia, N America). BM, CGE, E, NMW, OXF, RNG. 1, 8, 13, 14, 15, 19, 20*, 25*, 30*, 500*.

montana L.   Perennial Cornflower
•••• An established garden escape; naturalised on roadsides, railway banks and waste places in widely scattered localities. Europe. 1, 20*, 22*, 23*, 27*, 28*, 34*.

napifolia L.
Pre-1930 only. Medit. BM, SLBI. 1, 3, 21, 46*, 116(p.229), 546*.

nicaeensis All.
Pre-1930 only. Medit. NMW, OXF, SLBI. 2, 21, 30*, 46*, 156, 300(1905:100), 354(6:132; 9:358). *Carthamus nicaeensis* sphalm.

[*nigrescens* Willd. Dubious records. 1, 151, 354(5:230; 6:131).]

orientalis L.
- A casual at Cardiff docks (Glam), vector unknown. Formerly a grain and ballast alien. No modern records. SE Europe, SW Asia. BM, NMW, OXF. 7, 21, 31*, 166, 354(4:202; 7:578; 11:485), 623a*, 688*, 1317*.

[*pallescens* Del.   Unconfirmed; probably all records are referable to *C. hyalolepis*.]

paniculata L.   Jersey Knapweed
- Established in Jersey (Channel Is) since 1851, probably now gone; elsewhere a wool casual. SW Europe; (Australia). BM, JSY, LANC, LIV, OXF, RNG. 2, 17, 20*, 23*, 201, 354(11:33), 546*. *C. gallica* Gugler, non Gouan.

patula DC.
Pre-1930 only. Turkey, N Iran. RNG. 39, 524, 554, 1249. *C. calliacantha* Fischer & C.Meyer; *C. ramosissima* Freyn & Sint.

pullata L.
- A casual, vector unknown. No modern records. W Medit. 21, 30*, 45*, 46*, 258, 546*.

rhenana Boreau   Panicled Knapweed
Pre-1930 only. C & SE Europe. OXF. 7, 21, 26, 354(5:37), 623a*, 688*, 1265*. *C. stoebe* L. pro parte.

solstitialis L.   Yellow Star-thistle
●●●● A grain, bird-seed, lucerne seed and wool casual; decreasing. Formerly persistent in lucerne fields. Medit.; (Australasia, N America). 1, 14, 19, 20*, 22*, 24*, 32*, 209. Incl. *C. adamii* Willd.

spinosa L.
Pre-1930 only.   Aegean, W Turkey.   OXF.   7, 21, 39, 160, 722*, 860*(43:146).

cf. **stenolepis** A.Kerner
- A casual collected from Splott, Cardiff (Glam), vector unknown. No modern records. C & SE Europe. RNG. 21, 683*, 688*, 715*, 1265* .

trichocephala M.Bieb. ex Willd.
Pre-1930 only.   Russia, W Asia.   OXF.   21, 354(5:36), 579*, 1104*. Illustrations labelled as *C. trichocephala* from Romania (e.g. 1265*) are now considered to belong to a separate taxon, *C. trichocephala* subsp. *simonkaiana* (Hayek) Dostál (*C. simonkaiana* Hayek).

triumfetti All.
- A grain casual. No modern records. Medit. BM, LCN, OXF. 7, 21, 28*, 31*, 160, 281, 289, 354(9:23), 546*, 623a. *C. axillaris* Willd., nom. illeg.; incl. *C. seussana* Chaix; *C. variegata* Lam.

verutum L.
- A grain casual. No modern records. E Medit. OXF. 1, 11, 44a*, 139, 184, 354(4:492), 528*, 667*, 1095*.

CRUPINA (Pers.) DC.
crupinastrum (Moris) Vis.
- An esparto casual. No modern records. Medit. RNG. 21, 32*, 546*, 683*, 737*. *C. morisii* Boreau; *Centaurea crupinastrum* Moris.

vulgaris Cass. False Saw-wort
Pre-1930 only. Medit.; (Australia). 7, 21, 31*(p.446), 46*, 313(16:215), 359(1:40), 623a*, 683*, 737*. *C. crupina* (L.) Karsten; *Centaurea crupina* L.

CNICUS L.
benedictus L. Blessed Thistle
• A bird-seed casual. Formerly a wool casual. Medit., SW Asia. BM, LCN, LIV, OXF. 1, 3, 20, 21, 24*, 31*, 44a*, 185, 206, 371(352:27). *Carbenia benedicta* (L.) Arcang.

CARTHAMUS L.
dentatus (Forsskål) Vahl
Pre-1930 only. SE Europe, SW Asia. E. 3*, 21, 24*, 31*.
lanatus L. Downy Safflower
••• A wool, grain and bird-seed casual. Eurasia, N Africa; (Australasia, N America). BM, CGE, E, NMW, OXF, RNG. 1, 11, 14, 19, 20*, 22*, 25*, 32*, 500*. Incl. *C. baeticus* (Boiss. & Reuter) Nyman; *C. creticus* L.; *Kentrophyllum baeticum* Boiss. & Reuter.
oxyacanthus M.Bieb.
• A casual at Avonmouth docks (W Gloucs). No modern records. SW Asia. 354(12:46), 660*, 690.
tenuis (Boiss. & Blanchet) Bornm.
Pre-1930 only. E Medit. OXF. 39, 44a*, 354(9:23).
tinctorius L. Safflower
•••• A bird-seed and grain casual; also a relic of cultivation. Origin obscure; (widespread). 2, 13, 19, 20*, 25*, 68*, 71*, 500*.

CARDUNCELLUS Adans.
caeruleus (L.) C.Presl
• A casual on refuse tips, probably from bird-seed. Medit. 19, 21, 24*, 30*, 45*, 46*, 458. *Carthamus caeruleus* L.

PEREZIA Lagasca
multiflora (Humb. & Bonpl.) Less.
• A casual at Southampton docks (S Hants), vector unknown. No modern records. S America. BM. 51, 354(13:31,112), 650*, 1151*. *Chaetanthera multiflora* Humb. & Bonpl.

SCOLYMUS L.
grandiflorus Desf.
Pre-1930 only. W Medit. 7, 21, 46*, 354(4:419), 546*, 556*, 633*, 683*.
hispanicus L. Golden Thistle
•• A bird-seed casual. Medit. BM, LIV, NMW, OXF, RNG. 1, 19, 20*, 22*, 24*, 31*, 737*.
maculatus L.
• A wool and bird-seed casual. Medit. RNG, HbEJC. 2, 14, 20, 21, 29*, 44a*, 45*, 46*, 331(42:9), 400, 737*.

CICHORIUM L.
endivia L.   **Endive**
••• A wool and bird-seed casual; also a relic of cultivation. Medit., Asia.
BM, E, LIV, LSR, NMW, RNG. 1, 14, 16, 19, 21, 44a*, 54*, 80*, 185,
826*(12:97-102). Incl. *C. divaricatum* Schousboe; *C. pumilum* Jacq.

CATANANCHE L.
caerulea L.   **Blue Cupidone**
• A garden escape persistent since 1988 on a road verge at Priddy
(N Somerset). W Medit. 21, 29*, 51*, 312*(t.293), 371(422:88), 423, 546*,
594*, 650*, 683*, 1207*.

LAPSANA L.
communis L. subsp. **intermedia** (M.Bieb.) Hayek   **Large Nipplewort**
•• Well established on a railway bank near Totternhoe (Beds); abundant in
Frosterley quarries (Co Durham) and on grassy slopes in the Isle of Dogs
(Middlesex); also recorded from Great Ormes Head (Caerns) and near Cilcain
(Flints). SE Europe, SW Asia. BM, CGE, E, OXF, RNG, SUN. 16, 20,
156, 166, 303(20:24; 34:34), 370(1:234-237; 12:175,196; 13:299-302), 623a*.
L. grandiflora auct., non Bieb.; *L. intermedia* M.Bieb.

HEDYPNOIS Miller
cretica (L.) Dum.-Cours.   **Scaly Hawkbit**
•• A wool, bird-seed and esparto casual. Medit., SW Asia; (Australia,
N America). BM, K, LIV, OXF, RNG, SLBI. 2, 14, 15, 19, 20*, 29*, 32*,
44a*, 160, 500*. *H. polymorpha* DC.; *H. rhagadioloides* (L.) F.W.Schmidt;
H. tubiformis Ten.; *Rhagadiolus creticus* (L.) All.; *R. hedypnois* (L.) All., non
Fischer & C.Meyer ex Fischer, C.Meyer & Trautv.

RHAGADIOLUS Scop.
[*angulosus* (Jaub. & Spach) Kupicha (*R. hedypnois* Fischer & C.Meyer ex
Fischer, C.Meyer & Trautv., non (L.) All.; *Garhadiolus hedypnois* Jaub. &
Spach) In error for *R. stellatus*. 182.]
stellatus (L.) Gaertner   **Star Hawkbit**
•• A bird-seed, grain and wool casual. Medit., SW Asia. BM, CGE, NMW,
OXF, RNG, SLBI. 1, 11, 17, 19, 20, 21, 24*, 32*, 39*, 44a*. Incl.
R. edulis Gaertner.

UROSPERMUM Scop.
dalechampii (L.) Scop. ex F.W.Schmidt
Pre-1930 only. Medit. OXF. 21, 24*, 29*, 546*, 737*.
picroides (L.) Scop. ex F.W.Schmidt
• A wool and esparto casual. No modern records. Medit.; (Australia).
K, OXF, RNG. 3, 14, 15, 21, 29*, 39*, 44a*, 251, 354(9:759), 546*, 737*.

HYPOCHAERIS L.
achyrophorus L.
Pre-1930 only. Medit. OXF. 7, 21, 44a*, 46*, 354(6:28). *H. aethnensis* (L.) Benth. & Hook.; *Seriola aethnensis* L.

LEONTODON L.
hispidus L. subsp. danubialis (Jacq.) Simonkai
Pre-1930 only. C & E Europe. LIV. 2, 21, 623a*, 683, 1140*. *L. danubialis* Jacq.; *L. hastilis* L.; *L. hispidus* var. *glabratus* (Koch) Bisch. Perhaps in error, since glabrous forms of native subsp. *hispidus* are virtually inseparable. 16.
muelleri (Schultz-Bip.) Fiori
• An esparto casual. No modern records. Sardinia, Sicily, N Africa. RNG. 21, 683*.
salzmannii (Schultz-Bip.) Ball
• An esparto casual. No modern records. Spain, Morocco. 15, 21, 737*.

PICRIS L.
altissima Del.
• An esparto and bird-seed casual. No modern records. Medit., SW Asia. CGE, E, OXF, RNG. 7, 9, 15, 19, 21, 44a*, 160, 683*, 825*(3:53-55). *P. sprengeriana* (L.) Poiret, nom. illeg.
[*arvalis* Jordan Now considered to be conspecific with native *P. hieracioides*. 376(1:187).]
cupuligera (Durieu) Walpers
• A bird-seed casual. N Africa. 45*, 46*, 371(346:28).
hieracioides L. subsp. spinulosa (Bertol. ex Guss.) Arcang.
Pre-1930 only. S Europe. SLBI. 1, 20, 21, 126, 241, 354(11:178,223,404), 623*, 683. *P. hieracioides* var. *umbellata* Vis.; *P. spinulosa* Bertol. ex Guss. Plants found near Longfield (W Kent) in the 1930s were probably the native subsp.
pilosa Del.
Pre-1930 only. N Africa. 1, 555*, 729. *P. radicata* (Forsskål) Less. var. *pilosa* (Del.) Asch. & Schweinf.

SCORZONERA L.
hispanica L.
•• A persistent garden escape. Eurasia. BM, E, LSR, RNG. 20, 21, 22*, 30*, 54*, 80*, 349(3:413), 371(334:21), 623a*, 1156*.

TRAGOPOGON L.
crocifolius L.
Pre-1930 only. Medit. BM, NMW, OXF. 1, 5, 21, 30*, 354(6:133), 546*, 683*.

hybridus L. Slender Salsify
●●● A bird-seed, spice and grain casual. Medit., SW Asia. BM, NMW, OXF, RNG. 7, 20*, 21, 44a*, 182, 303*(24:1,18), 546*, 683*, 729*. *Geropogon glaber* L.; *G. hybridus* (L.) Schultz-Bip.

porrifolius L. Salsify
●●●● An established escape from cultivation; naturalised on sea-walls, cliffs and roadsides, mainly in SE England. Medit. 1, 20, 22*, 24*, 33*, 45*, 191, 236, 245. A hybrid with native *T. pratensis* (*T.* × *mirabilis* Rouy) has been recorded. MANCH. 20.

pratensis L. subsp. **orientalis** (L.) Čelak.
Pre-1930 only. C & E Europe. BIRM, BM, E, LIV, OXF, SLBI. 1, 3, 20, 129, 139, 354(7:197), 623a*. *T. orientalis* L.; *T. undulatus* Wulfen. The 1960 record for St Margaret's Bay (E Kent) was in error. 371(331:19; 334:21). The European *T. dubius* Scop. may have been overlooked. 825*(12:23-25).
[*pratensis* L. subsp. *pratensis* Accepted, with reservations, as native.]

REICHARDIA Roth
tingitana (L.) Roth
● An esparto casual. No modern records. Medit. 15, 21, 24*, 30*, 44a*, 46*. *Picridium tingitanum* (L.) Desf.; *Scorzonera tingitana* L.

AETHEORHIZA Cass.
bulbosa (L.) Cass. Tuberous Hawk's-beard
● A persistent weed in a few gardens in eastern Ireland. Medit., W Europe. BM, CGE, E, RNG. 13, 20*, 22*, 44a*, 46*, 354(8:618), 371(374:9), 737*. *Crepis bulbosa* (L.) Tausch; *Intybus bulbosus* (L.) Fr.; *Leontodon bulbosus* L.; *Prenanthes bulbosa* (L.) DC.

SONCHUS L.
maritimus L.
Pre-1930 only. Medit. BM. 21, 22*, 44a*, 45*, 46*, 303(24:18). The record for Suffolk was in error. 369(13:412; 14:136).

tenerrimus L. Slender Sow-thistle
● (but confused with native *S. oleraceus*). An esparto and bird-seed casual. No modern records. Medit. E. 2, 15, 19, 21, 44a*, 45*, 46*, 354(10:532).

LACTUCA L.
[*dubia* Jordan (*L. scariola* L. var. *dubia* (Jordan) Rouy) Now considered to be conspecific with native *L. serriola*. 7, 245.]

perennis L. Mountain Lettuce
1930 only. A casual on shingle near Greatstone (E Kent), vector unknown. Europe. RNG. 21, 22*, 23*, 25*, 26*, 623a*.

sativa L. Garden Lettuce
●●●●● A relic of cultivation and a casual garden escape; probably also a bird-seed casual. Origin obscure. 3, 19, 20*, 68*, 80*.

tatarica (L.) C.Meyer Blue Lettuce
●●● Possibly a grain alien, established in a few places on dunes, by railways and at ports; known since 1923 on the shore east of Galway (NE Galway) and

since 1963 on dunes near Llandudno (Caerns). Eurasia, N America. BM, CGE, DBN, NMW, OXF, RNG. 1, 20*, 22*, 31*, 156, 275, 303*(26:12-13), 1138*. *Mulgedium tataricum* (L.) DC.; *Sonchus tataricus* L.; incl. *L. pulchella* (Pursh) DC.

viminea (L.) J.S. & C.Presl   Pliant Lettuce
• A casual at Galashiels (Roxburghs), vector unknown. Eurasia. E. 21, 24*, 25*, 29*, 44a*, 133, 457, 623*.

CICERBITA Wallr.

bourgaei (Boiss.) Beauverd   Pontic Blue-sow-thistle
••• An established garden escape; naturalised by roadsides in a few widely scattered localities; known since 1958 at the top of Bug Hill at Warlingham (Surrey). Turkey. BM, CGE, E, OXF, RNG. 13, 20, 39*, 370*(16:121-129), 1275*. *Mulgedium bourgaei* Boiss.; *Lactuca bourgaei* (Boiss.) Irish & Norman Taylor.

macrophylla (Willd.) Wallr.   Common Blue-sow-thistle
•••• An established garden escape; naturalised on roadsides in scattered localities throughout most of the British Is; increasing. E Europe, SW Asia. 7, 20, 22*, 34*, 370*(121-129), 1275*. *Lactuca macrophylla* (Willd.) A.Gray; *Mulgedium macrophyllum* (Willd.) DC. All records are referable to subsp. *uralensis* (Rouy) Sell.

plumieri (L.) Kirschl.   Hairless Blue-sow-thistle
•• (but over-recorded for *C. macrophylla*). An established garden escape; known since 1950 at Tighnabruich, near Invermoriston (Easterness) and since 1957 in a wood at Bothwell (Lanarks). Europe. E, OXF. 7, 20, 22*, 30*, 370*(16:121-129), 1275*. *Lactuca plumieri* (L.) Gren. & Godron; *Mulgedium plumieri* (L.) DC.

PRENANTHES L.

purpurea L.   Purple Lettuce
• A casual on a building site in York, vector unknown. Europe, SW Asia. 1, 21, 22*, 23*, 27*, 29*, 197, 371(367:29), 1149.

TARAXACUM Wigg.

The status, nomenclature, distribution and origin of the species (or microspecies) of *Taraxacum* are very uncertain. A provisional list is given here; some are probably better regarded as of dubious alien status, especially *T. pectinatiforme*. Most are probably native in N Europe. Few records have been localised in print, so vice-county numbers only are quoted for all species.

aequisectum M.Christiansen
(Vcs. 23, 60, 99). 91*, 370(11:80).

altissimum Lindb.f.
(Vcs. 89, 112). 253, 370(15:68).

angustisquameum Dahlst. ex Lindb.f.
(Vc. 89). 370(15:68).

austriacum Soest
   (Vc. 66).   91*, 166.
christiansenii Hagl.
   (Vcs. 5, 9, 12, 16, 60, 62, 95, 98, 109).   E.   91*, 277, 370(10:431; 11:401; 12:359).
copidophyllum Dahlst.
   (Vcs. S, 12, 15, 22, 38, 46, 61).   NMW.   91*, 156, 236, 370(11:80,401).
disseminatum Hagl.
   (Vcs. 17, 41).   91*.
falcatum Brenner
   (Vcs. 16, 24, 29, 50, 55).   OXF.   91*, 156, 370(13:195-201).   *T. canulum* auct., non Hagl. ex Markl.
fasciatum Dahlst.
   Widespread throughout Britain.   91*.
hamiferum Dahlst.
   (Vcs. 40, 50, 51, 57-59, 90).   183, 187, 370(13:195-201).   *T. atrovirens* auct., non Dahlst.
insigne Ekman ex M.Christiansen & Wiinst.
   (Vcs. 50, 57, 95, 112).   E, OXF.   183, 253, 277, 370(13:195-201).   *T. ordinatum* Hagend., Soest & Zevenb.
kernianum Hagend., Soest & Zevenb.
   (Vcs. 1, 12, 50, 51, 58).   E, NMW.   156, 211, 370(13:195-201).
lacerifolium Hagl.
   (Vcs. 15, 55, 58, 59).   LSR.   236, 370(13:195-201).   *T. lacinulatum* Markl.
laciniosum Dahlst.
   Widespread in England.   OXF.   91*, 370(11:401).
laetiforme Dahlst.
   (Vcs. 2, 28, 57, 89, 95, 101).   91*, 211, 370(15:68).
laticordatum Markl.
   (Vcs. 15, 55, 89).   LSR.   236, 240, 370(15:68).   *T. uncosum* Hagl.
linguatum Dahlst. ex M. Christiansen & Wiinst.
   Widespread in England.   91*, 370(10:431; 11:79,400; 12:359).   Records for Wales were in error.   156.
lividum (Waldst. & Kit.) Peterm.
   (Vcs. 20, 23).   21, 354(11:224; 660).   *T. lissocarpum* (Dahlst.) Dahlst.
monochroum Hagl.
   (Vc. 96).   E.   277.
obliquilobum Dahlst.
   (Vcs. 15, 17, 22, 24, 55, 57).   LSR.   91*, 183, 236, 240.
obtusilobum Dahlst.
   (Vcs. 15, 41, 66, 106).   BM.   91*, 153, 156, 166, 236.   A hybrid with native *T. hamatiforme* has been recorded.   236.
ostenfeldii Raunk.
   (Vc. 23).   91*.   *T. biforme* sensu A.Richards et auct., non Dahlst.; *T. parvuliceps* sensu A.Richards, non Lindb.f.

pachymerum Hagl.
   (Vcs. 12, 15, 55).   LSR, HbAB.   236, 240.
parvuliceps Lindb.f.
   (Vcs. 3, 22, 29, 79, 95, 99).   E.   91*, 277.   Records for Wales were in
   error. 156.
pectinatiforme Lindb.f.
   (Vcs. 5, 6, 12, 29, 34, 57, 60, 66, 95, 106, 112).   BM, HbAB.   91*.
piceatum Dahlst.
   (Vcs. 46, 50, 51).   NMW.   156.
polyhamatum Oellg.
   (Vcs. 15, 16, 40, 50, 51, 55, 58, 112).   E, LSR, NMW.   156, 236, 240, 253,
   370(13:195-201).
procerisquameum Oellg.
   (Vcs. 1, 12, 23, 50, 51, 55-58, 60, 66).   LSR.   91*, 156, 166, 183, 211,
   240, 370(10:430; 11:79; 13:195-201).   *T. procerum* sensu A.Richards et auct.,
   non Hagl.
ramphodes Dahlst.
   (Vcs. 11, 55, 60, 89).   HbAB.   240, 303(57:10), 370(15:68).
sagittipotens Dahlst. & Ohlsén ex Hagl.
   (Vcs. 41, 46, 55, 60, 66).   NMW.   156, 166, 240, 303(57:10).
undulatiflorum M.Christiansen
   (Vcs. 50, 51, 55, 60, 66, 89).   166, 240, 370(13:195-201; 5:68).

CHONDRILLA L.
juncea L.   Skeletonweed
   Pre-1930 only.   Eurasia, N Africa.   OXF.   2, 21, 22*, 25*, 31*, 44a*, 148,
   151, 201.

CREPIS L.
alpina L.
   Pre-1930 only.   Crimea, SW Asia.   2, 21, 300(1905:100), 695*, 1007*.
   Anthochytrum alpinum (L.) Reichb.f.
aurea (L.) Cass.   Golden Hawk's-beard
   Probably pre-1930 only.   C & SE Europe.   21, 23*, 26*, 28*, 197, 546*.
foetida L. subsp. **commutata** (Sprengel) Babc.
   Pre-1930 only.   SE Europe.   1, 21, 1007*.   *Rodigia commutata* Sprengel.
foetida L. subsp. **rhoeadifolia** (M.Bieb.) Schinz & Keller
   • A casual at Bristol (W Gloucs), vector unknown.   Eurasia. 21, 350(30:243),
   683*, 1007*.
nicaeensis Balbis   French Hawk's-beard
   •• A bird-seed and agricultural seed casual; decreasing, few modern records
   being correct.   S Europe; (N America).   BM, LANC, LIV, OXF, RNG.
   1, 19, 20*, 22*, 191, 715*.
[*oporinoides* Boiss. ex Froelich   In error for native *C. biennis*.   354(10:43;
   11:223).]

[*praemorsa* (L.) F.Walther (*Hieracium praemorsum* L.)   Accepted, with reservations, as native. LANC. 20*.]

pulchra L.   Small-flowered Hawk's-beard
    • A casual by the railway at Hadleigh (E Suffolk), vector unknown. Eurasia, N Africa. CGE, LINN. 2, 21, 22*, 159, 187, 258, 546*, 683*, 715*. *Phaecasium pulchrum* (L.) Reichb.; *P. lampsanoides* Cass.; *Prenanthes hieracifolia* Willd.

rubra L.   Pink Hawk's-beard
    Pre-1930 only. SE Europe. 1, 16, 21, 24*, 32*, 195.

sancta (L.) Babc.
    Pre-1930 only. Eurasia, N Africa. 7, 21, 44a*, 354(4:419; 6:27,386), 546*, 683*, 1007*. *Hieracium sanctum* L.; *Lagoseris nemausensis* (Gouan) M.Bieb.; *L. sancta* (L.) K.Malý; *Phaecasium sancta* (L.) K.Koch; *Pterotheca sancta* (L.) K.Koch. This species is steadily spreading across N Europe and approaching our shores. 825*(3:13-16).

setosa Haller f.   Bristly Hawk's-beard
    ••• A persistent agricultural seed, grass seed and grain alien. Eurasia; (N America). BM, E, LANC, NMW, OXF, RNG. 1, 20*, 22*, 192, 234, 264, 500*. *Barkhausia setosa* (Haller f.) DC.

suberostris Cosson & Durieu ex Battand.
    • A bird-seed casual. Algeria. RNG. 46, 1007*, 1018.

tectorum L.   Narrow-leaved Hawk's-beard
    ••• A persistent grass seed and grain alien. Europe; (N America). BM, E, LANC, LSR, OXF, RNG. 1, 20*, 22*, 26*, 303(13:24; 42*:16-17), 350(47:xxxi), 370(13:143). Early records were mostly in error for native *C. capillaris* (*C. tectorum* Smith, non L.).

vesicaria L.   Beaked Hawk's-beard
    ••••• An established alien; first recorded in 1713 and now widespread in S and E England, Wales and southern Ireland; spreading northwards. Medit., SW Asia. 1, 20*, 22*, 25*, 33*. Incl. *C. polymorpha* Pourret; *C. taraxacifolia* Thuill.; *Barkhausia taraxacifolia* (Thuill.) DC. All records are referable to subsp. *taraxacifolia* (Thuill.) Thell. ex Schinz & R.Keller (subsp. *haenseleri* (Boiss. ex DC.) Sell). A hybrid with a *Taraxacum* sp. was recorded pre-1930, probably in error. 1311.

zacintha (L.) Babc.
    • A grain casual. No modern records. Medit. RNG. 9, 21, 39*, 160, 683*, 1007*. *Lapsana zacintha* L.; *Zacintha verrucosa* Gaertner.

TOLPIS Adans.
barbata (L.) Gaertner
    • A wool, bird-seed and granite casual. Medit.; (New Zealand). BM, LCN, NMW, OXF, RNG. 1, 17, 19, 20, 21, 31*(p.475), 32*, 44a*, 45*, 303(10:16), 371(379:26). *Crepis barbata* L.

ANDRYALA L.
integrifolia L.
- A wool casual. Medit. ABD, CGE, E, NMW, OXF, RNG. 4, 10, 17, 21, 29*, 32*, 45*, 354(4:419; 7:43). Incl. *A. tenuifolia* DC. This plant is easily confused with *Hieracium* spp.

AGOSERIS Raf.
grandiflora (Nutt.) E.Greene
- A grass seed casual. Western N America. BM. 41*, 42, 371(331:22), 500*.

PILOSELLA Hill
aurantiaca (L.) F.Schultz & Schultz-Bip.   Fox-and-cubs
●●●●● An established garden escape; widely naturalised on roadsides and railway banks throughout the British Is. Europe. 1, 20, 22*, 26*, 33*, 34*, 1251, 1266*, 1275. Incl. *Hieracium aurantiacum* L.; *H. brunneocroceum* Pugsley (*H. aurantiacum* subsp. *carpathicola* Naeg. & Peter); *H. claropurpureum* Naeg. & Peter. A hybrid with native *P. officinarum* (*P. × stoloniflora* (Waldst. & Kit.) F.Schultz & Schultz-Bip.) has been recorded. ABD, BM, CGE, E. 20, 741*, 1218, 1275.

caespitosa (Dumort.) Sell & C.West   Yellow Fox-and-cubs
●●● An established garden escape; naturalised by railways, on river banks and fixed dunes, and by roadsides in a few widely scattered localities. Europe, SW Asia. BM, CGE, E, NMW, OXF, RNG. 2, 20, 36c*, 39, 277, 320*(11:353), 683*, 1251, 1266. *Hieracium caespitosum* Dumort.; *H. pratense* Tausch; incl. *H. colliniforme* (Naeg. & Peter) Roffey.

flagellaris (Willd.) Sell & C.West subsp. flagellaris
●●● An established garden escape; naturalised on roadsides and railway banks in a few scattered localities, mainly in C and S England; abundant by railways in the Forth area (Midlothian). Europe. BM, E, LANC, LSR, OXF, RNG. 20, 546*, 1251, 1266. *Hieracium flagellare* Willd.; *H. stoloniflorum* auct., non Waldst. & Kit.

× floribunda (Wimmer & Grab.) Arv.-Touv.
- Established on a road verge near Pig Bush in the New Forest (S Hants), vector unknown. Formerly naturalised in a limestone quarry at Cave Hill, Belfast (Co Antrim). N & C Europe. BM, CGE. 20, 21, 370(19:187-188), 589*, 1251, 1266, 1275. *P. caespitosa × P. lactucella*; *P. lactucella* subsp. *helveola* (Dahlst.) Sell & C.West; *Hieracium × floribundum* Wimmer & Grab.; *H. × helveolum* (Dahlst.) Pugsley.

lactucella (Wallr.) Sell & C.West
- A persistent weed at Harrow (Middlesex), vector unknown. Formerly established in pasture at Keevil (S Wilts). Eurasia. BM, CGE, LANC, OXF, RNG, SLBI. 2, 20, 172, 192, 683*, 1251, 1266, 1275. *Hieracium auricula* auct., non L.; *H. lactucella* Wallr. 20, 589*, 1275.

praealta (Villars ex Gochnat) F.Schultz & Schultz-Bip.   Tall Mouse-ear-hawkweed
••• An established garden escape; naturalised in a few widely scattered localities, mainly on railway banks in southern England. Europe. BM, E, LTN, OXF, RNG, SLBI. 1, 20, 119, 519*, 1251, 1266, 1275. *Hieracium praealtum* Villars ex Gochnat; incl. *H. arvorum* (Naeg. & Peter) Pugsley; *H. bauhinii* Besser; *H. pilosella* subsp. *thaumasium* (Peter) Sell; *H. spraguei* Pugsley.

HIERACIUM L.

[*alfvengrenii* Dahlst.   In error for native *H. acuminatum*. 354(9:560), 1266.]
amplexicaule L.   Sticky Hawkweed
• (but confused with *H. pulmonarioides* and *H. speluncarum*). An established garden escape; known for 200 years on the walls of Oxford Botanic Garden (Oxon). Europe. BM, HAMU, LANC, OXF. 1, 16, 33*, 257, 303*(39:23-25), 354(10:440), 370(2:403), 1251, 1310*. *H. amplexicaule* group.

cardiophyllum Jordan ex Sudre
• Established on a railway bank near Otford (W Kent), on a riverside near Llangattock (Brecs) and at Rhu (Dunbarton). S Europe. CGE, MNE. 156, 236, 1251, 1321*. *H. murorum* group.

[*cerinthoides* L.   In error for native *H. anglicum*. 159, 361(7:262).]
chondrillifolium Fries
Pre-1930 only. S Europe. 1, 21, 546a, 683, 1281. Probably of *H. bifidum* Kit. × *H. glaucum* All. × *H. villosum* origin. *H. chondrilloides* Villars, non L., nec Jacq., nec All.; *H. glaucopsis* Gren. & Godron var. *chondrillifolium* Rouy.

[*dentatum* sphalm.   A dubious record. 354(9:252).]
[*dubium* sphalm.   Dubious records. 106, 136, 203.]
gougetianum Gren. & Godron
• Established about Dublin, especially on walls near Mountjoy Prison and on the banks of the river Liffey, (Co Dublin); also recorded from a pit bing at Hallside (Lanarks). SW & C Europe. BM, GL. 370(17:480), 371(374:9), 695*, 1251, 1266. *H. maculosum* auct., non Dahlst. ex Stenström. *H. glaucinum* group.

grandidens Dahlst.
••• Established in widely scattered localities throughout the British Is; increasing. C Europe. BM, CGE, LSR, NMW. 16, 126, 153, 156, 199, 248, 564*, 1217*, 1251, 1266. *H. exotericum* Jordan ex Boreau forma *grandidens* (Dahlst.) Pugsley. *H. murorum* group.

hjeltii Norrlin ex T.Saelán, W.Nylander & T.S.Nylander
Pre-1930 only. Scandinavia. 589, 1251. *H. murorum* group.

lanatum (L.) Villars   Woolly Hawkweed
• An established garden escape at Canterbury (E Kent) and on the North Denes at Great Yarmouth (E Norfolk); also reported from Littleferry (E Sutherland). Europe. MNE, RNG. 21, 27*, 28*, 408, 546*, 683. *H. tomentosum* (L.) L. *H. lanatum* group.

[*maculatum* Smith   Accepted, with reservations, as native.]

oblongum Jordan
- Established on grassy banks of the reservoir at Hallington (S Northumb) and on walls near Alnwick (Cheviot). Europe. 1251, 1321*. *H. murorum* group.

patale Norrlin
- Established in abundance on roadsides, railway banks and in woods on the southern border of Dartmoor (S Devon); also on an old station platform at Torrington (N Devon) and on walls at Onich (Westerness). Scandinavia. CGE. 21, 370(16:448), 1251. *H. murorum* group.

pilosum Schleicher ex Froelich
- Established on the beach at Dungeness (E Kent), probably a garden escape. Europe. MNE. 21, 236, 683*. *H. morisianum* Reichb. *H. pilosum* group.

pulmonarioides Villars
- (but confused with *H. amplexicaule* and *H. speluncarum*). An established garden escape; naturalised on old walls in a few places in C and N England. Pyrenees, Alps. BM. 7, 16, 143, 185, 303*(39:23-25), 1251, 1266. *H. amplexicaule* group.

scotostictum N.Hylander
- •• Established in the London area and a few widely scattered localities in England and Wales; increasing since the first record in 1920. SW & C Europe. BM, LANC, NMW. 126, 156, 199, 370(17:194), 1251, 1266*. *H. praecox* auct., non Schultz-Bip. *H. glaucinum* group.

severiceps Wiinst.
- •• Established in a few widely scattered localities; abundant in Bishop Middleham quarry (Co Durham). Probably N Europe. CGE, SUN. 166, 1251. *H. murorum* group.

speluncarum Arv.-Touv.
- •• (but confused with *H. amplexicaule* and *H. pulmonarioides*). An established garden escape in a few widely scattered localities; long known on walls at Ham House (Surrey), Mells (N Somerset) and Richmond Hill in Bristol (W Gloucs). C Europe. BM, CGE, LANC, LIV, OXF, RNG. 16, 209, 248, 303*(39:23-25), 370(15:425), 1251, 1279*. *H. amplexicaule* group.

villosum Jacq.
- Recorded from the ruins of St Augustine's Abbey in Canterbury (E Kent) and, without locality, from a limestone cliff in Yorkshire. Europe. 21, 365(3:23), 525*, 546*, 594*, 595b*, 623*, 1016*. Other records, e.g. from Scotland, are in error for native *H. anglicum*. 195, 255, 354(5:775). *H. villosum* group.

zygophorum N Hylander
- Established at Weston Crossing, near Beccles (E Suffolk). Sweden. CGE. 370*(13:27-29), 916*(7:1-432). *H. murorum* group.

ARCTOTIS L.
breviscapa Thunb.
- A casual garden escape. S Africa. 51, 331(57:71), 593*, 594*, 655*.

stoechadifolia Bergius   White Arctotis
- An established garden escape; naturalised since 1978 in a quarry on Sark (Channel Is). S Africa. RNG. 21, 36c, 37*, 52*, 371(385:34), 583*. Incl. *A. grandis* Thunb.

ARCTOTHECA Wendl.
calendula (L.) Levyns   Plain Treasureflower
- ••• A wool and granite casual. S Africa; (widespread). BM, CGE, E, NMW, OXF, RNG. 2, 14, 20, 30*, 37*, 303(10:16), 533*, 583*, 737*. *Arctotis calendula* L.; *Cryptostemma calendula* (L.) Druce; *C. calendulacea* sphalm.

GAZANIA Gaertner
rigens (L.) Gaertner   Treasureflower
- •• A persistent garden escape in a few places near the sea in Scilly and the Channel Is; perhaps established on a cliff-face near Yellow Rock on St Martins (Scilly). S Africa. BM, OXF, RNG. 7, 20, 207, 220, 312*(t.2270), 349(1:330), 354(4:201), 594*, 650*, 657*, 835(3:370-375). *G. splendens* hort. ex Henderson & A.Henderson, non Moore; *Meridiana rigens* auct.; *M. splendens* auct.; incl. *G. leucolaena* DC.; *G. uniflora* (L.f.) Sims.

BERKHEYA Ehrh.
heterophylla (Thunb.) O.Hoffm.
- A wool casual. S Africa. RNG. 17, 145, 500, 565*, 592, 835(3:191-194). *Stobaea biloba* DC.; *S. heterophylla* Thunb.
pinnatifida (Thunb.) Thell.
- A wool casual. S Africa. E, RNG. 3, 592, 835(3:207-211), 1144. *Stobaea pinnatifida* Thunb.; incl. *B. stobaeoides* Harvey.
rigida (Thunb.) Ewart, J.White & Rees   African Thistle
- A wool casual. S Africa; (Australia). BM, E, RNG. 17, 37c*, 501, 583*, 628*, 655*, 835(3:423), 1222*. *Stobaea rigida* Thunb.

CUSPIDIA Gaertner
cernua (L.f.) B.L.Burtt
- Pre-1930 only. S Africa. 7, 354(6:611), 592, 835(3:316-319), 1145*. *Berkheya cernua* (L.f.) R.Br.

HETERORHACHIS Schultz-Bip. ex Walp.
cf. **aculeata** (Burm.f.) Roessler
- Pre-1930 only. S Africa. OXF. 7, 51, 312*(t.1788), 354(5:36), 592, 835(3:313-315). *Berkheya pinnata* (Thunb.) Less.; *Stobaea pinnata* Thunb.

EPALTES Cass.
australis Less.   Spreading Nut-heads
- A casual in 1962 on a slag heap between Windyridge and Braidwood Burn, Carluke (Lanarks), vector unknown. Formosa, Australia. BM. 37*, 514, 590*, 743.

FILAGO L.
arvensis L.   Field Cudweed
- A casual at Kingston North Common (S Hants) in 1986, and at Avonmouth docks (W Gloucs) and Laughton (N Lincs) in the 1930s, vector unknown. Eurasia.   LCN, LIV.   7, 9, 20, 21, 22*, 26*, 39*, 160, 354(6:24). *Logfia arvensis* (L.) Holub; incl. *F. lagopus* Stephan ex Willd.

duriaei Cosson ex Lange
- An esparto casual. S Spain, NW Africa.   RNG.   21, 737*, 1189*.

gallica L.   Narrow-leaved Cudweed
- Formerly established as a weed in a few sandy fields, mainly in SE England; not recorded in Britain since 1955, but still in Sark (Channel Is), vector unknown. Medit., Atlantic Is.   BM, IPS, LANC, OXF, RNG.   2, 20, 22*, 29*, 33*, 35*, 201, 303(17:5) in error?.   *Gnaphalium gallicum* L.; *Logfia gallica* (L.) Cosson & Germ.   A very recent, unpublished report suggests that this is a native species.

IFLOGA Cass.
verticillata (L.f.) Fenzl
- A wool casual.   S Africa.   E, K, RNG.   17, 145, 327(82:293-312), 344(31:22), 501, 592. *Gnaphalium verticillatum* L.f.; *Trichogyne verticillata* (L.f.) Less.

BOMBYCILAENA (DC.) Smoljan.
erecta (L.) Smoljan.   Micropus
Pre-1930 only.   Europe, SW Asia, N Africa.   7, 21, 22*, 25*, 39*, 354(4:199). *Micropus erectus* L.

MICROPSIS DC.
spathulata (Pers.) Cabrera
- A wool casual.   Uruguay, NE Argentina.   E, K, RNG.   17, 534*. *M. herleri* Beauverd; *Evax spathulata* Pers.

CHEVREULIA Cass.
sarmentosa (Pers.) Blake
- A wool casual.   S America.   HbEJC.   534*, 535*.   *C. stolonifera* Cass.

FACELIS Cass.
retusa (Lam.) Schultz-Bip.
- A wool casual.   S America; (S Africa, Australia).   LIV, RNG.   514, 534*, 590*, 598*, 682*. *Gnaphalium retusum* Lam.

STUARTINA Sonder
hamata Philipson   Hooked Cudweed
- A wool casual.   Australia.   BFT, BM, E, K, LTR, RNG.   14, 37c*, 236, 533*, 590*, 743.

muelleri Sonder   Spoon-leaved Cudweed
Pre-1930 only.   Australasia.   E.   1, 3, 36c*, 37*, 195, 533*, 583*, 590*.

ANAPHALIS DC.
margaritacea (L.) Benth.   Pearly Everlasting
●●●● An established garden escape; widely naturalised on railway banks, river
banks, hillsides and roadsides, especially in S Wales; known since 1698 in the
Rhymney Valley (Mons & Glam). E Asia, N America. 1, 20*, 22*, 28*, 55*,
156, 270. *Antennaria margaritacea* (L.) Gray; *Gnaphalium margaritaceum* L.
Some early records are referable to var. *subalpina* A.Gray.
triplinervis (Sims) C.B.Clarke
● (but perhaps overlooked as *A. margaritacea*). A casual or persistent
garden escape; recorded in 1986 from Wortham Common, near Diss
(E Suffolk) and in the 1970s from Glentarbuck (Dunbarton) and a tip at
Tholthorpe (NE Yorks). Himalayas. 51, 53*, 57*, 312*(t.2468), 369(24:100),
371(376:25), 454, 604, 685*. Incl. *A. nubigena* DC.

GNAPHALIUM L.
albescens Sw.
Pre-1930 only. Jamaica. 49, 197.
japonicum Thunb.   Japanese Cudweed
Pre-1930 only. E Asia, Australasia. E, K. 3*, 37, 43, 51, 312*(t.2582),
500*, 514, 649*. *G. collinum* Labill.; *G. involucratum* Forster f. The record
for Blackmoor (N Hants) was in error for *Ifloga verticillata*. 14.
[*polycephalum* Michaux   In error for *G.undulatum*. 201, 354(1:374).]
purpureum group   Purple Cudweeds
● An escape from the Royal Botanic Gardens, established in the churchyard of
St Ann's on Kew Green (Surrey); also a wool and granite casual. Tropical
America; (widespread). BM, CGE, E, K, OXF, RNG. 3, 17, 20, 38*, 41*,
48*, 182, 199, 303*(38:20-21), 331(58:65), 501, 541, 825(5:39),
843*(45:479), 1247*. Incl. *G. americanum* Miller; *G. pensylvanicum* Willd.,
nom. illeg.; *G. peregrinum* Fern.; *G. purpureum* L.; *G. purpureum* sensu
Hook.f.(1881), non L.; *G. spathulatum* Lam.; *G. spicatum* Lam.; *Gamochaeta
purpurea* (L.) Cabrera.
undulatum L.   Cape Cudweed
●●● An established alien; long naturalised, widespread and locally abundant in
the Channel Is, perhaps as an extension of its range from NW France, and
recently established on cliffs at Seaton (E Cornwall); also a wool casual and
garden escape. S Africa. BM, E, LANC, LTR, OXF, RNG. 2, 14, 20*, 22*,
117*, 201, 220, 236. *G. polycephalum* auct., non Michaux; *Pseudognaphalium
undulatum* (L.) Hilliard & B.L.Burtt.
viscosum Kunth   Sticky Cudweed
● A casual on a farm at Great Wakering (S Essex), vector unknown.
N America, Mexico. 38*, 38a*, 40, 41, 191, 500*, 707, 1210*, 1256*.
G. gracile Kunth; *G. hirtum* Kunth; *G. leptophyllum* DC.; *G. splendens* Willd.
Incl. *G. decurrens* Ives, non L.; *G. ivesii* Nelson & Macbr.; *G. macounii*
E.Greene.

TROGLOPHYTON Hilliard & B.L.Burtt
parvulum (Harvey) Hilliard & B.L.Burtt
- A wool casual. SW Africa. E. 3, 133, 327(82:208-209), 501, 592, 1092*.
Gnaphalium parvulum Harvey; *Helichrysum capillaceum* var. *erectum* DC.

VELLEREOPHYTON Hilliard & B.L.Burtt
dealbatum (Thunb.) Hilliard & B.L.Burtt   White Cudweed
- A wool casual. S Africa; (Australasia). BM, RNG. 17, 327(82:209-211),
541*, 590*, 592, 1078*, 1263*. *Gnaphalium candidissimum* Lam., nom.
illeg.; *G. dealbatum* Thunb.; *G. maculatum* Thunb.; *G. micranthum* Thunb.

HELICHRYSUM Miller
apiculatum (Labill.) D.Don sensu lato   Yellow-buttons
Pre-1930 only. Australia. E. 3, 37*, 51, 301*(1817:t.240), 569*, 583*,
666*. *Gnaphalium apiculatum* Labill. Incl. *H. ramosissimum* Hook.
bellidioides (Forster f.) Willd.   New Zealand Everlastingflower
- Introduced, or a garden escape, established in turf at Tagon, Voe (Shetland).
New Zealand. BM, OXF, ZCM. 20*, 36, 52*, 54*, 253, 303(14:14), 519*.
bracteatum (Vent.) Andrews   Strawflower
- ••• A garden escape and wool casual. Australia. OXF, RNG. 7, 14, 20, 21,
37*, 52*, 55*, 583*. Incl. *H. monstrosum* hort.
capillaceum (Thunb.) Less.
- A wool casual. S Africa. E, K, RNG. 17, 501, 628*.
expansum (Thunb.) Less.
- A wool casual. S Africa. RNG. 17, 501, 628*.
italicum (Roth) Don   Curry-plant
- A casual garden escape. Medit. BM. 24*, 21, 39*, 52*, 156, 371(409:40).
H. angustifolium (Lam.) DC.; incl. *H. microphyllum* (Willd.) Cambess.
odoratissimum (L.) Less.
- A wool casual. S Africa. BM, E, OXF, RNG. 3*, 17, 47*, 48*, 51, 145.
Gnaphalium odoratissimum L.
petiolare Hilliard & B.L.Burtt   Silver-bush Everlastingflower
- Introduced and persistent on dunes at Pentle Bay on Tresco and in a pit above
Tremelethen on St Mary's (both Scilly); also a casual garden escape. S Africa.
21, 53*, 207, 303(41:14), 562*. *H. petiolatum* auct., non (L.) DC.
stoechas (L.) Moench
Pre-1930 only. Medit. NMW, OXF. 7, 21, 22*, 24*, 32*, 39*, 354(7:878;
10:928).

RHODANTHE Lindley
charsleyae (F.Muell.) Paul G.Wilson
- A wool casual. Australia. E. 37, 133, 583*. *Helipterum charsleyae*
F.Muell.

corymbiflora (Schldl.) Paul G.Wilson
- A wool casual. No modern records. Australia. E, JDFC, RNG. 3*, 14, 37*, 51, 300(1909:42), 590*. *Helipterum corymbiflora* Schldl.

floribunda (DC.) Paul G.Wilson   Flowery Sunray
- A wool casual. No modern records. Australia. E, RNG. 3*, 14, 37*, 583*, 590*. *Helipterum floribundum* DC.

manglesii Lindley   Pink Sunray
- A casual garden weed, perhaps an impurity in horticultural seed. W Australia. NMW. 51, 54*, 55*, 156, 303(36:28), 543*, 583*. *Argyrocome manglesii* (Lindley) Kunth; *Helichrysum manglesii* (Lindley) Baillon; *Helipterum manglesii* (Lindley) F.Muell. ex Benth.; *Roccardia manglesii* (Lindley) Voss.

moschata (Cunn. ex DC.) Paul G.Wilson   Musk Sunray
- A wool casual. Australia. E, K, RNG. 17, 37c*, 590*, 743, 1074*. *Gnaphalium moschatum* Cunn. ex DC.; *Helipterum moschatum* (Cunn. ex DC.) Benth.

TRIPTILODISCUS Turcz.

pygmaeus Turcz.   Common Sunray
Pre-1930 only. Australia. E. 3, 37*, 317*(t.856), 514, 533, 583*, 590*. *Helipterum australe* (A.Gray) Druce; *H. dimorpholepis* Benth.; *H. pygmaeum* (Turcz.) Druce, non (DC.) F.Muell.

HYALOSPERMA Streetz

glutinosa Streetz   Golden Sunray
Pre-1930 only. Australia. E. 3*, 37*, 300(1911:101), 583*, 590*, 743. *Helipterum glutinosum* (Streetz) Druce; *H. hyalospermum* F.Muell. ex Benth.; *H. variabile* (Sonder) Ostenf.

LEPTORHYNCHOS Less.

squamatus (Labill.) Less.   Scaly-buttons
- A casual recorded without date, probably a wool alien. Australia. LIV. 37, 251, 514, 533*, 549, 590*. *Chrysocoma squamata* Labill.

MILLOTIA Cass.

muelleri (Sonder) P.S.Short   Common Bow-flower
Pre-1930 only. Australia. E. 3*, 37*, 743. *Toxanthes muelleri* (Sonder) Benth.

myosotidifolia (Benth.) Streetz   Broad-leaved Millotia
Pre-1930 only. Australia, Tasmania. K (holotype). 3, 37c*, 328(1910:22-23), 590*, 1074*, 1078*. *M. depauperata* Stapf.

perpusilla (Turcz.) P.S.Short   Tiny Bow-flower
- A wool casual. Australia. BM, RNG. 17, 37, 583*, 590*, 743, 1074*. *Toxanthes perpusilla* Turcz.

tenuifolia Cass.   Soft Millotia
Pre-1930 only. Australia. E. 3, 37c*, 549*, 583*, 1078*. *M. hispidula* Gand.

AMMOBIUM E.A.Brown
alatum R.Br.
- A casual garden escape. Australia. 371(425:7), 565*, 566*, 590*, 594*, 674*.

MYRIOCEPHALUS Benth.
rhizocephalus (DC.) Benth.   Tufted Woolly-heads
- A wool casual. Australia. HbEJC. 37c*, 583*, 590*, 743, 1074*.

CALOCEPHALUS R.Br.
knappii (F.Muell.) Ewart & J.White
- A wool casual. Australia. RNG, HbTBR. 17, 37c*, 909*(n.s. 22:11). *Eriochlamys knappii* F.Muell.

CASSINIA R.Br.
fulvida Hook.f.   Golden-bush
- Introduced or a garden escape, persistent or established by the side of the main road at Crianlarich (Mid Perth) and in a field on Raasay (N Ebudes). New Zealand. 36c, 51, 62*, 63*, 303(36:28), 458, 686*. Perhaps conspecific with *C. leptophylla* (Forster f.) R.Br.

LEONTOPODIUM (Pers.) R.Br. ex Cass.
alpinum Cass.   Edelweiss
- Introduced on chalk grassland at Pebblecombe (Surrey). Europe. 21, 22*, 23*, 26*, 52*, 199.

LEYSERA L.
capillifolia (Desf.) Sprengel
- A wool casual. S Spain, N Africa, SW Asia. RNG. 17, 21, 44a*, 45*, 46*, 810*(131:369-383). *L. leyseroides* (Desf.) Maire, nom. illeg.; *Asteropterus leyseroides* (Desf.) Rothm.

tenella DC.
- A wool casual. S Africa. E, RNG. 17, 133, 810*(131:369-383). *Asteropterus tenellus* (DC.) Rothm.

PODOLEPIS Labill.
gracilis (Lehm.) Graham   Slender Podolepis
- A wool casual. Australia. RNG. 17, 51, 583*.

longipedata Cunn. ex DC.   Tall Copper-wire-daisy
- A wool casual. Australia. K, RNG. 14, 37c*, 590*.

INULA L.
britannica L.   Meadow Fleabane
- Long established at Cropstone Reservoir (Leics), possibly brought in by waterfowl, but now extinct. Formerly a grain alien. Eurasia. BM, CGE, LANC, LIV, OXF, RNG. 1, 20, 21, 22*, 26*, 29*, 184, 354(5:110).

ensifolia L.
- • A garden escape reported from waste ground at Wisley (Surrey). Europe, SW Asia. OXF. 7, 21, 22*, 53*, 57*, 342(Feb89), 354(5:562).

helenium L.   Elecampane
- •••• An established escape from cultivation; naturalised in widely scattered localities throughout the British Is. Eurasia. 1, 20, 22*, 25*, 29*, 35*.

orientalis Lam.
- • A persistent garden escape on Box Hill Common (N Wilts) and on a railway embankment at Tunbridge Wells (W Kent). SW Asia. HbEJC. 39, 51*, 236, 303(29:12). *I. glandulosa* Mussin-Puschkin ex Willd.

DITTRICHIA Greuter

graveolens (L.) Greuter   Stinking Fleabane
- •• A wool casual. Medit., SW Asia; (Australia). BM, CGE, E, LIV, OXF, RNG. 3*, 14, 20, 22*, 44a*, 590*, 826*(16:96-99). *Erigeron graveolens* L.; *Inula graveolens* (L.) Desf.

viscosa (L.) Greuter   Woody Fleabane
- • Established on roadside waste ground at Landguard Common, Felixstowe (E Suffolk); elsewhere persistent for a few years at ports, or a casual; vector unknown. Medit. BM, NMW, OXF, RNG. 1, 10, 20*, 24*, 31*, 44a*, 303(21:16), 370(15:401), 371(399:39), 414. *Inula viscosa* (L.) Aiton.

PULICARIA Gaertner

arabica (L.) Cass.
- • A wool casual. Egypt, SW Asia. E, RNG. 3, 14, 39, 44a*, 45*, 729*. *Inula arabica* L. Possibly conspecific with *P. paludosa*.

paludosa Link
- Pre-1930 only. Spain, Portugal, NW Africa. BM, OXF. 7, 21, 192, 354(7:37), 737*. *P. arabica* var. *hispanica* Boiss.; *P. uliginosa* Hoffsgg. & Link, non Steven, nec S.F.Gray.

BUPHTHALMUM L.

salicifolium L.   Willow-leaved Yellow-oxeye
- • A persistent garden escape at the edge of Wandsworth Common (Surrey); casual elsewhere. Europe. 20, 21, 22*, 25*, 26*, 27*, 199, 230.

TELEKIA Baumg.

speciosa (Schreber) Baumg.   Yellow-oxeye
- ••• (but confused with *Inula helenium* and *Ligularia dentata*). An established garden escape; well naturalised and widespread on moist soils, especially in Scotland; known for many years near Lovat Bridge at Beauly (Easterness) and on Woodwalton Fen (Hunts). Europe, SW Asia. BM, CGE, E, NMW, OXF, RNG. 7, 16b, 20, 21, 22*, 29*, 57*, 162, 277, 349(5:343). *Buphthalmum speciosum* Schreber.

ASTERISCUS Miller
aquaticus (L.) Less.
- An esparto casual. No modern records. Medit. OXF, SLBI. 1, 15, 21, 24*, 29*, 44a*. *Bubonium aquaticum* (L.) Hill; *Buphthalmum aquaticum* L.; *Odontospermum aquaticum* (L.) Schultz-Bip.

maritimus (L.) Less.
- An esparto casual. W Medit., Canary Is. NWM. 4, 21, 24*, 29*, 30*, 32*, 836(5:307). *Odontospermum maritimum* (L.) Schultz-Bip.

GRINDELIA Willd.
chiloensis (Cornel.) Cabrera   Shrubby Gumweed
 Pre-1930 only. Argentina, Chile. K. 51, 61, 312*(t.9471), 354(11:261), 841*(33:217-219,249), 871*(6:59), 1050*, 1066*. *G. speciosa* Lindley & Paxton.

cf. **decumbens** E.Greene
 Pre-1930 only. SW USA. OXF. 7, 354(4:199), 706, 805*(21:505).

robusta Nutt.   Californian Gumweed
- Probably a garden escape, established abundantly on sandstone cliffs at Whitby (NE Yorks), known there since 1961; also a casual on shingle at Galashiels (Roxburghs). Western N America. E. 7, 71*, 354(5:32), 371(374:8), 650*. *G. rubricaulis* DC. var. *robusta* Steyerm. The name for these plants may need revision. The plant at Whitby is considered in 20 to be *G. stricta* DC., but the phyllaries of this species as described in 42 differ from those drawn in 303*(21:16-17).

squarrosa (Pursh) Dunal   Curly-cup Gumweed
 Pre-1930 only. N America. BM, CLE, NWM, OXF, SLBI. 1, 9, 38*, 41*, 42, 51. Incl. *G. grandiflora* Hook.

CALOTIS R.Br.
cuneifolia R.Br.   Bur-daisy
 •• A wool casual. Australia. BFT, BM, CGE, E, OXF, RNG. 1, 3*, 14, 20, 37*, 303*(31:13,15), 349(1:579), 590*, 591*.

dentex R.Br.   White Bur-daisy
- A wool casual. Australia. K, RNG. 14, 143, 514, 569*, 590*, 591*.

hispidula (F.Muell.) F.Muell.   Hairy Bur-daisy
 •• A wool casual. Australia. BM, CGE, E, LIV, OXF, RNG. 1, 3*, 14, 20, 37*, 583*, 590*. Incl. *C. squamigera* C.White.

lappulacea Benth.   Yellow Bur-daisy
 •• A wool casual. Australia; (New Zealand). BM, CGE, E, LIV, OXF, RNG. 14, 20, 37*, 569*, 583*, 590*.

AMELLUS L.
microglossus DC.
- A wool casual. S Africa. HbEJC. 370(16:168), 835*(13:579-729).

strigosus (Thunb.) Less.
- A wool casual. S Africa. BM, E, LIV, RNG. 3, 17, 51, 370(16:168), 835*(13:579-729), 892*(10:16-17). *A. annuus* hort., non Willd.; *Kaulfussia amelloides* hort., non Nees.

BOLTONIA L'Hér.
asteroides (L.) L'Hér.   False Chamomile
Pre-1930 only. Eastern N America. NMW. 38*, 40, 51, 55*, 57*, 156. Incl. *B. glastifolia* L'Hér.

BELLIUM L.
bellidioides L.   False Daisy
- A casual garden escape. W Medit. 21, 30*, 236, 546*, 650*, 683*.

SOLIDAGO L.
calcicola (Fern.) Fern.
- A garden escape established in woodland at Kilmel Ford (Main Argyll). Eastern N America. ABRN. 38, 38a*, 40*, 436, 1298. *S. virgaurea* L. var. *calcicola* Fern.

canadensis L. sensu lato   Canadian Goldenrod
●●●●● An established garden escape; naturalised in many places throughout the British Is. N America. 2, 20, 21, 22*, 38*, 56*, 1279*. Incl. *S. altissima* L.; *S. elongata* Nutt. A hybrid with native *S. virgaurea* (*S. × niederederi* Khek) has been recorded. 20, 370(13:123). According to 436, most British material is referable to the hybrid *S. canadensis × S. rugosa*. See also 826*(27:7-12).

flexicaulis L.   Zig-zag Goldenrod
- A casual in a garden at Wareham (Dorset), vector unknown. N America. 38*, 40*, 51, 371(409:38), 1210*, 1256*. *S. latifolia* L.

gigantea Aiton   Early Goldenrod
●●●●● An established garden escape; naturalised in many places throughout the British Is. N America. 3, 20, 21, 22*, 29*, 38*, 40*, 1113. Incl. *S. serotina* Aiton, non Retz. According to 436, British material is referable to subsp. *serotina* (Kuntze) McNeill.

graminifolia (L.) Salisb.   Grass-leaved Goldenrod
●● An established garden escape; naturalised in a wood near Barnstaple (N Devon) and on roadsides and riverbanks elsewhere. N America. BM, E, K, LANC, OXF, RNG. 1, 20*, 21, 22*, 38*, 189, 370(15:137), 1210*. *S. graminea* sphalm.; *S. lanceolata* L.; *Euthamia graminifolia* (L.) Elliot.

nemoralis Aiton   Grey Goldenrod
- Recorded from Bromley (W Kent). N America. 38*, 40*, 51, 126, 1113*, 1210*.

odora Aiton   Anise-scented Goldenrod
- A casual or persistent garden escape. Eastern N America. 38*, 40*, 51, 223, 1113*.

rugosa Miller   Rough-stemmed Goldenrod
- A well-established colony near Bowling Harbour and possibly still in Glen Fruin (both Dunbarton), probably a garden escape. N America. E, K. 20, 38*, 40*, 370(14:428;15:69), 1113*, 1210*. *S. altissima* Aiton, non L.

sempervirens L.   Salt-marsh Goldenrod
Pre-1930 only. Eastern N America. 7, 38*, 40*, 51, 354(4:416), 1113*, 1210*.

speciosa Nutt.   Showy Goldenrod
Pre-1930 only. N America. 38*, 40*, 267, 1113*, 1268*. *S. conferta* Miller; *S. sempervirens* Michaux, non L.

tortifolia Ell.   Twisted-leaf Goldenrod
- A persistent garden escape found in 1991 near the bus station in Leicester (Leics). SE USA, Texas. BM. 38*, 371(427:35), 545, 699, 1088*, 1113, 1268*.

HETEROTHECA Cass.

subaxillaris (Lam.) Britton & Rusby   Camphorweed
- A wool casual. USA, Mexico. K. 38*, 40. *Inula subaxillaris* Lam.

DICHROCEPHALA L'Hér. ex DC.

chrysanthemifolia DC.
- A wool casual. Tropical Africa & Asia. BM, RNG. 17, 48*, 604, 676, 892*(10:6-10).

integrifolia (L.f.) Kuntze
- A wool casual. Tropical Africa & Asia. E, K, RNG. 14, 21, 48*, 590*, 676, 892*(10:6-10), 1195*, 1247*. *D. bicolor* (Roth) Schldl.; *D. latifolia* DC.

BRACHYSCOME Cass.
The generic spelling *Brachyscome* Cass. (1816) has priority over *Brachycome* Cass. (1825).

ciliaris (Labill.) Less.   Variable Daisy
- A wool casual. Australia. BM, RNG. 14, 37*, 583*, 590*, 743.

graminea (Labill.) F.Muell.
- A wool casual. Australia. RNG. 14, 37*, 514, 533, 549*, 590*.

iberidifolia Benth.   Swan River Daisy
Pre-1930 only. Australia. BM. 37*, 51*, 583*.

perpusilla (Steetz) J.Black   Tiny Daisy
Pre-1930 only. Australia. E. 3, 37*, 583*, 590*, 743. Incl. *B. collina* (Sonder) Benth.

ASTER L.

ageratoides Turcz.
- A casual garden escape. No modern records. India, E Asia. 43, 51, 217, 617*, 649*, 685, 724*, 1195*. *A. leiophyllus* Franchet & Savat. Incl. *A. microcephalus* (Miq.) Franchet & Savat.; *A. trinervius* Roxb. ex D.Don; *A. viscidulus* Makino.

amellus L.   European Michaelmas-daisy
   • A garden escape established in some quantity on waste ground at a former
   railway station at South Wigston (Leics). Europe, W Asia. 20, 21, 22*, 23*,
   26*, 57*, 430, 458.
ascendens Lindley   Long-leaved Aster
   Pre-1930 only. N America. NMW, OXF, RNG. 7, 38*, 40, 41*, 42,
   354(7:997), 500*. *A. adscendens* Lindley ex DC.
[*carneus* Nees   The correct application of this name defeats us; it is probably a
   colour variant of *A. praealtus*. BM. 2, 354(9:718), 1333.]
concinnus Willd.   Narrow-leaved Smooth-aster
   • Established on waste ground by the railway, Hob Moor, near York
   (MW Yorks); also recorded, with some doubt, in 1991 from near the sea in
   Cards (without locality), possibly as a garden escape. Eastern N America. 20,
   38*, 432, 301*(t.1619), 305(53:19), 1088*, 1314*. Not to be confused with
   A. concinnus Colla, which appears to be a synonym of *A. novae-angliae*.
cordifolius L.   Blue Wood-aster
   • A persistent garden escape near Whittlesey (Cambs) and a garden relic at
   Denver (W Norfolk). Eastern N America. 38*, 40*, 51, 264, 460, 1297*.
   A. paniculatus Willd.
dumosus L.   Bushy Aster
   1930 only. Eastern N America. 20, 38*, 40*, 51, 51a, 354(9 :271), 1297*.
   A. foliosus Pers. A hybrid with *A. novi-belgii* has been recorded. 264.
ericoides L.   Heath Aster
   • (but confused with *A. pilosus*). A casual garden escape. Eastern N America.
   BM, LIV, OXF. 3, 12, 21, 38*, 40*, 192, 336(19:67), 1297*. *A. multiflorus*
   Aiton; *Virgulus ericoides* (L.) Reveal & Keener.
[*foliaceus* Lindley ex DC., non L.   In error for *A. novi-belgii*. 253, 349(6:44).]
junceus Aiton   Rush Aster
   Pre-1930 only. N America. OXF. 7, 38*, 51, 354(5:562).
laevis L.   Glaucous Michaelmas-daisy
   •• An established garden escape in a few widely scattered localities; long
   naturalised by Lough Neagh (Tyrone). N America. BM, E, LANC, RNG,
   SLBI. 1, 20*, 22*, 38*, 40*, 354(9:714), 1297*. Incl. *A. geyeri* (A.Gray)
   Howell. A hybrid with *A. lanceolatus* has been recorded. 135.
lanceolatus Willd.   Narrow-leaved Michaelmas-daisy
   •••• An established garden escape; naturalised in widely scattered localities
   throughout most of the British Is. Eastern N America. 1, 20*, 22*, 126*,
   1297*. *A. lamarckianus* auct., non Nees; *A. paniculatus* auct., non Lam.;
   A. simplex Willd.; *A. tradescantii* auct., non L. Hybrids with *A. laevis* and
   A. puniceus have been recorded, as well as *A. laevis* × *A. lanceolatus* ×
   A. novi-belgii. 336(19:67).
laterifiorus (L.) Britton   Starved Aster
   Pre-1930 only. Eastern N America. 38*, 40, 51, 53*, 354(9:270), 1297*.
   A. diffusus Aiton.
[*longifolius* Lam.   Unconfirmed; probably all records are referable to *A. novi-*
   belgii or *A. × salignus*. 1297*]

macrophyllus L.   Large-leaved Aster
• A garden escape persistent for a few years at Swanley (W Kent), now gone.
Eastern N America. K. 12, 21, 22*, 38*, 40*, 126, 1297*. The record for
Lochside station (Renfrews) was in error for *A. schreberi* Nees. 16,
349(7:510).

novae-angliae L.   Hairy Michaelmas-daisy
•••• An established garden escape; naturalised in scattered localities, mainly
in the south. Eastern N America. 3, 12, 20, 22*, 38*, 40*, 126*. *Virgulus
novae-angliae* (L.) Reveal & Keener.

novi-belgii L.   Confused Michaelmas-daisy
••••• An established garden escape; widely naturalised, but over-recorded.
Eastern N America. 1, 16, 20*. 21, 22*, 25*, 34*, 38*, 126*, 1297*.
A. brumalis Nees; *A. floribundus* auct., ?Willd.; *A. longifolius* auct. pro parte,
non Lam. A hybrid with *A. dumosus* has been recorded. 264.

pilosus Willd.   Frost Aster
• A casual garden escape. Eastern N America. RNG. 22*, 38a*, 40*, 51a,
199, 371(374:9), 1297*. *A. ericoides* auct., non L.

praealtus Poiret
Pre-1930 only. Eastern N America. BM. 38a*, 40*, 51, 51a, 151,
354(5:658), 1297*. *A. salicifolius* Aiton, non Lam.

prenanthoides Muhlenb.   Crooked-stem Aster
• A garden escape established by Annamoe river (Co Wicklow). No modern
records. Eastern N America. OXF. 7, 38*, 40*, 51, 124, 354(4:416), 1297*.

puniceus L.   Red-stalk Aster
•• An established garden escape; naturalised over an extensive area about
Clachan (Kintyre) and by the rivers Greeba and Neb (Man). Eastern
N America. BM, CGE, DUE, LANC, LIV, OXF. 3, 20, 21, 22*, 38*, 40*,
103, 137, 354(9:714), 1297*. *A. firmus* Nees; *A. hispidus* Lam. A hybrid
with *A. lanceolatus* has been recorded. 264.

× **salignus** Willd.   Common Michaelmas-daisy
••• (but under-recorded as *A. novi-belgii*). An established garden escape;
naturalised by streams in widely scattered localities. Originated in cultivation.
BM, CGE, E, NMW, OXF, RNG. 2, 20*, 22*, 33*, 234. *A. lanceolatus* ×
A. novi-belgii; *A.* × *lanceolatus* auct., non Willd.; *A. longifolius* auct. pro
parte, non Lam.

schreberi Nees   Nettle-leaved Michaelmas-daisy
• Introduced, or a garden escape, established since 1931 at Lochside station
(Renfrews). Eastern N America. BM. 20, 22*, 38*, 40*, 349(7:501), 1297*.
A. macrophyllus auct., non L.

sedifolius L.
• A garden escape persistent for a few years by a track near the Blackwater
river, Ash Vale (Surrey). Europe. HbACL. 20, 31, 370(16:447),
371(409:39).

spectabilis Aiton   Showy Aster
- A garden escape persistent for a time on Hayes Common (W Kent). Eastern
  N America. LIV. 12, 38*, 40*, 51, 126.

subulatus Michaux   Annual Saltmarsh-aster
  Recorded in 1992-3 on a wall in St Peter Port, Guernsey (Channel Is).
  N, C & S America; (widespread). K. 21, 37c*, 38*, 450, 533*, 546a*,
  864(36:463-466), 1297*, 1347*. Our plant is var. *sandwicensis* (A.Gray)
  A.G.Jones (*A. squamatus* (Sprengel) Hiern. ex Sodiro) from Florida,
  C & S America; var. *subulatus* is from eastern N America.

[*tradescantii* L. (*A. leucanthemus* Desf., non Raf.)   Unconfirmed; probably
  always in error for *A. lanceolatus*. 156.]

umbellatus Miller   Tall White-aster
- A persistent garden escape near Newton Stewart (Kirkcudbrights); also
  collected from a roadside at Limpsfield (Surrey). Eastern N America. RNG.
  38*, 40*, 51, 371(355:25), 1297*.

× **versicolor** Willd.   Late Michaelmas-daisy
  ••• (but perhaps under-recorded). A garden escape established in widely
  scattered localities. Originated in cultivation. BM, DUE, E, LANC, OXF.
  2, 20*, 22*, 126*, 370(2:45). *A. laevis × A. novi-belgii*; *A. novi-belgii* subsp.
  laevigatus (Lam.) Thell.

vimineus Lam.   Small White-aster
  Pre-1930 only. Eastern N America. OXF. 7, 38*, 40*, 51, 354(7:997).
  Considered by 1297* as conspecific with *A. lateriflorus*.

FELICIA Cass.

abyssinica Schultz-Bip. ex A.Rich.
- A wool casual. Africa. BM. 48*, 51, 676.

amelloides (DC.) Voss   Blue Marguerite
  A casual recorded only in 1930. S Africa. OXF. 51, 54*.

amoena (Schultz-Bip.) Levyns
- A casual, probably a garden escape. S Africa. RNG. 51*, 51a, 312*(n.s.
  t.239). *F. pappei* (Harvey) Hutch.; *Aster pappei* Harvey.

angustifolia (Jacq.) Nees
  Pre-1930 only. S Africa. BM. 354(5:228), 592, 889*(6:700), 1131*, 1162*,
  1201*, 1207*. *Aster angustifolius* Jacq., non Lindley. Perhaps conspecific
  with *F. hyssopifolia* (Berger) Nees.

bergeriana (Sprengel) O.Hoffm.   Kingfisher Daisy
- A casual garden escape. S Africa. 51, 53*, 66*, 220, 354(9:270), 562*,
  655*. *Aster bergerianus* (Sprengel) Harvey.

heterophylla (Cass.) Grau
- A wool casual. S Africa. K. 51*, 51b, 301*(t.490), 312*(t.2177), 655*,
  1274*. *Charieis heterophylla* Cass.; *Kaulfussia amelloides* Nees.

[*petiolata* (Harvey) N.E.Br. (*Aster petiolatus* Harvey)   In error for *Erigeron
  karvinskianus*. 119, 354(11:31).]

tenella (L.) Nees
- A casual garden escape. Formerly a wool alien. S Africa. E. 3*, 51, 143, 312*(t.33), 561*, 650*. *F. fragilis* Cass.; *Aster tenellus* L.

VITTADINIA A.Rich.
triloba (Gaudich.) DC. sensu lato   Fuzzweed
- A wool casual. Australasia. BFT, BM, CGE, E, LIV, RNG. 14, 36, 36c*, 37*, 145, 569*, 583*. Incl. *V. australis* A.Rich.

CHRYSOCOMA L.
coma-aurea L.   Shrub Goldilocks
- A garden escape established at Porth Seal on St Martin's, and in the dunes at Pentle Bay on Tresco (both Scilly). S Africa. BM, E, RNG. 3*, 20, 212, 312*(t.1972), 458, 1117*. Pre-1930 wool alien records may be referable to *C. tenuifolia*.

tenuifolia P.Bergius   Fine-leaved Goldilocks
- • A wool casual. S Africa. BFT, BM, CGE, E, RNG, TCD. 14, 20, 145, 577a, 578a*, 892*(10:14-16), 1019*.

ERIGERON L.
annuus (L.) Pers.   Tall Fleabane
- • • An established garden escape; naturalised on rough ground, as on Longmoor airfield (N Hants) and in a quarry on the Isle of Portland (Dorset); elsewhere a weed of newly sown grass. N America; (widespread). BM, CGE, E, LANC, OXF, RNG. 2, 20*, 22*, 26*, 38*, 303(13:23), 353(13:27), 370(11:399), 371(391:34), 683*, 825(2:9-12; 12*:51-56), 1261*. *Phalacroloma annuum* (L.) Dumort.; incl. *E. septentrionalis* Fern. & Wieg.; *E. strigosus* Mühlenb. ex Willd.

caucasicus Steven
Pre-1930 only. SW Asia. 7, 39, 51, 354(4:416; 5:32), 842*(1884:112), 922*(9:318), 1110. Incl. *E. amphilobus* Ledeb.; *E. pulchellus* (Willd.) DC., non Michaux f.; *Aster pulchellus* Willd.

divaricatus Michaux
Recorded without date or locality as a casual in Yorkshire. No modern records. N America. 38*, 40*, 197.

glaucus Ker Gawler   Seaside Daisy
- • • An established garden escape; naturalised in a few places, mainly on cliffs on the south coast of England; increasing. Western N America. BM, CGE, E, LANC, RNG. 13, 20, 41*, 212, 354(13:29).

karvinskianus DC.   Mexican Fleabane
- • • • A garden escape well established on walls in southern England, Wales and Ireland; known for over 100 years at St Peter Port on Guernsey, and now abundant in the Channel Is; increasing. Mexico. 2, 20, 22*, 30*, 207, 220, 1279*. *E. mucronatus* DC.; *Vittadinia triloba* hort., non (Gaudich.) DC.

philadelphicus L.   Robin's-plantain
●●● An established garden escape; naturalised in quantity on banks in a few
places in the south; long established on walls at Llandaf (Glam) and Tintern
(Mons). N America. BM, CGE, EXR, NMW, OXF, RNG. 7, 20, 38*, 41*,
156, 350*(47:43).
[*pseudoelongatus* Rouy   Now considered conspecific with native *E. acer*.
354(10:803), 1281.]
speciosus (Lindley) DC.   Garden Fleabane
● A garden escape established on an old wall in Knole Park (W Kent); also
recorded from Pett Level (E Sussex), Shalford and Godalming (both Surrey)
and from dunes between Ainsdale and Birkdale (S Lancs).   Western
N America. E, LANC, OXF. 7, 20, 41*, 54*, 236, 251, 371(385:34), 650*.
Incl. *E. macranthus* Nutt.   Various *Erigeron* cultivars are listed under this
species and may have been overlooked; most of them are of hybrid origin.
subtrinervis Rydb.
Pre-1930 only. Western N America. OXF. 38*, 41*.
[*uniflorus* L.   Dubious records.   155, 354(5:767), 361(7:263).]

CONYZA Less.
blakei (Cabrera) Cabrera
● A wool casual. Brazil, Uruguay, Argentina. BM, RNG. 17, 534*, 535,
546a*, 1027, 1050*. *Erigeron blakei* Cabrera; *E. coronopifolius* Sennen, non
Conyza coronipifolia Kunth.
bonariensis (L.) Cronq.   Argentine Fleabane
●●● (but confused with *C. sumatrensis*). A wool casual. Tropical America;
(widespread). BM, CGE, E, NMW, OXF, RNG. 3*, 14, 20, 40*, 44a*,
370(17:145-148), 546a*, 1247*.   *C. ambigua* DC.; *Erigeron bonariensis* L.;
E. crispus Pourret; *E. linifolius* Willd.   The hybrid with *C. canadensis*
(*C.* × *mixta* Fouc. & Neyraut; *C.* × *flahaultiana* Sennen) may have been
overlooked. 1332.
canadensis (L.) Cronq.   Canadian Fleabane
●●●●● Abundantly established on cultivated and waste ground, dunes and
walls, especially in the southeast; first recorded in the London area in 1690, but
still almost unknown in Ireland. N America; (widespread). 1, 20, 22*, 25*,
26*, 35*, 268. *Erigeron canadensis* L.   A hybrid with native *Erigeron acer*
(× *Conyzigeron huelsenii* (Vatke) Rauschert; *E.* × *huelsenii* Vatke) has been
recorded. HbEJC. 20, 354(8:117).
japonica (Thunb.) Less.   Japanese Fleabane
● A wool casual. E Asia. RNG, HbEJC. 17, 43, 649*, 617*.
sumatrensis (Retz.) E.Walker   Guernsey Fleabane
●●●● Well established and spreading in the Channel Is, about Yeovil
(S Somerset), along the Thames estuary and in the London area, probably as
an extension of its range in N France; also a wool casual. Also, since 1990 at
Dublin port (Co Dublin). Peru; (widespread). 17, 20, 220*, 303*(37:16-17
as *C. bonariensis*; 62:38-39), 331(64:113; 65:193), 370(17:145-148), 533*,
546a*, 1247*.   *C. albida* Willd. ex Sprengel; *C. bonariensis* auct., non (L.)

Cronq.; *C. floribunda* Kunth; *C. naudinii* Bonnet; *Erigeron floribundus* (Kunth) Schultz-Bip.; *E. sumatrensis* Retz. Intermediates between this species and *C. canadensis* have been recorded. 303(64:22).

CALLISTEPHUS Cass.
chinensis (L.) Nees   China Aster
●●● A casual garden escape. China. BM, LIV, OXF, RNG. 2, 20, 54*, 55*, 56*. *C. hortensis* Cass.; *Aster chinensis* L.; *Callistemma chinensis* (L.) Druce.

OLEARIA Moench
× **haastii** Hook.f.   Daisybush
• A garden escape introduced and self-sown in Darenth Wood (W Kent) and self-sown on Weymouth harbour wall (Dorset). New Zealand. RNG. 20*, 54*, 61, 63*, 236, 253, 303(26:16). *O. avicenniifolia* (Raoul) Hook.f. × *O. moschata* Hook.f. The record for Lerwick (Shetland) was in error for *Ligustrum ovalifolium*. 253.

ilicifolia (Hook.f.) Hook.f.   Maori-holly
• A persistent garden escape at Beachy Head (E Sussex). New Zealand. 36, 52*, 61, 371(343:29), 560*, 686*, 709*.

macrodonta Baker   New Zealand Holly
●● An established garden escape; naturalised in a few places in the south and west, as in scrub in a coastal ravine at Knock Bay (Wigtowns), on dunes at Llandudno (Caerns) and on roadsides at Cahersiveen (Kerry). New Zealand. BM, GLAM, RNG. 13, 20, 52*, 54*, 61*, 63*, 102, 126, 220, 415, 458.

nummulariifolia (Hook.f.) Hook.f.   Sticky Daisybush
• A persistent garden escape high on the hills south of Carlingford Lough (Co Louth). New Zealand. BM, RNG. 36, 53*, 55*, 61, 371(352:26), 686*.

paniculata (Forster & Forster f.) Druce   Forster's Daisybush
• A garden escape persistent in an old quarry in Guernsey, and used as a hedging plant in Guernsey and Alderney (both Channel Is), where occasional seedlings occur. New Zealand. 20*. 36, 61, 117*, 220, 370(8:183-194), 629*, 686*. *O. forsteri* (Hook.f.) Hook.f.

solandri Hook.f.   Coastal Daisybush
• A garden escape persistent for at least 10 years on Hayle Towans (W Cornwall); seedlings have been recorded from Penzance (W Cornwall). New Zealand. 20, 36, 213, 371(403:39-40), 686*, 709*.

traversii (F.Muell.) Hook.f.   Chatham Island Daisybush
• A persistent garden escape or relic in Scilly and the Channel Is, much used as a windbreak; though it seldom flowers, small plants thought to be self-sown have been recorded for Scilly. Chatham Is. RNG. 20*, 36, 61, 207, 220, 629*, 1117*.

BACCHARIS L.
halimiifolia L.   Tree Groundsel
• An established garden escape; known since 1924 on the shore at Mudeford (S Hants); also recorded from the shore at Hamworthy (Dorset). Eastern

N America. BM, OXF. 7, 20*, 38*, 61, 349(4:169), 354(7:774; 13:59), 371(391:33), 586*, 590*.

BELLIS L.
annua L.
Pre-1930 only. Medit. OXF. 7, 21, 24*, 29*, 44a*, 354(6:24).

ISOETOPSIS Turcz.
graminifolia Turcz.   Grass-cushion
Pre-1930 only. Australia. E. 3, 7, 37*, 354(5:34), 533*, 549*, 583*.

TANACETUM L.
balsamita L.   Costmary
• An established garden escape; naturalised on waste land at Southport (S Lancs) and Gosford Bay (E Lothian); also recorded from Mitcham Junction station (Surrey). SW Asia. OXF, LIV. 2, 20, 21, 39, 80*, 88*, 199, 251, 349(7:638), 546a*, 650*, 695*. Much confused nomenclaturally, the taxon without ray florets being known as *T. balsamita* L. sensu stricto (*Balsamita major* Desf.) and the rayed form as *Chrysanthemum balsamita* L. (*Pyrethrum balsamita* (L.) Willd.; *T. balsamitoides* Schultz-Bip.).
cinerariifolium (Trev.) Schultz-Bip.   Dalmatian Pyrethrum
Pre-1930 only. Yugoslavia, Albania. E. 21, 31*, 68*, 112, 617*, 695*. *Chrysanthemum cinerariifolium* (Trev.) Vis.; *Pyrethrum cinerariifolium* Trev.
coccineum (Willd.) Grierson   Garden Pyrethrum
• A garden escape. SW Asia. 39, 51, 54*, 55*, 199. *Chrysanthemum coccineum* Willd.; *C. roseum* Adams; *Pyrethrum coccineum* (Willd.) Vorosch.; *P. roseum* (Adams) Bieb.
corymbosum (L.) Schultz-Bip.   Scentless Feverfew
Pre-1930 only. Europe, Caucasus. LIV, NMW, OXF, SLBI. 2, 21, 22*, 26*, 39*, 156, 234, 248, 286, 354(4:80). *Chrysanthemum corymbosum* L.; *Pyrethrum corymbosum* (L.) Willd.; incl. *C. italicum* L. Some early records are in error for native *T. vulgare*.
leucophyllum Regel
• A casual on a refuse tip at Dartford (W Kent), vector unknown. Turkestan. 51, 236, 824*(t.1064).
macrophyllum (Waldst. & Kit.) Schultz-Bip.   Rayed Tansy
•• An established garden escape; recorded from widely scattered localities; known since 1912 at Jervaulx Abbey (NW Yorks). SE Europe, SW Asia. BM, JSY, OXF. 2, 20*, 22*, 258, 303(53:33), 371(400:36), 695*. *Chrysanthemum macrophyllum* Waldst. & Kit.
parthenium (L.) Schultz-Bip.   Feverfew
••••• An established garden escape throughout the British Is, but rarely far from gardens; an ancient introduction as a medicinal herb. Eurasia. 1, 20, 22*, 33*, 35*. *Chrysanthemum parthenium* (L.) Bernh.; *Matricaria parthenium* L.

NEOPALLASIA Polj.
pectinata (Pallas) Polj.
- • A wool casual. Asia. HbEJC. 370(15:111), 617*, 630*. *Artemisia pectinata* Pallas.

SERIPHIDIUM (Besser ex Hook.) Fourr.
compactum (Fischer ex DC.) Polj.
Pre-1930 only. C Asia. 2, 3, 630, 1349. *Artemisia compacta* Fischer ex DC.

ARTEMISIA L.
abrotanum L.   Southernwood
- •• A persistent garden escape. Origin obscure; (Europe). BM, OXF. 2, 20, 38*, 71*, 80*, 236, 253, 595b*, 623*. *A. paniculatum* Lam.
afra Jacq.
- • A wool casual. Tropical & S Africa. BM, E, K, LTN, RNG. 3, 14, 17, 145, 370(15:110), 742*, 1162*, 1294*.
anethifolia Weber
- • A wool casual. Asia. BM, HbEJC. 370(15:111), 579*, 617, 630*, 1313. Not to be confused with the allied Asian species *A. anethoides* Mattf. 617*.
annua L.   Annual Mugwort
- •• A wool and bird-seed casual. Europe, SW Asia; (N America). BM, JSY, K, OXF, RNG. 2, 14, 20*, 38*, 39*, 201, 251, 617*, 826(25:28-36).
[*artemisiifolia* sphalm. In error for *Ambrosia artemisiifolia*.]
biennis Willd.   Slender Mugwort
- ••• An established wool, grain and oil-seed alien; naturalised on the banks of the Chew Valley Reservoir (N Somerset) and Arlington Reservoir (E Sussex), and persistent for at least 40 years by the river at Bath (N Somerset). Asia, N America; (widespread). BM, CGE, E, NMW, OXF, RNG. 1, 3*, 20*, 38*, 39*, 41*, 303(37*:26-27; 45:25), 331*(63:69-70), 350(31:30), 370(15:111), 617*. Incl. *A. tournefortiana* Reichb.
[*caerulescens* L., incl. *A. gallica* Willd. Dubious records. 1, 123, 354(5:769), 358(1874:136,163).]
chamaemelifolia Villars
Pre-1930 only. Eurasia. 21, 39*, 239, 525*, 546*, 573*, 1317*.
dracunculus L.   Tarragon
- •• A persistent garden escape; also a wool casual. Eurasia, N America. BM, HTN, LIV, OXF, RNG, TRU. 2, 14, 20, 22*, 80*, 88*, 325*(110:237-240), 579*, 617*, 715*. Incl. *A. dracunculoides* Pursh; *A. glauca* Pallas ex Willd.
herba-alba Asso   Pyrenean Wormwood
- • An esparto casual. Medit., SW Asia. E. 21, 23*, 39*, 44a*, 370(15:110), 528*.
longifolia Nutt.
Pre-1930 only. N America. SLBI. 2, 38*, 40, 192, 300(1907:40).

ludoviciana Nutt.   Western Mugwort
  • (but confused with *A. stelleriana*). A garden escape and possibly an oil-seed
  alien, casual or persistent. N America. GL, NMW. 4, 10, 38*, 51, 53*, 54*,
  55*, 192, 354(4:16). Incl. *A. gnaphalodes* Nutt.; *A. purshiana* Besser.
macrantha Ledeb.
  • A casual at Hull docks (SE Yorks), vector unknown. No modern records.
  E Europe, Asia. 21, 285, 636*.
pontica L.   Roman Wormwood
  • A garden escape and wool casual. Europe, SW Asia; (N America). BM,
  CGE, LANC, LIV, NMW, OXF. 1, 16, 21, 22*, 38*, 156, 192, 199,
  331(70:160), 370(15:110), 525*, 595b*, 623*.
[*rupestris* Willd.   In error for *Chenopodium multifidum*.   354(7:593),
  362(1922:48).]
scoparia Waldst. & Kit.
  • A wool casual. Eurasia. BM, E, K, OXF, RNG, SLBI. 1, 3, 16, 17, 21,
  39*, 212, 370(15:111), 617*, 623a*, 715*, 741*.
sericea Weber
  Pre-1930 only. E Europe, Asia. 2, 21, 300(1907:40), 579*, 1313.
stelleriana Besser   Hoary Mugwort
  •• (but confused with *A. ludoviciana*).   An established garden escape;
  naturalised in a few places on coastal dunes in southern England and
  SW Scotland; known for about 90 years on Marazion beach (W Cornwall), but
  now gone. NE Asia, Western N America. BM, E, K, LANC, OXF, RNG.
  1, 20, 21, 22*, 33*, 38*, 212, 304(4:6-9; 5*:1,7), 354(5:229).
  A. ludoviciana hort., non Nutt.
tschernieviana Besser
  Pre-1930 only. E Europe. 2, 21, 300(1896:237), 715*. *A. arenaria* DC.
verlotiorum Lamotte   Chinese Mugwort
  •••• Established, as an extension of its spread through Europe; naturalised and
  locally abundant on roadsides and waste ground, mostly in SE England,
  especially in the London area; increasing. First found in 1908. SW China;
  (widespread). 12, 20, 22*, 33*, 35*, 191, 370(1:209-223). A hybrid with
  native *A. vulgaris* has been recorded. 20, 331(69:140).

LASIOSPERMUM Lagasca
bipinnatum (Thunb.) Druce   Royal Down-flower
  • A wool casual. S Africa. BM, E, LTN, RNG. 14, 51, 133, 143, 549,
  565*, 578a*, 628*, 1294*. *L. pedunculare* Lagasca; *L. radiatum* Trev.

ATHANASIA L.
crithmifolia L.
  • A bird-seed and wool casual. S Africa. BM, RNG. 17, 19, 51, 432, 628*.
trifurcata L.
  • A wool casual. S Africa. CGE, E, K, RNG. 17, 133, 592, 628*, 1064*,
  1274*.

SANTOLINA L.

chamaecyparissus L. Lavender-cotton
•• An established garden escape; naturalised on sandy shores at Upton Towans and Constantine Bay (both W Cornwall). Medit. BM, CGE, LANC, NMW, OXF, RNG. 1, 20, 23*, 39*, 52*, 212, 1279*. Subsp. *tomentosa* (Pers.) Arcang. (*S. tomentosa* Pers.) has been recorded from Leckwith (Glam). NMW. 305(54:26).

rosmarinifolia L. Holy-flax
Pre-1930 only. SW Europe. RNG. 7, 21, 30*, 61*, 354(4:200).

ACHILLEA L.

ageratum L. Sweet Yarrow
• A casual or persistent garden escape. W Medit. BM, RNG. 21, 29, 45*, 303(22:12), 331(68:151).

alpina L.
• A casual on dockland at Hull (SE Yorks) and Bristol (W Gloucs), vector unknown. No modern records. Russia, Asia. BM. 43, 51, 285, 354(10:471), 638*. Incl. *A. sibirica* Ledeb.

asplenifolia Vent.
Pre-1930 only. C & SE Europe. BM, OXF. 1, 21, 192, 623a*, 715*.

cartilaginea Ledeb. ex Reichb.
Pre-1930 only. E Europe, Caucasus, Siberia. 21, 116, 623a*, 715*. *A. ptarmica* L. var. *cartilaginea* (Ledeb. ex Reichb.) J.F.Bevis & W.H.Griffin; *Ptarmica cartilaginea* (Ledeb.) Ledeb. ex Reichb. Not to be confused with *A. cartilaginea* Schur, which is synonymous with native *A. ptarmica*.

cretica L. Chamomile-leaved Lavender-cotton
Pre-1930 only. Crete, Aegean Is, Turkey. 2, 4, 21, 263, 320(1907:45), 658*. *Santolina anthemoides* L.

crithmifolia Waldst. & Kit.
Pre-1930 only. SE Europe. 7, 21, 39, 267, 623a*, 715*, 741*.

decolorans Schrader Serrated Yarrow
• A casual garden escape. Originated in cultivation, perhaps of hybrid origin. BM, CGE, LANC, LCN, OXF, SLBI. 1, 51, 1301, 1310*. *A. ptarmica* var. *decolorans* (Schrader) ined.; *A. serrata* Smith.

distans Waldst. & Kit. ex Willd. Tall Yarrow
•• (but confused with *Tanacetum macrophyllum*). A persistent garden escape. C Europe. BM, DUE, OXF, SLBI. 1, 20, 27*, 248, 549*, 1310*. *A. tanacetifolia* All., non Miller. Modern records from Edinburgh, Herts, W Kent and Devon are in error for *Tanacetum macrophyllum*. 371(343:27; 352:27; 367:30).

filipendulina Lam. Fern-leaf Yarrow
•• A persistent garden escape. SW & C Asia. BM, OXF, RNG. 7, 12, 20, 39*, 51, 51a, 53*, 55*, 56*, 370(13:167). *A. eupatorium* M.Bieb. A putative hybrid with *A. clypeolata* hort., non Sibth. & Smith, has been collected. BM.

[*grandiflora* Friv. In error for *Tanacetum macrophyllum*. 371(367:30; 388;36).]

ligustica All.   Southern Yarrow
  • Established for 30 years at Newport docks (Mons), vector unknown. Medit.
  BM, LCN, NMW, OXF. 1, 20*, 21, 156, 270, 370(1:119; 12:287), 546a*.
[*macrophylla* L.   In error for *Tanacetum macrophyllum*.   4, 371(334:22; 367:30).]
[*magna* L. (*A. compacta* Lam.) Now considered to be conspecific with native
  A. millefolium, and not worthy of even varietal rank. 1, 354(4:15-16).]
micrantha Willd.
  Pre-1930 only. Russia, Asia. LCN, OXF. 7, 21, 160, 354(6:25), 518, 524,
  579*, 1371*. *A. aegyptiaca* S.Gmelin; *A. gerberi* Willd. Our plant is
  A. micrantha Willd.(1789). *A. micrantha* Willd.(1803) and *A. micrantha* Bieb.
  are synonyms of *A. biebersteinii* Afan. 21, 39, 546*.
nobilis L.
  •• A casual or persistent wool and grain alien, possibly also a garden escape.
  Eurasia. BM, CGE, E, LANC, OXF, RNG. 1, 17, 20, 21, 22*, 31*, 51,
  145, 182, 546a*, 623a*.
ochroleuca Ehrh.
  Pre-1930 only. E Europe. 7, 21, 354(3:164), 623a*. *A. pectinata* Willd.,
  non Lam.
pannonica Scheele
  Pre-1930 only. C & SE Europe. 9, 21, 623a*. *A. millefolium* L. subsp.
  collina (Becker ex Reichb.) E.Weiss var. *lanata* Koch.
santolina L.
  Pre-1930 only. E Medit. 7, 44a*, 46, 300(1905:100, 1906:103).
[*stricta* (Koch) Schleicher ex Gremli (*A. lanata* Sprengel) A dubious record.
  354(2:294).]
tomentosa L.   Yellow Milfoil
  • A casual or persistent garden escape. SW Europe. BM, CGE, GLR, LIV,
  OXF, SLBI. 1, 20, 21, 26*, 29*, 55*, 403.

CHAMAEMELUM Miller
mixtum (L.) All.
  •• A bird-seed, grain, wool and granite casual. Medit. BM, E, NMW, OXF,
  RNG, SLBI. 1, 16, 17, 19, 21, 39*, 44a*, 45*, 46*, 182, 303(16:16),
  323(25:434). *Anthemis mixta* L.; *Ormenis mixta* (L.) Dumort.

ANTHEMIS L.
abrotanifolia (Willd.) Guss.
  Pre-1930 only. Crete. BM. 2, 21.
altissima L.
  • A grain and carrot seed casual. Formerly a wool casual. Eurasia. ABD,
  BM, CGE, E, OXF, RNG. 1, 3, 11, 21, 39*, 133, 277, 331(63:143). *A. cota*
  auct., ?an L.; *Cota altissima* (L.) Gay.
austriaca Jacq.   Austrian Chamomile
  • A wool, grass seed and wild flower seed casual; possibly also a bird-seed
  alien. C & E Europe, SW Asia. BM, CGE, K, LIV, OXF, RNG. 1, 17, 21,
  26*, 39*, 251, 303(19:12; 59:47; 60:44), 331(70:160).

[*chamomilla* Willd.   Claimed as an alien from S Europe, but it is a synonym of native *Matricaria recutita*.   7, 139.]

cretica L.   Mountain Dog-daisy
  • A casual in a lawn, perhaps from grass seed.   No modern records.   Medit.
  BM.   21, 23*, 27*.   Incl. *A. montana* L., nom. illeg.

hyalina DC.
  • A grain casual.   No modern records.   SW Asia.   39*, 44a*, 350(28:173),
  370(2:107), 826*(11:45-51).   *A. crassipes* Boiss.

leucanthemifolia Boiss. & Blanchet
  Pre-1930 only.   Palestine, S Lebanon.   1, 44a*, 524, 528*, 667*.

[*macrantha* Heuffel   In error for *A. punctata*.   156.]

maritima L.
  Pre-1930 only.   W Medit.   OXF.   2, 21, 112, 300(1904:110), 737*.

muricata (DC.) Guss.
  Pre-1930 only.   Sicily.   BM, OXF.   7, 21, 313(22:40), 354(4:200,295), 683*.

punctata Vahl   Sicilian Mayweed
  ••• An established garden escape; naturalised on a few cliffs, railway banks
  and roadsides in the south, especially in Cornwall and the Isle of Wight; known
  since 1923 on cliffs near Fishguard (Pembs).   Sicily, NW Africa.   BM, E,
  NMW, OXF, RNG, SLBI.   20*, 21, 54*, 156, 212, 436, 594*, 657*.   Incl.
  A. cupaniana Tod. ex Nyman.   Probably all records are referable to subsp.
  cupaniana (Tod. ex Nyman) Ros.Fernandes.

ruthenica M.Bieb.
  •• A grain and bird-seed casual.   E Europe, Caucasus.   BM, LIV, OXF.   19,
  21, 39, 182, 683, 695*, 715*.   *A. neilreichii* Ortm.   The record for Fishguard
  (Pembs) was in error for *A. punctata*.   156.

tinctoria L.   Yellow Chamomile
  •••• A persistent garden escape; possibly also a bird-seed alien.   Europe,
  SW Asia.   1, 20*, 22*, 26*, 34*.   *Cota tinctoria* (L.) Gay; incl. *A. discoidea*
  (All.) Willd.

tomentosa L.   Woolly Chamomile
  Pre-1930 only.   Medit.   OXF.   7, 21, 24*, 109, 286, 354(7:38), 615*.   Incl.
  A. peregrina L.

triumfetti (L.) All.   Southern Dog-daisy
  • Recorded in 1989 from waste ground at Kew (Surrey).   S Europe, SW Asia.
  RNG.   9, 21, 23*, 39*, 342(Feb90).   *Cota triumfetti* (L.) Gay.

wiedemanniana Fischer & C.Meyer
  • A casual at Spirthill (N Wilts) and at Ashton Gate, Bristol (N Somerset),
  vector unknown.   No modern records.   Turkey.   BM, BRIST, K, LIV, OXF,
  RNG.   7, 39*, 172, 354(13:297).   *A. ormenioides* Boiss.; *Cota ormenioides*
  (Boiss.) Boiss.

CHRYSANTHEMUM L.
coronarium L.   Crown Daisy
  ●●● A wool, grain and bird-seed casual; also a garden escape. Medit. BM,
  CGE, E, NMW, OXF, RNG. 1, 14, 19, 20*, 30*, 44a*, 55*, 182.
segetum L.   Corn Marigold
  ●●●●● An established grain alien, plentiful throughout the British Is, with a
  preference for sandy, arable fields. Eurasia; (widespread). 1, 20, 22*, 33*,
  35*, 44a*.

ISMELIA Cass.
carinata (Schousboe) Schultz-Bip.   Tricolor Chrysanthemum
  ●● A casual garden escape and bird-seed alien. N Africa. OXF, RNG. 16,
  19, 20, 21, 51*, 53*, 54*, 220, 258, 354(8:618). *Chrysanthemum carinatum*
  Schousboe; *C. tricolor* Andrews.   This genus may be better sunk into
  Chrysanthemum.

LEUCANTHEMELLA Tzvelev
serotina (L.) Tzvelev   Autumn Oxeye
  ●●● An established garden escape; naturalised in damp areas on heaths,
  commons and dunes in the south. E Europe. BM, OXF, RNG. 7, 13, 20*,
  31*, 55*, 57*, 199, 209, 826*(16:92-95). *Chrysanthemum serotinum* L.;
  C. uliginosum (Waldst. & Kit. ex Willd.) Pers.; *Leucanthemum uliginosum*
  (Pers.) ined.; *Pyrethrum uliginosum* Waldst. & Kit. ex Willd.; *Tanacetum
  serotinum* (L.) Schultz-Bip.

COLEOSTEPHUS Cass.
myconis (L.) Reichb.f.
  ● A wool and granite casual. Medit. BM, CGE, E, NMW, OXF, RNG. 2,
  17, 21, 29*, 30*, 44a*, 45*, 147, 303(10:15-17). *Chrysanthemum myconis* L.

LEUCANTHEMUM Miller
atratum (Jacq.) DC.   Saw-leaved Moon-daisy
  Pre-1930 only. Europe. 7, 21, 23*, 26*. *Chrysanthemum atratum* Jacq.; incl.
  L. coronopifolium (Villars) Gren. & Godron; *C. coronopifolium* Villars.
lacustre (Brot.) Samp.   Portuguese Swamp-daisy
  ●● A persistent garden escape. Portugal. BM, OXF, RNG. 7, 21,
  315*(5:589), 371(331:20; 340:28; 376:25), 510*. *Chrysanthemum lacustre*
  Brot.; *C. latifolium* DC.
maximum (Ramond) DC.
  ● Recorded for S Lancs. Pyrenees. 370(18:89), 1290*. *Chrysanthemum
  maximum* Ramond. Other records are probably all in error for *L. × superbum*.
  370(18:89).
monspeliense (L.) Coste
  Pre-1930 only. SW Europe. SLBI. 7, 21, 300(1908:102), 546*, 1290*,
  1293*. *Chrysanthemum monspeliense* L.; *L. cebennense* DC.
paludosum (Poiret) Bonnet & Barratte
  Pre-1930 only. SW Europe, NW Africa. LANC. 21, 46*, 556*, 737*,
  826(26:107-109), 1027*. *L. murcicum* Gay, nom. nud.; *L. setabense* DC.;

Chrysanthemum glabrum Poiret; *C. paludosum* Poiret; *Hymenostemma fontanesii* Willk. A polymorphic species.

× **superbum** (Bergmans ex J.Ingram) Kent   **Shasta Daisy**
●●●● An established garden escape; on roadsides, railway banks and in disused quarries, in widely scattered localities. Originated in cultivation. BM, OXF, SLBI. 7, 20*, 22*, 53*, 54* (all except 20 as *L. maximum*). *L. lacustre* × *L. maximum*; *L. maximum* auct., non (Ramond) DC.; *Chrysanthemum maximum* auct., non Ramond. A fertile hybrid mis-named, until recently, as *L. maximum*.

ARGYRANTHEMUM Webb
frutescens (L.) Schultz-Bip.   **Paris Daisy**
● A casual garden escape. Canary Is. OXF. 7, 36c, 51, 66*, 151, 300(1908:102), 303(50:27), 530*, 617*. *Chrysanthemum frutescens* L.; *C. grandiflorum* (Willd.) DC.

DENDRANTHEMA (DC.) Des Moul.
× **grandiflorum** group   **Florists' Chrysanthemums**
●● (but probably under-recorded). Casual garden escapes. Originated in cultivation. OXF. 7, 51*, 55*, 66*, 67*, 236, 331(37:203), 617*. Incl. *D.* × *grandiflorum* (Ramat.) Kitam.; *D. indicum* (L.) Des Moul.; *D. morifolium* (Ramat.) Tzvelev; *Chrysanthemum indicum* hort., non L., nec Thunb.; *C. morifolium* Ramat.; *C. sinense* Sabine; *C. vestitum* (Hemsley) Stapf.

Korean Hybrids   **Korean Chrysanthemums**
● (but probably under-recorded). Casual or persistent garden escapes. Originated in cultivation. 55, 475, 732a, 1163, 1237*, 1334. Of hybrid origin between *D.* × *grandiflora* group and *D. zawadskii* group; *Chrysanthemum coreanum* hort., non (Lév. & Vaniot) Vorosch. Incl. *D.*'Wedding Sunshine'.

zawadskii group   **Rubellum Chrysanthemums**
● (but probably under-recorded). A casual garden escape. Originated in cultivation. 21, 51*, 53*, 236. Incl. *D. zawadskii* (Herbich) Tzvelev; *D.* × *rubellum* (Sealy) Philp; *Chrysanthemum erubescens* Stapf; *C.* × *rubellum* Sealy; *C. zawadskii* Herbich.

LONAS Adans.
annua (L.) Vines & Druce   **African Daisy**
Pre-1930 only. C Medit. BM, OXF. 7, 21, 46*, 51, 52*, 354(5:287). *L. inodora* (L.) Gaertner; *Athanasia annua* L.

ANACYCLUS L.
clavatus (Desf.) Pers.
● A wool and grain casual. No modern records. Medit. BM, NMW, OXF, RNG. 1, 8, 14, 16, 21, 24*, 30*, 39*, 45*, 182, 350(28:386), 354(10:710). *A. tomentosus* (Gouan) DC.; incl. *A. marginatus* Guss.

radiatus Lois.
• A grain and wool casual. No modern records. Medit. BM, CGE, LCN,
NMW, OXF, RNG. 1, 9, 14, 21, 31*(p.446), 44a*, 45*, 182, 354(10:710).
A. medians Murb.; *A. officinarum* Hayne; *A. pallescens* Guss.;
A. purpurascens DC.
valentinus L.
• An esparto and grain casual. W Medit. BM, CGE, NMW, OXF, SLBI.
2, 15, 16, 21, 30*, 45*, 46*, 182. Incl. *A. dissimilis* Pomel;
A. prostratus Pomel; *Anthemis valentinus* (L.) Willd.

CLADANTHUS Cass.
arabicus (L.) Cass.
• A casual in a siding at Duns (Berwicks), vector unknown. S Spain,
N Africa. OXF. 2, 21, 30*, 32*, 45*, 54*, 55*, 300(1905:100),
312*(n.s. t.490), 354(8:400), 371(328:20). *Anthemis arabicus* L.; incl.
C. canescens Sweet; *C. proliferus* DC., nom. illeg.

PENTZIA Thunb.
calcarea Kies
• A wool casual. S Africa. BM, RNG. 17, 830*(11:118-119).
cooperi Harvey
• A wool casual. S Africa. ABD, BFT, BM, CGE, E, RNG. 17,
328*(1916:241-254), 592, 1294*.
incana (Thunb.) Kuntze
• A wool casual. S Africa. BM, RNG. 17, 328*(1916:241-254), 592,
831*(35:283-287), 875*(57:14), 1145*. *P. virgata* Less.; *Chrysanthemum
incanum* Thunb.

ONCOSIPHON Källersjö
grandiflorum (Thunb.) Källersjö
•• A persistent wool alien. S Africa. BM, CGE, E, OXF, RNG, TCD.
3, 17, 20, 51b, 145, 303(18:10), 327*(96:299-322), 328*(1916:241-254), 592,
1343*. *Matricaria grandiflora* (Thunb.) Fenzl ex Harvey, non Poiret; *Pentzia
grandiflora* (Thunb.) Hutch.; *Tanacetum grandiflorum* Thunb.
piluliferum (L.f.) Källersjö   **Stink-net**
• A wool casual. S Africa. ABD, BM, CGE, E, LIV, RNG. 3, 6, 17, 51b,
133, 327*(96:299-322), 592, 743, 830*(6:45), 1096*. *Cotula globifera*
Thunb.; *C. pilulifera* L.f.; *Matricaria globifera* (Thunb.) Fenzl ex Harvey;
Matricaria pilulifera (L.f.) Druce; *Pentzia globifera* (Thunb.) Hutch. Not to
be confused with its S African congener *P. globosa* Less.
suffruticosum (L.) Källersjö
• A wool casual. S Africa; (Australia). BFT, BM, CGE, E, OXF, RNG.
3*, 8, 17, 37, 51b, 133, 327*(96:299-322), 328*(1916:241-254), 354(4:16;
12:492), 592, 1343*. *Cotula tanacetifolia* L.; *Matricaria multiflora* Fenzl ex
Harvey; *M. suffruticosa* (L.) Druce; *Pentzia suffruticosa* (L.) Druce;
P. tanacetifolia (L.) Hutch.; *Tanacetum suffruticosum* L.

MATRICARIA L.
aurea (Loefl.) Schultz-Bip.
- A casual on a refuse tip at Yeovil (S Somerset) and at Highweek, Newton Abbot (S Devon), vector unknown. Medit., SW Asia. CGE, OXF. 1, 21, 39, 44a*, 45*, 248, 390(85:178-188), 528*, 683*, 729, 737*. *Chamomilla aurea* (Loefl.) Gay ex Cosson & Kralik; *Cotula aurea* Loefl., non Sibth. & Smith; *Perideraea aurea* (Loefl.) Willk.

decipiens (Fischer & C.Meyer) K.Koch
- •• A grain and wool alien; established in a gravel pit at Stanton Harcourt (Oxon), perhaps now gone. SW Asia. BFT, CGE, E, LTR, OXF, RNG. 1, 17, 39*, 160, 303(22:12), 354(13:298). *Cotula aurea* Sibth. & Smith, non Loefl.; *Pyrethrum decipiens* Fischer & C.Meyer; *Tripleurospermum decipiens* (Fischer & C.Meyer) Bornm.

disciformis (C.Meyer) DC.
- A casual at Worsley (S Lancs) and by a canal at Stourton (SW/MW Yorks), vector unknown. No modern records. Formerly probably a grain alien. SW Asia. OXF. 7, 16, 21, 39*, 228, 251, 354(2:504). *M. corymbifera* DC.

discoidea DC.   Pineappleweed
- ••••• An established weed throughout the British Is, probably originally a grain alien; locally abundant. NE Asia; (widespread). 1, 20, 21, 22*, 29*, 33*, 35*. *M. matricarioides* (Less.) Porter pro parte, nom. illeg.; *M. occidentalis* Greene; *M. suaveolens* (Pursh) Buch., non L.; *Chamomilla suaveolens* (Pursh) Rydb.

oreades Boiss.   Lawn Pyrethrum
Pre-1930 only. SW Asia. K, OXF. 7, 39, 44a*, 209, 354(6:291). *Tripleurospermum oreades* (Boiss.) Rech.f.; incl. *M. tschihatchewii* (Boiss.) Voss.

cf. **trichophylla** (Boiss.) Boiss.
- A casual at Hull docks (SE Yorks), vector unknown. No modern records. C & SE Europe, Turkey. 21, 31, 39, 285, 695*. *Chamaemelum uniglandulosum* Vis.; *Tripleurospermum tenuifolium* (Kit.) Freyn.

COTULA L.
anthemoides L.
Pre-1930 only. Africa, India, China. OXF. 7, 46*, 354(5:288), 527*, 592, 604, 617*, 676, 677*, 729*.

australis (Sieber ex Sprengel) Hook.f.   Annual Buttonweed
- •• Established since 1946 on the wall of a leat at Newton Abbot (S Devon); elsewhere a wool casual. Australasia. BFT, BM, CGE, E, OXF, RNG. 3*, 14, 20*, 371(328:19; 367:29), 500*, 519*, 534*, 583*.

bipinnata Thunb.   Ferny Buttonweed
- A wool casual. S Africa; (Australia). BM, RNG. 17, 37, 371(382:25), 583*, 590*, 743, 1074*.

coronopifolia L.   Buttonweed
●●● An established wool alien and garden escape; naturalised in a few widely
scattered, mostly coastal marshes; known since about 1880 on saline marsh near
Birkenhead (Cheshire); increasing. S Africa, Australia; (widespread). BM,
CGE, E, LANC, OXF, RNG. 1, 3*, 14, 20, 21, 22*, 25*, 30*, 33*, 230,
303*(64:42-46), 345(22:114), 534*. Incl. *C. integrifolia* Hook.f.

dioica (Hook.f.) Hook.f.   Hairless Leptinella
●● A garden escape well established in short grass in a few widely scattered
localities, perhaps originally planted as a grass substitute. New Zealand.
BM, E, K, OXF, RNG. 13, 20, 21, 36, 201, 209, 370(2:170), 1170*.
Leptinella dioica Hook.f.

filifolia Thunb.
● A casual at Sharpness docks (W Gloucs), vector unknown. No modern
records. Formerly a wool alien. S Africa. E, K. 3, 592, 830*(7:132).

macroglossa Bolus ex Schldl.
Pre-1930 only. S Africa. E. 3, 861(27:209-210).

pusilla Thunb.
● A wool casual. S Africa. E. 3, 475, 592.

sororia DC.
● A wool casual. S Africa. E, LIV, RNG. 3, 17, 592.

squalida (Hook.f.) Hook.f.   Leptinella
●●● An established garden escape and wool alien; naturalised in turf on
roadsides near Cat Firth, Dales Voe and Tofts Voe (all Shetland) and in turf
under bracken above Gairloch beach (W Ross); elsewhere a persistent lawn
weed. New Zealand. BM, CGE, E, LANC, OXF, RNG. 14, 20, 36, 253,
368(41:547), 371(352:27), 651*, 709*, 1170*. *Leptinella squalida* Hook.f.

turbinata L.
●● A wool casual. S Africa; (Australasia). BM, CGE, E, K, OXF, RNG.
3*, 14, 133, 303*(31:13-14), 578a*, 583*, 592. *Cenia turbinata* (L.) Pers.

villosa DC.
Pre-1930 only. S Africa. OXF. 3, 592. Possibly conspecific with
C. australis.

zeyheri Fenzl ex Harvey
● A wool casual. S Africa. BM, K, OXF, RNG. 3, 14, 17, 335(880:27),
592.

CENTIPEDA Lour.
minima (L.) A.Braun & Asch.   Spreading Sneezeweed
Pre-1930 only. Asia, Australasia; (N America). E. 3, 36c*, 37*, 583*, 590*,
617*, 649*, 1170*. *C. orbicularis* Lour.

SOLIVA Ruíz Lopez & Pavón
anthemifolia (Juss.) Sweet   Button Onehunga-weed
● A wool casual. S America; (widespread). BM, CGE, E, LTR, OXF, RNG.
3*, 14, 17, 20, 36c, 133, 145, 514, 534*, 590*, 617*, 905*(14:123-139).

aff. **macrocephala** Cabrera
- ● A wool casual. Uruguay, NE Argentina. K, HbEJC. 17, 534*, 905*(14:123-139).

pterosperma (Juss.) Less. Jo-jo
- ● A wool and granite casual. America; (widespread). E, K, LIV, RNG. 14, 17, 21, 182, 303(10:16), 500*, 533*, 534*, 583*, 905*(14:123-139). *S. sessilis* auct., non Ruíz Lopez & Pavón.

sessilis Ruíz Lopez & Pavón Onehunga-weed
- ● A wool casual. S America; (Australasia). E, K. 3, 17, 21, 36c*, 133, 500*, 590*, 905*(14:123-139).

stolonifera (Brot.) R.Br. ex Don Carpet Burweed
Pre-1930 only. Uruguay, NE Argentina. E, OXF. 3, 21, 354(5:34), 534*, 590*, 905*(14:123-139). *Gymnostyles stolonifera* (Brot.) Tutin; *Soliva lusitanica* Less.; *S. nasturtiifolia* (Juss.) DC. pro parte.

SENECIO L.

aegyptius L.
Pre-1930 only. NE Africa. RNG. 2, 3, 300(1904:239; 1905:101), 676, 729*. *S. arabicus* L.

arenarius Thunb.
- ● A wool casual. No modern records. S Africa. LTN. 14, 143, 511, 628*, 823*(32:t.1254), 1002*, 1283*.

bipinnatisectus Belcher Australian Fireweed
- ● A wool casual. No modern records. Australia. LTN. 14, 36c*, 145, 514, 519*, 590*.

brasiliensis Less.
- ● A wool casual. S America. BM, RNG. 1, 3, 17, 534*, 535, 554, 1199*. *S. canabinifolius* Hook. & Arn., non *S. cannabifolius* Less.; incl. *S. tripartitus* DC.

[*burchellii* DC. In error for *S. inaequidens*. This S African shrub is unlikely to occur. 1144.]

[*cambrensis* Rosser Alien status dubious. Derived from the alien *S.* × *baxteri*, but 20 treats it as an endemic.]

caracasanus Klatt
Pre-1930 only. Venezuela. BM. 635, 800(15:331). *S. cucullatus* Klatt. An endemic of the Cordillera de la Costa.

cineraria DC. Silver Ragwort
- ●●●● An established garden escape; naturalised, often abundantly, especially on sea cliffs in southern England and at Dalkey, Killiney Bay (Co Dublin). Medit. 1, 20, 21, 22*, 25*, 33*, 52*. *S. bicolor* (Willd.) Tod. subsp. *cineraria* (DC.) Chater; *S. maritimus* (L.) Reichb.; *Cineraria maritima* (L.) L. Hybrids with native *S. erucifolius* (*S.* × *thurettii* Briq. & Cavill.; *S.* × *patersonianus* Burton) and *S. jacobaea* (*S.* × *albescens* Burb. & Colgan) have been recorded. RNG, SLBI. 20.

doria L.   Golden Ragwort
•• (but confused with *S. fluviatilis*). An established garden escape; naturalised
in a few places in the north, as at Hillberry, Fistard and near Fleshwick
(all Man). C & S Europe. BM, E, OXF, RNG. 1, 20, 103, 683*, 1279*.

doronicum (L.) L.   Chamois Ragwort
• A garden escape recorded from Balnaknock, near Mulben (Banffs) and from
the banks of the river Lyon in Glen Lyon (Mid Perth). C & S Europe. 20, 21,
23*, 27*, 28*, 29*, 303(32:19), 371(407:40), 683*.

elegans L.   Purple Groundsel
• A casual or persistent garden escape in N Devon, the Channel Is and Scilly.
S Africa. JSY, LIV. 7, 21, 52*, 55*, 201, 371(427:35), 519*, 562*, 590*.
Jacobaea elegans (L.) Moench.

fluviatilis Wallr.   Broad-leaved Ragwort
•••• (but confused with *S. doria*). An established garden escape; naturalised
by streams and on damp ground in widely scattered localities; known since
1680 by a stream between Wells and Glastonbury (N Somerset). Europe.
1, 20, 21, 22*, 33*, 34*, 35*, 248.   *S. sarracenicus* L. pro parte;
S. salicetorum Godron.

gallicus Chaix   French Groundsel
• A casual at the Crown Wallpaper tip, Darwen (S Lancs), vector unknown.
No modern records. Medit. RNG. 7, 21, 44a*, 45*, 46*, 300(1909:42),
683*.   *S. coronopifolius* Desf., non Burm.f.; *S. desfontainei* Druce;
S. subdentatus Ledeb., non Phil.

glastifolius L.f.   Holly-leaved Groundsel
• A casual escape from cultivation on Tresco (Scilly). S Africa. BM, RNG.
36c, 51, 301*(t.1342), 303(29:12; 36:23), 717*. *S. lilacinus* Schrader. Not
to be confused with *S. glastifolius* Hook.f., a synonym of the New Zealand
S. kirkii Hook.f. ex Kirk.

glomeratus Desf. ex Poiret   Annual Fireweed
Pre-1930 only. Australasia. E. 3, 36, 37c*, 42, 500*, 1078*. *S. argutus*
A.Rich.; *S. multicaulis* auct., non A.Rich.; *Erechtites arguta* (A.Rich.) DC.

glossanthus (Sonder) Belcher   Slender Groundsel
• A wool casual. Australia. BM, E, K, RNG. 3*, 14, 17, 37*, 133, 583*,
590*, 1078*. *S. brachyglossus* F.Muell., non Turcz.; *Erechtites glossanthus*
Sonder.

grandiflorus P.Bergius   Purple Ragwort
• A garden escape established in grass between Hauteville and Fermain in
Guernsey (Channel Is). S Africa. BM. 20, 220, 301*(t.901), 370(9:185),
501, 1159*. *S. venustus* Aiton. Not to be confused with *S. grandiflorus*
Hoffsgg. & Link, a synonym of the Portuguese *S. lopezii* Boiss.

hieracioides DC.
Pre-1930 only. S Africa. OXF. 3, 554, 592.

inaequidens DC.   Narrow-leaved Ragwort
••• A wool alien persistent or established in a few places; plants found at
Walmer (E Kent) may be an extension of its range from Calais, France, where
it is abundantly naturalised. S Africa; (Europe). ABD, BM, CGE, E, OXF,

RNG. 3*, 14, 20, 22*, 133, 277, 303*(31:13,15), 370(12:392), 546a*, 598*, 683*, 826*(11:98-99), 1144. *S. burchellii* auct. pro parte, non DC.; *S. lautus* auct., non Sol. ex Willd. The closely related S African *S. harveianus* MacOwan (*S. vimineus* auct., non DC.), and *S. madagascariensis* Poiret (*S. burchellii* auct. pro parte, non DC.) may both have been overlooked. 1050, 1144.

juniperinus L.f.
- A wool casual. S Africa. E, OXF. 3, 133, 578a*, 1144.

[*lautus* Sol. ex Willd. (*S. australis* Willd.) Unconfirmed; probably always in error for *S. inaequidens*. This variable Australian species, mostly from coastal habitats, is unlikely to occur.]

leucanthemifolius Poiret
- A casual, perhaps a grain alien. No modern records. Medit. IPS. 1, 21, 45*, 46*, 149, 180, 244, 354(10:22), 683*, 1149. Incl. *S. crassifolius* Willd.

linearifolius A.Rich.   Fireweed Groundsel
Pre-1930 only. Australia, Tasmania. E, OXF. 3, 533, 540*, 590*, 743, 1150*. *S. australis* A.Rich., non Willd.; *S. cinerarioides* A.Rich., non Kunth; *S. dryadeus* F.Muell.; *S. macrodontus* DC. A variable species.

minimus Poiret   Toothed Fireweed
- A wool casual. Australasia. BM, CGE, E, K, OXF, RNG. 3*, 17, 36c*, 37*, 519*, 583*, 590*, 1042a.   *Erechtites minima* (Poiret) DC.; *E. prenanthoides* DC.

[*nebrodensis* L. In error for *S. squalidus*. 4, 156.]

nudiusculus DC.
- A wool casual. S Africa. RNG. 17, 554.

oederiifolius DC.
- A wool casual. S Africa. RNG. 17, 554, 592.

ovatus (P.Gaertner, Meyer & Scherb.) Willd.   Wood Ragwort
- Introduced, or a garden escape, established at Holden Clough and Arncliffe (both MW Yorks). Eurasia. 7, 20, 21, 22*, 23*, 25*, 320(1913:305), 371(340:24; 355:28), 1016*. *S. fuchsii* C.Gmelin; *S. nemorensis* L. subsp. *fuchsii* (C.Gmelin) Čelak. Modern records are referable to subsp. *alpestris* (Gaudin) Herborg.

pinnulatus Thunb.
- A wool casual. S Africa. ABD, BM, CGE, E, RNG, TCD. 17, 133, 371(382:25), 592, 1273*. Incl. *S. scabriusculus* DC.

pterophorus DC.   Shoddy Ragwort
- •• A wool casual. S Africa; (Australia). ABD, BM, E, OXF, RNG. 3, 14, 20, 37c*, 145, 303(37:14), 590*, 1078*, 1144. Incl. *S. subserratus* DC.

quadridentatus Labill.   Cotton Fireweed
- A wool casual. Australasia. CGE, E, K, RNG. 3, 17, 36, 36c*, 37, 133, 519*, 583*, 590*, 1078*. *Erechtites quadridentata* (Labill.) DC.

smithii DC.   Magellan Ragwort
- ••• An established garden escape or relic; naturalised by streams and in ditches in many places in Scotland, especially in Caithness, Orkney and

Shetland. Southern S America. BM, CGE, LANC, OXF, RNG, SLBI. 7, 20, 21, 253, 303*(32:16-17), 669*.

squalidus L.   Oxford Ragwort
●●●●● An escape from Oxford Botanic Garden, first recorded in 1794, now widely and abundantly established, especially in England. C & S Europe. 1, 20, 21, 22*, 23*, 25*, 33*, 35*, 349(5:210-219). *S. chrysanthemifolius* DC.; *S. rupestris* Waldst. & Kit.   Hybrids with native *S. viscosus* (*S.* × *subnebrodensis* Simonkai; *S.* × *londinensis* Lousley; *S.* × *pseudonebrodensis* sphalm.) and with *S. vulgaris* (*S.* × *baxteri* Druce) have been recorded. E, LSR, NMW, RNG, SUN. 20, 370(15:36). Pre-1930 records of native *S. jacobaea* × *S. squalidus* from Ireland are in error. 1311. [*subdentatus* Phil.   A dubious record. 354(2:349).]

vernalis Waldst. & Kit.   Eastern Groundsel
●●● (but confused with *S. squalidus* × *S. vulgaris* L.). A grass seed casual and possibly a wool casual. Eurasia. BM, CGE, E, NMW, OXF, RNG. 1, 17, 20*, 21, 22*, 26*, 303*(41:12-13). *S. leucanthemifolius* var. *vernalis* (Waldst. & Kit.) J.Alexander. A hybrid with native *S. vulgaris* (*S.* × *helwingii* Beger ex Hegi; *S.* × *pseudovernalis* Zabel) has been recorded. 20, 370(8:90).

PERICALLIS D.Don

hybrida R.Nordenstam   Cineraria
● A garden escape established in the extreme southwest of England; abundant on walls and roadsides at St Ives (W Cornwall) and Garrison Hill, St Mary's (Scilly); casual elsewhere. Originated in cultivation. BM, RNG. 20, 52*, 54*, 56*, 67*, 212, 303(36:23). *Senecio cruentus* auct., non Roth, nec (L'Hér.) DC.; *S. hybridus* N.Hylander, nom. inval. Probably of hybrid origin from *P. cruenta* (L'Hér.) Bolle and other Canary Is species.

CINERARIA L.

aspera Thunb.
● A wool casual. S Africa. ABD, BM, CGE, E, LTR, RNG. 17, 554, 592, 1294*. *Senecio aspera* (Thunb.) ined.

burkei Burtt Davy
● A wool casual. S Africa. RNG. 17, 328(1936:80). Not to be confused with *Senecio burkei* Greenm.

geifolia L.
● A wool casual. S Africa. RNG. 17, 735*. *C. erodioides* DC.

lobata L'Hér.
● A wool casual. S Africa. BM, CGE, E, RNG. 17, 51, 565*, 578*, 592, 640*, 838*(14:156). Not to be confused with *Senecio lobatus* Pers.

lyrata DC.
● A wool casual. S Africa. ABD, CGE, E, K, LTN. 3, 145, 598*. *Senecio lyratus* L.f., non Forsskål.

DELAIREA Lemaire
odorata Lemaire   German-ivy
●●● An established garden escape; naturalised, sometimes abundantly, in a few
places in SW England, the Channel Is and Ireland; increasing. S Africa. BM,
CGE, LANC, OXF, RNG, TOR. 7, 20*, 21, 22*, 37c*, 201, 207*, 212, 275,
354(3:475), 683*. *Senecio mikanioides* Otto ex Walp.

BRACHYGLOTTIS Forster & Forster f.
monroi (Hook.f.) R.Nordenstam   Monro's Ragwort
● A persistent garden escape on dunes at West Shore, Llandudno (Caerns).
New Zealand. 20*, 36, 55*, 61*, 63*, 458. *Senecio monroi* Hook.f.
repanda Forster & Forster f.   Hedge Ragwort
● A relic of cultivation in Scilly; seedlings have been reported from St Martin's
(Scilly). New Zealand. E, RNG. 20*, 36, 53*, 61, 207*, 458, 1117*.
'Sunshine'   Shrub Ragwort
●● A garden escape persistent on dunes at West Shore, Llandudno (Caerns),
at the Mumbles (Glam), Meadowside, Glasgow (Lanarks) and Shoreham
(W Sussex); also collected from Laxey (Man). Originated in cultivation. BM.
20*, 52*, 61, 61a, 62*, 63*, 370(18:430), 445, 458. Probably *B. compacta*
(Kirk) R.Nordenstam × *B. laxifolia* (J.Buch.) R.Nordenstam; *Senecio greyi*
hort., non Hook.f. All records are tentatively referred to this taxon; other
cultivars in the Dunedin Hybrid group may occur, but have not yet been
detected. The illustrations quoted may not all apply strictly to *B.*'Sunshine'.

SINACALIA H.Robinson & Brettell
tangutica (Maxim.) R.Nordenstam   Chinese Ragwort
●●● An established garden escape; naturalised for more than 30 years by the
river North Esk (Angus). China. BM, E, K, OXF, RNG, TOR. 7, 13, 20*,
51*, 187. *Ligularia tangutica* (Maxim.) Bergmans ex Mattf.; *Senecio
tanguticus* Maxim.

LIGULARIA Cass.
dentata (A.Gray) H.Hara   Leopard-plant
●● (but confused with *Buphthalmum speciosum* and *Inula helenium*). An
established garden escape or relic; naturalised in woods and damp places in
widely scattered localities. E Asia. CGE, OXF, RNG. 7, 20, 21, 52*, 55*,
57*, 234, 325*(106:122-125), 371(352:28). *L. clivorum* Maxim.; *Senecio
clivorum* (Maxim.) Maxim.
fischeri (Ledeb.) Turcz.
Pre-1930 only. E Asia. 7, 354(5:288), 685*, 724*. *L. sibirica* (L.) Cass. var.
racemosa (DC.) Kitam.; *Senecio ligularia* Hook.f.
japonica (Thunb.) Less.
● Details not known to us. E Asia. BM. 43, 51, 594*. *Senecio japonicus*
Schultz-Bip., non Thunb.

przewalskii (Maxim.) Diels   Przewalski's Leopard-plant
• A garden escape persistent by the river Tyne (S Northumb) and in a wild bog
garden at Witherslack (Westmorland). N China. 20, 53*, 371(428:8). *Senecio
przewalskii* Maxim.   *L.*'The Rocket', probably of *L. przewalskii* ×
L. stenocephala (Maxim.) Matsum. & Koidz. origin, may have been
overlooked. 51b.

PARASENECIO W.Smith & Small
hastata (L.) ined.
Pre-1930 only.   Siberia.   21, 227, 281, 579*, 678*, 1104*.   *Cacalia
hastata* L.; *Senecio hastatus* (L.) ined.   The plant from E Asia sometimes
described under this name may have been overlooked; it is probably best
separated as *P. hastata* subsp. *orientalis* (Kitam.) ined. 43, 638*, 649*, 1324*.

EMILIA (Cass.) Cass.
javanica (Burm.f.) C.Robinson sensu lato   Tassel-flower
Pre-1930 only.   Origin obscure; (tropics).   1, 47*, 48*, 51, 52*, 55*,
346(32:33), 1042a, 1316*.   *E. coccinea* (Sims) G.Don; *E. flammea* Cass.;
E. sagittata (Vahl) DC.; *Cacalia coccinea* Sims; *C. sagittata* Vahl; *Senecio
sagittatus* Schultz-Bip.; incl. *E. fosbergii* Nicolson.

CRASSOCEPHALUM Moench
sarcobasis (DC.) S.Moore
• A casual garden or greenhouse escape. Tropical Africa. 47*, 371(376:25).

OTHONNA L.
angustifolia DC.
• A wool casual. Iran. E, K, RNG. 17, 524, 554, 622*. *Othonnopsis
angustifolia* (DC.) Jaub. & Spach.

DORONICUM L.
austriacum Jacq.   Austrian Leopard's-bane
Pre-1930 only. C & S Europe. 21, 22*, 23*, 151, 546*, 683*.
columnae Ten.   Eastern Leopard's-bane
• A garden escape on the bank of a reservoir at Barnes and a pathside on White
Hill at Chipstead (both Surrey).   C & S Europe.   20, 21, 31*, 199,
331(68:149), 683*. *D. cordatum* auct., non Lam.
orientale Hoffm.
• A casual, probably a garden escape. SE Europe, SW Asia. NMW. 21,
24*, 156, 683*. *D. caucasicum* M.Bieb.; *D. eriorhizon* Guss.
pardalianches L.   Leopard's-bane
•••• An established escape from cultivation; naturalised in woods, plantations
and by roadsides in scattered localities throughout Britain, especially in
Scotland.   Formerly cultivated as a medicinal plant. W Europe. 1, 20, 21,
22*, 25*, 303*(27:22-23), 1275*. *D. cordatum* Lam., nom. illeg.
plantagineum group   Plantain-leaved Leopard's-banes
••• Established garden escapes; naturalised in woodland in a few widely
scattered localities; the hybrid *D.* × *willdenowii* known since 1800 on a ditch

bank and hedgerow near Widdington (N Essex). W Europe. BM, CGE, E,
NMW, OXF, RNG. 1, 20, 21, 22*, 33*, 35*, 303*(27:22-23), 1275*.
D. plantagineum L. has been over-recorded in error for its hybrids
D. × *willdenowii* (Rouy) A.W.Hill (probably *D. pardalianches* ×
D. plantagineum; *D. plantagineum* var. *willdenowii* (Rouy) A.B.Jackson) and
D. × *excelsum* (N.E.Br.) Stace (probably *D. columnae* × *D. pardalianches* ×
D. plantagineum; *D.* 'Harpur Crewe'; *D.* × *hybridum* hort.; *D. plantagineum*
var. *excelsum* N.E.Br.). LANC. 20.

ARNICA L.
angustifolia Vahl   Arctic Arnica
Pre-1930 only. N Scandinavia, N Siberia, Labrador, Greenland. 21, 38*,
362(1912:24-31), 589*, 1026*. *A. alpina* (L.) Olin, non Salisb.;
A. montana L. var. *alpina* L.; incl. *A. iljinii* (Maguire) Iljin.

PETASITES Miller
albus (L.) Gaertner   White Butterbur
●●●● An established garden escape; naturalised in woods, by streams and on
roadsides in widely scattered localities throughout the British Is, especially in
eastern Scotland; increasing. Eurasia. 1, 20, 21, 22*, 23*, 29*, 303*(26:1).
Tussilago alba L.

fragrans (Villars) C.Presl   Winter Heliotrope
●●●●● An established garden escape; naturalised on roadsides and by streams
throughout the British Is, especially in the south and west; locally abundant.
C Medit. 1, 20, 21, 22*, 25*, 201*, 258*. *P. odorata* sphalm.;
P. ?pyrenaicus (L.) G.Lopez; *Nardosmia fragrans* (Villars) Reichb.; *Tussilago
fragrans* Villars.

japonicus (Siebold & Zucc.) Maxim.   Giant Butterbur
●●●● An established garden escape; naturalised, sometimes abundantly, on
damp ground in widely scattered localities. E Asia. 12, 20, 21, 22*,
303*(38:1), 354(12:282). Most records are referable to subsp. *giganteus*
Kitam. 1275.

palmatus (Aiton) A.Gray   Palmate-leaved Butterbur
● A garden relic at Tunbridge Wells (W Kent). N America. MNE. 38*, 40,
41*, 42, 303(29:12). *P. frigidus* (L.) Fries var. *palmatus* (Aiton) Cronq.

HOMOGYNE Cass.
alpina (L.) Cass.   Purple Colt's-foot
● Probably introduced; established on rocks since about 1800 on the Clova
mountains (Angus); also recorded but never refound on Beinn Mhor, South Uist
(Outer Hebrides). Europe. BM, CGE, E, GL, K, RNG. 7, 20, 21, 22*, 23*,
25*, 27*, 370(2:237-238), 1246. *Tussilago alpina* L.

CALENDULA L.
arvensis L.   Field Marigold
●●● An established garden escape and esparto alien; abundant as a weed in a
few fields in Scilly and the Channel Is. Medit. BM, LSR, NMW, OXF,

RNG, SLBI. 1, 15, 20, 21, 29*, 32*, 44a*, 117*, 212, 220. Incl.
C. aegyptiaca Desf.; *C. persica* C.Meyer.
officinalis L. Pot Marigold
●●●●● A persistent or established garden escape and bird-seed alien; perhaps
naturalised in a few places in the south. Origin obscure. 1, 19, 20, 21, 22*,
25*, 54*, 66*, 207.

DIMORPHOTHECA Moench
pluvialis (L.) Moench Cape Rain-daisy
● A casual garden escape. No modern records. S Africa. 1, 51, 53*, 235,
312*(n.s. t.532), 562*, 655*. *D. annua* Less.; *Calendula pluvialis* L.;
Gattenhoffia pluvialis (L.) Druce.
sinuata DC. Namaqualand Daisy
● A casual in a carrot field at Houghton (W Norfolk), vector unknown.
No modern records. S Africa. RNG. 51, 52*, 55*, 56*, 66*. *D. aurantiaca*
hort., non DC.

OSTEOSPERMUM L.
jucundum (Phillips) Norlindh
● A garden plant reported from a number of walls in W Cornwall. S Africa.
51, 53*, 54*, 458, 543*. *O. barberae* auct., non (Harvey) Norlindh;
Dimorphotheca barberae auct., non Harvey; *D. jucunda* Phillips.

AMBROSIA L.
acanthicarpa Hook. Hooker's Bur-ragweed
Pre-1930 only. N America. 2, 38*, 40, 41*, 42, 279. *Franseria acanthicarpa*
(Hook.) Cov.; *Gaertneria acanthicarpa* (Hook.) Britton.
artemisiifolia L. Ragweed
●●●● A persistent bird-seed, grain, oil-seed and agricultural seed alien.
N America; (Europe, Australia). 1, 19, 20*, 21, 22*, 41*, 303*(22:13), 1285.
Artemesia artemesiifolia sphalm.; *Franseria artemisioides* Willd.; *Gaertneria
artemisioides* Britton; incl. *Ambrosia elatior* L.
maritima L.
● A casual at Crown Wallpaper tip, Darwen (S Lancs), vector unknown.
No modern records. Medit. BM, NMW, RNG. 1, 21, 44a*, 139, 156, 165,
258, 263, 683*, 1285*.
peruviana Willd.
Pre-1930 only. Tropical America. BM. 7, 49, 150, 354(4:416).
psilostachya DC. Perennial Ragweed
●● A grain alien established in a few widely scattered localities; known for
many years on the edge of Broughton Gifford Common (N Wilts), now gone,
and on coastal dunes in S and W Lancs. N America. BM, NMW, OXF.
7, 20*, 21, 22*, 37c*, 38*, 41*, 172, 251, 349(6:172), 590*. Incl.
A. coronopifolia Torrey & A.Gray. There seems little justification for
recognising two species within this variable taxon, as in 683*.
tenuifolia Sprengel Lacy Ambrosia
Pre-1930 only. S America. 21, 37c*, 266, 508*, 590*, 653*, 683*, 1074*.

trifida L. Giant Ragweed

●●● A persistent oil-seed and grain alien. N America. BM, E, LANC, LSR, NMW, OXF, RNG. 1, 8, 20*, 22*, 41*, 236, 303*(25:12-13), 683*, 868*(62:1003-1010). Incl. *A. integrifolia* Muhlenb.; *A. aptera* DC.

IVA L.

ambrosiifolia (A.Gray) A.Gray

Pre-1930 only. SW USA, Mexico. 7, 545, 652.

microcephala Nutt.

● A casual on a field border at Harefield (Middlesex), vector unknown. No modern records. SE USA. 12, 580, 646.

xanthiifolia Nutt. Marsh-elder

●● A persistent bird-seed, grain, carrot seed and wool alien; perhaps established on the beach at Greenhithe (W Kent) and by the river Gipping at Ipswich (E Suffolk). N America; (Europe). BM, CGE, LTN, NMW, OXF, RNG. 1, 19, 20*, 21, 38*, 236, 258, 1138*. *Cyclachaena xanthiifolia* (Nutt.) Fresen.

XANTHIUM L.

ambrosioides Hook. & Arn. Argentine Cocklebur

●● A persistent wool alien; abundant for a few years in one arable field near Flitwick (Beds). Argentina, Patagonia. BM, E, K, LTN, NMW, RNG. 3, 14, 20*, 303*(29:1,13), 894*(5:67). *Acanthoxanthium ambrosioides* (Hook. & Arn.) ined.

spinosum L. Spiny Cocklebur

●●●● A persistent wool, grain and bird-seed alien; rarely locally abundant for a short time. S America; (widespread). 2, 14, 19, 20*, 21, 22*, 34*, 191, 303*(29:1), 458.

strumarium L. group Rough Cockleburs

●●● Persistent grain, wool and oil-seed aliens; in scattered localities in England and Wales, sometimes abundant for a time; frequent on the estuarine shores of the river Thames from the London border as far as the east coast of Grain (W Kent), sometimes self-sown. America; (widespead). BM, LANC, LTN, NMW, OXF, RNG. 1, 14, 20*, 21, 22*, 24*, 236, 303*(29:1,13), 349(1:70), 825*(9:300-303). Incl. *X. canadense* Miller; *X. cavanillesii* Schouw ex Didr.; *X. chinense* Miller; *X. echinatum* Murray; *X. italicum* Moretti; *X. macrocarpum* DC.; *X. occidentale* Bert.; *X. orientale* L.; *X. pungens* Wallr.; *X. saccharatum* Wallr.

ACANTHOSPERMUM Schrank

australe (Loefl.) Kuntze Spiny-bur

● A wool casual. Tropical America, Uruguay. E, LIV, LTN, RNG. 14, 38*, 40, 48, 145, 371(397:34), 598*, 892(6:5-8), 1144. Incl. *A. brasilum* Schrank.

glabratum (DC.) Wild

● A wool casual. S America. RNG. 17, 48, 892(6:5-8), 1144*. *A. australe* auct., non (Loefl.) Kuntze.

ZINNIA L.
elegans Jacq.   Youth-and-age
- A casual garden escape. Mexico. 16, 20, 51, 52*, 53*, 54*, 126, 199.

peruviana (L.) L.
- A wool casual. Arizona, Mexico, S America. K. 17, 51, 312*(t.149), 511, 627, 707, 710*, 1316*. *Z. multiflora* L.; *Z. pauciflora* L.; *Z. tenuiflora* Jacq.; *Z. verticillata* Andrews.

SANVITALIA Lam.
procumbens Lam.
Pre-1930 only. C America. BM. 1, 16, 20, 51, 52*, 53*, 54*, 300(1905:101).

GUIZOTIA Cass.
abyssinica (L.f.) Cass.   Niger
•••• A bird-seed, grain, oil-seed and wool casual. E Africa. 1, 19, 20*, 21, 22*, 68*, 251, 303*(21:1), 546a*, 826*(25:52). *G. oleifera* DC. Some records may be referable to the similar, Ethiopian, *G. scabra* (Vis.) Chiov. subsp. *schimperi* (Schultz-Bip.) J. Baagøe (*G. schimperi* Schultz-Bip.). 826*(25:113-119), 862*(69:1-39).

SIGESBECKIA L.
microcephala DC.   Pale Indian-weed
- A wool casual. Australia. BM, CGE, E, K, LTN, RNG. 21, 37c*, 349(7:21), 590*, 1074*.

orientalis L.   Eastern St Paul's-wort
•• A wool casual. Africa, Asia; (Australasia). BM, CGE, K, LTN, OXF, RNG. 14, 20, 21, 37c*, 349(7:20), 583*, 1074*, 1247*.

serrata DC.   Western St Paul's-wort
•• (but confused with *S. orientalis*). An established grain alien; known since 1928 at Freshfield (S Lancs), casual or sporadically recurrent elsewhere. C & S America. BM, K, LIV, RNG. 20*, 21, 22*, 349(7:19-20). *S. cordifolia* auct., non Kunth; *S. jorullensis* auct., non Kunth.

ECLIPTA L.
prostrata (L.) L.
Pre-1930 only. Tropics. 2, 21, 38*, 44a*, 184, 544*, 1247*. *E. alba* (L.) Hassk.; *E. erecta* L.; *Verbesina alba* L.

VERBESINA L.
australis (Hook. & Arn. ex DC.) Baker
Pre-1930 only. Brazil, Uruguay. 2, 181(pp.81,356), 508*, 653. *Ximenesia australis* Hook. & Arn. ex DC.

encelioides (Cav.) Benth. & Hook.f. ex A.Gray   Golden Crownbeard
•• A wool casual. N America; (Australia). BM, CGE, E, K, LTN, RNG. 7, 14, 38*, 40, 44a, 51, 354(6:291), 500*, 590*. *Ximenesia encelioides* Cav.

SILPHIUM L.
perfoliatum L.   Cup-plant
• Persistent in scrub by a roadside at Freckenham (W Suffolk), now gone; also for a few years on a tip at Guildford (Surrey), perhaps a garden escape. Eastern N America. 21, 22*, 38*, 51b, 369(22:37-47), 370(8:187), 1110*.

CHRYSOGONUM L.
virginianum L.   Golden-knee
• A casual garden escape. Eastern N America. BM. 38*, 51, 55*, 57*, 354(12:470). Recorded as var. *dentatum* A.Gray.

PARTHENIUM L.
hysterophorus L.   Santa-Maria
Pre-1930 only. Tropical America. 1, 38*, 40, 49, 197, 245, 247, 590*, 1247*.

RUDBECKIA L.
amplexicaulis Vahl   Clasping-leaved Coneflower
• A wool casual and garden escape. USA, Mexico. BM, CGE, E, RNG. 7, 17, 38*, 40*, 51, 354(7:439), 371(416:8), 620*. *Dracopsis amplexicaulis* (Vahl) Cass.

fulgida Aiton   Perennial Coneflower
• A casual or persistent garden escape. N America. RNG. 38*, 51a, 52*, 53*, 55*, 312*(n.s. t.238), 353(35:37), 354(9:23), 403. Incl. *R. speciosa* Wender.

hirta L.   Bristly Coneflower
••• A persistent garden escape and wool casual. N America. BM, LTN, NMW, OXF, RNG. 7, 17, 20*, 21, 22*, 52*, 55*, 236. Incl. *R. bicolor* Nutt.; *R. serotina* Nutt., non Sweet.

laciniata L.   Coneflower
••• An established garden escape; naturalised on the banks of Lunan Water, near Friockheim (Angus) and by the river Conon (E Ross); known since 1898, and now abundant, on the banks of the river Tay below Perth (E Perth). N America. ABD, BM, CGE, E, NMW, OXF. 1, 20, 21, 22*, 34*, 38*, 57*, 153, 187.

triloba L.   Thin-leaved Coneflower
• Probably a casual garden escape. Eastern N America. 38*, 40*, 51, 371(364:29).

RATIBIDA Raf.
columnifera (Nutt.) Wooton & Standley   Prairie Coneflower
• A wool casual. N America. E, OXF. 7, 9, 38*, 41*, 51, 354(5:32). *Lepachys columnaris* (Pursh) Torrey & A.Gray; *Rudbeckia columnaris* Sims.

HELIANTHUS L.
angustifolius L.   Swamp Sunflower
• A casual at Hull docks (SE Yorks), vector unknown; also collected from Notts. Eastern N America. OXF, RNG. 7, 38*, 40, 51, 285, 354(5:286,660), 1319*. Incl. *H. giganteus* L.; *H. nuttallii* Torrey & A.Gray.

annuus L.   Sunflower
••••• A persistent or established garden escape, bird-seed, oil-seed and wool alien; possibly also from food for small mammals; abundant and reproducing along the shores of the river Thames between Erith and Sheppey (W Kent). N America; (widespread). 1, 14, 19, 20, 21, 38*, 52*, 53*, 54*, 192, 236. Incl. *H. lenticularis* Douglas; *H. macrocarpus* DC.

argophyllus Torrey & A.Gray   Silver-leaf Sunflower
Pre-1930 only. Texas. 2, 40, 51, 267, 842*(t.431), 1137*. *H. argyrophyllus* sphalm. Possibly conspecific with *H. annuus* L.

bolanderi A.Gray   Bolander's Sunflower
Pre-1930 only. California, SW Oregon. 7, 42, 354(5:33), 500*.

debilis Nutt.
Pre-1930 only. Eastern USA. 7, 40, 51, 312*(t.7432), 354(3:164; 5:111), 1137*. Incl. *H. cucumerifolius* Torrey & A.Gray.

decapetalus group   Thin-leaved Sunflowers
•••• Persistent garden escapes. Eastern N America; hybrids originating in cultivation. 2, 20, 52*, 53*, 54*, 57*. Incl. *H. decapetalus* L. (*H. decapitatus* sphalm.; *H. multiflorus* hort.) and hybrids with *H. annuus* (*H.* × *multiflorus* L.). Most records are referable to double-flowered forms.

divaricatus L.   Rough Sunflower
Pre-1930 only. Eastern N America. 7, 38*, 40, 51, 354(5:286).

hirsutus Raf.   Bristly Sunflower
• Persistent on a roadside at Bagillt (Flints), probably a garden escape, now gone. Eastern N America. NMW. 38*, 40, 51, 156.

× **laetiflorus** group   Perennial Sunflowers
•••• Persistent or established garden escapes. N America; hybrids originating in cultivation. 2, 20, 21, 22*, 36c*, 38*, 41*, 156, 199, 1137*. Incl. *H. diffusus* Sims; *H.* × *laetiflorus* Pers. (*H. pauciflorus* × *H. tuberosus*); *H. pauciflorus* Nutt.; *H. rigidus* (Cass.) Desf.; *H.* × *scaberrimus* Elliott.

maximilianii Schrader   Maximilian's Sunflower
• A casual in Black Rock Quarry in the Avon Gorge (W Gloucs), vector unknown. No modern records. N America. 7, 38*, 40, 41*, 42, 267, 300(1909:42), 350(29:431).

petiolaris Nutt.   Lesser Sunflower
•• A persistent oil-seed, grain and wool alien; perhaps established by the river Thames at Erith (W Kent). N America. BM, MNE, OXF, RNG. 2, 20, 38*, 41*, 126, 182, 331(58:63), 371(388:37), 403, 1137*.

strumosus L.   Pale-leaved Sunflower
Pre-1930 only. Eastern N America. OXF. 7, 38*, 40, 51, 354(5:33), 1319. Incl. *H. mollis* Willd., non Lam.; *H. tomentosus* Michaux.

tuberosus L.   Jerusalem Artichoke
•••• A persistent or established garden escape or relic of cultivation; perhaps
also from bird-seed. N America. BM, LNHS, OXF, RNG, SLBI. 1, 19, 20,
21, 22*, 36c*, 41*, 56*, 1137*. *H. serotinus* Tausch.

HELIOPSIS Pers.
helianthoides (L.) Sweet   Rough Oxeye
• A garden escape established on waste ground near Swanscombe (W Kent).
N America. BM, OXF. 7, 20, 21, 22*, 38*, 52*, 55*, 126, 236, 354(6:290).
Incl. *H. laevis* Pers.; *H. scabra* Dunal.

SPILANTHES Jacq.
decumbens (Smith) A.Moore
• A casual at Hull docks (SE Yorks), vector unknown. No modern records.
S America. BM, OXF. 7, 134, 354(5:33; 7:439), 833*(3:373), 841*(n.s.
4:223), 1066*. *S. arnicioides* DC.; *S. leptophylla* DC.
[*oleracea* L. Unconfirmed. No localised record found to justify the entries in
16 and 20.]

SIMSIA Pers.
foetida (Cav.) Blake
Pre-1930 only. Mexico, C America. 1, 281, 575*, 710. *Encelia mexicana*
Martius ex DC.

GALINSOGA Ruíz Lopez & Pavón
parviflora Cav.   Gallant-soldier
••••• Originally an escape from cultivation at the Royal Botanic Gardens,
Kew (Surrey), now an established weed in many places in Britain, especially
in SE England; also a wool alien; increasing. S America; (widespread). 1, 14,
20*, 21, 22*, 29*, 33*, 209, 546a*.

quadriradiata Ruíz Lopez & Pavón   Shaggy-soldier
••••• Perhaps originally brought in with ornamental plants, now an
established weed in many places in Britain, especially in the London Area; also
a wool and bird-seed alien; increasing, and only recently recorded from Ireland.
Tropical America; (widespread). 12, 14, 19, 20*, 21, 22*, 33*, 192, 209,
546a*, 1285*. *G. aristulata* E.Bickn.; *G. ciliata* (Raf.) S.F.Blake.

LAYIA Hook. & Arn. ex DC.
chrysanthemoides (DC.) A.Gray   Smooth Layia
Pre-1930 only. California. BM. 42, 500*, 699, 1166*. *L. calliglossa*
A.Gray.

MADIA Molina
glomerata Hook.   Mountain Tarweed
• A casual at Kirkcaldy harbour (Fife), vector unknown. No modern records.
Western N America. BM, E. 1, 9, 16, 40, 41*, 42, 113, 195, 318(15:147).

sativa Molina   Coast Tarweed
•• A wool and flax seed casual. No modern records. Western N & S
America; (Australia). BM, CGE, E, NMW, OXF, RNG. 1, 12, 14, 16, 41*,
42, 500*, 546a*, 665*. Incl. *M. capitata* Nutt.; *M. gracilis* (Smith) Keck &
J.Clausen ex Appleg.; *M. racemosa* Torrey & A.Gray. The Tweedside
(Selkirks) wool alien record is an error for *Inula graveolens*.

HEMIZONIA DC.
fasciculata (DC.) Torrey & A.Gray   Fascicled Spikeweed
Pre-1930 only. California. 1, 42, 195, 500*.

fitchii A.Gray   Fitch's Spikeweed
• A casual at Bristol (W Gloucs), vector unknown. No modern records.
Western N America. CGE. 7, 42, 136, 144, 354(11:32), 500*. *Centromadia
fitchii* (A.Gray) E.Greene.

kelloggii E.Greene   Kellogg's Spikeweed
• A grain casual. No modern records. Western N America. BM, E, NMW,
OXF, SLBI. 1, 8, 42, 258, 500*.

pungens Torrey & A.Gray   Common Spikeweed
•• A grain and agricultural seed casual; decreasing. Western N America.
BM, E, K, LCN, NMW, OXF. 1, 8, 16, 41*, 42, 235, 500*. *Centromadia
pungens* (Torrey & A.Gray) E.Greene.

ramosissima Benth.   Slender Spikeweed
Pre-1930 only. California. LIV. 42, 168, 500*.

CALYCADENIA DC.
multiglandulosa DC.   Sticky Calycadenia
Pre-1930 only. California. 7, 42, 197, 500*. *Hemizonia multiglandulosa*
(DC.) A.Gray; incl. *H. cephalotes* E.Greene.

LASTHENIA Cass.
glabrata Lindley   Yellow-rayed Lasthenia
Pre-1930 only. California. 2, 42, 51, 300(1907:40), 301*(t.1780), 500*,
650*.

platycarpha (A.Gray) E.Greene   Alkali Goldfields
Pre-1930 only. California. BM, NMW. 1, 42, 354(4:416). *Baeria carnosa*
E.Greene; *B. platycarpha* A.Gray.

FLAVERIA Juss.
bidentis (L.) Kuntze sensu lato   Speedy-weed
• A casual at Avonmouth docks (W Gloucs), vector unknown. S America;
(widespread). BM, E. 37c*, 350(30:19), 646, 729, 805(65:590-636),
815*(45:235-239), 1074*. *F. contrayerba* (Cav.) Pers.; incl. *F. australasica*
Hook.

VILLANOVA Lagasca
dissecta (Hook.) DC.
Pre-1930 only. Peru. 1, 247, 554, 605. *Unxia dissecta* Hook.

BIDENS L.
aristosa (Michaux) Britton   Western Tickseed
- A casual at Avonmouth docks (W Gloucs), vector unknown. No modern records. Eastern N America. 1, 9, 38*, 40*, 350(30:19), 545, 721*. *Coreopsis aristata* Muhlenb. ex Willd.; *C. aristosa* Michaux.

bipinnata L.   Spanish-needles
- ••• A wool and bird-seed casual. Formerly a cotton alien. S America; (widespread). BM, CGE, E, LIV, OXF, RNG. 3, 14, 20*, 21, 38*, 40*, 303*(31:13,15), 371(357:24), 545, 546, 546a*, 590*, 825*(2:140-143), 1247*. *Kerneria bipinnata* (L.) Gren. & Godron.

biternata (Lour.) Merr. & Sherff
- A wool casual. Africa, S Asia. 17, 43, 47, 48*, 649*, 721*. *Coreopsis biternata* Lour.

camporum (Hutch.) Mesfin
- A bird-seed casual, perhaps also from wool. Northern tropical Africa. 19, 371(364:29), 676, 916*(24:1-138), 1302. *B. schimperi* auct., non Schultz-Bip.; *Coreopsis abyssinica* Schultz-Bip. ex Walp. Not to be confused with the Ethiopian endemic *B. carinata* Cufod. ex Mesfin (*Coreopsis schimperi* O.Hoffm.), nor with the widespread African *B. schimperi* Schultz-Bip., nor *B. setigera* (Schultz-Bip.) Sherff (*Microlecane abyssinica* (Schultz-Bip.) Benth. ex Oliver & Hiern). 328(48:437-516). All African species of *Coreopsis* are now considered to be best referred to the genus *Bidens*; the same may be appropriate for the American species.

connata Muhlenb. ex Willd.   London Bur-marigold
- •• Well established by the Grand Union Canal and adjoining waterways in S Essex, Middlesex, Herts and Bucks, and by the river Thames near Cross Ness (W Kent), vector unknown; increasing. Eastern N America; (W & C Europe). K, LTN, OXF. 20*, 21, 22*, 303*(18:15-16; 37:24-25), 331(58:12-13; 64:113), 546a*.

ferulifolia (Jacq.) DC.   Fern-leaved Beggarticks
Pre-1930 only. Arizona, Mexico. OXF. 7, 51a, 301*(t.684), 312*(t.2059), 354(6:25), 721*. *B. procera* D.Don; *Coreopsis ferulifolia* Jacq.

frondosa L.   Beggarticks
- ••• Established by canals, especially in the Midlands and at ports, vector unknown; also a wool casual; increasing. N & S America; (New Zealand). BM, E, K, NMW, OXF, RNG. 7, 14, 20*, 21, 22*, 25*, 36c*, 331(58:10), 546a*. *B. melanocarpa* Wieg. Var. *anomala* Porter ex Fernald has been recorded. 354(12:282).

laevis (L.) Britton, Sterns & Pogg.   Larger Bur-marigold
Pre-1930 only. N & S America. 2, 38*, 40*, 42, 267, 545, 721*, 1065*. *B. chrysanthemoides* Michaux.

pilosa L.   Black-jack
- ••• A wool, cotton and bird-seed casual. S America; (widespread). BM, CGE, E, LIV, OXF, RNG. 3*, 14, 20*, 21, 36c*, 37c*, 44a*, 303(18:11; 31*:15), 371(361:26), 545, 546a*, 1247*. *B. leucantha* (L.) Willd.; *Coreopsis coronata* L.

subalternans DC.
- (but probably overlooked as *B. bipinnata*). A wool casual. S America; (Australia). RNG, HbALG. 17, 21, 303(37:14), 331(58:9), 546a*, 590*, 721*, 825*(2:140-143), 1074*.

tenuisecta A.Gray
- A wool casual. Western USA, Mexico. HbEJC. 331(58:9), 652, 627, 699, 721*. *B. cognata* E.Greene.

vulgata E.Greene   Tall Beggarticks
- •• A wool casual. N America. BM, K, OXF, RNG. 14, 20, 38*, 40*, 41*, 145, 349(2:240), 546a*.

COREOPSIS L.
grandiflora Hogg ex Sweet   Large-flowered Tickseed
- •• A persistent or established garden escape. Eastern N America. RNG, SLBI. 7, 38*, 51*, 54*, 56*, 126, 192, 199, 545.

nuecensis A.A.Heller
- A casual garden escape. SE Texas. 51b, 236, 312*(t.3460), 371(343:30), 545, 1042a, 1302(p.358). *C. coronata* sensu Hook., non L.

tinctoria Nutt.   Garden Tickseed
- •• A garden escape and wool casual. N America. BM, E, LIV, NMW, OXF, RNG. 1, 17, 20, 38*, 51b, 55*, 236, 545. *C. bicolor* hort.; *Calliopsis bicolor* Reichb. Incl. *Coreopsis atkinsoniana* Douglas ex Lindley.

verticillata L.   Whorled Tickseed
- A garden escape established for a time on acid heathland at Church Crookham (N Hants), now gone. USA. HbARGM. 20, 38*, 51b, 53*, 55*, 303(39:9).

COSMOS Cav.
bipinnatus Cav.   Mexican Aster
- •••• A persistent garden escape and a bird-seed, wool and cotton alien. SW USA, Mexico; (Africa, New Zealand). 14, 19, 20, 36c*, 55*, 56*, 126, 191, 892(6:32-33), 1247*. *Bidens formosa* (Bonato) Schultz-Bip.

THELESPERMA Less.
trifidum (Poiret) Britton   Fine-leaved Thelesperma
Pre-1930 only. N America. 38*, 40, 51, 251. *Coreopsis trifida* Poiret.

TRIDAX L.
trilobata (Cav.) Hemsley
Pre-1930 only. Mexico. CGE. 51, 312*(t.1895), 333(16:118), 650. *Galinsoga trilobata* Cav.

MELAMPODIUM L.
sp.
Pre-1930 only. America. 179.

DAHLIA Cav.

× **cultorum** hort. group   Garden Dahlias
•• Casual garden escapes. Mexico, or originated in cultivation. 20, 54*, 55*, 199, 236. A complex of hybrids involving *D. coccinea* Cav.; *D. jaurezii* hort. ex Sas., nom. nud.; *D. merckii* Lehm.; *D. pinnata* Cav. (*D. variabilis* (Willd.) Desf.) and *D. rosea* Cav.

TAGETES L.
erecta L.   African Marigold
• (but confused with *T. patula*). A casual garden escape. Mexico, C America. BM, OXF. 20*, 53*, 54*, 55*, 199, 354(24:45).
erythrocephala Rusby
• A wool casual. Colombia, Ecuador, Bolivia, Argentina. HbEJC. 17, 812(8:133). *T. multiflora* Kunth.
[*foetidissima* hort. ex DC. Unconfirmed; probably all records are referable to *T. gracilis*. 434.]
gracilis DC.
• A wool casual. Peru, Bolivia, Argentina. E, RNG. 17 (as *T. foetidissima*), 554.
micrantha Cav.
Pre-1930 only. SW USA, Mexico. 1, 16, 538*, 652*, 707, 710*. The 1926 Avonmouth (W Gloucs) record was in error for *T. minuta*.
minuta L.   Southern Marigold
••• A wool, grain, cotton and bird-seed casual. S America; (widespread). ABD, BM, CGE, E, OXF, RNG. 7, 14, 20*, 21, 37c*, 42, 47*, 303*(18:1,10), 1247*. *T. minima* sphalm.
patula L.   French Marigold
••• A casual garden escape. Mexico, Guatemala. LIV, RNG. 20*, 53*, 54*, 55*, 236, 354(10:471). Probably not specifically distinct from *T. erecta* in spite of traditional separation in horticultural literature.
tenuifolia Cav.   Signet Marigold
• (but confused with *T. patula*). A casual garden escape. Mexico, C America. BM. 51, 55*, 135, 235, 594*, 650*, 710*. *T. signata* Bartlett; incl. *T. pumila* L.

SCHKUHRIA Roth
pinnata (Lam.) Thell.   Dwarf Marigold
••• A wool and bird-seed casual. Tropical America; (widespread). BFT, BM, CGE, E, OXF, RNG. 1, 14, 20*, 21, 354(12:359), 565*, 841*(n.s. 4:245), 892*(10:119-121). *Pectis pinnata* Lam.; incl. *S. abrotanoides* Roth; *S. advena* Thell.; *S. bonariensis* Hook. & Arn.; *S. isopappa* Benth.; *S. pedata* A.Gray.

BAHIA Lagasca
dissecta (A.Gray) Britton
Pre-1930 only. SW USA, Mexico. LSR. 42, 184, 500*, 652*, 699*.
Villanova dissecta (A.Gray) Rydb., non (Hook.) DC.; *Amauria dissecta*
A.Gray; *Amauriopsis dissecta* (A.Gray) Rydb.
cf. **neomexicana** (A.Gray) A.Gray   **New Mexico Bahia**
Pre-1930 only. SW USA, Mexico. E, OXF. 3, 42, 354(5:34), 500*, 699.
Schkuhria neomexicana A.Gray.

GAILLARDIA Foug.
aristata group   **Blanket-flowers**
•• Garden escapes established on light soils in southern England; naturalised
on heathland on Witley Common (Surrey) and on dunes at Greatstone-on-Sea
(E Kent). N America. CGE, LIV, OXF, RNG. 20*, 38*, 54*, 55*, 209,
331(37:203), 458, 1110*. Incl. *G. aristata* Pursh; *G. pulchella* Foug.; and
hybrids. Most records are referable to *G.* × *grandiflora* hort. ex Van Houtte
(*G. aristata* × *G. pulchella*).

HELENIUM L.
autumnale group   **Sneezeweeds**
•• Persistent garden escapes, perhaps established in a chalk pit near Reigate
and at the edge of a plantation at Croydon (both Surrey). N America. BM.
20, 38*, 52*, 54*, 56*, 199, 349(5:26-27). Incl. *H. autumnale* L.;
H. pumilum Schldl.; and putative hybrids with other species.
quadridentatum Labill.
• A wool casual. Texas, Mexico. BM, E, RNG. 17, 545, 661*, 699*,
803*(1:22). *H. elegans* DC.

AGERATUM L.
houstonianum Miller   **Flossflower**
•• A casual garden escape. Mexico, C America. LIV, OXF, RNG. 3, 20,
21, 54*, 56*, 199, 1247*. *A. mexicanum* Sims.

LIATRIS Gaertner ex Schreber
ligulistylis (Nelson) Nelson
• A casual garden escape. N America. 40, 223, 673*.

STEVIA Cav.
paniculata Lagasca
Pre-1930 only. Mexico. OXF. 312*(t.1861), 674. *S. hyssopifolia* Sims, non
Cav., nec Kunth.

MONOCOTYLEDONAE
LIMNOCHARITACEAE
HYDROCLEYS Rich.
nymphoides (Willd.) Buchenau   Water-poppy
  Pre-1930 only. Tropical S America. 36b*, 50, 171, 544*, 545(p.100), 1065*,
  1170*, 1230*. *Limnocharis humboldtii* Rich.

ALISMATACEAE
SAGITTARIA L.
graminea Michaux   Grass-leaved Arrowhead
  • Recorded from a gravel pit near Sandhurst (Berks). N America. 38*, 40*,
  50, 87*, 303(26:18), 646*. Incl. *S. platyphylla* (Engelm.) J.G.Smith.
latifolia Willd.   Duck-potato
  •• An established garden escape and relic; naturalised in a few ponds and on
  streamsides in southern England and Jersey (Channel Is). N & C America.
  BM, JSY, LANC, RNG. 12, 20, 22*, 38*, 41*, 50, 122, 126, 201, 209,
  826*(27:55-58).
rigida Pursh   Canadian Arrowhead
  • An established water-garden escape; naturalised since 1898 in the Exeter
  Canal (S Devon); also recorded from Clowance Lake, near Camborne
  (W Cornwall). Eastern N America. BM, LANC, OXF, RAMM, RNG, TOR.
  7, 20, 22*, 33*, 38*, 189, 212, 217, 320*(1908:273-278), 354(6:858; 7:785).
  S. heterophylla Pursh, non Schreber; incl. var. *iscana* Hiern.
subulata (L.) Buchenau   Narrow-leaved Arrowhead
  • An escape from aquaria established since 1962 in a pond on Shortheath
  Common (N Hants). Eastern N America. BM, RNG. 20, 38*, 40*, 50,
  303(36:29), 370(10:411,431), 1230*.

CALDESIA Parl.
parnassifolia (L.) Parl.   Parnassus-leaved Water-plantain
  • An aquatic plant established in woods at Kinfauns (E Perth). Europe. 21,
  22*, 25*, 51, 79*, 371(374:8), 546*. *Alisma parnassifolium* L. Contrary to
  much literature, this species is endemic to Europe; extra-European plants
  appear to be referable to the closely allied *C. reniformis* (D.Don) Makino.

HYDROCHARITACEAE
EGERIA Planchon
densa Planchon   Large-flowered Waterweed
  •• An established escape from aquaria; naturalised in canals at Failsworth and
  Droylesden, Manchester (both S Lancs) and in the canal and river system in the
  Huddersfield-Brighouse-Elland area (SW Yorks); also reported from Amberley
  (W Sussex). Argentina; (widespread). BM, BON, LANC, NMW, RNG. 13,

20, 22*, 25*, 33*, 87*, 122, 251, 349(1:322), 1065*.   *Elodea densa*
(Planchon) Caspary.

ELODEA Michaux

callitrichoides (Rich.) Caspary   South American Waterweed
●●● (but confused with *E. nuttallii*).   An apparently short-lived escape from
aquaria, found in several waterways in southern England and Wales; first
recorded in 1948.   Southern S America.   BM, CGE, E, LANC, OXF, RNG.
12, 20, 22*, 192, 349(1:321), 370(15:1-9; 16*:1-14), 826*(14:33-56), 1275*.
E. ernstiae H.St John; *Anacharis callitrichoides* Rich.

canadensis Michaux   Canadian Waterweed
●●●●● An established escape from garden pools and aquaria; first recorded
in 1836, now long naturalised in many canals, streams and ponds throughout
the British Is.   N America; (widespread).   1, 20, 22*, 25*, 303*(12:15),
370(15:1-9; 16*:1-14), 1275*.   *Anacharis alsinastrum* Bab.; *A. canadensis*
(Michaux) Planchon; *Udora canadensis* (Michaux) Nutt. pro parte.   A putative
hybrid with *E. nuttallii* has recently been recorded from Littlebrook Marshes
(W Kent).   303(60:35).

nuttallii (Planchon) H.St John   Nuttall's Waterweed
●●●● An established escape from aquaria; first recorded in 1966, now widely
naturalised in canals, streams and ponds in England, and recorded from a few
localities in Wales, Scotland and Ireland; increasing, displacing *E. canadensis*
in many places.   N America; (Europe).   20, 22*, 25*, 303*(12:14-16),
370(11:427; 15:1-9,402; 16*:1-14), 1275*.   *E. occidentalis* (Pursh) H.St John,
nom. illeg.; *Anacharis nuttallii* Planchon; *Hydrilla lithuanica* Dandy pro parte.
The name of this species has been widely misapplied to native *Hydrilla
verticillata*.   16, 377, 319(18:327-331).

LAGAROSIPHON Harvey

major (Ridley) Moss   Curly Waterweed
●●●● An established escape from aquaria; naturalised in canals, ponds and
quarry pools in widely scattered localities.   S Africa; (widespread).   12, 20,
22*, 25*, 87*, 143, 182, 251, 349(1:322).   *L. muscoides* Harvey var. *major*
Ridley; *Elodea crispa* hort.

VALLISNERIA L.

spiralis L.   Tapegrass
●●● An established aquarium escape or introduction; abundantly naturalised
in the Lea Navigation Canal (S Essex, N Essex, Herts, Middlesex) and the
Reddish Canal (S Lancs); also recorded from canals in W Gloucs and
SW Yorks.   Tropics, subtropics.   BM, BON, E, LANC, OXF, RNG.   7, 20,
22*, 25*, 44a*, 87*, 191, 251, 370(9:253-256).

APONOGETONACEAE
APONOGETON L.f.
distachyos L.f.   Cape-pondweed
••• An established water-garden escape; naturalised in ponds and canals in widely scattered localities; known since 1924 at Keston Ponds (W Kent). S Africa; (widespread). BM, DBY, E, OXF, SLBI. 1, 20, 22*, 87*, 126, 544*, 600*.

[*POTAMOGETONACEAE*
POTAMOGETON L.
foliosus Raf. In error. 354(6:632,860; 10:35).]

NAJADACEAE
NAJAS L.
graminea Del.
• A cotton alien established between 1883 and 1947 in the Reddish Canal (S Lancs). Old World tropics and subtropics. BM, LIV, LANC, OXF, RNG. 1, 20, 251, 320*(22:262), 544*, 571*, 1107*. Recorded as var. *delilei* Magnus.

ARECACEAE
Palmae
TRACHYCARPUS H.A.Wendl.
fortunei (Hook.) H.A.Wendl.   Chusan Palm
• Self-sown from gardens at Bosahan, near Helford, and at Trelowarren on the Lizard peninsula (both W Cornwall) and at Abbotsbury (Dorset). China. 20, 21, 50, 54*, 58*, 371(403:40), 403, 505*. *Chamaerops fortunei* Hook.

PHOENIX L.
dactylifera L.   Date
•••• A food refuse casual on tips. N Africa, SW Asia. 2, 20, 21, 1156*.

ARACEAE
ACORUS L.
calamus L.   Sweet-flag
••••• A long-established escape from cultivation; naturalised by ponds, rivers and canals in many places throughout the British Is, especially in England. E Asia; (widespread). 1, 20, 22*, 29*, 33*, 79*. *Acorus vulgaris* sphalm.
gramineus Aiton   Slender Sweet-flag
• Recorded from Mytchett Lake (Surrey). E Asia. 20, 43, 50, 371(409:38-41), 638*.

LYSICHITON Schott
americanus Hultén & H.St John   American Skunk-cabbage
●●●● An established garden escape, freely reproducing by seed; naturalised by streams and ponds in widely scattered localities; increasing. Western N America. 12, 20, 22*, 28*, 50, 54*, 209, 1039*. *L. camtschatcensis* auct., non Schott.

camtschatcensis (L.) Schott   Asian Skunk-cabbage
●● (but probably over-recorded for *L. americanus*). An established garden escape; by streams near Virginia Water and on Fairmile Common (both Surrey) and by a lakeside at Margam (Glam). E Asia. BM. 20, 50, 51*, 52*, 54*, 57*, 156, 319(13:120,182), 349(7:632). *L. japonicus* hort. A hybrid with *L. americanus* occurs in gardens and may have been overlooked. 442. Records of this species as naturalised and spreading rapidly in Ireland are referable to *L. americanus*. 1292.

CALLA L.
palustris L.   Bog Arum
●●● Introduced, or a garden escape, in ponds, bogs and by streams in widely scattered localities in Britain, sometimes long established; known for at least 70 years at Boldermere (Surrey), now gone. Europe, N Asia, N America. BM, DBY, LANC, OXF, RNG. 1, 20, 22*, 25*, 26*, 209, 1242*.

AMORPHOPHALLUS Blume ex Decne.
[*abyssinicus* N.E.Br.   In error for *Dracunculus vulgaris*. 303(19:16).]
rivieri Durieu   Umbrella Arum
● Introduced in an orchard at Harwell (Berks). SE Asia. 50, 67*, 119, 1119*, 1229*. According to 854(7:1,7-8) this plant is conspecific with the food plant *A. konjac* K.Koch, a name which has priority. 1039*.

ZANTEDESCHIA Sprengel
aethiopica (L.) Sprengel   Altar-lily
●●● An established garden escape; naturalised in ditches and damp places in the Channel Is, SW England, W Kent, Glam and S Kerry; known for over 50 years in Fermain Valley, Guernsey (Channel Is). S Africa; (widespread). RNG. 20, 21, 52*, 54*, 57*, 66*, 207, 212, 220. *Calla aethiopica* L.

DIEFFENBACHIA Schott
seguine (Jacq.) Schott   Dumb-cane
● A casual greenhouse escape. Tropical America. 50, 54*, 458, 503*. *D. maculata* Sweet; *D. picta* Schott.

COLOCASIA Schott
esculenta (L.) Schott   Taro
● A food refuse casual on tips near London. SE Asia; (widespread). 21, 50, 66*, 68*, 69*, 80*, 236. Incl. *C. antiquorum* Schott.

ARUM L.

italicum Miller subsp. **canariense** (Webb & Berth.) Boyce
  Pre-1930 only. Canary Is, Madeira. 320(1:25), 1040, 1218. Recorded in
  1863 as being naturalised in Guernsey (Channel Is), with no records since then;
  possibly overlooked here or elsewhere. *A. canariense* Webb & Berth.

italicum Miller subsp. **italicum**
  •••• (but much confused with its congeners). A garden escape established in
  widely scattered localities. S Europe. 20, 22*, 79*, 201, 213, 370(16:196),
  1040*, 1122. A hybrid with native *A. maculatum* has been recorded. 1311.

DRACUNCULUS Miller

vulgaris Schott   Dragon Arum
  ••• An established garden escape; naturalised in hedges and on grassy slopes
  in several places in the south. Medit. BM, RNG. 7, 20, 24*, 29*, 79*, 207,
  212, 236. *D. dracunculus* (L.) Voss; *Arum dracunculus* L.

ARISARUM Miller

proboscideum (L.) Savi   Mouse-tail Arum
  • A garden escape established in a hedgerow near Dorchester (Dorset), on
  Harrow Weald Common (Middlesex) and near Salford Priors (Warks); a garden
  relic in W Cornwall. Spain, Italy. BM. 20, 50, 57*, 66*, 79*, 213,
  331(63:146), 349(6:125), 1039*. *Arum proboscideum* L.

PISTIA L.

stratiotes L.   Water-lettuce
  • A casual escape from aquaria. Tropics, subtropics. HbEJC. 49, 50, 66*,
  87*, 303(36:29), 582*, 825*(7:28-29), 1230*.

LEMNACEAE

LEMNA L.

minuta Kunth   Least Duckweed
  •••• An established escape from aquatic nurseries; widely naturalised,
  sometimes abundantly, in ponds, rivers and canals, mainly in southern England;
  increasing rapidly since its discovery in 1977. N & S America. 20, 38*,
  370(14*:243-248; 17:483), 826*(25:86-98). *L. minima* Philippi; *L. minuscula*
  Herter, nom illeg.; *L. valdiviana* auct., non Philippi.
  [*valdiviana* Philippi In error for *L. minuta*. 370(14:243-248), 371(391:32).]

COMMELINACEAE

TRADESCANTIA L.

fluminensis Vell.Conc.   Wandering-Jew
  •• A persistent greenhouse or garden escape; known for well over 30 years on
  rough ground at La Tertre, Câtel, Guernsey (Channel Is), but now gone; casual
  elsewhere in the south. S America. BM. 20, 50, 66*, 199, 331(62:109),
  371(376:25), 503*, 519*, 593*, 719*. Incl. *T. albiflora* Kunth; *T. tricolor*
  hort. ex C.B.Clarke; *T. viridis* hort. ex Vilm., in syn.

pallida (Rose) Hunt   Purple-heart
* A casual greenhouse escape. Mexico. 50, 54*, 66*, 303(29:12), 503*, 680*. *Setcreasea pallida* Rose; *S. purpurea* Boom.

virginiana group   Spiderworts
••• Garden escapes established in a few places in southern England. Originated in cultivation. LANC, OXF, RNG. 20, 50, 52*, 54*, 199, 331(46:34), 369(22:48), 506*, 680*. *T.* × *andersoniana* W.Ludwig & Rohw., nom. inval.; *T. virginiana* L. Some records are probably referable to modern garden spiderworts, which are complex hybrids between *T. virginiana* and *T. canaliculata* Raf., *T. ohiensis* Raf. and *T. subaspersa* Ker Gawler.

zebrina hort. ex Bosse   Inch-plant
* A casual greenhouse escape. Mexico. OXF. 7, 20, 50, 66*, 331(44:27), 354(8:319), 371(394:36), 593*, 600*, 680*. *Zebrina pendula* Schnitzl.

COMMELINA L.

coelestis Willd.   Blue Spiderwort
* A garden escape and oil-seed casual. Mexico, C & S America. OXF. 50, 51*, 51b*, 66*, 199, 354(9:571,841), 370(9:189).

communis L.
* A casual at Avonmouth docks (W Gloucs), vector unknown. No modern records. Asia; (Europe, N America). OXF. 7, 9, 21, 38*, 50, 51b*, 354(6:318), 571*, 804*(t.20), 1138*, 1261*, 1356*.

diffusa Burman f.   Creeping Spiderwort
* A casual greenhouse escape. Tropics, subtropics. NMW, OXF. 7, 38*, 50, 51b*, 156, 192, 354(8:319), 1226*. *C. nudiflora* auct., non L.

elliptica Kunth   Peruvian Spiderwort
Pre-1930 only. Tropical America. OXF. 7, 50, 51, 312*(t.3047), 354(6:629), 705*. *C. gracilis* Ruíz & Pavón. Possibly conspecific with *C. tuberosa* L.

virginica L.
* A greenhouse plant reported from Trafford Park (S Lancs). Eastern N America. BON. 38*, 51, 469, 1226*, 1319*. *C. caerulea* Salisb.

JUNCACEAE

JUNCUS L.

aridicola L.A.S.Johnson   Tussock Rush
•• A wool alien persistent in orchards at Barming (W Kent). Australia. BM, RNG. 17, 37c*, 192, 236, 371(382:25), 590*, 666*.

australis Hook.f.   Austral Rush
* A wool alien persistent in orchards at Barming (W Kent). Australasia. BM, LTN, RNG. 17, 36a, 36b, 37c*, 145, 236, 533, 590*, 603*, 1170*. Hybrids between various native and alien species, including *J. australis*, have occurred, but none appear to have been precisely identified.

continuus L.A.S.Johnson
* A wool casual. No modern records. Australia. BM, RNG. 36b, 37c*, 192, 569*, 590*, 720*, 725*.

distegus Edgar
- A wool alien persistent in orchards at Barming (W Kent).   New Zealand. RNG. 17, 36a, 36b, 236, 1170*.

dudleyi Wieg.   Dudley's Rush
- Naturalised in marshy ground by a roadside near Crianlarich (Mid Perth) and recorded from Rhum (N Ebudes), but not recorded recently; also a wool casual. N America. BM, E, K, OXF, RNG. 16, 20, 21, 38*, 40*, 354(9:251,283), 1279*. Possibly conspecific with *J. tenuis.*

ensifolius Wikström   Sword-leaved Rush
- A casual on a canal bank at Chester (Cheshire), vector unknown. No modern records. Japan, Western N America; (N Europe, New Zealand). LIV, RNG. 21, 41*, 43, 73*, 349(3:49), 825*(8:77-80), 826*(21:86-88), 1170*.   The eastern N American *J. canadensis* J.Gay ex Laharpe may have been overlooked. 38*, 825*(7:17-20).

flavidus L.A.S.Johnson
- A wool casual. Australia; (New Zealand). BFT, BM, CGE, E, K, RNG. 17, 36b, 37c*, 1170. *J. flavescens* L.A.S.Johnson, nom. nud., non Host.

gregiflorus L.A.S.Johnson
- A wool casual. Australia. BM, LTN, RNG. 17, 145, 514, 590*, 819(3:243-244), 1170*, 1363*. The New Zealand plant under this name (36b*) is distinct and is now known as *J. edgariae* L.A.S.Johnson, nom. nud. The identity of our plants needs confirmation.

hybridus Brot.
  Pre-1930 only. Medit. BM.   21, 39*, 370(12*:113-128; 14:263-272). *J. mutabilis* Savi, non Lam.

imbricatus Laharpe
- A wool casual. Temperate S America; (New Zealand). E, RNG. 17, 21, 36b, 514, 535*, 884*(4:107), 911*(4:141).

involucratus Steudel
- A wool casual. Andes. BM. 564, 820*(10:438), 1066*.

ochrocoleus L.A.S.Johnson
- A wool casual. Australia. BM. 590*, 919(5:312-313). *J. ochrolepis* L.A.S.Johnson, nom. nud. Originally misidentified as *J. effusus* L. × *J. pallidus*; all records of this hybrid probably belong here.

oxycarpus E.Meyer
- A wool casual. Africa. E, K, RNG. 17, 592, 676, 736*, 801*(4:431).

pallidus R.Br.   Great Soft-rush
- • A wool and bird-seed alien established in a few gravel pits in Middlesex and Beds, probably now gone, but still present in gravel pit ditches near Evesham (Worcs). Australasia. BM, E, LTN, OXF, RNG, SLBI. 14, 20*, 36a*, 37b*, 143, 192, 354(13:172), 371(419:15), 520*, 590*, 1170*. A hybrid with native *J. inflexus* has been recorded. 20, 1311. Records of a hybrid with native *J. effusus* are probably referable to *J. ochrocoleus.*

cf. **pauciflorus** R.Br.   Loose-flowered Rush
- A wool casual. Australasia. MNE. 36a, 37c*, 251, 520*, 590*, 837*(95:21-22), 1170*.

[*pelocarpus* E.Meyer   In error for native *J. articulatus*. 272.]

planifolius R.Br.   Broad-leaved Rush
- Well established and colonising wet habitats over an area of about 16 square miles in the Carna-Glink peninsula (W Galway); vector unknown. Australasia, S America. BEL, BM, DBN, OXF, RNG. 20*, 21, 36b*, 37c*, 73*, 319*(17:307-309), 370(10:418), 590*, 1170*, 1347*.

[*polyanthemus* Buchenau Unconfirmed; probably always in error for *J. continuus* or other species. BM.]

procerus E.Meyer
- A wool alien persistent in orchards at Barming (W Kent). S America; (Australasia). LTN, RNG. 14, 17, 36b, 37c*, 145, 236, 514, 590*, 669*, 1170*, 1363*.

radula Buchenau   Hoary Rush
- A wool casual. Australia. OXF, RNG. 17, 37, 37c*, 590*, 743.

sarophorus L.A.S.Johnson
- A wool casual. Australasia. BM, LTN, RNG. 17, 36a, 36b, 37c*, 145, 590*, 819(3:242-243), 1170*, 1363*.   *J. luxurians* auct., non Colenso; *J. polyanthemos* sensu J.Black pro parte; *J. polyanthemus* 'southern race'. 485.

subsecundus N.A.Wakef.   Fingered Rush
- A wool alien persistent in orchards at Barming (W Kent). Australia; (New Zealand). 36b, 37c*, 236, 533, 590*, 720*, 725*. *J. vaginatus* sensu J.Black pro parte, non R.Br.

subulatus Forsskål   Somerset Rush
- Established, possibly brought in by birds; naturalised in a salt marsh at Berrow (N Somerset) and on reclaimed land at Grangemouth (Kirkcudbrights). Medit. BM, E, LANC, OXF, RNG. 20*, 39*, 44a*, 73*, 304(9:12), 350(35:16), 370*(4:211-217).

tenuis Willd. sensu lato   Slender Rush
- ••••• Long established, possibly originally imported with hay from America; naturalised in woodland rides, by paths and on roadsides in many places throughout the British Is, sometimes abundantly; increasing. N America; (widespread). 2, 20*, 35*, 73*, 255, 283, 300(1897:32,197,260), 354(9:282). Incl. *J. bicornis* Michaux; *J. gesneri* Smith; *J. gracilis* Smith, non Roth; *J. macer* Gray. Some records are referable to var. *anthelatus* Wieg. 40*, 349(2:23).

cf. **uruguensis** Griseb.
Pre-1930 only. S America. E, OXF. 3, 354(5:54), 820*(10:389). Possibly in error for *J. imbricatus*.

usitatus L.A.S.Johnson
- A wool alien persistent in orchards at Barming (W Kent). Australasia. BM, CGE, E, K, LTN, RNG. 17, 36a, 36b, 37c*, 145, 192, 236, 569*, 590*, 720*, 725*, 819(3:241-242), 1170*. *J. polyanthemus* auct., non Buchenau.

vaginatus R.Br.   Clustered Rush
  • A wool casual. Australia. BM, E, LTN, MNE, OXF, RNG. 3, 145, 251, 354(5:53), 514, 533, 576*, 590*. This name has been much misapplied, especially to *J. australis* and less so to other species including *J. subsecundus*.

LUZULA DC.
luzuloides (Lam.) Dandy & Wilm.   White Wood-rush
  •••• Long established, probably originally introduced as a landscape plant, perhaps also a grass seed and grain alien; naturalised in woods, on roadsides and on railway banks, sometimes abundantly, in scattered localities throughout Britain. Europe. 1, 20, 33*, 35*, 39*, 73*. *L. albida* (Hoffm.) DC.; *L. cuprina* Rochel ex Asch. & Graebner; *L. nemorosa* (Pollich) E.Meyer, non Hornem.; *Juncoides nemorosum* (Pollich) Kuntze; *J. rubellum* Hoppe.
nivea (L.) DC.   Snow-white Wood-rush
  Pre-1930 only. Europe. BM, K, LIV, OXF, SLBI. 1, 20, 27*, 29*, 73*, 354(5:683), 361(6:286), 1279*. *Juncoides niveum* (L.) Kuntze.

CYPERACEAE

ELEOCHARIS R.Br.
nodulosa (Roth) Schultes
  • A wool casual. S America. HbEJC, HbTBR. 17, 49, 370(11:69), 866*(ser.2, 13:256), 907*(9:375), 1050*.

SCIRPUS L.
cf. **inundatus** (R.Br.) Poiret   Swamp Club-rush
  • A wool casual. Malaysia, Australasia, S America. E, K. 36b, 37, 37b, 520*, 720*, 1170*. *Isolepis inundata* R.Br.
nodosus Rottb.   Knotted Club-rush
  • A wool casual. Southern hemisphere. BM, K, RNG. 17, 36b, 37b*, 335(904:24), 370(11:69), 520*, 540*, 1170*, 1363*. *Isolepis nodosa* (Rottb.) R.Br.

CYPERUS L.
albostriatus Schrader
  • A casual greenhouse escape. S Africa. HbACL. 36b, 37c*, 50, 199, 331(70:160), 720*, 725*. *C. diffusus* hort., non Vahl.
brevifolius (Rottb.) Endl. ex Hassk.   Globe Kyllinga
  • A wool casual. Asia, Tropical America; (widespread). HbTBR. 21, 37c*, 370(11:69), 599*, 649*, 720*. *Kyllinga brevifolia* Rottb.; *K. colorata* (L.) Druce.
clarus S.T.Blake
  • A wool casual. Australia. K, HbTBR. 37, 37b, 370(11:69), 720*, 725*.

congestus Vahl   Dense Flat-sedge
- A wool casual. S Africa; (widespread). BM, CGE, E, LTR, OXF, RNG. 3*, 17, 21, 36b*, 39*, 354(4*:215; 5:528), 370(11:69), 371(409:39), 578a*, 1170*, 1194*. *C. strigosus* Willd., non L.; *Mariscus congestus* (Vahl) C.B.Clarke.

cyperinus (Retz.) Sur.
    Pre-1930 only. Tropical Asia, Polynesia, N Australia. 7, 354(5:583), 604, 720, 1215*, 1347. *Kyllinga cyperina* Retz.; *Mariscus cyperinus* (Retz.) Vahl.

dactylotes Benth.
- A wool casual. Australia. LIV, HbEJC, HbTBR. 37, 37b, 370(11:69), 720*, 1169*.

difformis L.   Variable Flat-sedge
- A casual near Bristol (W Gloucs), possibly brought in with rice. No modern records. Origin obscure; (widespread). 9, 21, 37, 37b*, 39*, 44a*, 720*, 1130*, 1169*, 1247*, 1347*.

eragrostis Lam.   Pale Galingale
- ●●● A garden escape and wool alien established in a few places in the south; well naturalised and spreading in the Channel Is, perhaps carried here by wildfowl from colonies in N France. N & S America; (widespread). BM, CGE, E, NMW, OXF, RNG. 7, 14, 20, 21, 36b*, 50, 73*, 117*, 220, 720*, 826*(18:101-104), 1170*. *C. declinatus* Moench; *C. limbatus* C.B.Clarke; *C. vegetus* Willd.

erectus (Schum.) Mattf. & Kük.
- A wool casual. Africa. K, RNG. 17, 370(11:69), 676, 1130*. *Kyllinga erecta* Schum.

esculentus L.   Yellow Nutsedge
- A food refuse casual on tips; also a wool casual. Medit.; (widespread). BM, LTN, RNG. 21, 38*, 39*, 40*, 46*, 50, 145, 349(1:158), 720*, 1130*. Although a serious weed in maize crops in N Europe, this species has not yet been recorded from arable fields in the British Is. 825(12*:135-138; 19:65-73).

flavus (Vahl) Boeckeler   Cayenne Cyperus
- A casual at Bristol (W Gloucs), vector unknown. W Indies; (Australia, N America). BM, RNG. 38*, 40*, 303(33:10; 38*:27), 370(14:229), 720*, 1199*. *C. cayennensis* (Lam.) Britton; *Mariscus flavus* Vahl.

globulosus Aublet   Baldwin's Cyperus
- A wool casual. N & S America. HbTBR. 17, 38*, 40*.

gunnii Hook.f.   Flecked Flat-sedge
- A wool casual. Australia. K, HbTBR. 17, 36b, 37*, 37b*, 370(11:69), 720.

involucratus Rottb.
- Recorded from the banks of Seven Kings Water, near Little Heath (S Essex), probably a garden escape, and from long-derelict nursery walls at Friern Barnet (Middlesex). Africa; (tropics, subtropics). RNG. 36b*, 50, 53*, 87*, 303(64:41), 581*, 720*, 1130*. *C. alternifolius* L. subsp. *flabelliformis* (Rottb.) Kük.; *C. flabelliformis* Rottb.

longus L. subsp. **tenuiflorus** (Rottb.) Kük.
- A wool casual. E & S Africa. RNG. 520*, 1130. *C. badius* Desf. var. *tenuiflorus* (Rottb.) Husnot; *C. longus* var. *tenuiflorus* (Rottb.) Boeckeler; *C. tenuiflorus* Rottb.

luzulae (L.) Retz.
- A wool casual. Mexico, W Indies, S America. K, RNG. 49, 370(11:69), 654*, 1199*. *C. luzuloides* sphalm.

ovularis (Michaux) Torrey   Globose Cyperus
- A wool casual. N America. HbTBR. 21, 38*, 40*, 370(11:69).

reflexus Vahl
- A wool casual; persistent for several years on a fruit farm at Blackmoor (N Hants). Subtropical N & S America. BM, LIV, HbEJC. 303*(37:18-19), 514.

rigidifolius Steudel
- A wool casual. E Africa. K, LIV, RNG. 370(11:69), 676, 1130*.

rotundus L.   Purple Nutsedge
- A persistent food-refuse and wool alien.   Widespread in tropics and subtropics. BM, K, LTN, RNG. 17, 21, 37b*, 38*, 39*, 44a*, 192, 349(1:159), 370(11:69), 720*, 1130*.

rutilans (C.B.Clarke) Maiden & Betche
- A wool casual. Australia. BM, K, RNG. 17, 37, 37b, 370(11:69), 514, 564*, 743. *Mariscus rutilans* C.B.Clarke.

sesquiflorus (Torrey) Mattf. & Kük.
- A wool casual. Africa, America; (Australia). RNG, HbTBR. 49, 370(11:69), 720*, 1130*. *Kyllinga odorata* Vahl, non *Cyperus odoratus* L.; *K. sesquiflorus* Torrey.

sporobolus R.Br.
- A wool casual. Australia. LIV, HbTBR. 370(11:69), 516, 720*.

tenuis Sw.
- A wool casual. Tropical & S Africa, Mexico, S America. K, RNG. 17, 49, 370(11:69), 828*(26:70). *Mariscus tenuis* (Sw.) C.B.Clarke.

ustulatus A.Rich.   Giant Umbrella-sedge
- A wool casual. New Zealand. HbTBR. 36a, 370(11:69), 1170*.

vaginatus R.Br.   Stiff-leaved Flat-sedge
- A wool casual. Australia. BM, E, RNG. 14, 17, 37*, 37b*, 370(11:69), 520*, 540*, 720*, 725*, 1169*. Incl. *C. gymnocaulos* Steudel.

CAREX L.
appressa R.Br.   Tall Sedge
- A wool casual. Australasia. BM, RNG. 17, 36a, 37b, 370(11:70), 520*, 564*, 569*, 720*.

bicolor All.   Bicoloured Sedge
- Introduced on Rhum (N Ebudes), now gone. N & C Europe, Greenland, Iceland. 21, 81, 337(163:105), 354(12:475), 368(34:270-277), 546*, 1026*, 1203*, 1242*.

[*brizoides* L. A dubious record. 1, 81.]

brunnea Thunb.
- Established, possibly brought in with bamboos, in the Kinloch grounds, Rhum (N Ebudes). No modern records. S Asia, Australasia. RNG. 43, 354(13:39), 368(34:276), 514, 526*, 564*, 720*, 725*, 842*(1892:383). *C. gracilis* R.Br. The record is referable to var. *minor* Boott.

buchananii Berggren   Leatherleaf Sedge
- A garden escape recorded in 1990 from waste ground at Cessnock, Glasgow (Lanarks). New Zealand. HbEJC, HbPM. 36a, 50, 53, 316(22:194), 445, 1170*, 1362*.

capitata L.
- Introduced on South Uist (Outer Hebrides), now gone. N Europe, Greenland, Iceland. 21, 81, 368(34:270-277), 1026*, 1203*, 1242*.

[*cespitosa* L. In error for native *C. elata*. 81, 354(5:793).]

crawfordii Fern.   Crawford's Sedge
- A fodder or agricultural seed alien established for many years in W Kent and Surrey, now gone. N America. RNG. 12, 20, 38*, 81*, 354(11:515; 12:767), 825*(2:21-22), 826*(21:77-79).

devia Cheeseman
- A wool casual. New Zealand. HbTBR. 17, 36a*, 370(11:70).

deweyana Schwein.
- A wool casual. N America. HbEJC. 38*, 40*, 428.

flagellifera Colenso
- A wool casual. New Zealand. RNG, HbTBR. 36a, 370(11:70), 1170*. *C. lucida* Boott.

[*frigida* All. In error for native *C. binervis*. 361(3:20; 9:258).]

glacialis Mackenzie
- Introduced on Rhum (N Ebudes), now gone. Scandinavia, Iceland, Greenland. 21, 81, 155, 368(34:270-277), 1026*, 1203*, 1242*.

[*hordeistichos* Villars   A dubious record. BM. 81, 159. *C. hordeiformis* Wahlenb.; *C. secalina* Smith.]

hubbardii Nelmes
- A wool casual. Australia. HbTBR. 370(11:70), 720*.

inversa R.Br.   Knob Sedge
- A wool casual. Australasia. BM, LTN, RNG. 17, 81, 145, 370(11:70), 514*, 520*, 564*, 720*, 1170*.

[*laxa* Wahlenb. A dubious record. 354(5:792), 361(7:266).]

[*ligerica* Gay   A dubious record. 354(5:793).]

longebrachiata Boeckeler   Drooping Sedge
- A wool casual. Australia; (New Zealand). ABD, BM, CGE, E, K, RNG. 17, 81, 526*, 564*, 569*. *C. longifolia* R.Br., non Host.

secta Boott
- A wool casual. New Zealand. ABD, BM, CGE, E, K, RNG. 17, 36a, 51a, 370(11:70), 1170*. *C. appressa* var. *secta* (Boott) Kük.

solandri Boott
- A wool casual. New Zealand. HbTBR. 36a, 370(11:70), 1170*.

[*straminea* Willd. ex Schkuhr.   Unconfirmed; probably in error for native
C. *remota*. 187.]
tereticaulis F.Muell.
  • A wool casual. Australia. E, RNG. 17, 37, 37b, 370(11:70), 514, 564*.
virgata Sol. ex Boott
  • A wool casual. New Zealand. BM, CGE, E, K, RNG. 17, 36a,
370(11:70), 1170*. C. *appressa* var. *virgata* (Sol. ex Boott) Kük.
vulpinoidea Michaux   American Fox-sedge
  •• A fodder or agricultural seed alien long persistent on Banstead Heath
(Surrey) and for some years at Green Street Green, near Farnborough
(W Kent); also a wool casual, and recorded from three localities in the Glasgow
area in the 1980s. N America; (Europe, New Zealand). BM, E, K, LANC,
OXF, RNG. 1, 17, 20*, 38*, 40*, 41*, 81, 209, 370(11:70; 18:435),
825(14:29-340), 1170*.

BULBOSTYLIS Kunth
humilis (Kunth) C.B.Clarke
  • A wool casual. S Africa. BM, E, K, RNG. 17, 370(11:69), 828*(25:25),
835*(10:534).
striatella C.B.Clarke
  • A wool casual. Africa. RNG, HbEJC. 17, 370(11:70), 676, 1130*.
Abildgaardia striatella (C.B.Clarke) K.Lye.   Possibly conspecific with
B. *humilis*.

SCLERIA P.Bergius
bracteata Cav.
  Pre-1930 only. S America. LCN. 7, 160, 354(4:215), 538*.

TYPHACEAE
TYPHA L.
laxmannii Lepechin
  • Introduced in a lake near Hextable (W Kent). SE Europe, Asia. 21, 39*,
51, 458, 564*, 565*.
[*minima* Funck   Dubious records. 16, 234, 268.]

BROMELIACEAE
FASCICULARIA Mez
[*littoralis* (Philippi) Mez   In error for *Ochagavia carnea*. 370(10:289).]
pitcairniifolia (Verlot) Mez   Rhodostachys
  •• (but confused with *Ochagavia carnea*). A garden escape established by the
sea in a few places in Scilly, W Cornwall and Guernsey (Channel Is). Chile.
BM.   20, 21, 50, 66, 207, 212, 220, 312*(t.8087), 370(10:289),
858*(1876:t.10-11), 1243*.   *Bilbergia joinvillei* Van Houtte ex Morris;
Rhodostachys pitcairniifolia (Verlot) Baker.

OCHAGAVIA Philippi
carnea (Beer) Lyman B.Smith & Looser   **Tresco Rhodostachys**
- (but confused with *Fascicularia pitcairniifolia*). Introduced and established on Appletree Banks on Tresco, and near Normandy on St Mary's (both Scilly). Chile. 20, 51*, 61, 370(10:290), 371(370:29), 458, 566*, 1232*, 1269*. *O. lindleyana* (Lemaire) Mez.

TILLANDSIA L.
crocata (C.J.Morren) Baker
- A casual greenhouse escape. Brazil, Argentina. 51, 371(407:41), 1081*, 1158*, 1269*.

CANNACEAE
CANNA L.
indica L.   **Indian-shot**
Pre-1930 only. Tropical America. OXF. 21, 50, 301*(t.776), 565a*, 1218, 1260*, 1347*.

PONTEDERIACEAE
PONTEDERIA L.
cordata L.   **Pickerelweed**
- • An established water-garden escape or introduction; naturalised in a few ponds and gravel pits in the south; probably increasing. N & S America. BM, LANC, OXF, RNG. 16, 20, 38*, 50, 51*, 52*, 54*, 303(26:18), 1065*.

EICHHORNIA Kunth
crassipes (C.Martius) Solms-Laub.   **Water-hyacinth**
- A casual, introduced or a greenhouse escape. Brazil; (tropics). 21, 36b*, 50, 54*, 128, 199, 588*, 680*.

LILIACEAE
Incl. *Alliaceae, Alstroemeriaceae, Amaryllidaceae, Asparagaceae, Asphodelaceae, Colchicaceae, Convallariaceae, Hemerocallidaceae, Hyacinthaceae, Melanthiaceae, Ruscaceae, Trilliaceae*

VERATRUM L.
viride Aiton   **Green False-helleborine**
- Introduced, or a garden escape, persistent in woodland at Douglaston, Milngavie (Stirlings). Eastern N America. 20, 38*, 50, 51*, 300(1899:169), 316(20:478), 347*(11:35-61), 543*.

ASPHODELUS L.
albus Miller   **White Asphodel**
- A garden escape persistent on a grassy bank at the foot of Mont Rossignol, Jersey (Channel Is) and on a railway bank near Tunbridge Wells West station (W Kent). S Europe. 20, 22*, 23*, 24*, 26*, 50, 201, 409.

fistulosus L.   Hollow-stemmed Asphodel
•• A grain and wool casual; possibly also a garden escape. Medit.; (widespread). BM, E, LANC, NMW, OXF, RNG. 1, 8, 9, 14, 21, 24*, 30*, 44a*, 50, 79*. Incl. *A. tenuifolius* Cav.

ramosus L.
• Collected from a hedge east of Durret (Caithness). S Europe, N Africa, Turkey. RNG. 21, 50, 57*, 695*. *A. cerasiferus* Gay.

ASPHODELINE Reichb.
lutea (L.) Reichb.   Yellow Asphodel
Pre-1930 only. Medit. 7, 21, 24*, 26*, 44a*, 50, 79*, 197. *Asphodelus luteus* L.

CHLOROPHYTUM Ker Gawler
comosum group   Spider-plants
• Casual greenhouse escapes. S Africa. 50, 67*, 199, 416, 503*, 719*. Incl. *C. capense* (L.) Kuntze; *C. comosum* (Thunb.) Jacques; *C. elatum* (Aiton) R.Br.; *Anthericum comosum* Thunb.

ANTHERICUM L.
ramosum L.   Branched St Bernard's-lily
• A casual garden escape. Europe. 21, 22*, 23*, 26*, 27*, 31*, 50, 223.

HOSTA Tratt.
elata N.Hylander
• Introduced, or a garden escape, reported from a laneside verge at Rigton Moor, near Otley (MW Yorks). Japan. 50, 51a, 77*, 347*(7:9-10), 432. *H. fortunei* (Baker) L.Bailey var. *gigantea* (Koidz.) L.Bailey.

fortunei (Baker) L.Bailey
• A garden escape by the road north of Inverary (Main Argyll); also reported from Gourock (Renfrews), Helensburgh and Bearsden (both Dunbarton). Originated in cultivation, probably of *H. sieboldiana* ancestry. 50, 52*, 57*, 77*, 371(334:21), 454. Incl. var. *albopicta* (Miguel) N.Hylander; *H.*'Picta'.

lancifolia (Thunb.) Engl.   Narrow-leaved Plantain-lily
• A casual garden escape. Japan. 50*, 77*, 342(Feb89). *H. lanceolata* hort.; *Funkia japonica* (Thunb.) Druce; *F. lanceolata* Siebold ex Miq.

cf. nakaiana (Maek.) Maek.
• A garden escape persistent on the site of the Glasgow Garden Festival (Lanarks). Korea, Japan. 50, 51b, 370(19:175), 890*(5:419), 891*(11:688), 1257*.

sieboldiana (Hook.) Engl.
• Introduced and persistent near a pond at Gregynog Hall, near Newtown (Salop). Japan. 50, 51*, 51a, 52*, 57*, 156, 371(385:34). *H. glauca* (Miq.) Stearn; *H. sieboldii* hort., non (Paxton) J.Ingram; *Funkia sieboldii* Lindley; *Funkia sieboldiana* Hook.

ventricosa (Salisb.) Stearn   Blue Plantain-lily
- Introduced, or a garden escape, near Loch Oich (Easterness). China.
50, 51a, 52*, 57*, 77*, 371(349:29). *Funkia latifolia* hort., non Miq.
ex Matsum.; *F. ovata* hort., non Sprengel; *Hemerocallis caerulea* (Andrews)
Tratt.

HEMEROCALLIS L.
citrina Baroni   Citron Day-lily
- A persistent garden escape at Helensburgh (Dunbarton). China. 50, 51, 53*,
223, 347*(7:68-89).

fulva group   Orange Day-lilies
●●●● Established garden escapes; naturalised on cliffs, in quarries and on
dunes in widely scattered localities. E Asia. 7, 20, 50, 52*, 55*,
347*(7:68 89). *H. fulva* (L.) L. (*H. lilioasphodelus* var. *fulva* L.) and garden
hybrids. Most records are probably referable to a sterile triploid.

lilioasphodelus L.   Yellow Day-lily
●● An established garden escape; on roadsides, railway banks and waste ground
in scattered localities; naturalised for at least 25 years at the old lead-mine
workings at Charterhouse (N Somerset). China; (Europe). BM, LANC,
NMW, OXF, RNG. 1, 20, 29*, 50, 53*, 56*, 340(4:2). *H. flava* L., nom.
illeg.; *H. lutea* Gaertner.

KNIPHOFIA Moench
rufa Leichtlin ex Baker
- An established garden escape; naturalised in abundance on the dunes at Point
of Ayr (Flints). S Africa. 50, 312*(t.7706), 347(7:129-160), 432.

uvaria group   Red-hot-pokers
●●● Established garden escapes; naturalised on dunes, beaches, railway
embankments and grassy slopes in widely scattered coastal areas. S Africa.
BM. 20, 50, 53*, 54*, 56*, 201, 202, 212, 347*(7:129-160), 350(35:22). A
much confused group incl. *K. bruceae* Codd; *K. linearifolia* Baker; *K. uvaria*
(L.) Oken and garden hybrids. *K.* × *praecox* Baker is no longer regarded as
a species in this group and is now used to cover both *K. bruceae* ×
K. linearifolia and *K. bruceae* × *K. uvaria* hybrids. 51b.

COLCHICUM L.
agrippinum hort. ex Baker
- A garden plant recorded in 1952 as established in orchards and pastures in
Suffolk and Surrey, but apparently now gone. Origin obscure. 21, 50, 54*,
371(346:28), 529, 700*. Possibly of hybrid origin between native
C. autumnale and *C. variegatum*; *C. tessellatum* hort.

byzantinum Ker Gawler
- (but possibly in error for *C. agrippinum*). A garden escape at Tenterden
(E Kent). Origin obscure. RNG. 50, 51*, 54*, 371(340:28; 346:28), 529,
700*. Possibly of ancient hybrid origin between native *C. autumnale* and
C. cilicicum (Boiss.) Dammer.

variegatum L.
 Pre-1930 only. SE Europe, Turkey. 7, 21, 31*, 50, 79*, 206, 320(45:411),
 513*, 529, 615*. *C. parkinsonii* Hook.f.

ERYTHRONIUM L.
dens-canis L.   Dog's-tooth-violet
 •• A persistent garden escape and relic. Eurasia. RNG. 20, 23*, 26*, 30*,
 50, 79*, 113, 119, 277.

TULIPA L.
gesneriana group   Tall Garden Tulips
 ••• Persistent garden escapes and relics. Originated in cultivation. RNG. 20,
 50, 53*, 79*, 126, 201, 458. Incl. *T. acuminata* Vahl ex Hornem., from
 Wisley Common (Surrey); *T. didieri* Jordan, from the bank of a country lane
 west of Wilmington (W Kent); *T. elegans* hort. ex Baker, from Swanley
 by-pass (W Kent); and *T. gesneriana* L.
greigii Regel
 • A casual garden escape on waste ground at Wisley (Surrey). C Asia. 50,
 54*, 199, 700*.
saxatilis Sieber ex Sprengel   Cretan Tulip
 • A garden escape established in stony ground on Tresco (Scilly).   Crete,
 Turkey. 20, 24*, 50, 79*, 303(16:19), 615*, 700*.
sylvestris L.   Wild Tulip
 ••• A long-established garden escape or relic; naturalised in meadows,
 orchards and woods, mainly in S and E England; decreasing. Medit. BM,
 CGE, DEE, DUH, E, OXF. 1, 20, 21, 22*, 23*, 26*, 50, 79*.

FRITILLARIA L.
pyrenaica L.   Pyrenean Snake's-head
 • A garden escape persistent in one place in the New Forest (S Hants).
 SW Europe. OXF. 1, 21, 23*, 30*, 50, 79*, 315(178:40), 354(4:75; 8:421).

LILIUM L.
bulbiferum L.   Orange-lily
 • A garden escape or relic at Lambhill, Bride (Man). C Europe. 7, 21, 22*,
 23*, 26*, 50, 79*, 103, 354(3:34). *L. aurantiacum* Weston; incl. *L. croceum*
 Chaix.
canadense L.   Canada Lily
 Pre-1930 only. Eastern N America. 7, 38*, 50, 52*, 54*, 354(5:53).
candidum L.   Madonna Lily
 •• A persistent garden escape; recently recorded from St Ouen in Jersey
 (Channel Is), near Cuddington Golf Course at Sutton (Surrey), at Tattingstone
 and near Pond Hall, Ipswich (both E Suffolk). E Medit. 20, 44a*, 50, 52*,
 54*, 79*, 119, 201, 331(69:140), 369(24:67).
× **hollandicum** Bergmans ex Woodcock & Stearn   Hybrid Orange-lily
 • A persistent garden escape on a railway embankment near Glamis (Angus).
 Originated in cultivation. HbUKD. 20, 50, 54*, 187, 371(357:23), 594*.

L. bulbiferum × *L. maculatum* Thunb.; *L.* × *umbellatum* hort. ex Baker, non Pursh.

× **imperiale** E.Wilson
- A casual garden escape. Originated in cultivation. 50, 54*, 458.
L. regale × *L. sargentiae* E.Wilson.

martagon L.   Martagon Lily
•••• An established garden escape; long naturalised in woods in scattered localities in England, Wales & southern Scotland; sometimes occurring in abundance, as in the valley of the river Wye below Madgett (W Gloucs); known for well over 100 years at Headley (Surrey) and Bourton Woods (E Gloucs). Eurasia. 1, 20, 22*, 23*, 26*, 28*.

monodelphum M.Bieb.   Caucasian Lily
- A garden relic long persistent on the site of Warley Place gardens (S Essex); also recorded without locality from Yorkshire. Crimea, SW Asia. 21, 50, 54*, 313(31:370), 349(1:377), 700*.

pomponium L.
Pre-1930 only. France, Italy. 21, 50, 79*, 193(1869:237), 700*.

pyrenaicum Gouan   Pyrenean Lily
•••• An established garden escape; naturalised in woodlands and on roadsides in widely scattered localities in Britain; known since 1853 on hedgebanks between South Molton and Molland (N Devon). Pyrenees. 1, 20, 22*, 23*, 34*, 50, 79*, 156, 193(1869:237), 217, 329*(1:36-43).

regale E.Wilson   Royal Lily
- A casual or persistent garden escape. China. 20, 50, 52*, 54*, 126, 135, 199, 506*, 680*.

POLYGONATUM Miller
biflorum (Walter) Elliott
- A garden relic established at Virginia Water (Surrey). Eastern N America. RNG. 38*, 50, 51, 699*, 1268*, 1319*. *P. commutatum* (Schultes) D.Dietr.; *P. giganteum* D.Dietr.; incl. *P. canaliculatum* (Muhlenb.) Pursh.

[× *hybridum* Bruegger Alien status dubious; both parents (*P. multiflorum* and *P. odoratum*) are native, but the hybrid may be of garden or European origin. 20, 1311.]

SMILACINA Desf.
stellata (L.) Desf.   Star-flowered Lily-of-the-valley
Pre-1930 only. N America. BM. 7, 21, 22*, 38*, 41*, 50. 300(1909:43), 1242*. *Convallaria stellata* L.; *Vagnera stellata* (L.) Morong.

MAIANTHEMUM Weber
kamtschaticum (Cham.) Nakai   False Lily-of-the-valley
- A garden escape established since 1983 in woodland at Porlock (S Somerset). Western N America, E Asia. 41*, 50, 51, 423, 500*, 856*(14:50-59), 1057*, 1240*. *M. dilatatum* (Wood) Nelson & Macbride.

ASPIDISTRA Ker Gawler
elatior Blume   Cast-iron-plant
- A casual greenhouse escape. Japan. 51a, 67*, 303(52:30), 503*, 551*, 719*.

REINECKEA Kunth
carnea (Andrews) Kunth   Reineckea
- Introduced and established; extensively naturalised in woodland borders at Bosahan, near Helford, Lizard (W Cornwall). China, Japan. 20, 50, 66*, 371(403:40), 617*, 649*, 1119*. *Sansevieria carnea* Andrews.

ORNITHOGALUM L.
arabicum L.
- A persistent garden escape on a sandy bank near Vale Pond, Guernsey (Channel Is). Medit. STP. 21, 44a*, 50, 79*, 303(57:52), 401, 648*, 737*, 1021*(t.3).
narbonense L.
- A casual garden escape. No modern records. Medit., SW Asia. 21, 24*, 44a*, 50, 79*, 261, 737*.
nutans L.   Drooping Star-of-Bethlehem
•••• An established garden escape and relic; naturalised in grassy places, mainly in S and E England. SE Europe, Turkey. 1, 20, 22*, 25*, 31*, 32*, 144, 264. *Myogalum nutans* (L.) Kunth.
cf. thyrsoides Jacq.   Chincherinchee
- A garden escape persistent or established in a wood in Kincardineshire. No modern records. S Africa. RNG. 50, 51*, 54*, 66*, 593*, 1091*.

CAMASSIA Lindley
quamash (Pursh) E.Greene   Common Camass
- A casual garden escape. Western N America. 41*, 50, 223, 543*, 698*. *C. esculenta* Lindley.

SCILLA L.
bifolia L.   Alpine Squill
•• An established garden escape and introduction; naturalised in Abbey Wood (W Kent), on a roadside bank near Benenden (E Kent) and well established in Kirdford churchyard (W Sussex). C & S Europe, SW Asia. BM, MNE, NMW, RNG. 7, 20*, 22*, 23*, 26*, 50, 236, 311(25:197), 406.
bithynica Boiss.   Bithynian Squill
- A garden escape established by the river Cray at St Mary Cray (W Kent) and reported from Milford, Salisbury (Wilts); also a long-persistent relic on the site of Warley Place gardens (S Essex). Bulgaria, Turkey. HbEJC. 20, 21, 79*, 303*(58:40), 331(69:141), 371(397:35).
liliohyacinthus L.   Pyrenean Squill
•• An established garden escape or introduction; naturalised in woodlands at Coleshill (Berks), Abbotsford and Jedburgh (both Roxburghs) and Longformacus House and Dryburgh Abbey (both Berwicks); first recorded in 1964 from a bank of the river Ure at Mickley, near Ripon (MW Yorks).

SW Europe. E, LANC, RNG. 20, 22*, 23*, 30*, 50, 79*, 119, 335(90:27), 370(17:481), 1041.

messeniaca Boiss.   Greek Squill
- A garden escape or introduction established at the edge of Smallcombe Wood, Bath (N Somerset). Greece. 20, 31*, 50, 303(32:19), 350(41:89), 370(14:197), 543*, 700*.

peruviana L.   Portuguese Squill
- •• An established garden escape; naturalised in a few widely scattered localities, mainly in the Channel Is and SW England; known for over 60 years on a railway bank near Pontac, in Jersey (Channel Is). W Medit. JSY, OXF. 20, 24*, 32*, 50, 79*, 201, 354(9:281), 325*(117:440-442).

puschkinioides Regel   Russian Squill
- A casual garden escape. No modern records. W Asia. 50, 220, 700*.

siberica Haw.   Siberian Squill
- ••• An established garden escape and relic, mainly in SE England; naturalised on Headley Heath and Banstead Heath (both Surrey), and on Dartford Heath (W Kent). Russia, Siberia. LANC, OXF, RNG. 20*, 22*, 50, 54*, 79*, 126, 236.

HYACINTHOIDES Heister ex Fabr.

hispanica (Miller) Rothm.   Spanish Bluebell
- ••••• An established garden escape; naturalised in many places in Britain and the Channel Is, sometimes abundantly. SW Europe, NW Africa. 7, 20, 22*, 25*, 29*, 30*, 50. *Endymion hispanicus* (Miller) Chouard; *Scilla campanulata* Aiton; *S. hispanica* Miller. Fertile intermediates between *H. hispanica* and native *H. non-scripta* are apparently now more common than pure *H. hispanica*. 303(46:9), 825(12:91-104).

italica (L.) Rothm.   Italian Bluebell
- A garden escape established in abundance on the site of Warley Place gardens (S Essex); casual elsewhere. SW Europe. BM, OXF. 7, 20*, 50, 79*, 119, 191, 261, 354(4:429). *Scilla italica* L.

PUSCHKINIA Adams

scilloides Adams   Striped Squill
- A casual garden escape. SW Asia. 39, 50, 53*, 312*(t.2244), 369(29:44), 513*, 667*, 700*. *P. hyacinthoides* Baker; *P. libanotica* Zucc.

HYACINTHUS L.

orientalis L.   Hyacinth
- •• A persistent garden escape on roadsides in a few scattered localities in southern England. E Medit. RNG. 12, 20, 29*, 32*, 44a*, 54*, 172, 236.

CHIONODOXA Boiss.

luciliae group   Glory-of-the-snow
- ••• Established garden escapes and relics; mainly in SE England, abundant on a grassy roadside near Mickley (MW Yorks) and on the site of Warley Place gardens (S Essex). Turkey. 20, 50, 54*, 119, 126, 199, 371(355:27), 513*,

594*, 700*. Incl. *C. luciliae* Boiss. (*C. gigantea* Whittall); *C. forbesii* Baker (*C. luciliae* hort., non Boiss.; incl. *C. siehei* Stapf and *C. tmoli* hort.).

nana (Schultes & Schultes f.) Boiss. & Heldr.
- Recorded without status or locality. Crete. 7, 21, 31*, 50, 354(5:580), 371(340:28), 615*, 700*. *C. cretica* Boiss. & Heldr.

sardensis Drude
•• (but perhaps overlooked). An established garden escape; recorded from Surrey, S Essex and Kirkcudbrights. Turkey. E, RNG. 20, 50, 54*, 199, 312*(n.s. t.50), 331(61:26), 371(400:37), 594*, 700*. *Scilla sardensis* (Whittall ex Barr & Sugden) Speta.

BELLEVALIA Lapeyr.

romana (L.) Reichb.   **Roman Squill**
- A garden escape recorded from South Meadow, Eton (Bucks). No modern records. S Europe. OXF. 21, 29*, 50, 79*, 354(9:571), 546*, 683*, 700*. *Hyacinthus romanus* L.

MUSCARI Miller

armeniacum Leichtlin ex Baker   **Garden Grape-hyacinth**
••• (but overlooked as native *M. neglectum*). An established garden escape; mainly in SE England and East Anglia. SE Europe, SW Asia. LANC, OXF, RNG. 20*, 50, 79*, 119, 135*, 826(15:75-77), 1275*. Known, incorrectly, in horticulture as *M. botryoides* 'Heavenly Blue'.

azureum Fenzl
- A garden escape established in rough grass on the site of old parkland at Curry Rivel (S Somerset). Turkey. 39*, 50, 51*(fig. as *Hyacinthus azureus*), 423, 700*, 824*(t.1199). *M. praecox* Siehe; *Bellevalia azurea* (Fenzl) Boiss.; *Hyacinthella azurea* (Fenzl) Chouard; *Hyacinthus azureus* (Fenzl) ined., non Baker; *Pseudomuscari azureum* (Fenzl) Garb. & Greuter.

botryoides (L.) Miller   **Compact Grape-hyacinth**
•• (but confused with its congeners). An established garden escape; naturalised in sand or gravel pits and on roadsides in the south. C & SE Europe. 20, 22*, 25*, 50, 79*, 135*, 199, 371(334:23), 1275*. *Hyacinthus botryoides* L.

comosum (L.) Miller   **Tassel Hyacinth**
••• An established garden escape; naturalised on dunes in several places in S Wales, SW England and the Channel Is. Europe, N Africa, SW Asia. JSY, LANC, NMW, OXF, RNG, SLBI. 1, 20, 22*, 26*, 44a*, 50, 79*, 156, 207. *Hyacinthus comosus* L.; *Leopoldia comosa* (L.) Parl.

ALLIUM L.

carinatum L.   **Keeled Garlic**
•••• An established garden escape; naturalised in widely scattered localities throughout the British Is. Europe, Turkey. 1, 20, 22*, 25*, 26*, 31*, 50, 370(18:381-385). Some early illustrations, e.g. 1308*, are in error for native *A. oleraceum* (*A. carinatum* Smith, non L.).

cepa L.   Onion, incl. **Shallot**
•••• A casual garden escape or relic of cultivation. Origin obscure;
(widespread as a crop). 3, 16, 20, 21, 22*, 50, 68*, 80*, 546*. Incl.
A. ascalonicum auct., non L.

cyrilli Ten.
• A casual at the edge of a golf course at Grouville and on a refuse tip at
Mont Mado, both in Jersey (Channel Is), vector unknown. Italy, Greece. 21,
39*, 50, 201. *A. nigrum* L. var. *cyrilli* (Ten.) Fiori.

ericetorum Thore
• A garden escape recorded without locality from Westmorland. S Europe.
21, 50, 79*, 175, 546*. Incl. *A. ochroleucum* Waldst. & Kit. Possibly an
error for *A. moly.* 484.

fistulosum L.   Welsh Onion
• A casual garden escape. E Asia; (widespread as a crop). BM. 7, 20, 22*,
25*, 50, 80*, 215, 354(6:49), 546*, 1242*.

moly L.   Yellow Garlic
•• An established garden escape; naturalised in fields and on roadsides in a few
widely scattered localities; known for well over 20 years at Grève d'Azette in
Jersey (Channel Is). SW Europe. BM, JSY, K, OXF. 7, 20*, 30*, 50, 54*,
79*, 201.

neapolitanum Cirillo   Neapolitan Garlic
••• (but confused with *A. subhirsutum*). An escape and relic of cultivation
established the Channel Is, Scilly and most coastal counties of southern
England; known for over 20 years at Abbotsbury (Dorset), Chichester
(W Sussex) and Norwich (E Norfolk). Medit. BM, JSY, OXF, RNG, TOR.
7, 20*, 32*, 44a*, 50, 79*, 117*, 207, 220, 303(23:10).

nigrum L.   Broad-leaved Onion
•• A long-persistent or established garden escape and relic in a few scattered
localities in southern England, as on the airfield at Farnborough (N Hants).
Medit. OXF, RNG. 1, 20, 24*, 44a*, 50, 79*, 119, 191, 303(39:9),
370(16:195). Incl. *A. multibulbosum* Jacq. Early records were misnamed as
native *A. ampeloprasum*.

paradoxum (M.Bieb.) G.Don   Few-flowered Garlic
•••• An established garden escape; naturalised and locally abundant in woods
and on riverbanks and roadsides in England and Scotland, especially in the
Edinburgh, Glasgow and Cambridge areas; known since 1863 in woods near
Edinburgh; increasing rapidly. Caucasus, Iran. 1, 20*, 22*, 50, 209,
329*(4:194-199), 370(8:379-384), 1083*.

pendulinum Ten.   Italian Garlic
• A long-persistent garden relic on the site of E.A.Bowles' garden at Enfield
(Middlesex) and on the site of Warley Place gardens (S Essex). C Medit. 20,
50, 303(26:14), 546*, 645*. *A. triquetrum* var. *pendulinum* (Ten.) Regel.

porrum L.   Leek
•• A casual garden escape or relic. Origin obscure; (widespread as a crop).
BM, OXF. 20, 22*, 50, 80*, 174, 187, 192, 354(9:671), 546*.
A. ampeloprasum var. *porrum* (L.) Gay.

roseum L.   Rosy Garlic
●●●● An established garden escape; naturalised in several places in southern England and Wales, especially on the cliffs of the Avon Gorge (W Gloucs) and on St Mary's (Scilly).   Medit.   1, 20*, 22*, 30*, 34*, 50, 79*, 207, 350(47:43), 370(12:177; 18:381-385).   Incl. *A. ambiguum* Sibth. & Smith, non DC.

sativum L.   Garlic
●● An established garden escape and food refuse alien; naturalised on the shore at Port Dinllaen (Caerns) and in a salt-marsh beside the river Lune at Lancaster (W Lancs); elsewhere a relic of cultivation or a casual.   Origin obscure; (widespread as a crop).   BM, K, LIV, OXF, RNG.   7, 20, 22*, 50, 68*, 80*, 191, 303(36:28; 60:11), 331(18:72), 546*.   Incl. *A. ophioscorodon* G.Don.

subhirsutum L.   Hairy Garlic
●● (but under-recorded for *A. neapolitanum*).   A garden escape persistent in a few places in the Channel Is and SW England, as at Exwick and Teignmouth (both S Devon).   Medit., NE Africa.   BM, RNG.   20*, 24*, 32*, 50, 79*, 201, 303(23:10, 25:14).

triquetrum L.   Three-cornered Garlic
●●●●● An established garden escape; naturalised and locally abundant in many places near the sea; mainly in SW England, Scilly and the Channel Is, but with scattered localities reaching as far north as Ayrs; increasing.   W Medit.   2, 20,22*, 30*, 33*, 50, 79*, 117*, 201*, 207, 303(23:10), 368(40:328).

tuberosum Rottler ex Sprengel   Chinese Chives
● A casual, probably a garden escape.   SE Asia.   43, 50, 223, 312*(n.s. t.386), 700*.   *A. uliginosum* G.Don, non Ledeb.   This species is widely naturalised from cultivation in eastern Asia, obscuring its true native area which is probably China.

unifolium Kellogg   American Onion
● A persistent garden escape in damp woodland behind Cardross Park (Dunbarton), now gone.   California.   20, 41*, 42, 50, 370(17:195), 700*.   *A. uniflorum* sphalm.

vineale L. **alien variants**
Var. *purpureum* H.P.G.Koch (*A. kochii* Lange) from the Baltic has been recorded without locality.   21, 50, 458.

NECTAROSCORDUM Lindley
siculum (Ucria) Lindley   Honey Garlic
●●● An established introduction or garden escape; naturalised in a few places in southern England; known for many years at Abbotsbury (Dorset) and St Vincent's Rocks, Clifton (W Gloucs).   S Europe, Turkey.   NMW, RNG.   7, 9, 20*, 31*, 50, 54*, 79*, 312*(n.s. t.257), 350(47:44), 370(13:84; 14:106; 18:381-385).   *Allium siculum* Ucria; *A. dioscoridis* auct.; incl. *N. bulgaricum* Janka (*A. bulgaricum* (Janka) Prodán).

NOTHOSCORDUM Kunth
borbonicum Kunth   Honey-bells
●●● An established garden escape; well naturalised about Mont Cambrai and
Rue de Haut in Jersey (Channel Is); also on roadsides and a persistent garden
weed in a few scattered localities in southern England and Ireland. S America.
BM, CGE, E, RNG. 20*, 36b*, 50, 201, 207, 220, 650*, 700*. *N. fragrans*
(Vent.) Kunth; *N. gracile* sensu Stearn, non (Aiton) Stearn; *N. inodorum* auct.,
non (Aiton) Nicholson; *Allium fragrans* Vent.; *A. gracile* auct., non Aiton;
A. inodorum auct., non Aiton.

AGAPANTHUS L'Hér.
praecox Willd.   African Lily
●● A persistent garden escape or introduction in Scilly and the Channel Is,
perhaps established on consolidated dunes on Tresco (Scilly). S Africa. RNG.
13, 20, 22*, 50*, 201, 207, 220, 562*, 681*, 1117*. Incl. *A. orientalis*
F.M.Leighton.

TULBAGHIA L.
natalensis Baker
● A casual greenhouse escape. S Africa. HbACL. 50, 199, 577a*, 735*,
823*(t.979).

TRISTAGMA Poeppig
uniflorum (Lindley) Traub   Spring Starflower
●●● An established garden escape; abundantly naturalised in a few sandy places
in Scilly and the Channel Is, and recorded from widely scattered localities in
the south; increasing. Argentina, Uruguay. BM, JSY, NMW, OXF, RNG.
7, 20*, 22*, 50, 51*, 54*, 201, 207, 312*(n.s. t.185). *Brodiaea uniflora*
(Lindley) Engler; *Ipheion uniflorum* (Graham) Raf.; *Milla uniflora* Graham;
Triteleia uniflora Lindley.

NERINE Herbert
bowdenii Will.Watson
● Introduced or a garden escape; recently recorded from roadsides at
Farningham (W Kent) and Wherstead (E Suffolk). S Africa. 50, 54*,
303(60:35), 369(26:59), 700*, 1046*.
sarniensis (L.) Herbert   Guernsey Lily
Pre-1930 only. S Africa. 50, 54*, 117*, 214(p.31), 220, 357(19:269-286),
506*, 593*. *Amaryllis sarniensis* L. The claim of its accidental introduction
by shipwreck has been shown to be false; it was a very early garden escape.

AMARYLLIS L.
belladonna L.   Jersey Lily
●● A long-persistent escape or relic of cultivation in hedges, thickets and on
cliffs in Scilly and the Channel Is; recently recorded from the mainland as a
relic at Poldhu (W Cornwall). S Africa. BM. 13, 21, 50, 54*, 117*, 201,
212, 220, 593*, 594*, 700*, 1046*, 1117*.

CYRTANTHUS Aiton
elatus (Jacq.) Traub   Scarborough Lily
- A casual garden escape. S Africa. 50, 51*, 54*, 192, 650*, 719*.
C. purpureus (Aiton) Herbert; *Amaryllis purpurea* Aiton; *Vallota purpurea* (Aiton) Herbert; *V. speciosa* (L.f.) T.Durand & Schinz.

STERNBERGIA Waldst. & Kit.
lutea (L.) Ker Gawler ex Sprengel   Winter Daffodil
- An established garden escape, apparently now gone from Guernsey, but long naturalised and abundant by the sea at Gorey Castle in Jersey (Channel Is). Medit., SW Asia. JSY, OXF. 7, 20, 24*, 31*, 50, 51*, 201, 220, 347*(5:1 16), 354(5:580). *Amaryllis lutea* L.

CRINUM L.
× **powellii** Baker   Swamp Lily
- A garden escape in Jersey (Channel Is), and persistent for 30 years in a derelict garden at Farnham, near Saxmundham (E Suffolk). Originated in cultivation.   51, 54*, 66*, 220, 369(22:48), 458, 593*, 700*.
C. bulbispermum (Burman f.) Milne-Redh. & Schweick. × *C. moorei* Hook.f.

LEUCOJUM L.
aestivum L. subsp. **pulchellum** (Salisb.) Briq.
- ●●● A garden escape long established in Scilly and the Channel Is; increasingly recorded in southern England. W Medit. BM. 2, 20, 50, 94, 199, 207, 212, 1021*(t.3), 1030*, 1318*. *L. hernandezii* Cambess.; *L. pulchellum* Salisb.
[*vernum* L.   Accepted, with reservations, as native.]

GALANTHUS L.
cf. **allenii** Baker
- An established garden escape; naturalised on a laneside and in a churchyard at Copgrove (MW Yorks). Origin obscure, probably Caucasus. 50, 94*, 371(379:25), 700*.
caucasicus (Baker) Grossh.   Caucasian Snowdrop
- A persistent garden escape or relic on waste ground near Henley Park House (Surrey). SW Asia. HbACL. 20, 50, 94*, 199, 700*. *G. nivalis* subsp. *caucasicus* Baker.
elwesii Hook.f.   Greater Snowdrop
- ●● An established garden escape and introduction; naturalised, sometimes abundantly, in parks and churchyards in the south; a garden escape on Chislehurst Common, Dartford Heath and a trackside at Darenth (all W Kent). SE Europe, Turkey. OXF, RNG. 13, 20*, 50, 79*, 94*, 119, 199, 209, 458, 700*, 825(11;114-119). A hybrid with *G. nivalis* has been recorded. HbACL. 370(15:403).
ikariae Baker
- ●● A garden escape recently found near the beach at Swanage (Dorset), in a churchyard at Burgates, near Liss (N Hants), in woodland on Reigate Heath (Surrey), in Eaglefield Park (Berks) and at Whittlesford (Cambs). Aegean Is,

Turkey. RNG, HbKWP. 21, 39, 79*, 94*, 312*(t.9474), 336(35:65-66), 342(Feb93:5), 700*.

[*nivalis* L. Accepted, with reservations, as native.]

plicatus M.Bieb.   Pleated Snowdrop

•• An established garden escape and relic; naturalised in quantity in woods, plantations and under hedges at Great and Little Bealings (E Suffolk), Barton Park (W Suffolk) and Adwell (Oxon). E Europe, Turkey. RNG, HbACL. 20*, 50, 79*, 94*, 199, 258, 371(357:22; 376:25), 700*. Incl. *G. byzantinus* Baker. A hybrid with *G. nivalis* and backcrosses have been recorded. 122, 258, 371(357:22).

NARCISSUS L.

In addition to the species and hybrids listed below, many taxa of more complex derivation or uncertain parentage have occurred as persistent escapes from cultivation, especially in Scilly and the Channel Is. 207, 220.

bicolor L.

• An escape from cultivation recorded from Guernsey (Channel Is). Pyrenees. 20, 21, 30*, 50, 220. *N. abscissus* (Haw.) Schultes & Schultes f.; incl. *N. horsfieldii* hort.

bulbocodium L.   Hoop-petticoat Daffodil

• Introduced and established; long naturalised on Jethou (Channel Is); also naturalised at Hextable (W Kent) and by seed at Virginia Water (Berks). W Medit. BM. 1, 20, 30*, 50, 51*, 54*, 79*, 202, 207, 353(37:31-36), 458. *N. conspicuus* D.Don.

cyclamineus DC.   Cyclamen-flowered Daffodil

• Introduced and established; on a roadside at Wisley and naturalised at Henley Park (both Surrey); also recorded without locality from Yorkshire. SW Europe. 20, 50, 51*, 54*, 79*, 199, 371(374:9).

× **incomparabilis** Miller sensu lato

••• A garden escape persistent in a few widely scattered localities. France; or partly originated in cultivation. BM, LANC, NMW, OXF, RNG. 1, 20*, 21, 50, 79*, 117*, 192, 700*. *N. poeticus* group × *N. pseudonarcissus* group; incl. *N. aurantius* Schultes f.; *N.* × *barrii* Baker; *N.* × *burbidgei* Baker; *N.* × *leedsii* Baker.

jonquilla L.   Jonquil

• Introduced or a garden escape; recorded without locality from Kent. SW Europe. 20, 50, 54*, 79*, 236, 594*, 1023*.

× **medioluteus** Miller sensu lato

•••• A garden escape or relic long persistent or established in widely scattered localities throughout most of the British Is. France, or originated in cultivation. 1, 20, 21, 22*, 25*, 50, 79*, 207*. *N. poeticus* group × *N. tazetta*; *N.* × *biflorus* Curtis; *N. poetaz* hort.; incl. *N.* 'Primrose Peerless'.

minor L.

• An established garden escape or relic; naturalised in the Ballaugh Curraghs (Man) and known from 1885 to 1945 at Charles, near South Molton (N Devon). Pyrenees, N Spain. BM. 2, 20, 23*, 50, 54*, 79*, 103, 142,

217. Incl. *N. eystettensis* hort.; *N. nanus* hort. The South Molton plant was
N. minor 'Flore Pleno'.

× **odorus** L. sensu lato   **Hybrid Jonquil**
•• A long-persistent or established escape from cultivation in Scilly and the
Channel Is, and known for over 100 years in a field about two miles south of
St Austell (E Cornwall). Originated in cultivation. BM, OXF, RNG, SLBI.
2, 20, 21, 50, 79*, 207, 220. *N. jonquilla* × *N. pseudonarcissus* group;
N. campernelli hort. ex Haw.; *N.* × *conspicuus* Salisb.; *N.* × *heminalis*
(Salisb.) Schultes f.; *N.* × *infundibulum* Poiret.

papyraceus Ker Gawler
• A persistent escape from or relic of cultivation in Guernsey (Channel Is)
and Scilly. Medit. 20, 30*, 32*, 50, 79*, 207, 354(5:401), 1218. Incl.
N.'Paper White'.

poeticus group   **Pheasant's-eye Narcissi**
•••• Persistent garden escapes or introductions in widely scattered localities.
S Europe. 1, 20*, 23*, 26*, 50*, 79*, 129, 156. Incl. *N.*'Actaea';
N. angustifolius Curtis ex Haw.; *N. majalis* Curtis; *N. ornatus* Haw.;
N. patellaris Haw.; *N. poeticus* L.; *N. radiiflorus* Salisb.

pseudonarcissus L. subsp. **major** (Curtis) Baker   **Spanish Daffodil**
•••• A persistent garden escape or introduction in widely scattered localities.
SW Europe. BM, NMW, OXF. 2, 20, 50, 79, 700*. *N. grandiflorus* Salisb.;
N. hispanicus Gouan; *N. major* Curtis; incl. *N. spurius* Haw.

pseudonarcissus L. subsp. **moschatus** (L.) Baker
Pre-1930 only. SW Europe. OXF. 2, 50, 51, 79*. *N. moschatus* L.

pseudonarcissus L. subsp. **obvallaris** (Salisb.) Fernandes   **Tenby Daffodil**
• An established garden escape or introduction; long naturalised about Tenby
(Pembs); also recorded from Scilly, and from Newcastle Emlyn and near
Whitland (both Carms). Originated in cultivation. BM, NMW, OXF, RNG.
2, 20, 33*, 50, 140*, 1279*. *N. bromfieldii* Syme; *N. concolor* Bromf.;
N. lobularis (Haw.) Schultes & Schultes f.; *N. obvallaris* Salisb.; *N. sibthorpii*
Haw.; *Ajax lobularis* Haw.

pseudonarcissus L. subsp. *pseudonarcissus* **alien variants**
Subsp. *gayi* (Hénon) Fernandes (*N. gayi* (Hénon) Pugsley), incl. *N.*'Princeps',
has been recorded, but is probably not worthy of recognition at subspecific
rank. 103, 207, 1218. *N.*'Van Sion' (var. *plenus* hort.; *N. telamonius* Link;
N. tenuiflorus sphalm., non Schultes f.) is widely naturalised. BM. 220.
N. serratus Haw., a cultivar, was recorded in error; not from a wild habitat.
BM. 192.

tazetta group   **Bunch-flowered Narcissi**
••• Long-persistent escapes from or relics of cultivation, especially in
Scilly and the Channel Is. S Europe, Asia. NMW, OXF, RNG. 1, 20, 24*,
32*, 50*(1:292), 79*, 117*, 201, 207, 220. Incl. *N. aureus* Lois.;
N.'Grand Monarque'; *N.*'Grand Primo'; *N.*'Grand Soleil d'Or'; *N. italicus*
Ker Gawler; *N. ochroleucus* Lois.; *N. orientalis* L.; *N.*'Scilly White';
N. tazetta L.; *N. tenuiflorus* Schultes f.

triandrus L.   Angel's-tears Narcissus
- Introduced and established on Herm (Channel Is). SW Europe. 1, 21, 30*, 50, 54*, 79*, 202, 1023*.

ASPARAGUS L.

aethiopicus L.
- A casual greenhouse escape. S Africa. BM. 50, 56*, 66*, 67*, 347(7:249). *A. densiflorus* (Kunth) Jessop 'Sprengeri'; *A.* 'Sprengeri'.

officinalis L. subsp. **officinalis**   Garden Asparagus
- •••• An established garden escape, especially in dry, sandy places near the sea. Eurasia; (widespread as a crop). 1, 20, 21, 546*. *A. altilis* L.; *A. campestris* Reich.; *A. caspius* Hohen.; *A. polyphyllus* Steven.

setaceus (Kunth) Jessop   Asparagus-fern
- A casual garden escape. S Africa. 50, 371(391:34), 503*, 582*, 719*. *A. plumosus* Baker.

RUSCUS L.

hypoglossum L.   Spineless Butcher's-broom
- Introduced and long established in a quarry at Craigmillar, Edinburgh, now gone, and at the edge of a wooded slope at Kinver Edge (Staffs). C & SE Europe, Turkey. E. 20, 29*, 50, 61*, 79*, 303(32:19; 34:22), 344*(28:334), 370(16:229). *R. hypophyllum* L. var. *hypoglossum* (L.) Baker.

ALSTROEMERIA L.

aurea Graham   Peruvian Lily
- ••• An established garden escape; well naturalised in damp places in the Greeba Curraghs, near St John's (Man) and in several places in Scotland. Chile. BM. 12, 20, 50, 51*, 54*, 102, 199, 331(37:203), 371(394:35), 594*. *A. aurantiaca* D.Don ex Sweet.

SMILAX L.

aspera L.   Common Smilax
- A persistent garden weed in Guernsey (Channel Is). Medit., W Asia. 21, 24*, 30*, 31*, 32*, 50, 220, 546*.

excelsa L.   Larger Smilax
- A garden relic long persistent on the site of Warley Place gardens (S Essex). SE Europe, SW Asia. BM. 21, 51*, 313(31:370), 630*, 1085*.

IRIDACEAE

LIBERTIA Sprengel

caerulescens Kunth
- A garden escape persistent or established on the gravelly bed of the river Flesk near Killarney (N Kerry). Chile. RNG. 50, 616*, 1014.

elegans Poeppig   Lesser Chilean-iris
- A garden escape persistent on a railway embankment at Helensburgh (Dunbarton), now gone. Chile. 20, 303(55:16), 316(22:91-92), 467, 616, 1014, 1100*. According to 1066*, it occurs in Argentina and is probably conspecific with *L. formosa*. Both names were published in Oct. 1833.

formosa R.Graham   Chilean-iris
•• An established garden escape; naturalised near Loch Sween (Kintyre), near
the sea at the Towans (W Cornwall), and at Caragh Lake (S Kerry). Chile.
E, OXF, RNG. 7, 20, 50, 52*, 57*, 66*, 303*(19:12-13), 312*(t.3294),
354(4:27), 1014, 1117*. *L. chilensis* (Molina) Klotzsch ex Baker, nom. inval.;
L. grandiflora hort., non (R.Br.) Sweet.

HOMERIA Vent.
collina (Thunb.) Vent.   Cape-tulip
• A persistent weed in Abbey Gardens on Tresco (Scilly) and in a plant nursery
in Guernsey (Channel Is), very rarely escaping. S Africa. 20, 36b*, 37b*, 50,
66*, 207, 220, 371(416:7), 458, 616*. *H. breyniana* G.Lewis, non *Tulipa
breyniana* L.

SISYRINCHIUM L.
[*bermudiana* L. Accepted, with reservations, as native.]
californicum (Ker Gawler) Dryander   Yellow-eyed-grass
•• An established garden escape; abundantly naturalised by Loch Corrib,
NW of Oughterard (W Galway) and recorded from a few localities widely
scattered throughout the British Is, mainly on damp soils.   At one time
naturalised in abundance over many acres of marshy meadows north of Rosslare
(Co Wexford), perhaps now gone. Western N America. BM, OXF, RNG.
1, 20, 22*, 33*, 50, 52*, 238(para.160), 274, 318(16:151), 319(20:470).
Hydastylus californicus (Ker Gawler) Salisb.; *Marica californica* Ker Gawler;
incl. *S. boreale* (Bickn.) J.K.Henry (*Hydastylus borealis* Bickn.); *S. brachypus*
(Bickn.) J.K.Henry.
[*chilense* Hook. Dubious records. BM. 7, 354(4:429; 6:149; 7:598).]
iridifolium group   Veined Yellow-eyed-grasses
• A garden escape established for a time at the Royal Aerospace Establishment,
Farnborough (N Hants), site now destroyed; also a persistent weed in a garden
and churchyard in Jersey and a casual in Guernsey (both Channel Is).
S America. 36b*, 50, 201, 220, 226, 370(9:186; 16:229), 519*, 1014, 1066*.
Incl. *S. iridifolium* Kunth; *S. laxum* Otto ex Sims; *S. valdivianum* Philippi.
According to 1218, *S. iridifolium* subsp. *valdivianum* (Philippi) Ravenna is a
synonym of *S. chilense*.
montanum E.Greene   American Blue-eyed-grass
••• (but much confused with *S. bermudiana*). An established garden escape;
naturalised on waste ground in widely scattered localities in Britain.
N America. BM, NMW, RNG. 20, 22*, 50. *S. angustifolium* Miller pro
parte; *S. bermudiana* sensu Coste, non L. All records are referable to var.
crebrum Fern.
platense Johnston
• A wool casual. S America. RNG. 535a*, 616, 827(19:395), 1050*.
striatum Smith   Pale Yellow-eyed-grass
••• A persistent or established garden escape; in a chalkpit, on heaths and
waste ground, in a few places in the south. Chile, Argentina. BM, OXF,
RNG. 7, 20, 50, 52*, 54*, 56*, 199, 207, 258, 654*(t.1870). *S. iridifolium*

auct., non Kunth; *S. lutescens* G. Lodd.; *Marica striata* (Smith) Ker Gawler; *Phaiophleps nigricans* (Philippi) R.C.Foster.

ARISTEA Sol. ex Aiton

ecklonii Baker   Blue Corn-lily
- A garden escape established on Tresco (Scilly). S Africa. BM, HbEJC. 20, 36b*, 50, 371(370:29), 519*, 562*. First recorded in error as *A.* cf. *caerulea* (Thunb.) Vahl. 370(11:289).

HERMODACTYLUS Miller

tuberosus (L.) Miller   Snake's-head Iris
- •• An established garden escape; naturalised in several localities in SW England; known for over 50 years in Scorrier Woods and near Ludgvan (both W Cornwall) and on sandhills at Woolacombe (N Devon); also known since 1950 at Sand Point, Kewstoke (N Somerset). Medit. BM, NMW, OXF, RAMM, RNG. 1, 20, 24*, 29*, 33*, 50, 79*, 189, 212, 217, 354(7:215). *Iris tuberosa* L.

IRIS L.

cf. **chrysographes** Dykes
- Introduced and long persistent by Loch Moy (Moray); casual elsewhere. China. E. 50, 55*, 223, 277, 559*.

danfordiae (Baker) Boiss.
- A garden escape at Darenth (W Kent). Turkey. 50, 54*, 82*, 303(55:32), 371(416:8), 559*, 700*.

ensata Thunb.   Japanese Iris
- A garden escape persistent in a swamp on Dartford Heath (W Kent). E Asia. 20, 50, 54*, 82*, 303(50:27; 55:32), 559*. *I. kaempferi* Siebold ex Lemaire.

germanica group   Flag Irises
- ••••• Established garden escapes; naturalised on railway banks and roadsides, mainly in the south; known for over 50 years on railway banks between East Ham and Bromley-by-Bow (S Essex). Originated in cultivation. 1, 20, 22*, 23*, 50, 71*, 79*, 191. Incl. *I. germanica* L. and hybrid Bearded Irises derived from *I. pallida* Lam., *I. variegata* and other related species.

× **hollandica** hort.   Dutch Iris
- A casual or persistent escape from cultivation in the Channel Is. Originated in cultivation. 20, 201, 220, 594*, 700*. A complex of hybrids involving *I. filifolia* Boiss., *I. latifolia*, *I. tingitana* Boiss. & Reuter and *I. xiphium*. Incl. *I.* 'Wedgwood'.

latifolia (Miller) Voss   English Iris
- •• A persistent garden escape or relic of cultivation; in several grassy places in Shetland, and abundant in a hayfield at Fawkham (W Kent). Pyrenees. 1, 20, 23*, 30*, 50, 54*, 79*, 142, 159(p.306), 236, 253. *I. xiphioides* Ehrh.

orientalis Miller   Turkish Iris
- •• A garden escape established in several places in the south; known since the early 1950s in scrub at Sand Point, Kewstoke (N Somerset) and for many years in a field at Abbotsbury (Dorset); abundant about Northfleet and

Swanscombe (W Kent). Turkey. E, OXF, RNG. 20, 39*, 50, 54*, 122, 303(28:16; 50*:27-29; 55:32; 59:45). *I. ochroleuca* L.; *I. spuria* subsp. *ochroleuca* (L.) Dykes.

pumila L.   Pygmy Iris
Pre-1930 only. Europe, SW Asia. 1, 21, 26*, 32*, 50, 79*, 82*, 184.

reticulata M.Bieb.
• A casual or persistent garden escape recorded in 1990 from Mitcham Common (Surrey). SW Asia. 39*, 50, 342(Feb91), 700*, 1354*.

sibirica L.   Siberian Iris
••• A garden escape persistent or established in widely scattered localities in England and Scotland. Europe, SW Asia. E, RNG. 7, 20, 22*, 26*, 50, 79*, 82*, 209, 303(18:13). *I. orientalis* Thunb., non Miller.

sofarana Foster   Mourning Iris
Pre-1930 only. Lebanon. 2, 50, 54*, 82*, 1235*. Incl. *I. susiana* L.

[*spuria* L.   Accepted, with reservations, as native.]

unguicularis Poiret   Algerian Iris
• A garden escape persistent since 1989 in a drove at Long Load (S Somerset). E Medit., N Africa. 21, 31*, 51b, 423, 594*, 615*. *I. stylosa* Desf.; incl. *I. cretensis* Janka.

variegata L.
• Reported from a grassy bank at Hook Green, near Meopham (W Kent). C & SE Europe. 21, 50, 79*, 82*, 458, 623*.

versicolor L.   Purple Iris
••• An established garden escape; naturalised by lakes, ponds and streams in a few widely scattered localities in Britain; known for many years in a reed-swamp in Ullswater (Cumberland) and by a pond at High Beach (S Essex). Eastern N America. BM, E, K, LANC, RNG. 12, 20, 22*, 25*, 50, 82*. A hybrid with the very similar, and possibly conspecific, *I. virginica* L. (*I.* × *robusta* E.S.Anderson) has been recorded. LANC. 20, 50, 1311, 1359.

xiphium L.   Spanish Iris
•• A garden escape and relic of cultivation persistent in Scilly and the Channel Is. W Medit. RNG. 1, 20, 30*, 32*, 50, 79*, 82*, 207, 236, 349(3:62). Incl. *I. lusitanica* Ker Gawler.

WATSONIA Miller

beatricis Mathews & L.Bolus   Beatrice Watsonia
• A garden escape on the foreshore at Rhu (Dunbarton). S Africa. 50, 51*, 52*, 66*, 223, 454. The 1964 record from Helensburgh (Dunbarton) may have been in error. 454.

borbonica (Pourret) Goldblatt   Bugle-lily
• Introduced and established on Appletree Banks and possibly elsewhere on Tresco (Scilly). S Africa. E. 20, 50, 52*, 66, 370(11:289), 371(370:29), 519*, 1117*. Incl. *W. ardernei* Sander (*W. meriana* auct., non (L.) Miller). Hybrids have also been introduced and persist in the same area. 412.

ROMULEA Maratti

[*bulbocodium* (L.) Sebast. & Mauri (*Trichonema bulbocodium* (L.) Ker Gawler; *Ixia bulbocodium* (L.) L.)   In error for *R. rosea.* 220.]

rosea (L.) Ecklon   Onion-grass
- An established garden escape; naturalised in quantity at Cobo in Guernsey (Channel Is). S Africa. 20, 22*, 50, 117*, 220, 1082*. *R. longifolia* (Salisb.) Baker; *Ixia rosea* L. All records are referable to var. *australis* (Ewart) de Vos.

CROCUS L.

biflorus Miller   Silvery Crocus
- A garden escape and relic naturalised for many years in old pastures and on the sites of old gardens in E and W Suffolk, probably now gone; also naturalised in marshes near Friarton Bridge (E Perth). E Medit. BM, CGE, IPS, LIV, OXF, RNG.   1, 20, 50, 54*, 79*, 84*, 258*, 303*(60:1), 363(16:215-219), 1307. *C. argenteus* Sabine; *C. praecox* Haw.

chrysanthus (Herbert) Herbert   Golden Crocus
- (but probably under-recorded). A persistent or established garden escape at Weymouth (Dorset) and in a few places in Surrey, and probably W Sussex. SE Europe, Turkey. 20, 50, 54*, 79*, 84*, 199, 303*(60:1), 371(425:31; 428:6). *C. annulatus* Herbert var. *chrysanthus* Herbert. Some records may be referable to hybrids with *C. biflorus*.

flavus Weston   Yellow Crocus
- ••• An established garden escape and relic; naturalised in quantity on Hayes Common (W Kent); known for well over 100 years as a relic of cultivation at Barton Park near Bury St Edmunds (W Suffolk). SE Europe, Turkey. BM, CGE, IPS, LANC, OXF, RNG. 1, 20*, 31*, 32*, 50, 79*, 84*, 188, 258, 458. *C. aureus* Smith; *C. luteus* Lam. Most records are probably referable to the hybrid with *C. angustifolius* Weston (*C.* 'Dutch Yellow'; *C.* 'Golden Yellow'). 371(425:30).

kotschyanus K.Koch   Kotschy's Crocus
- An established garden escape; naturalised in abundance in a small meadow in Shrubland Park (E Suffolk), by a trackside on Wisley Common (Surrey) and in rough grass at Stowmarket (W Suffolk). Turkey, Syria, Lebanon. 20, 50, 54*, 84*, 199, 209, 369(20:76-86), 700*. *C. zonatus* Gay ex Klatt.

longiflorus Raf.
A garden escape found in 1992 on a grassy roadside verge at Milford (Surrey). Italy, Sicily, Malta. HbACL. 21, 79*, 84*, 342(Feb93:5), 700*.

nudiflorus Smith   Autumn Crocus
- •••• An established escape from cultivation; formerly commercially grown as a substitute for saffron crocus, long naturalised in abundance in meadows and pastures mainly in NW England, now much reduced; also a persistent garden escape. SW Europe. 1, 20, 29*, 50, 79*, 84*, 303(58:33), 312*(n.s. t.169), 335(835:133-141).

pulchellus Herbert   Hairy Crocus
- Introduced and naturalised in the churchyard at Bildeston (W Suffolk). Balkans, W Turkey. 20, 50, 79*, 84*, 369(20:76-86), 700*.

sativus L.   Saffron Crocus
Pre-1930 only. Formerly grown commercially about Saffron Walden (N Essex) for culinary purposes and recently re-planted there. Origin obscure. 1, 20, 50, 80*, 84*, 191, 234, 1156*.

serotinus Salisb.
- A garden escape, found in 1992, in thousands on the edge of a river cliff at Wisley (Surrey). Portugal. HbACL. 21, 79*, 84*, 342(Feb93:5).

sieberi Gay   Sieber's Crocus
- A garden escape on a trackside on Wisley Common (Surrey). SE Europe. 20, 50, 54*, 79*, 84*, 199, 312*(n.s. t.340), 700*. *C. sieberianus* Herbert.

speciosus M.Bieb.   Bieberstein's Crocus
- • (but probably overlooked as *C. nudiflorus*). An established garden escape in a few places in SE England; abundantly naturalised in churchyards at Cold Ash, near Newbury (Berks) and at Chiddingfold (Surrey). Greece, Crimea, SW Asia. RNG, HbACL. 20, 50, 51*, 54*, 79*, 84*, 199, 370(16:195), 371(367:30).

tommasinianus Herbert
- • • (but probably overlooked as *C. vernus*). An established garden escape; naturalised in widely scattered localities in England, extensively so in fields north of North Ash and on Hayes Common (both W Kent); increasing. SE Europe. DBY, RNG. 20*, 50, 54*, 79*, 84*, 199, 209, 371(388:35). A hybrid with *C. vernus* has been recorded. 342(1993:5).

vernus (L.) Hill   Spring Crocus
- • • • An established garden escape and relic of cultivation, mainly in southern England; long naturalised in abundance in fields at Warley Place (S Essex) and Inkpen (Berks). C & S Europe. 1, 20*, 23*, 28*, 50, 79*, 84*, 119, 191. *C. albiflorus* Kit.; *C. albifrons* Kit.; *C. officinalis* Hudson pro parte; *C. purpureus* Weston; *C. vernus* All.; incl. Large Dutch cultivars. Records for *C. vernus* Miller, a synonym of *C. angustifolius* Weston, are in error for this species.

versicolor Ker Gawler
- A garden escape persistent on Dartford Heath (W Kent). SE France. 50, 79*, 84*, 458, 700*.

GLADIOLUS L.

× **colvillei** Sweet
- A garden escape persistent in an old quarry at Mont Cuet in Guernsey (Channel Is); also recorded from near Dartford (W Kent). Originated in cultivation. 50, 52*, 56*, 220, 236, 458, 596*, 700*. *G. cardinalis* Curtis × *G. tristis* L.

communis L.   Eastern Gladiolus
- • • An established escape from cultivation; long persistent in old bulbfields, hedges, on dunes and railway banks in southern England and the Channel Is; locally naturalised since the 1920s in Scilly. Medit. 1, 20, 29*, 50, 54*, 79*, 117*, 207, 220, 1117*. Incl. *G. byzantinus* Miller. Records are mostly referable to subsp. *byzantinus* (Miller) A.P.Ham.

× **hortulanus** Bailey   **Large-flowered Gladiolus**
- A casual garden escape. Originated in cultivation. 50, 53*, 54*, 204, 223.
A complex of hybrids involving *G. natalensis* (Ecklon) Hook., *G. oppositiflorus*
Herbert and other species.
[*italicus* Miller (*G. segetum* Ker Gawler)   In error for *G. communis*. 220.]

SCHIZOSTYLIS Backh. & Harvey ex Hook.f.
coccinea Backh. & Harvey ex Hook.f.
- An escape from cultivation in a semi-natural habitat, recently recorded
without locality from Wilts. S Africa. 50, 54*, 312*(t.5422), 594*, 700*,
1118. *S. pauciflora* Klatt.

IXIA L.
campanulata group   **Red Corn-lilies**
- • Persistent or established escapes from cultivation in a few places, mainly on
St Martin's and St Mary's (both Scilly). S Africa; hybrids originated in
cultivation. BM, E, RNG. 50, 54*, 207, 312*(t.594), 458, 1117*(as Pink
Ixia). Incl. *I. campanulata* Houtt.; *I. crateroides* Ker Gawler; *I. speciosa*
Andrews and hybrids.
maculata L.
- An escape from cultivation, probably persistent, in Scilly. S Africa.
50, 51b, 312*(t.539), 562*, 823*(t.329), 830*(28:45-195), 1014, 1046*, 1143.
The typical plant has not been recorded although undoubtedly present in
cultivation; all records are referable to var. *nigroalbida* (Klatt) Baker,
which appears to be the plant illustrated in 1117*(as White Ixia), in
616*(as *I. viridiflora*) and in 1000*(t.159 as *I. capitata* var.).
paniculata Delaroche   **Tubular Corn-lily**
- A persistent garden escape on Tresco (Scilly). S Africa. K. 20, 22*, 50,
207, 312*(t.256, t.1502), 616*, 655*, 1046*, 1164*.   *I. longiflora* Bergius;
Tritonia longiflora (Bergius) Ker Gawler.

SPARAXIS Ker Gawler
grandiflora (Delaroche) Ker Gawler   **Plain Harlequinflower**
- • An escape from cultivation persistent in a few fields and by roadsides in
Scilly. S Africa. RNG. 20, 50, 51*, 54*, 207, 312*(t.779), 458, 1091*.
Ixia grandiflora Delaroche.

FREESIA Ecklon ex Klatt
× **hybrida** L.Bailey   **Freesia**
- An escape from cultivation persistent in the Channel Is and Scilly.
Originated in cultivation. 20, 51*, 55*, 56*, 220, 371(334:22). *F. refracta*
auct., non (Jacq.) Ecklon ex Klatt.   A complex of hybrids involving
F. armstrongii Watson, *F. corymbosa* (Burman) N.E.Br., *F. lactea* Klatt
(*F. alba* Foster), *F. refracta* (Jacq.) Ecklon ex Klatt, *F. sparrmannii* (Thunb.)
N.E.Br. and *F. xanthospila* Klatt.

CROCOSMIA Planchon

× crocosmiiflora (Lemoine) N.E.Br. Montbretia
••••• An established garden escape; naturalised throughout most of the British Is, especially in SW England, Ireland and western Scotland. Originated in cultivation. 3, 20*, 22*, 34*, 50, 347*(5:246-253), 370(19:265-267). *C. aurea* (Pappe ex Hook.) Planchon × *C. pottsii*; *Montbretia crocosmiiflora* Lemoine; *Tritonia* × *crocosmiiflora* (Lemoine) Nicholson.

masoniorum (L.Bolus) N.E.Br. Giant Montbretia
• A persistent garden escape in three unpublished localities in Dunbarton, now gone from them all. S Africa. 20*, 50, 57*, 304(8:15-17), 347*(5:246-253), 467. *Tritonia masoniorum* L.Bolus.

paniculata (Klatt) Goldblatt Aunt-Eliza
••• An established garden escape; on roadsides, in quarries and on waste ground in widely scattered localities, especially in Dunbarton and Kintyre. S Africa. LANC, NMW, RNG. 20*, 50, 54*, 156, 304(8:16), 349(5:344), 562*, 596*. *Antholyza paniculata* Klatt; *Curtonus paniculatus* (Klatt) N.E.Br.

pottsii (MacNab ex Baker) N.E.Br. Potts' Montbretia
•• (but probably overlooked as *C.* × *crocosmiiflora*). An established garden escape; long naturalised in a few damp places in Kirkcudbrights, abundantly so by the river Dee; also recorded from Wigtowns, Main Argyll and Clyde Is. S Africa. E. 20*, 50, 124, 303(54:54), 304(6:13), 347*(5:246-253), 370(18:433), 732*, 1275*. *Montbretia pottsii* MacNab ex Baker; *Tritonia pottsii* (MacNab ex Baker) Baker. Many early records were confused with *C.* × *crocosmiiflora*.

ANTHOLYZA L.

ringens L.
• A garden escape persistent in woodland on Tresco (Scilly). S Africa. 51, 412, 566*, 628a*, 655*, 1274*.

CHASMANTHE N.E.Br.
[*aethiopica* L. In error for *C. bicolor*. 20, 830*(51:253-261).]
bicolor (Gasp. ex Ten.) N.E.Br. Chasmanthe
• A garden escape persistent or established in a few places on Tresco, on St Agnes, and at Rose Hill, St Mary's (all Scilly). S Africa. RNG, HbEJC. 20, 36b*, 51b, 207, 371(340:25). *Antholyza aethiopica* L. var. *bicolor* (Gasp. ex Ten.) Baker; *A. bicolor* Gasp. ex Ten. All records are tentatively assigned to this species. As a result of the varied circumscriptions used by authors, the name *C. aethiopica* has covered three separate species. The closely related *C. floribunda* (Salisb.) N.E.Br. may have been overlooked. 36b*, 51b, 576*.

AGAVACEAE

FURCRAEA Vent.
bedinghausii K.Koch
• Introduced on Appletree Banks, Tresco (Scilly). Mexico. 50, 312*(t.7170), 325*(93:336), 371(340:25).

YUCCA L.

aff. filamentosa L.   Adam's-needle
• Introduced and long persistent at Dawlish Warren (S Devon). Eastern USA.
21, 50, 51*, 52*, 54*, 371(400:36), 699*, 1181*.

gloriosa L.   Spanish-dagger
• Introduced; also a casual garden escape. SE USA. 7, 21, 50*, 54*, 61*,
62*, 142(p.40), 156, 370(2:306), 699*.

recurvifolia Salisb.   Curved-leaved Spanish-dagger
• A garden escape persistent on dunes at Crymlyn Burrows, near Swansea
(Glam), at Dawlish (S Devon), and in a gravel pit at Broadway (Worcs).
SE USA. 20, 50, 52*, 54*, 61*, 303(45:24), 370(15:139), 446. *Y. gloriosa*
var. *recurvifolia* (Salisb.) Engelm.

rupicola Scheele   Twisted-leaf Spanish-dagger
• A casual on a refuse tip on Crayford Marshes (W Kent), probably a garden
escape. Texas. 50, 312*(t.7172), 458, 699*.

AGAVE L.

americana L.   Centuryplant
• Introduced, or a garden escape, persistent in a few places in the Channel Is
and on Tresco (Scilly). Mexico; (widespread). 20, 24*, 32*, 50, 202, 220,
221, 458, 684*, 702*.

CORDYLINE Comm. ex Adr.Juss.

australis (Forster f.) Endl.   Cabbage-palm
•• Introduced; seedlings have been reported from a few places in W Cornwall,
Scilly and the Channel Is; originally planted and long persistent at Abbotsbury
(Dorset). New Zealand. 20, 50, 54*, 60*, 117*, 207, 220, 303(36:23),
370(13:84), 1181*. *Dracaena australis* Forster f.

PHORMIUM Forster & Forster f.

cookianum Le Jolis   Lesser New Zealand Flax
• An established escape from former cultivation; first recorded in 1920, now
naturalised in abundance on cliffs and dunes on St Martin's (Scilly).
New Zealand. 13, 20, 36a, 50, 52*, 207*, 560*, 686*. *P. colensoi* Hook.f.;
P. hookeri Gunn ex Hook.f.

tenax Forster & Forster f.   New Zealand Flax
••• (but probably over-recorded for hybrids). An established escape from
cultivation; long naturalised in Scilly and Man, where formerly grown as a
fibre crop; elsewhere a garden escape, or planted as windbreaks, mainly in
SW England, SW Ireland and the Channel Is. New Zealand. BM, E, K,
RNG. 7, 20, 36a, 50, 52*, 62*, 103, 207, 560*, 686*.

ORCHIDACEAE
CHAMORCHIS Rich.
alpina (L.) Rich.   False Orchid
   • Reported from the New Forest area (S Hants). N Europe, Alps, Carpathians.
   21, 22*, 23*, 25*, 27*, 1188. *Herminium alpinum* (L.) Lindley.

[*GYMNADENIA* R.Br.
odoratissima (L.) Rich. (*Habenaria odoratissima* (L.) Benth.)   A dubious record.
   166.]

ORCHIS L.
pallens L.   Pale-flowered Orchid
   Pre-1930 only. C & S Europe, Turkey, Caucasus. 7, 21, 22*, 23*, 79*, 98*,
   354(3:33).
palustris Jacq.
   Pre-1930 only.   Europe, E Medit.   7, 22*, 98*, 354(2:508), 552*, 675*,
   1149. *O. laxiflora* Lam. subsp. *palustris* (Jacq.) Bonnier & Layens. The 1872
   record from Guernsey was in error for native *O. laxiflora*. 220, 320(11:209;
   13:377).

SERAPIAS L.
lingua L.   Tongue-orchid
   Found in 1992 in a meadow with native *Orchis laxiflora* at Vagon, Guernsey
   (Channel Is), status uncertain. Medit. 21, 24*, 32*, 303(63:53), 546*, 615*.
neglecta De Notaris   Scarce Tongue-orchid
   Pre-1930 only. Medit. OXF. 7, 21, 29*, 32*, 50, 79*, 98*, 354(5:308).
parviflora Parl.   Small-flowered Tongue-orchid
   • Probably introduced; reported in 1989 without locality from E Cornwall.
   Medit. 20, 21, 213*, 303*(52:11-12), 546*, 1049*, 1079*. *S. occultata* Gay.

OPHRYS L.
bertolonii Moretti   Bertoloni's Bee-orchid
   • Introduced in 1976 on calcareous grassland in Purbeck (Dorset), now gone.
   Medit. 21, 24*, 79*, 98*, 165, 303(54:15), 370(11:430), 1188.

CALYPSO Salisb.
bulbosa (L.) Oakes   Calypso
   • Probably introduced; reported without locality from W Sutherland.
   N temperate regions. 21, 22*, 26*, 50, 98*, 441, 675*, 1188.

BLETILLA Reichb.f.
striata (Thunb.) Reichb.f.   Hyacinth Orchid
   • Introduced in the 1960s at Kingsdown (W Kent), probably now gone.
   E Asia. 43, 50, 54*, 57*, 66*, 310(13:42). *B. hyacinthina* (Smith) Reichb.f.;
   Bletia hyacinthina (Smith) R.Br.

APPENDIX: DRIFT ALIENS

Fruits or seeds of the following alien species have been found washed up on beaches; there are no records of germination in the wild.

Amblygonocarpus andogensis (Welw. ex Oliver) Exell & Torre MIMOSACEAE

Astrocaryum sp. ARECACEAE

Baillonella heckelii (Pierre ex A.Chev.) Baehni (*Dumoria heckelii* Pierre ex A.Chev.; *Mimusops heckelii* (Pierre ex A.Chev.) Hutch. & Dalz.; *Tieghemella heckelii* Pierre ex A.Chev., nom. nud.) SAPOTACEAE

Caesalpinia bonduc (L.) Roxb. CAESALPINIACEAE

Canavalia nitida (Cav.) Piper FABACEAE

C. rosea (Sw.) DC. (*C. maritima* (Aublet) Urban; *Dolichos maritimus* Aublet) FABACEAE

Cocos nucifera L. ARECACEAE

Dioclea hexandra (Ralph) Mabb. (*D. reflexa* Hook.f.) FABACEAE

Entada gigas (L.) Fawcett & Rendle MIMOSACEAE

Erythrina sp. FABACEAE

Ipomoea cf. *alba* L. (*Calonyction aculeatum* (L.) House) CONVOLVULACEAE

Merremia discoidesperma (F.D.Smith) O'Don. (*Ipomoea discoidesperma* F.D.Smith) CONVOLVULACEAE

Mucuna sloanei Fawcett & Rendle FABACEAE

Sacoglottis amazonica Mart. HUMIRIACEAE

Trachycarpus sp. ARECACEAE

303*(34:10-13; 44:16-17), 370(12:104-107). Other species have probably been overlooked. 825*(4:95-98; 8:156-157; 13:44-45).

REFERENCES

MAJOR REFERENCES FOR RECORDS

These (**1-19**) are numbered in order of date. The earliest is quoted in the Catalogue for each species, indicating the period in which it was first recognised as a British alien.

For pre-1930 records

Dunn, S.T. (1905). *Alien flora of Britain*. London.
Druce, G.C. (1908). *List of British plants*. Oxford.
Hayward, I.M. & Druce, G.C. (1919). *The adventive flora of Tweedside*. Arbroath.
Wade, A.E. & Smith, R.L. (1926). The adventive flora of the port of Cardiff. *Rep. Bot. Soc. Brit. Isles*, **7**:999-1027.
Wade, A.E. & Smith, R.L. (1927). Additions to the flora of the port of Cardiff. *Rep. Bot. Soc. Brit. Isles*, **8**:181-183.
Melville, R. & Smith, R.L. (1928). Adventive flora of the metropolitan area. *Rep. Bot. Soc. Brit. Isles*, **8**:444-454.
Druce, G.C. (1928). *British plant list*. Arbroath.
Curtis, R. (1931). Adventive flora of Burton-upon-Trent. *Rep. Bot. Soc. Brit. Isles*, **9**:465-469.
Sandwith, C.I. (1933). The adventive flora of the port of Bristol. *Rep. Bot. Soc. Brit. Isles*, **10**:314-363.

For 1931-1960 records

10 Smith, R.L. & Wade, A.E. (1939). Notes on the adventive flora of the Cardiff district. *Rep. Bot. Soc. Brit. Isles*, **12**:72-83.
11 Burges, R.C.L. (1946). Adventive flora of Burton-on-Trent. *Rep. Bot. Soc. Brit. Isles*, **12**:815-819.
12 Kent, D.H. & Lousley, J.E. (1951-57). A hand list of the plants of the London area. Parts 1-7. Suppls to *London Naturalist*, **30-36**.
13 McClintock, D. (1957). *Supplement to The pocket guide to wild flowers*. Privately published. Platt, Kent.
14 Lousley, J.E. (1961). A census list of wool aliens found in Britain, 1946-1960. *Bot. Soc. Brit. Isles Proc.*, **4**:221-247.
15 Beattie, E.P. (1962). Esparto grass aliens in Fife (v.c.85). *Bot. Soc. Brit. Isles Proc.*, **4**:404-406.

16   Clapham, A.R., Tutin, T.G. & Warburg, E.F. (1962). *Flora of the British Isles*, 2nd ed. Cambridge.
  16a   Clapham, A.R., Tutin, T.G. & Warburg, E.F. (1952). *Flora of the British Isles*. Cambridge.
  16b   Clapham, A.R., Tutin, T.G. & Moore, D.M. (1987). *Flora of the British Isles*, 3rd ed. Cambridge.

For 1961-1990 records

17   Copy of **14** annotated by J.E.Lousley, 1961-1975, in possession of E.J.Clement.
18   Palmer, J.R. (1977). Oil-milling adventive plants in north-west Kent 1973-6. *Trans. Kent Field Club*, **6**:85-90.
19   Hanson, C.G. & Mason, J.L. (1985). Bird seed aliens in Britain. *Watsonia*, **15**:237-252.

MAJOR REFERENCES FOR DESCRIPTIONS AND ILLUSTRATIONS

European floras and wild flower books

✛   20   Stace, C.A. (1991). *New Flora of the British Isles*. Cambridge.
21   Tutin, T.G. *et al.*, eds. (1964-1980). *Flora Europaea*, **1-5**. Cambridge.
  21a   Tutin, T.G. *et al.*, eds. (1993). *Flora Europaea*, **1**, 2nd ed. Cambridge.
✷   22   Blamey, M. & Grey-Wilson, C. (1989). *The illustrated flora of Britain and northern Europe*. London.
✷   23   Grey-Wilson, C. & Blamey, M. (1979). *The alpine flowers of Britain and Europe*. London.
24   Schönfelder, I. & Schönfelder, P. (1990). *Collins photoguide to the wildflowers of the Mediterranean*. London.
✷   25   Fitter, R., Fitter, A. & Blamey, M. (1985). *The wild flowers of Britain and northern Europe*, 4th ed. London.
✷   26   Schauer, T. (1982). *A field guide to the wild flowers of Britain and Europe*. London.
27   Huxley, A. (1986). *Mountain flowers in colour*, 2nd rev. ed. Poole, Dorset.
28   Moggi, G. (1985). *The Macdonald encyclopedia of alpine flowers*. London.
29   Polunin, O. (1969). *Flowers of Europe: a field guide*. London. (The colour plates of this guide are reproduced in Polunin, O. (1972). *The concise flowers of Europe*. Oxford; and in Polunin, O. (1988). *A concise guide to the flowers of Britain and Europe*. Oxford.)
30   Polunin, O. & Smythies, B.E. (1973). *Flowers of south-west Europe: a field guide*. London.
31   Polunin, O. (1980). *Flowers of Greece and the Balkans: a field guide*. Oxford.
✷   32   Polunin, O. & Huxley, A. (1965). *Flowers of the Mediterranean*. London.
33   Butcher, R.W. (1961). *A new illustrated British flora*, **1-2**. London.
✷   34   Phillips, R. (1977). *Wild flowers of Britain*. London.
✷   35   Ross-Craig, S. (1948-1974). *Drawings of British plants*, **1-32**. London.

Non-European floras and wild flower books

36   Allan, H.H. (1961). *Flora of New Zealand*, **1**. Wellington.
  36a Moore, L.B. & Edgar, E. (1971). *Flora of New Zealand*, **2**. Wellington.
  36b Healy, A.J. & Edgar, E. (1980). *Flora of New Zealand*, **3**. Wellington.
  36c Webb, C.J., Sykes, W.R. & Garnock-Jones, P.J. (1988). *Flora of New Zealand*, **4**. Christchurch.
37   Black, J.M. (1960-1965). *Flora of South Australia*, **1-4**, 2nd ed. Adelaide.
  37a Eichler, H. (1965). *Supplement to J.M.Black's Flora of South Australia*, *2nd ed.* Adelaide.
  37b Black, J.M. (1978). *Flora of South Australia*, **1**, 3rd ed., rev. & ed. Jessop, J.P. Adelaide.
  37c Jessop, J.P. & Toelken, H.R. (1986). *Flora of South Australia*, **1-4**, 4th ed. Adelaide.
38   Britton, N.L. & Brown, A. (1970). *An illustrated flora of the northern United States and Canada*, **1-3**, Dover ed. New York.
  38a Gleason, H.A. (1968). *The new Britton and Brown illustrated flora of the northeastern United States and adjacent Canada*, **1-3**, rev. ed. New York.
39   Davis, P.H., ed. (1965-1988). *Flora of Turkey*, **1-10**. Edinburgh.
40   Fernald, M.L. (1950). *Gray's manual of botany*, 8th ed. New York.
41   Hitchcock, C.L. & Cronquist, A. (1973). *Flora of the Pacific northwest: an illustrated manual*. Seattle & London.
42   Munz, P.A. & Keck, D.D.(1959). *A California flora*. Berkeley & Los Angeles.
43   Ohwi, J. (Meyer, G.F. & Walker, E.H., eds.) (1965). *Flora of Japan*, rev. ed. Washington, D.C.
44   Zohary, M. (1966-1972). *Flora Palaestina*, text **1-2**, plates **1-2**. Jerusalem.
  44a Feinbrun-Dothan, N. (1978-1986). *Flora Palaestina*, text **3-4**, plates **3-4**. Jerusalem.
45   Nègre, R. (1962-1963). *Petite flore des régions arides du Maroc occidental*, **1-2**. Paris.
46   Quézel, P. & Santa, S. (1962-1963). *Nouvelle flore de l'Algérie et des régions désertiques méridionales*, **1-2**. Paris.
47   Blundell, M. (1987). *Collins guide to the wild flowers of East Africa*. London.
48   Agnew, A.D.Q. (1974). *Upland Kenya wild flowers*. Oxford.
49   Adams, C.D. (1972). *Flowering plants of Jamaica*. Mona, Jamaica.

Garden plants

50   Walters, S.M. *et al.*, eds. (1984-). *The European garden flora*, **1**-. Cambridge.

51 Synge, P.M., ed. (1956). *The Royal Horticultural Society dictionary of gardening*, **1-4**, 2nd ed. Oxford.

   51a Synge, P.M., ed. (1969). *The Royal Horticultural Society Supplement to the dictionary of gardening*, 2nd ed. Oxford.

   51b Huxley, A., Griffiths, M. & Levy, M., eds. (1992). *The new Royal Horticultural Society dictionary of gardening*, **1-4**. London & New York.

52 Beckett, K.A. (1984). *The concise encyclopedia of garden plants*. London.

53 Brickell, C., ed. (1989). *The Royal Horticultural Society gardeners' encyclopedia of plants and flowers*. London.

54 Hay, R. & Beckett, K.A. (1987). *Reader's Digest encyclopedia of garden plants and flowers*, 4th ed. London.

55 Wright, M. (1984). *The complete handbook of garden plants*. London.

56 Perry, F., ed. (1980). *The Macdonald encyclopedia of plants and flowers*. London.

57 Bloom, A. (1990). *Hardy perennial plants including alpines*. Wisbech.

Trees, shrubs and climbers

58 Mitchell, A. & More, D. (1985). *The complete guide to trees of Britain and northern Europe*. Limpsfield, Surrey.

59 Phillips, R. (1978). *Trees in Britain, Europe and North America*. London.

60 Mitchell, A. (1974). *A field guide to the trees of Britain and northern Europe*. London.

61 Bean, W.J. (1970-1980). *Trees and shrubs hardy in the British Isles*, **1-4**, 8th ed. London.

   61a Clarke, D.L. (1988). *W.J.Bean. Trees and shrubs hardy in the British Isles. Supplement*. London.

62 Davis, B. (1987). *The gardener's illustrated encyclopedia of trees and shrubs*. Harmondsworth.

63 Phillips, R. & Rix, M. (1989). *Shrubs*. London.

64 Herklots, G. (1976). *Flowering tropical climbers*. Folkestone.

65 Grey-Wilson, C. & Matthews, V. (1983). *Gardening on walls*. London.

House plants and tropical plants

66 Beckett, K.A. (1987). *The RHS encyclopedia of house plants including conservatory plants*. London.

67 Přibl, J. (1987). *Hamlyn colour guides: indoor plants*. Twickenham, Middlesex.

68 Purseglove, J.W. (1968-1972). *Tropical crops*: **1** (Dicotyledons), **2** (Monocotyledons). London.

69 Lötschert, W. & Beese, G. (1983). *Collins guide to tropical plants*. London.

Monographs and special groups

70 Chalk, D. (1988). *Hebes and parahebes*. London.

71 Chiej, R. (1984). *The Macdonald encyclopedia of medicinal plants*. London.

72 Evans, R.L. (1983). *Handbook of cultivated sedums*. Northwood, Middlesex.

73 Fitter, R., Fitter, A. & Farrer, A. (1984). *Collins guide to the grasses, sedges, rushes and ferns of Britain and Northern Europe*. London.
74 Fretwell, B. (1989). *Clematis*. London.
75 Goodspeed, T.H. (1954). *The genus Nicotiana*. Waltham, Massachusetts.
76 Green, R. (1976). *Asiatic primulas*. Alpine Garden Society. Woking, Surrey.
77 Grenfell, D. (1990). *Hosta: the flowering foliage plant*. London.
78 Grey-Wilson, C. (1988). *The genus Cyclamen*. Kew.
79 Grey-Wilson, C. & Mathew, B. (1981). *Bulbs*. London.
80 Harrison, S.G. *et al.* (1969). *The Oxford book of food plants*. Oxford.
81 Jermy, A.C., Chater, A.O. & David, R.W. (1982). *Sedges of the British Isles*. B.S.B.I. handbook No.1. London.
82 Köhlein, F. (1987). *Iris*. Bromley, Kent.
83 Lousley, J.E. & Kent, D.H. (1981). *Docks and knotweeds of the British Isles*. B.S.B.I. handbook No.3. London.
84 Mathew, B. (1982). *The crocus*. London.
85 Mathew, B. (1989). *Hellebores*. Alpine Garden Society. Woking, Surrey.
86 Meikle, R.D. (1984). *Willows and poplars of Great Britain and Ireland*. B.S.B.I. handbook No.4. London.
87 Muhlberg, H. (1982). *The complete guide to water plants*. Wakefield.
88 Page, M. & Stearn, W.T. (1979). *Culinary herbs*, 2nd ed. Wisley handbook No.16. London.
89 Phillips, R. & Rix, M. (1988). *Roses*. London.
90 Rich, T.C.G. (1991). *Crucifers of Great Britain and Ireland*. B.S.B.I. handbook No.6. London.
91 Richards, A.J. (1972). The *Taraxacum* flora of the British Isles. Suppl. to *Watsonia*, 9.
92 Rose, P.Q. (1980). *Ivies*. Poole, Dorset.
93 Rowley, G.D. (1980). *Name that succulent*. Cheltenham.
94 Stern, F.C. (1956). *Snowdrops and snowflakes*. London.
95 Tutin, T.G. (1980). *Umbellifers of the British Isles*. B.S.B.I. handbook No.2. London.
96 Webb, D.A. & Gornall, R.J. (1989). *Saxifrages of Europe*. London.
97 Webb, W.J. (1984). *The Pelargonium family*. London.
98 Williams, J.G. *et al.* (1978). *A field guide to the orchids of Britain and Europe with North Africa and the Middle East*. London.
99 Yeo, P.F. (1985). *Hardy geraniums*. London.

LOCAL FLORAS

100   Allen, D.E. (1957). *The flora of the Rugby district*. Rugby.
101   Allen, D.E. (1961). *A first supplement to The flora of the Rugby district*. Rugby.
102   Allen, D.E. (1969). *The flowering plants and ferns of the Isle of Man*. Douglas.
103   Allen, D.E. *et al.* (1984). *Flora of the Isle of Man*. Douglas.
104   Amphlett, J. & Rea, C. (1909). *The botany of Worcestershire*. Birmingham.
105   Bagnall, J.E. (1891). *The flora of Warwickshire*. London.
106   Baker, J.G. (1885). *A flora of the English Lake District*. London.
107   Baker, J.G. (1906). *North Yorkshire: studies of its botany, geology, climate and physical geography*, 2nd ed. London. (Reprinted from *Trans. Yorkshire Naturalists' Union*, **1863**).
108   Baker, J.G. & Nowell, J. (1854). *A supplement to Baines' Flora of Yorkshire*. London.
109   Baker, J.G. & Tate, G.R. (1868). A new Flora of Northumberland and Durham. *Trans. Nat. Hist. Soc. Northumberland*, **2**: 1-316.
110   Balfour, F.R.S. (1925). *Botany*, in Buchan, J.W. & Paton, H., eds. *The history of Peeblesshire*, **1**:340-428.
111   Balfour, J.H. & Sadler, J. (1863). *Flora of Edinburgh*. Edinburgh.
112   Ballantyne, G.H. (1971). Ballast aliens in south Fife, 1820-1919. *Trans. & Proc. Bot. Soc. Edinburgh*, **41**:125-137.
113   Ballantyne, G.H. (1970). *The flowering plants of Kirkcaldy and district*. Kirkcaldy.
114   Ballantyne, G.H. (1985). *The flowering plants of Kinross*, 2nd ed. Kirkcaldy.
115   Barkham, J.P., *et al.* (1981). *Foula, Shetland. Volume 2. The flora of Foula*. Ambleside.
116   Bevis, J.F. & Griffin, W.H., eds. (1909). *Botany*, in Grinling, C.H., Ingram, T.A. & Polkinghorne, B.C., eds. *A survey and record of Woolwich and West Kent*, pp. 31-230. Woolwich.
117   Bichard, J.D. & McClintock, D. (1975). *Wild flowers of the Channel Islands*. London.
118   Booth, E.M. (1979). *The flora of County Carlow*. Dublin.
119   Bowen, H.J.M. (1968). *The flora of Berkshire*. Oxford.
120   Boydon Ridge, W.T. (1922-1929). The flora of North Staffordshire. Append. 1-8 to *Trans. N. Staffordshire Field Club*, **56-63**.
121   Brewer, J.A. (1863). *Flora of Surrey*. London.
122   Briggs, M. (1990). *Sussex plant atlas: selected supplement*. Brighton.
123   Bromfield, W.A. (Hooker, W.J. & Bell-Salter, T., eds.) (1856). *Flora vectensis*. London.
124   Brunker, J.P. (1950). *Flora of the County Wicklow*. Dundalk.
125   Burgess, J.J., ed. (1935). *Flora of Moray*. Elgin.
126   Burton, R.M. (1983). *Flora of the London area*. London Natural History Society. London.

127 Buxton, R. (1859). *A botanical guide to the flowering plants, ferns, mosses, and algae, found indigenous within eighteen miles of Manchester*, 2nd ed. London.

128 Byatt, J.I. (1985). *Checklist of the flora of the Croydon survey area*. South Croydon.

129 Cadbury, D.A., Hawkes, J.G. & Readett, R.C. (1971). *A computer-mapped flora. A study of the county of Warwickshire*. Birmingham.

130 Clapham, A.R., ed. (1969). *Flora of Derbyshire*. Derby.

131 Clitheroe, W. & Robinson, E.C., (1903). *Flora of Preston and neighbourhood*. Preston.

132 Colgan, N. (1904). *Flora of the County Dublin: flowering plants, higher cryptogams, and characeae*. Dublin.

133 Corner, R.W.M. (1985). *Flowering plants and ferns of Selkirkshire and Roxburghshire*. Lancaster.

134 Crackles, F.E. (1990). *Flora of the East Riding of Yorkshire*. Hull.

135 Crompton, G. & Whitehouse, H.L.K. (1983). *Annotated checklist of the flora of Cambridgeshire*. Cambridge.

136 Crump, W.B. & Crossland, C. (1904). *The flora of the parish of Halifax*. Halifax. (Reprinted from suppl. to *Halifax Naturalist*, **1-8** (1896-1904):1-120.

137 Cunningham, M.H. & Kenneth, A.G. (1979). *The flora of Kintyre*. Wakefield.

138 Davey, F.H. (1902). *A tentative list of the flowering plants, ferns, etc., known to occur in the county of Cornwall, including the Scilly Isles*. Penryn.

139 Davey, F.H. (1909). *Flora of Cornwall*. Penryn.

140 Davis, T.A.W. (1970). *Plants of Pembrokeshire*. Haverfordwest.

141 Dickinson, J. (1851). *The flora of Liverpool*. (Reprinted from *Proc. Liverpool Lit. Soc.*, **9**.)

142 Dillwyn, L.W. (1848). *Materials for a fauna and flora of Swansea and the neighbourhood*. Swansea.

143 Dony, J.G. (1953). *Flora of Bedfordshire*. Luton.

144 Dony, J.G. (1967). *Flora of Hertfordshire*. Hitchin.

145 Dony, J.G. (1969). Additional notes on the flora of Bedfordshire. *Bot. Soc. Brit. Isles Proc.*, **7**:523-553.

146 Dony, J.G. (1976). *Bedfordshire plant atlas*. Luton.

147 Dony, J.G. & Dony, C.M. (1986). Further notes on the flora of Bedfordshire. *Watsonia*, **16**:163-172.

148 Druce, G.C. (1886). *The flora of Oxfordshire*. Oxford.

149 Druce, G.C. (1897). *The flora of Berkshire*. Oxford.

150 Druce, G.C. (1926). *The flora of Buckinghamshire*. Arbroath.

151 Druce, G.C. (1927). *The flora of Oxfordshire*, 2nd ed. Oxford.

152 Druce, G.C. (1930). *The flora of Northamptonshire*. Arbroath.

153 Duncan, U.K. (1980). *Flora of East Ross-shire*. Edinburgh.

154 Easy, G.M. (1976). The flora and fauna of rubbish tips and waste places in Cambridgeshire. *Nat. Cambridgeshire*, **19**:23-31.

155   Eggeling, W.J. (1965).   Check list of the plants of Rhum, Inner Hebrides (v.c.104, North Ebudes).   Part 1.   Stoneworts, ferns and flowering plants. *Trans. & Proc. Bot. Soc. Edinburgh*, **40**:20-59.
156   Ellis, R.G. (1983). *Flowering plants of Wales*. Cardiff.
157   F[alconer], R.W. (1848). *Contributions towards a catalogue of plants indigenous to the neighbourhood of Tenby*. London.
158   Frazer, O.H., ed. (1978). *Flora of the Isle of Wight*. Newport.
159   Gardiner, W. (1848). *The flora of Forfarshire*. London & Edinburgh.
160   Gibbons, E.J. (1975). *The flora of Lincolnshire*. Lincoln.
161   Gibbons, E.J. & Weston, I. (1985). *Supplement to The flora of Lincolnshire*. Lincoln.
162   Gilbert, J.L. (1965). *Flora of Huntingdonshire: wildflowers*. Peterborough.
163   Good, R. (1948). *A geographical handbook of the Dorset flora*. Dorchester.
164   Good, R. (1970). Hand-list of the Dorset flora (second addendum). *Proc. Dorset Naturalists' Hist. Archaeol. Soc.*, **83**:1-10.
165   Good, R. (1984). *A concise flora of Dorset*. Dorchester.
166   Graham, G.G. (1988). *The flora and vegetation of County Durham*. Durham.
167   Graham, G.G., Sayers, C.D. & Gaman, J.H. (1972).   *A checklist of the vascular plants of County Durham*. Durham.
168   Green, C.T. (1933). *The flora of the Liverpool district*, 2nd ed.   Arbroath.
169   Grierson, R. (1931). Clyde casuals, 1916-1928. *Glasgow Naturalist*, **9**:5-51.
170   Griffith, J.E. (1895). *The flora of Anglesey and Carnarvonshire*. Bangor.
171   Grindon, L.H. (1859). *The Manchester flora*. London.
172   Grose, [J.]D. (1957). *The flora of Wiltshire*. Devizes.
173   Hadley, G., ed. (1985). *A map flora of mainland Inverness-shire*. Edinburgh & London.
174   Hall, P.C. (1980). *Sussex plant atlas*. Brighton.
175   Halliday, G. (1978). *Flowering plants and ferns of Cumbria*. Lancaster.
176   Hanbury, F.J. & Marshall, E.S. (1899). *Flora of Kent*. London.
177   Harrison, J.W.Heslop, ed. (1941).   A preliminary flora of the Outer Hebrides. *Proc. Univ. Durham Philos. Soc.*, **10**:228-273.
178   Harrison, J.W.Heslop, *et al.* (1941).   The flora of the isles of Coll, Tiree and Gunna (v.c. 110B). *Proc. Univ. Durham Philos. Soc.*, **10**:274-308.
179   Hart, H.C. (1887). *The flora of Howth*. Dublin.
180   Hind, W.M. [& Babington, C.] (1889). *The flora of Suffolk*. London.
181   Hodgson, W. (1898). *Flora of Cumberland*. Carlisle.
182   Holland, S.C. *et al.* (1986). *Supplement to the Flora of Gloucestershire*. Bristol.
183   Hollick, K.M. & Patrick, S.J. (1980). *Supplement to the Flora of Derbyshire 1969: additional records received 1974-1979*. Derby.
184   Horwood, A.R. & Noel, C.W.F. (1933). *The flora of Leicestershire and Rutland*. London.
185   Howitt, R.C.L. & Howitt, B.M. (1963). *A flora of Nottinghamshire*. Newark.
186   Hyde, H.A. & Wade, A.E. (1978). *Welsh ferns, clubmosses, quillworts and horsetails*, 6th ed., rev. Harrison, S.G.   Cardiff.

187   Ingram, R. & Noltie, H.J. (1981). *The flora of Angus (Forfar, v.c. 90)*. Dundee.

188   Irvine, A. (1838). *The London flora*. London.

189   Ivimey-Cook, R.B. (1984). *Atlas of the Devon flora*. Exeter.

190   Jermy, A.C. & Crabbe, J.A., eds. (1978). *The island of Mull: a survey of its flora and environment*. London.

191   Jermyn, S.T. (1975 [1974]). *Flora of Essex*. Fingringhoe.

192   Kent, D.H. (1975). *The historical flora of Middlesex*. London.

193   Keys, I.W.N. (1865-1871). Flora of Devon and Cornwall. Parts 1-5. *Rep. Trans. Plymouth Inst. Devon Cornwall Nat. Hist. Soc.*, **2-3**.

194   Lee, J.R. (1933). *The flora of the Clyde area*. Glasgow.

195   Lees, F.A. (1888). *The flora of West Yorkshire*. London.

196   Lees, F.A. (1939). The vegetation of Craven in Wharfedale with its adjacencies in Aire and Ribble. Suppl. to *N.W. Naturalist*, **12-14**.

197   Lees, F.A. (Cheetham, C.A. & Sledge, W.A., eds.) (1941). *A supplement to the Yorkshire floras*. London.

198   Leslie, A.C. (1971). A contribution to the alien flora of the Sherborne area. *Proc. Dorset Nat. Hist. Archaeol. Soc.*, **92**:45-47.

199   Leslie, A.C. (1987). *Flora of Surrey: supplement and checklist*. Privately published. Guildford.

200   Lester-Garland, L.V. (1903). *A flora of the island of Jersey*. London.

201   Le Sueur, F. (1984). *Flora of Jersey*. Jersey.

202   Le Sueur, F. & McClintock, D., comp. (1962). A check list of the flowering plants and ferns wild on Herm and its off-islets. *Rep. Trans. Soc. Guernésiaise*, **17**:303-318.

203   Linton, W.R. (1903). *Flora of Derbyshire*. London.

204   Livermore, L.A. & P.D. (1987). *The flowering plants and ferns of North Lancashire*. Lancaster.

205   Liverpool Naturalists' Field Club (1872-1876). *The flora of Liverpool*. Liverpool.

206   Lloyd, L.C. & Rutter, E.M. (1957). *Handlist of the Shropshire flora*. Shrewsbury.

207   Lousley, J.E. (1971). *Flora of the Isles of Scilly*. Newton Abbot.

208   Lousley, J.E. (1975). *Flowering plants and ferns in the Isles of Scilly*, rev. ed. Isles of Scilly Museum publication No.4.   Hugh Town, St Mary's.

208a   Lousley, J.E. (1983). *Flowering plants and ferns in the Isles of Scilly*, rev. ed., rev. Harvey, C.   Isles of Scilly Museum publication No.4. Hugh Town, St Mary's.

209   Lousley, J.E. (1976). *Flora of Surrey*. Newton Abbot.

210   Mackay, J.T. (1836). *Flora hibernica*. Dublin.

211   Margetts, L.J. (1988). *The difficult and critical plants of the Lizard district of Cornwall*. Bristol.

212   Margetts, L.J. & David, R.W. (1981). *A review of the Cornish flora 1980*. Redruth.

213   Margetts, L.J. & Spurgin, K.L. (1991). *The Cornish flora: supplement 1981-1990*. St Ives, Cornwall.

214   Marquand, E.D. (1901). *Flora of Guernsey and the lesser Channel Islands*. London.
215   Marshall, A. (1982). *The Rochdale flora*. Rochdale.
216   Martin, I.H., ed. (1927). *The field-club flora of the Lothians*. Edinburgh.
217   Martin, W.Keble & Fraser, G.T., eds. (1939). *Flora of Devon*. Arbroath.
218   Martyn, T. (1763). *Plantae cantabrigienses*. London.
219   May, R.F. (1968). *A list of the flowering plants and ferns of Carmarthenshire*. Haverfordwest.
220   McClintock, D. (1975). *The wild flowers of Guernsey*. London.
221   McClintock, D. & Marsden, M. (1979). *A revised checklist of flowering plants and ferns wild on Sark and its off-islets*. Privately published. Platt, Kent & Rondellerie, Sark.
222   Messenger, G. (1971). *Flora of Rutland*. Leicester.
223   Mill, R.R. (1967). *Flora of Helensburgh and district*. Helensburgh.
224   Morton, J.K. (1959). *The flora of Islay and Jura*. Suppl. to *Bot. Soc. Brit. Isles Proc.*, **3**.
225   Mott, F.T., *et al.* (1886). *The flora of Leicestershire*. London.
226   Mundell, A.R.G. (1991). *Royal Aerospace Establishment conservation group working paper CG 01/91*. Internal report. Farnborough, Hants.
227   Murray, R.P. (1896). *The flora of Somerset*. Taunton.
228   Nelson, G.A., (1963). A flora of Leeds and district. *Proc. Leeds Philos. Lit. Soc.,* **9**:113-170.
229   Nelson, G.A., comp. (1965). *Addendum II (1964-1965) to A flora of Leeds and district*. Duplicated manuscript. Leeds.
230   Newton, A. (1971). *Flora of Cheshire*. Chester.
231   P[amplin], W. & I[rvine], A. (1857). *A botanical tour in the highlands of Perthshire*. London. (Reprinted from *Phytologist*, **2**.)
232   Paton, J.A. (1968). *Wild flowers in Cornwall and the Isles of Scilly*. Truro.
233   Paulson, R. (1926). *The botanical investigation of Essex*, in Hutchings, G.E., ed. *Essex: an outline scientific survey*, pp.34-55. Colchester.
234   Perring, F.H., *et al.* (1964). *A flora of Cambridgeshire*. Cambridge.
235   Petch, C.P. & Swann, E.L. (1968) *Flora of Norfolk*. Norwich.
236   Philp, E.G. (1982). *Atlas of the Kent flora*. The Kent Field Club. Maidstone.
237   Praeger, R.L. (1909). *A tourist's flora of the west of Ireland*. Dublin.
238   Praeger, R.L. (1934). *The botanist in Ireland*. Dublin.
239   Praeger, R.L. & Megam, W.R., eds. (1938). *A flora of the north-east of Ireland, by S.A.Stewart & T.H.Corry*, 2nd ed. Belfast.
240   Primavesi, A.L. & Evans, P.A. (1988). *Flora of Leicestershire*. Leicester.
241   Pryor, A.R. (1887). *A flora of Hertfordshire*. London.
242   Purton, T. (1821). *An appendix to the Midland flora*. London.
243   Rees, J.S. (1970). *A flora of Oundle*. Oundle.
244   Riddelsdell, H.J. (1907). A flora of Glamorganshire. Suppl. to *J. Bot.*, **45**.
245   Riddelsdell, H.J., ed. (1948). *Flora of Gloucestershire*. Cheltenham.
246   Roberts, R.H. (1982). *The flowering plants and ferns of Anglesey*. Cardiff.
247   Robinson, J.F. (1902). *The flora of the East Riding of Yorkshire*. London.

248 Roe, R.G.B. (1981). *The flora of Somerset*. Taunton.
249 Salmon, C.E. (1931). *Flora of Surrey*. London.
250 Salter, J.H. (1935). *The flowering plants and ferns of Cardiganshire*. Cardiff.
251 Savidge, J.P. *et al.*, eds. (1963). *Travis's flora of South Lancashire*. Liverpool.
252 Scannell, M.J.P. & Synnott, D.M. (1972). *Census catalogue of the flora of Ireland*. Dublin.
253 Scott, S. & Palmer, R. (1987). *The flowering plants and ferns of the Shetland Islands*. Lerwick.
254 Scott Elliot, G.F. (1891). *Flora of Dumfriesshire and Dumfries district*. London.
255 Scully, R.W. (1916). *Flora of County Kerry*. Dublin.
256 Shepard, B. (1985). A supplement to the Flora of the Isle of Wight, 1978. *Proc. Isle of Wight Nat. Hist. Soc.*, 7:569-576.
257 Sibthorp, J. (1794). *Flora oxoniensis*. Oxford.
258 Simpson, F.W. (1982). *Simpson's flora of Suffolk*. Ipswich.
259 Sinker, C.A. *et al.* (1985). *Ecological flora of the Shropshire region*. Shrewsbury.
260 Sonntag, C.O. (1894). *A pocket flora of Edinburgh*. London.
261 Stearn, L.F. (1975). *Supplement to The flora of Wiltshire*. Devizes.
262 Stewart, R. (1860). *Handbook of the Torquay flora*. Torquay & London.
263 Storrie, J. (1886). *The flora of Cardiff*. Cardiff.
264 Swann, E.L. (1975). *Supplement to the Flora of Norfolk*. Norwich.
265 Thompson, R. (1980). *A new flora of Alnwick district*. Alnwick.
266 Thurston, E. & Vigurs, C.C. (1922). *A supplement to F.Hamilton Davey's Flora of Cornwall*. Truro.
267 Trail, J.W.H. (1923). Flora of the City parish of Aberdeen, in *James William Helenus Trail: a memorial volume*, pp. 57-331. Aberdeen University studies No.91. Aberdeen.
268 Trimen, H. & Thiselton-Dyer, W.T. (1869). *Flora of Middlesex*. London.
269 Trist, P.J.O., ed. (1979). *An ecological flora of Breckland*. Wakefield.
270 Wade, A.E. (1970). *The flora of Monmouthshire*. Cardiff.
271 Wade, W. (1804). *Plantae rariores in Hibernia inventae*. Dublin.
272 Warren, J.B.L. (Lord de Tabley) (1899). *The flora of Cheshire*. London.
273 Watts, N.S. & Watts, G.D. (1979). Norwich bird seed mixtures and the casual plants of Harford tip. *Trans. Norfolk Norwich Naturalists' Soc.*, 25:300-309.
274 Webb, D.A. (1959). *An Irish flora*, 2nd ed. Dundalk.
275 Webb, D.A. & Scannell, M.J.P. (1983). *Flora of Connemara and the Burren*. Cambridge.
276 Webster, M.McCallum (1968). *A check list of the flora of the Culbin State Forest*. Dyke, Moray.
277 Webster, M.McCallum (1978). *Flora of Moray, Nairn and East Inverness*. Aberdeen.
278 Wheldon, J.A. (1914). *Some alien plants of the Mersey province*. (Reprinted from *Lancash. Naturalist*, 5-6 and *Lancashire Cheshire Naturalist*, 7.)

279   Wheldon, J.A. & Wilson, A. (1907). *The flora of West Lancashire.* Liverpool.

280   White, F.B.W. (1898). *The flora of Perthshire.* Edinburgh.

281   White, J.W. (1912). *The flora of Bristol.* Bristol.

282   Whitehead, J. *et al.,* comps. (1888). *The district flora.* Ashton-under-Lyne.

283   Whitehead, L.E. (1976). *Plants of Herefordshire.* Hereford.

284   Wilson, A. (1938). *The flora of Westmorland.* Arbroath.

285   Wilson, A.K. (1938). *The adventive flora of the East Riding of Yorkshire.* Hull Scientific and Field Naturalists' Club occasional paper No.1. Hull.

286   Winch, N.J. (1831). Flora of Northumberland and Durham. Newcastle-upon-Tyne. (Reprinted from *Trans. Nat. Hist. Soc. Northumberland,* **1831**.)

287   Winch, N.J., Thornhill, J. & Waugh, R. (1805). *The botanist's guide through the counties of Northumberland and Durham,* **1**. Newcastle-upon-Tyne.

288   Wolley-Dod, A.H. (1937). *Flora of Sussex.* Hastings.

289   Woodruffe-Peacock, E.A. (1909). *A check-list of Lincolnshire plants.* Louth.

290   Zawadzki, J.J. (1978). *Bury metropolitan borough flora.* Bury.

BRITISH AND IRISH PERIODICALS

Two versions of abbreviations of titles of periodicals are given; the first in the traditional style of the *World list of scientific periodicals* (1044), the second in the increasingly used style of the *B-P-H Botanico-Periodicum-Huntianum* list (1192) and *Supplementum* (1192a).

World list	*B-P-H*
300 *Ann. Scot. nat. Hist.*	*Ann. Scott. Nat. Hist.*
- - - *(Bot. Mag.)* See **312**	
301 *Bot. Reg.*	*Bot. Reg.*
302 *BSBI Abstr.*	*B.S.B.I. Abstr.*
303 *BSBI News*	*B.S.B.I. News*
304 *BSBI Scott. Newsl.*	*B.S.B.I. Scott. Newslett.*
305 *Botl Soc. Br. Isl. Welsh Bull.*	*Welsh Bull. Bot. Soc. Brit. Isles*
306 *Bull. alp. Gdn Soc.* continued as *Q. Bull. alp. Gdn Soc.*	*Bull. Alpine Gard. Soc. Gr. Brit.*
307 *Bull. Flora Bucks.*	*Bull. Fl. Buckinghamshire*
308 *Bull. For. Commn, Lond.*	*Bull. Forest. Commiss.*
309 *Bull. hardy Pl. Soc.*	*Bull. Hardy Pl. Soc.*
310 *Bull. Kent Fld Club*	*Bull. Kent Field Club*
311 *Countryside*	*Country-Side*
312 *Curtis's bot. Mag.*	*Bot. Mag.*
313 *Essex Nat.*	*Essex Naturalist*
314 *Fern Gaz.*	*Fern Gaz.*
315 *Gdnrs' Chron.*	*Gard. Chron.*
316 *Glasg. Nat.*	*Glasgow Naturalist*
317 *Hooker's Icon. Pl.*	*Hooker's Icon. Pl.*
318 *Ir. Nat.*	*Irish Naturalist*
319 *Ir. Nat. J.*	*Irish Naturalists' J.*
320 *J. Bot., Lond.*	*J. Bot.*
321 *J. Cheltenham Distr. Nat. Soc.*	*Cheltenham Distr. Naturalists' Soc. J.*
322 *J. Ecol.*	*J. Ecol.*
323 *J. Gloucs. Nat. Soc.*	*J. Gloucestershire Naturalists' Soc.*
324 *Jl N. Gloucs. Nat. Soc.*	*N. Gloucestershire Naturalists' Soc. J.*
325 *Jl R. hort. Soc.*	*J. Roy. Hort. Soc.*
326 *Jl R. Instn Cornwall*	*J. Roy. Inst. Cornwall*
327 *J. Proc. Linn. Soc. (Bot.)* continued as *J. Linn. Soc. (Bot.)* and *Bot. J. Linn. Soc.*	*J. Proc. Linn. Soc., Bot.* continued as *J. Linn. Soc., Bot.* and *Bot. J. Linn. Soc.*
328 *Kew Bull.* preceded by *Bull. misc. Inf. R. bot. Gdns, Kew*	*Kew Bull.* preceded by *Bull. Misc. Inform. Kew*
329 *Kew Mag.*	*Kew Mag.*

	World list	*B-P-H*
330	*Living Countryside*	*Liv. Countryside*
331	*Lond. Nat.*	*London Naturalist*
332	*Middle-Thames Nat.*	*Middle-Thames Naturalist*
333	*Midl. Nat.*	*Midl. Naturalist (London)*
334	*Moorea*	*Moorea*
335	*Naturalist, Hull*	*Naturalist (Hull)*
336	*Nature Cambs.*	*Nat. Cambridgeshire*
337	*Nature, Lond.*	*Nature*
338	*Nature N.E. Essex* An alternative title for **356**	
339	*Nature Wales*	*Nat. Wales*
340	*Newsl. Bristol Nats. Soc. bot. Sect.*	*Newslett. Bristol Naturalists' Soc. Bot. Sect.*
341	*Newsl. Lond. nat. Hist. Soc.*	*Newslett. London Nat. Hist. Soc.*
342	*Newsl. Surrey Flora Comm.*	*Newslett. Surrey Fl. Committee*
343	*Newsl. Sussex Flora Comm.*	*Newslett. Sussex Fl. Committee*
344	*Notes R. bot. Gdn Edinb.*	*Notes Roy. Bot. Gard. Edinburgh*
345	*NWest. Nat.*	*N.W. Naturalist*
346	*Phytologist*	*Phytologist*
347	*Plantsman*	*Plantsman*
348	*Proc. Bgham nat. Hist. & Micr. Soc.* continued as *Rep. Trans. Bgham nat. Hist. & microsc. Soc.* Cf. **386**	*Proc. Birmingham Nat. Hist. Microscop. Soc.* continued as *Trans. Birmingham Nat. Hist. Microscop. Soc.*
349	*Proc. bot. Soc. Br. Isl.*	*Bot. Soc. Brit. Isles Proc.*
350	*Proc. Bristol Nat. Soc.*	*Proc. Bristol Naturalists' Soc.*
351	*Proc. Dorset nat. Hist. antiq. Fld Club* continued as *Proc. Dorset nat. Hist. archaeol. Soc.*	*Proc. Dorset Nat. Hist. Antiq. Field Club* continued as *Proc. Dorset Nat. Hist. Archaeol. Soc.*
352	*Proc. R. Ir. Acad.*	*Proc. Roy. Irish Acad.*
353	*Reading Nat.*	*Reading Naturalist*
354	*Rep. botl Soc. Exch. Club Br. Isl.* 1914-1947 and several variations on this title, 1862-1913	*Bot. Exch. Club Brit. Isles Rep.* continued as *Bot. Exch. Club Soc. Brit. Isles* and *Bot. Soc. Exch. Club Brit. Isles*
355	*Rep. Penzance nat. Hist. antiq. Soc.*	*Rep. Penzance Nat. Hist. Antiq. Soc.*
356	*Rep. Rec. Colchester Distr. nat. Hist. Soc. Fld Club*	*Rep. Rec. Colchester Distr. Nat. Hist. Soc. Field Club*
357	*Rep. Trans. Guernsey Soc. nat. Sci.* continued as *Trans. Soc. guernés.*	*Rep. Trans. Guernsey Soc. Nat. Sci.* continued as *Rep. Trans. Soc. Guernésiaise*
358	*Sci. Gossip*	*Sci.-Gossip*

World list	B-P-H
359 *Scott. bot. Rev.*	*Scott. Bot. Rev.*
360 *Scott. For.*	*Scott. Forest.*
361 *Scott. Nat.*	*Scott. Naturalist (Edinburgh)*
362 *SEast. Nat.*	*S.E. Naturalist*
363 *Suffolk nat. Hist.* An alternative title for **369**	*Suffolk Nat. Hist.*
364 *Trans. Herts. nat. Hist. Soc. Fld Club*	*Trans. Hertfordshire Nat. Hist. Soc.*
365 *Trans. Kent Fld Club*	*Trans. Kent Field Club*
366 *Trans. Leicester lit. phil. Soc.*	*Trans. Leicester Lit. Soc.*
367 *Trans. Penzance nat. Hist. antiq. Soc.*	*Trans. Penzance Nat. Hist. Antiq. Soc.*
368 *Trans. Proc. bot. Soc. Edinb.*	*Trans. & Proc. Bot. Soc. Edinburgh*
369 *Trans. Suffolk Nat. Soc.*	*Trans. Suffolk Naturalists' Soc.*
370 *Watsonia*	*Watsonia*
371 *Wild Flower Mag.*	*Wild Fl. Mag.*
372 *Wilts. Pl. Notes*	*Wiltshire Plant Notes*
373 *Xerophyte*	*Xerophyte*
374 *Yb. bot. Soc. Br. Isl.*	*Year Book Bot. Soc. Brit. Isles*
375 *A. Bull. Soc. jersiaise*	*Annual Bull. Soc. Jersiaise*
- - - *(Bot. Cabinet)* See **645**	
376 *Bot. (Locality) Rec. Club*	*Bot. (Locality) Rec. Club*
377 *Bull. imp. Bur. Breed. Genet. Camb.*	*Bull. Imp. Bur. Breed. Genet. Cambridge*
378 *Forth Nat. & Historian*	*Forth Naturalist & Historian*
379 *Garden, Lond.* (1871-1927)	*Garden (London 1871-1927)*
- - - *(Garden, Lond.* (1975-)) See **325**	*Garden (London 1975+)*
380 *Heredity, Lond.*	*Heredity*
381 *Leeds Nats. Club scient. Ass. Newsl.*	*Leeds Naturalists' Club Sci. Assoc. Newslett.*
- - - *(Loddiges' Bot. Cabinet)* See **645**	
382 *Lotus Newsl.*	*Lotus Newslett.*
383 *New Phytol.*	*New Phytol.*
384 *Newsl. Atlas Flora Somerset Proj.*	*Newslett. Atlas Fl. Somerset Proj.*
385 *Newsl. Bot. Cornwall*	*Newslett. Bot. Cornwall*
386 *Proc. Bgham nat. Hist. phil. Soc.* continued as *Proc. Bgham nat. Hist Soc.* Cf. **348**	*Proc. Birmingham Nat. Hist. Philos. Soc.* continued as *Proc. Birmingham Nat. Hist. Soc.*
387 *Proc. Bournemouth nat. Sci. Soc.*	*Proc. Bournemouth Nat. Sci. Soc.*
388 *Proc. Somerset. archaeol. nat. Hist. Soc.*	*Proc. Somersetshire Archaeol. Nat. Hist. Soc.*
389 *Pteridologist*	*Pteridologist*

World list	*B-P-H*
390 *Rep. Trans. Devon. Ass. Advmt Sci.*	*Rep. & Trans. Devonshire Assoc. Advancem. Sci.*
391 *Trans. Edinb. Fld Nat. microsc. Soc.*	*Trans. Edinburgh Field Naturalists' Soc.*
392 *Trans. Market Harborough. nat. Hist. Soc.*	*Trans. Market Harborough Nat. Hist. Soc.*
393 *Trans. Norfolk Norwich Nat. Soc.*	*Trans. Norfolk Norwich Naturalists' Soc.*
394 *Rep. Watson bot. Exch. Club*	*Rep. (Annual) Watson Bot. Exch. Club*

[**382** should have appeared under Foreign Periodicals; it is published in Quebec.]

PERSONAL COMMUNICATIONS

400	Ballantyne, G.H.	443	Lourteig, Dr A.
401	Barrett, R.A.	444	Lucas, Mrs M.J.
402	†Booth, Miss E.M.	445	Macpherson, Dr P.
403	Bowen, Dr H.J.M.	446	Margetts, L.J.
404	Bowman, R.P.	447	Marner, Ms S.K.
405	Brewis, Lady Anne B.M.	448	Martin, J.
406	Briggs, Mrs M.	449	Martin, Mrs M.E.
407	Brummitt, Dr R.K.	450	McClintock, D.
408	Bull, A.L.	451	McKean, D.R.
409	Bull, K.E.	452	†McLean, Mrs J.K.
410	Burton, R.M.	453	Meikle, R.D.
411	Chater, A.O.	454	Mill, Dr R.R.
412	Clough, P.	455	Mullin, J.M.
413	Coles, Dr S.M.	456	Mundell, A.R.G.
414	Copping, A.	457	Norman, Mrs E.
415	Dony, †Dr J.G. & Mrs C.M.	458	Palmer, J.R.
416	Easy, G.M.S.	459	Partridge, Dr J.W.
417	†English, R.D.	460	Perring, Dr F.H.
418	Flood, R.J.	461	†Pratt, H.M.
419	Fox, Prof. B.W.	462	Presland, J.L.
420	Gardner, M.F.	463	Radcliffe-Smith, A.
421	Graham, Rev. G.G.	464	Roe, Capt. R.G.B.
422	Gravestock, Miss I.F.	465	Roper, P.
423	Green, I.P. & P.R.	466	†Russell, Mrs J.M.L.
424	†Grenfell, A.L.	467	Rutherford, Miss A.
425	Gunn, J.M.	468	Ryan, Mrs P.
426	Hall, C.R.	469	Shaw, Rev. C.E.
427	Hall, P.C.	470	Simpson, F.W.
428	Hanson, C.G.	471	Smith, Mrs J.E.
429	Harron, W.J.	472	Southam, M.J.
430	Hesselgreaves, Mrs E.	473	Stace, Prof. C.A.
431	†Hollick, Miss K.M.	474	Stern, R.C.
432	†Houseman, Mrs F.	475	Stewart, Mrs O.M.
433	Hyde, Mrs E.M.	476	Swan, Prof. G.A.
434	Jury, Dr S.L.	477	†Swanborough, Mrs J.
435	Karley, S.L.M.	478	Trotman, M.J.
436	Kent, D.H.	479	Trueman, I.C.
437	†Kertland, Miss M.P.	480	Underhill, J.A.
438	Knipe, P.R.	481	Woodell, Dr S.R.J.
439	Lambley, P.W.	482	Wurzell, B.
440	Lancaster, C.R.	483	Fryer, Ms J.D.
441	Lang, D.C.	484	Halliday, Dr G.
442	Leslie, Dr A.C.	485	Johnson, Dr L.A.S.

418

ADDITIONAL BOOKS FOR IDENTIFICATION

500 Abrams, L. & Ferris, R.S. (1923-1960). *Illustrated flora of the Pacific states*, **1-4**. Stanford, California.
501 Adamson, R.S. & Salter, T.M., eds. (1950). *Flora of the Cape Peninsula.* Cape Town.
502 Ali, S.I. & Jafri, S.M.H. (1976-). *Flora of Libya*, **1**-. Tripoli.
503 Anon. (1979). *Reader's Digest Success with house plants.* New York.
504 Anon. (1981). *The Hillier colour dictionary of trees and shrubs.* Newton Abbot.
505 Anon. (1982). *The Macdonald encyclopedia of trees.* London.
506 Anon. (1983). *The Macdonald encyclopedia of flowers for balcony and garden.* London.
507 Apgar, A.C. (1910). *Ornamental shrubs of the United States (hardy, cultivated).* New York.
508 Arechavaleta, J. (1898-1911). Flora Uruguaya, **1-4**. *Anales Mus. Nac. Montevideo*, **3** & **5-7**.
509 Aubréville, A. *et al.*, eds. (1963-). *Flore du Cameroun*, **1**-. Paris.
510 Bailey, L.H. (1937). *The standard cyclopedia of horticulture*, **1-3**. New York.
511 Bailey, L.H. & Bailey, E.Z. (1976). *Hortus Third*, rev. the staff of the L.H.Bailey Hortorium. New York & London.
512 Baum, B.R. (1978). *The genus Tamarix.* Jerusalem.
513 Baytop, T. & Mathew, B. (1984). *The bulbous plants of Turkey.* London.
514 Beadle, N.C.W., Evans, O.D. & Carolin, R.C. (1972). *Flora of the Sydney region.* Sydney & London.
515 Beales, P. (1985). *Classic roses.* London.
516 Bentham, G. (1863-1878). *Flora Australiensis*, **1-7**. London.
517 Berhaut, J. (1971). *Flore illustrée du Sénégal.* Dakar.
518 Bieberstein, F.A.Marschall von (1808-1819). *Flora Taurico-Caucasica*, **1-2**. Char'kov.
519 Bishop, O. (1990). *Wild flowers of New Zealand.* Auckland & London.
520 Blackall, W.E. & Grieve, B.J. (1974). *How to know Western Australian wildflowers*, parts 1-3, new ed. Perth.
520a Blackall, W.E. & Grieve, B.J. (1980-1981). l.c., parts 3a & 3b, 2nd ed., rev. Grieve, B.J. Perth. See also **583**.
521 Bloom, A. (1980). *Alpines for your garden.* Nottingham.
522 Boissier, [P.]E. (1839-1845). *Voyage botanique dans le midi de l'Espagne pendant l'année 1837*, **1-2**. Paris.
523 Boissier, [P.]E. (1866). *Icones Euphorbiarum.* Geneva.
524 Boissier, [P.]E. (1867-1888). *Flora Orientalis*, **1-5**. Basle & Geneva, etc.
525 Bonnier, G.E.M. (1911-1935). *Flore complète illustrée en couleurs de France, Suisse et Belgique*, **1-13**. Paris.
526 Boott, F. (1858). *Illustrations of the genus Carex.* London.
527 Boulos, L., El-Hadidi, M.N. & El-Gohary, M. (1967). *Common weeds in Egypt.* Cairo.
528 Bouloumoy, L. (1930). *Flore du Liban et de la Syrie.* Paris.

529 Bowles, E.A. (1952). *A handbook of Crocus and Colchicum for gardeners*, rev. ed. London.

530 Bramwell, D. & Bramwell, Z.I. (1974). *Wild flowers of the Canary Islands*. London.

531 Brooker, M.I.H. & Kleinig, D.A. (1983). *Field guide to eucalypts*. Melbourne.

532 Bruggeman, L. (1957). *Tropical plants and their cultivation*. London.

533 Burbidge, N.T. & Gray, M. (1970). *Flora of the Australian Capital Territory*. Canberra.

534 Burkart, A., *et al.*, eds. (1969-). *Flora ilustrada de Entre Ríos (Argentina)*, **1-**. Buenos Aires.

535 Cabrera, A.L. & Zardini, E.M. (1978). *Manual de la flora de los alrededores de Buenos Aires*, 2nd ed. Buenos Aires.

535a Cabrera, A.L. & Zardini, E.M. (1953). l.c., 1st ed.

536 Castroviejo, S. *et al.* (1986-). *Flora Iberica*, **1-**. Madrid.

537 Cavanilles, A.J. (1785-1790). *Monadelphiae classis dissertationes decem*, **1-10**. Paris & Madrid.

538 Cavanilles, A.J. (1791-1801). *Icones et descriptiones plantarum*, **1-6**. Madrid.

539 Chinnock, R.J. & Heath, E. (1974). *Ferns and fern allies of New Zealand*. Wellington & London.

540 Cochrane, G.R. *et al.* (1968). *Flowers and plants of Victoria*. Sydney.

541 Codd, L.E.[W.] *et al.*, eds. (1970). *Flora of southern Africa*, **13**. Pretoria.

542 Collenette, S. (1985). *An illustrated guide to the flowers of Saudi Arabia*. London.

543 Compton, J. (1987). *Success with unusual plants*. London.

544 Cook, C.D.K. *et al.* (1974). *Water plants of the world*. The Hague.

545 Correll, D.S. & Johnston, M.C. (1970). *Manual of the vascular plants of Texas*. Renner, Texas.

546 Coste, H.[J]. (1900-1906). *Flore descriptive et illustrée de la France, de la Corse et des contrées limitrophes*, **1-3**. Paris.

546a Jovet, P. & Vilmorin, R. de (1972-1990). *Flore descriptive et illustrée de la France*, suppls **1-7**. Paris.

547 Costermans, L. (1981). *Native trees and shrubs of S.E. Australia*. Adelaide.

548 Crook, H.C. (1951). *Campanulas: their cultivation and classification*. London & New York.

549 Curtis, W.M. (1956-1967). *The student's flora of Tasmania*, **1-3**. Hobart.

550 Cusin, L.A. & Ansberque, E. (1868[1867]-1876). *Herbier de la flore Française*, **1-14** (in 25 parts). Lyons.

551 Davidson, W. (1983). *Illustrated dictionary of house plants*. London.

552 Davies, P. & Huxley, A. (1983). *Wild orchids of Britain and Europe*. London.

553 Davis, B. (1990). *The gardener's illustrated encyclopedia of climbers and wall shrubs*. Harmondsworth, Middlesex.

554 de Candolle, A.P. (1824-1874). *Prodromus systematis naturalis regni vegetabilis*, **1-17**. Paris.

555   Delile, A.R. (1813). *Flore d'Egypte*. Paris.
556   Desfontaines, R.L. (1798-1799). *Flora Atlantica*, 1-2. Paris.
557   Don, G.[fil.] (1831-1838). *A general system of gardening and botany*, 1-4 (incomplete). London.
558   Duncan, B.D. & Isaac, G. (1986). *Ferns and allied plants of Victoria, Tasmania and South Australia*. Melbourne.
559   Dykes, W.R. (1974). *The genus Iris*, Dover ed. Cambridge.
560   Eagle, A. (1975). *Eagle's trees and shrubs of New Zealand in colour*. Auckland & London.
561   Eliovson, S. (1972). *Namaqualand in flower*. Johannesburg.
562   Eliovson, S. (1980). *Wild flowers of southern Africa*, 6th ed. Johannesburg.
563   Elwes, H.J. & Henry, A. (1906-1913). *The trees of Great Britain and Ireland*, 1-7. Edinburgh.
564   Engler, H.G.A., ed. (1900-1953). *Das Pflanzenreich*, 1-107. Berlin.
565   Engler, H.G.A. & Prantl, K.A.E. (1887-1915). *Die natürlichen Pflanzenfamilien*, 1-4 (Teil), 1-4 (Nachträge). Leipzig & Berlin.
565a   Engler, H.G.A. & Prantl, K.A.E., eds. (1924-). *Die natürlichen Pflanzenfamilien*, 13-, 2nd ed. Leipzig & Berlin.
566   Everard, B. & Morley, B.D. (1970). *Wild flowers of the world*. London.
567   Everist, S.L. (1957). *Common weeds of farm and pasture*. Brisbane.
568   Fabian, A. & Germishuizen, G. (1982). *Transvaal wild flowers*. Johannesburg.
569   Fairley, A. & Moore, P. (1989). *Native plants of the Sydney district*. Sydney.
570   Fiori, A. (1923-1929). *Nuova flora analitica d'Italia*, 1-2. Florence.
571   Fiori, A. & Paoletti, G. (1933). *Iconographia florae italicae ossia flora italiana illustrata*, 3rd ed. Florence.
572   Fournet, J. (1978). *Flore illustrée des phanérogames de Guadeloupe et de Martinique*. Paris.
573   Fournier, P. (1961). *Les quatre flores de la France*, new ed. Paris.
574   Fryxell, P.A. (1988). Malvaceae of Mexico. *Syst. Bot. Monogr.*, 25.
575   Garcia, J.G.L. *et al.* (1975). *Malezas prevalentes de America Central. Prevalent weeds of Central America*. El Salvador.
576   George, A.S., ed. (1981-). *Flora of Australia*, 1-. Canberra.
577   Gibson, J.M. (1975). *Wild flowers of Natal (coastal region)*. Durban.
577a   Gibson, J.M. (1978). *Wild flowers of Natal (inland region)*. Durban.
578   Gledhill, E. (1969). *Eastern Cape veld flowers*. Cape Town.
578a   Gledhill, E. (1981). *Eastern Cape veld flowers*, 2nd ed. Cape Town.
579   Gmelin, J.G. (1747-1769). *Flora Sibirica*, 1-4. St Petersburg.
580   Godfrey, R.K. & Wooten, J.W. (1981). *Aquatic and wetland plants of southeastern United States. Dicotyledons*. Athens, Georgia.
581   Graf, A.B. (1978). *Tropica*. East Rutherford, New Jersey.
582   Graham, V.E. (1963). *Tropical wild flowers*. London.

583   Grieve, B.J. & Blackall, W.E. (1975). *How to know Western Australian wildflowers*, part 4. Perth. See also **520**.
584   Griffith, A.N. (1973). *Collins guide to alpines and rock garden plants*, 3rd ed. London.
585   Griffith, W. (1847-1854). *Icones plantarum asiaticum*, **1-4**. Calcutta.
586   Grimm, W.C. (1957). *The book of shrubs*. New York.
587   Guest, E., ed. (1966). *Flora of Iraq*, **1**. Baghdad. (Continued as **734**).
588   Häfliger, E. *et al.* (1982). *Monocot weeds 3. Monocot weeds excluding grasses*. Basle.
589   Hämet-Ahti, L. *et al.* (1986). *Retkeilykasvio [A field flora of Finland]*. Helsinki.
590   Harden, G.J. (1990-). *Flora of New South Wales*, **1-**. Kensington.
591   Harris, T.Y. (1979). *Wild flowers of Australia*, 8th ed. London.
592   Harvey, W.H. & Sonder, O.W. (1859-1865). *Flora Capensis*, **1-3**. (Continued by Thiselton-Dyer, W.T. *et al*, eds. (1896-1909), **4-10** (incomplete)). Cape Town.
593   Hay, R., *et al.* (1974). *The dictionary of indoor plants in colour*. London.
594   Hay, R. & Synge, P.M. (1969). *The dictionary of garden plants in colour with house and greenhouse plants*. London.
595   Hegi, G, ed. (1936-). *Illustrierte Flora von Mittel-Europa*, **1-**, 2nd ed. Munich.
       595a   Hegi, G., ed. (1966-). l.c., **1-**, 3rd ed. Munich.
       595b Hegi, G. (1906-1931). l.c., **1-7** (in 13). Munich.
596   Hellyer, A.G.L. (1958). *Garden plants in colour*. London.
597   Hellyer, A.G. (1965). *Shrubs in colour*. London.
598   Henderson, M. & Anderson, J.G. (1966). *Common weeds in South Africa*. Botanical Survey memorial No.37. Pretoria.
599   Herter, [W.]G. (1939-1957). *Flora ilustrada del Uruguay*, **1-2** (incomplete). Montevideo.
600   Heywood, V.H. *et al.*, eds. (1978). *Flowering plants of the world*. Oxford.
601   Heywood, V.H. & Chant, S.R., eds. (1982). *Popular encyclopedia of plants*. Cambridge.
602   Higgins, V. (1964). *Crassulas in cultivation*. London.
603   Hooker, J.D. (1855-1860). *Flora Tasmaniae*. London.
604   Hooker, J.D. (1872-1897). *The Flora of British India*, **1-7**. Ashford, Kent.
605   Hooker, W.J., ed. (1829-1833). *Botanical miscellany*, **1-3**. London.
606   Hora, B. (1981). *The Oxford encyclopedia of trees of the world*. Oxford.
607   Hultén, E. (1968). *Flora of Alaska and neighboring territories*. Stanford, California.
608   Humboldt, F.H.A.von, Bonpland, A.J.A. & Kunth, K.[C.]S. (1816-1825). *Nova genera et species plantarum*, **1-7**. Paris.
609   Hutchins, G. (1979). *Hebe and Parahebe*. Hornchurch, Essex.
610   Hutchinson, J. (1959). *The families of flowering plants*, ed. 2., **1** (*Dicotyledons*) - **2** (*Monocotyledons*). Oxford.
611   Hutchinson, J. & Dalziel, J.M. (1954-1972). *Flora of west tropical Africa*, **1-3**, 2nd ed., rev. Keay, R.W.J. & (later) Hepper, F.N. London.
612   Huxley, A., ed. (1970). *Garden perennials and water plants*. New York.

613 Huxley, A., ed. (1973). *Deciduous garden trees and shrubs.* Poole, Dorset.
614 Huxley, A., ed. (1973). *Evergreen trees and shrubs.* New York.
615 Huxley, A. & Taylor, W. (1977). *Flowers of Greece and the Aegean.* London.
616 Innes, C. (1985). *The world of Iridaceae.* Ashington, Sussex.
617 Institute of Botany, Academia Sinica. (1972-1976). *Iconographia cormophytorum sinicorum,* **1-5**. Beijing.
618 Ivens, G.W. (1971). *East African weeds and their control,* rev. ed. Nairobi.
619 Jacobsen, H. (1960). *A handbook of succulent plants,* **1-3**, trans. Raabe, H. London.
620 Jacquin, N.J.von (1781-1795). *Icones plantarum rariorum,* **1-3**. Vienna.
621 Jafri, S.M.H. (1966). *The flora of Karachi.* Karachi.
622 Jaubert, H.-F. & Spach, E. (1842-1857). *Illustrationes plantarum orientalium,* **1-5**. Paris.
623 Jávorka, S. & Csapody, V. (1929-1934). *Iconographia florae hungaricae.* Budapest.

    623a Jávorka, S. & Csapody, V. (1979). *Iconographia florae partis austro-orientalis Europae centralis / Ikonographie der Flora des Südöstlichen Mitteleuropa.* Stuttgart.

624 Johow, F. (1896). *Estudios sobre la flora de las Islas de Juan Fernandez.* Santiago, Chile.
625 Jones, D.L. (1987). *Encyclopaedia of ferns.* Melbourne & London.
626 Karsten, G. & Schenck, H. (1904-). *Vegetationsbilder.* Jena.
627 Kearney, T.H. *et al.* (1960). *Arizona Flora,* 2nd ed., rev. with suppl. by Howell, J.T. & McClintock, E. Berkeley, Los Angeles.
628 Kidd, M.M. (1950). *Wild flowers of the Cape peninsula.* London.

    628a Kidd, M.M. (1973). *Wild flowers of the Cape peninsula,* 2nd ed. London.

629 Kirk, T. (1889). *The forest flora of New Zealand.* Wellington.
630 Komarov, V.L. *et al.,* eds. (1934-1964). *Flora SSSR [Flora URSS],* **1-30**. Moscow & Leningrad.
631 Krüssmann, G. (Daniels, G.S., ed.) (1984). *Manual of cultivated broad-leaved trees and shrubs,* **1-3**, trans. Epp, M.E. London.
632 Krüssmann, G. (Warda, H.-D., ed.) (1985). *Manual of cultivated conifers,* trans. Epp, M.E. London.
633 Lamarck, J.B.A.P.de Monnet de (1783-1817). *Encyclopédie méthodique. Botanique,* **1-8**, Suppls **1-5**, & *Atlas.* Suppls and several articles in vols **1-4** by Poiret, J.L.M. Paris.
634 Lancaster, [C.]R. (1987). *Garden plants for connoisseurs.* London.
635 Lasser, T., ed. (1964-). *Flora de Venezuela,* **1-**. Caracas.
636 Ledebour, C.F.von (1829-1834). *Icones plantarum floram rossicam,* **1-5**. Riga.
637 Ledebour, C.F.von (1842[1841]-1853). *Flora Rossica,* **1-4**. Stuttgart.
638 Lee, T.B. (1989). *Illustrated flora of Korea.* Seoul.
639 L'Héritier de Brutelle, C.L. (1785-1805). *Stirpes novae aut minus cognitae.* Paris.

640 L'Héritier de Brutelle, C.L. (1789-1792). *Sertum Anglicum*. Paris.
641 Lid, J. (1963). *Norsk og Svensk flora*. Oslo.
642 Lind, E.M. & Tallantire, A.C. (1971). *Some common flowering plants of Uganda*, rev. ed. Nairobi.
643 Lloyd, C. (1979). *Clematis*, rev. ed. London.
644 Lloyd, F.E. (1976). *The carnivorous plants*. New York.
645 Loddiges, C., ed. (1818-1833). *The botanical cabinet*, 1-20. London.
646 Long, R.W. & Lakela, O. (1971). *A flora of tropical Florida*. Miami.
647 Lucas, A. & Pike, B. (1971). *Wild flowers of the Witwatersrand*. Cape Town.
648 Maire, R.[C.J.E.] (1952-1987). *Flore de l'Afrique du Nord*, 1-16 (incomplete). Paris.
649 Makino, T. (1961). *Makino's new illustrated flora of Japan*, 3rd ed., rev. Maekawa, F., Hara, H. & Tuyama, T. Tokyo.
650 Makins, F.K. (1957). *Herbaceous garden flora*. London.
651 Mark, A.F. & Adams, N.M. (1973). *New Zealand alpine plants*. Wellington & London.
652 Martin, W.C. & Hutchins, C.R. (1980-1981). *A flora of New Mexico*, 1-2. Vaduz.
653 Martius, C.F.P.von, *et al.* (1840-1906). *Flora Brasiliensis*, 1-15 (in 40). Munich.
654 Marzocca, A. (1957). *Manual de malezas*. Buenos Aires.
655 Mason, H., *et al.* (1972). *Western Cape Sandveld flowers*. Cape Town.
656 Mason, H.L. (1957). *A flora of the marshes of California*. Berkeley.
657 Meikle, R.D. (1963). *Garden flowers*. London.
658 Meikle, R.D. (1977-1985). *Flora of Cyprus*, 1-2. Kew.
659 Menninger, E.A. (1970). *Flowering vines of the world*. New York.
660 Migahid, A.M. (1978). *Migahid and Hammouda's Flora of Saudi Arabia*, 2nd rev. ed., 1-2. Riyadh.
661 Millspaugh, C.F. & Chase, M.A. (1903-1904). *Plantae yucatanae (regionis antillanae)*, fasc. 1-2. Chicago.
662 Mitchell, A. (1982). *The trees of Britain and northern Europe*. London.
663 Montasir, A.H. & Hassib, M. (1956). *Illustrated manual flora of Egypt*, 1 (incomplete). [Cairo.]
664 Moore, D.M. (1968). *The vascular flora of the Falkland Islands*. London.
665 Moore, D.M. (1983). *Flora of Tierra del Fuego*. Oswestry, Shropshire.
666 Morley, B.D. & Toelken, H.R., eds. (1983). *Flowering plants in Australia*. Adelaide.
667 Mouterde, P.S.J. (1966-1984). *Nouvelle flore du Liban et de la Syrie*, 1-3(*Texte*) & 1-3(*Atlas*). Beyrouth.
668 Mozaffarian, V. (1983). *The family of Umbelliferae in Iran*. Tehran.
669 Muñoz Schick, M. (1980). *Flora del Parque Nacional Puyehue*. Santiago, Chile.
670 Munz, P.A. (1972). *California spring wildflowers*. Berkeley.
671 Munz, P.A. (1974). *A flora of southern California*. Berkeley.
672 Muschler, R. (1912). *A manual flora of Egypt*. Berlin.

673   Neufeld, J.B. (1968). *Wild flowers of the prairies*. Saskatoon, Saskatchewan.

674   Nicholson, G., ed. (1884-1888). *The illustrated dictionary of gardening*, 1-4. London.
674a   Nicholson, G., ed. (1901). *The century supplement to the dictionary of gardening*. London.

675   Nilsson, S. & Mossberg, B. (1979). *Orchids of northern Europe*. Harmondsworth.

676   Oliver, D. *et al.*, eds. (1868-1877). *Flora of tropical Africa*, 1-3. (Continued by Thiselton-Dyer, W.T., ed. (1904-1934), 4-8.) Ashford, Kent.

677   Ozenda, P. (1977). *Flore du Sahara*, 2nd ed. Paris.

678   Pallas, P.S. (1784-1788[1831]). *Flora Rossica*, 1-2. St Petersburg.

679   Paul, W. (1848). *The rose garden*. London.

681   Perry, F. & Hay, R. (1982). *Tropical and subtropical plants*. London.

682   Phillips, E.P. (1938). *The weeds of South Africa*. Pretoria.

683   Pignatti, S. (1982). *Flora d'Italia*, 1-3. Bologna.

684   Polunin, O. & Everard, B. (1976). *Trees and bushes of Europe*. London.

685   Polunin, O. & Stainton, A. (1984). *Flowers of the Himalaya*. Oxford.

686   Poole, A.L. & Adams, N.M. (1964). *Trees and shrubs of New Zealand*. Wellington.

687   Praeger, R.L. (1932). *An account of the Sempervivum group*. London.

688   Prodán, J. (1930). *Centaureele României (Centaureae Romaniae), monographie*. Cluj.

689   Ralph, T.S. (1849). *Icones carpologicae*. London.

690   Rechinger, K.H. (1964). *Flora of lowland Iraq*. Weinheim.

691   Rehder, A. (1940). *Manual of cultivated trees and shrubs*, 2nd ed. New York.

692   Reichenbach, H.G.L. (1820-1821). *Monographia generis Aconiti*. Leipzig.

693   Reichenbach, H.G.L. (1823-1827). *Illustratio specierum Aconiti generis*. Leipzig.

694   Reichenbach, H.G.L. (1823-1832). *Iconographia botanica seu plantae criticae*, 1-10. Leipzig.

695   Reichenbach, H.G.L. *et al.* (1834-1912). *Icones florae germanicae et helveticae*, 1-25. Leipzig.

696   Reitz, P.R., ed. (1965-). *Flora ilustrada catarinense, I. parte. Plantas*, fasc. 1-. Itajaí, Brasil.

697   Retzius, A.J. (1779-1791). *Observationes botanicae*, fasc. 1-6. Leipzig.

698   Rickett, H.W. (1953). *Wild flowers of America*. New York.

699   Rickett, H.W. (1967-1971). *Wild flowers of the United States*, 1-6. New York.

700   Rix, M. & Phillips, R. (1981). *The bulb book*. London.

701   Robyns, W., ed. (1950-). *Flore générale de Belgique*. Brussels.

702   Rochford, T.C. (1983). *The Collingridge book of cacti and other succulents*. Feltham, Middlesex.

703   Rose, F. (1989). *Colour identification guide to the grasses, sedges, rushes and ferns of the British Isles and north-western Europe*. London.

704 Rowley, G. (1978). *The illustrated encyclopedia of succulents*. London.
705 Ruíz Lopez, H. & Pavón, J. (1798-1802). *Flora peruviana et chilensis*, 1-4. Madrid.
705a Ruíz Lopez, H. & Pavón, J. (1794). *Florae peruvianae et chilensis prodromus*. Madrid.
706 Rydberg, P.A. (1922). *Flora of the Rocky Mountains and adjacent plains*, 2nd ed. New York.
707 Rzedowski, J. & Rzedowski, G.C.de, eds. (1979-). *Flora fanerogámica del Valle de México*, 1-. Mexico City.
708 Saint-Hilaire, A.F.C.P.de (1825-1833). *Flora brasiliae meridionalis*, 1-3. Paris.
709 Salmon, J.T. (1970). *New Zealand flowers and plants in colour*, rev. ed. Wellington & London.
710 Sánchez Sánchez, O. (1969). *La flora del Valle de México*. Mexico City.
711 Sargent, C.S. (1891-1902). *The silva of North America*, 1-14. Boston & New York.
712 Sargent, C.S. (1902-1913). *Trees and shrubs*, 1-2. Boston & New York.
713 Saunders, D.E. (1975). *Cyclamen: a gardener's guide to the genus*, rev. ed. Alpine Garden Society. Woking, Surrey.
714 Saunders, H.N. (1958). *A handbook of West African flowers*. London.
715 Săvulescu, T. *et al.*, eds., Nyárády, E.J. *et al.*, comps. (1952-1976). *Flora republicii populare Romîne*, 1-10. (Retitled as *Flora republicii socialiste România*, 11-13.) Bucharest.
716 Schönfelder, I. & P. (1990). *Collins photoguide to the wild flowers of the Mediterranean*. London.
717 Schrader, H.A. (1827). *Blumenbachia, novum e Loasearum familia genus*. Göttingen.
718 Scopoli, J.A. (1772). *Flora Carniolica*, 2nd ed. Vienna.
719 Seabrook, P. & Rochford, T.C. [1975]. *Plants for your home*. [Nottingham.]
720 Sharpe, P.R. (1986). *Keys to Cyperaceae, Restionaceae & Juncaceae of Queensland*. Queensland botany bulletin No.5. Brisbane.
721 Sherff, E.E. (1937). *The genus Bidens*, parts 1-2. *Field Mus. Nat. Hist., Bot. Ser.*, 16:1-709. Publication Nos.388 & 389. Chicago.
722 Sibthorp, J. & Smith, J.E. (1806-1840). *Flora Graeca*, 1-10. London.
723 Spach, E. (1834-1848). *Histoire naturelle des végétaux. Phanérogames*, 1-14. Paris.
724 Stainton, A. (1988). *Flowers of the Himalaya: a supplement*. Oxford.
725 Stanley, T.D. & Ross, E.M. (1983-1989). *Flora of south-eastern Queensland*, 1-3. Brisbane.
726 Steyermark, J.A. (1963). *Flora of Missouri*. Ames, Iowa.
727 Steyermark, J.A. & Huber, O. (1978). *Flora del Ávila*. Caracas.
728 Sweet, R. (1823-1829). *The British flower garden*, 1-7. London.
728a Sweet, R. (1829-1838). *The British flower garden*, ser.2, 1-4.
729 Täckholm, V. (1974). *Students' flora of Egypt*, 2nd ed. Beirut.
730 Takhtajan, A.L., ed. (1954-). *Flora Armenii*, 1-. Erevan.
731 Tampion, J. (1977). *Dangerous plants*. Newton Abbot.

732   Thomas, G.S. (1982). *Perennial garden plants*, 2nd ed. London.
733   Tobler, F. (1912). *Die Gattung Hedera*. Geneva.
734   Townsend, C.C. & Guest, E., eds. (1966-). *Flora of Iraq*, 2- (incomplete).
         Baghdad. (Preceded by **587**).
735   Trauseld, W.R. (1969). *Wild flowers of the Natal Drakensburg*.
         Cape Town.
736   Turrill, W.B. *et al.*, eds. (1952-). *Flora of tropical East Africa*, fasc. 1-.
         London.
737   Valdés, B., Talavera, S. & Fernández-Galiano, E. (1987). *Flora vascular
         de Andalucía occidental*, **1-3**. Barcelona.
738   Vallentin, E.F. & Cotton, E.M. (1921). *Illustrations of the flowering plants
         and ferns of the Falkland Islands*. London.
739   Ventner, H.J.T. (1979). *A monograph of Monsonia L.* Wageningen.
740   Verboom, W.C., ed. (1973). *Common weeds of arable lands in Zambia*.
         Lusaka.
741   Waldstein-Wartemberg, F.A.von & Kitaibel, P. (1799-1812). *Descriptiones
         et icones plantarum rariorum hungariae*. Vienna.
742   Watt, U.M. & Breyer-Brandwijk, A. (1962). *The medicinal and poisonous
         plants of southern and East Africa*. Edinburgh.
743   Willis, J.H. (1970-1972). *A handbook to plants in Victoria*, **1-2**, 2nd ed.
         Carlton, Victoria.
744   Willmott, E.A. (1910-1914). *The genus Rosa*, **1-2**. London.
745   Wooton, E.O. & Standley, P.C. (1915). Flora of New Mexico. *Contr.
         U.S. Natl. Herb.*, **19**.
746   Yeo, P.F. (1973). *Acaena*, in Green, P.S., ed. *Plants: wild and
         cultivated*, pp. 51-55 & Appendix III (pp.193-221). B.S.B.I. conference
         report No.13. Hampton, Middlesex.
747   Zohary, M. & Heller, D. (1984). *The genus Trifolium*. Jerusalem.

FOREIGN PERIODICALS

Two versions of abbreviations of titles of periodicals are given; the first is in the traditional style of the *World list of scientific periodicals* (1044), the second in the increasingly used style of the *B-P-H Botanico-Periodicum-Huntianum* list (1192) and the *Supplementum* (1192a).

	World list	B-P-H
800	*Abh. naturforsch. Ges. Halle*	*Abh. Naturf. Ges. Halle*
801	*Abh. naturw. Ver. Bremen*	*Abh. Naturwiss. Vereine Bremen*
802	*Acta bot. fenn.*	*Acta Bot. Fenn.*
803	*Act. Soc. Hist. nat. Paris*	*Actes Soc. Hist. Nat. Paris*
804	*Addisonia*	*Addisonia*
805	*Ann. Mo. bot. Gdn*	*Ann. Missouri Bot. Gard.*
- - -	*(Arch. wiss. Bot.)* A subtitle of **838**	
806	*Ber. dt. bot. Ges.*	*Ber. Deutsch. Bot. Ges.*
807	*Biblthca bot.*	*Biblioth. Bot.*
808	*Blumea*	*Blumea*
809	*Boissiera*	*Boissiera*
810	*Bot. Notiser*	*Bot. Not.*
811	*Brunonia*	*Brunonia*
812	*Bull. N.Y. bot. Gdn*	*Bull. New York Bot. Gard.*
813	*Bull. Soc. Hist. nat. Afr. N.*	*Bull. Soc. Hist. Nat. Afrique N.*
814	*Bull. Soc. Nat. Moscou* or *Byull. Mosk. Obshch. Ispyt. Prir.*	*Byull. Moskovsk. Obshch. Isp. Prir., Otd. Biol.*
815	*Candollea*	*Candollea*
816	*Can. J. Bot.*	*Canad. J. Bot.*
817	*Commentat. Soc. Scient. götting.*	*Commentat. Soc. Regiae Sci. Gött.*
818	*Contr. Bolus Herb.*	*Contr. Bolus Herb.*
819	*Contr. N.S.W. natn. Herb.*	*Contr. New South Wales Natl. Herb.*
820	*Darwiniana*	*Darwiniana*
821	*Dep. agric. Qd Bot. Bull.*	*Dept. Agric. Queensland Bot. Bull.*
822	*Feddes Reprium* preceded by *Feddes Reprium Spec. nov. veg.; Reprium Spec. nov. Regni veg. and Reprium nov. Spec. Regni veg*	*Feddes Repert.* preceded by *Feddes Repert. Spec. Nov. Regni Veg.* and *Repert Spec. Nov. Regni Veg.*
823	*Flower. Pl. S. Afr.* continued as *Flower. Pl. Afr.*	*Fl. Pl. South Africa* continued as *Fl. Pl. Africa*
824	*Gartenflora*	*Gartenflora*
825	*Gorteria*	*Gorteria*
826	*Göttinger florist. Rundbr.* continued as *Florist. Rundbr.*	*Göttinger Florist. Rundbr.* continued as *Florist. Rundbr.*

World list	B-P-H

827 *J. Arnold Arbor.*   *J. Arnold Arbor.*

828 *Jl E. Africa nat. Hist. Soc. natn. Mus.*   *J. E. Africa Nat. Hist. Soc. Natl. Mus.*

829 *J. Proc. R. Soc. N.S.W.*   *J. Proc. Roy. Soc. New South Wales*

830 *Jl S. Afr. Bot.*   *J. S. African Bot.*

831 *Jl S. Afr. Sci.*   *J. S. African Sci.*

832 *K. svenska VetenskAkad. Handl.*   *Kungl. Svenska Vetenskapsakad. Avh. Naturskyddsärenden*

833 *Lilloa*   *Lilloa*

834 *Madroño*   *Madroño*

835 *Mitt. bot. StSamml. Münch.*   *Mitt. Bot. Staatssamml. München*

836 *Norw. J. Bot.*   *Norweg. J. Bot.*

837 *N.Z. Jl agric. Res.*   *New Zealand J. Agric. Res.*

838 *Planta*   *Planta*

839 *Proc. Linn. Soc. N.S.W.*   *Proc. Linn. Soc. New South Wales*

840 *Qd agric. J.*   *Queensland Agric. J.*

841 *Revta Mus. La Plata*   *Revista Mus. La Plata, Secc. Bot.*

842 *Revue hort.*   *Rev. Hort.*

843 *Rhodora*   *Rhodora*

844 *Vieraea*   *Vieraea*

845 *Abh. westf. Mus. Ges. Landeskunde*   *Abh. Westfälischen Mus. Naturk.*

846 *Acta Univ. upsal.*   *Acta Univ. Upsal.*

847 *Agronomia lusit.*   *Agron. Lusit.*

848 *Aliso*   *Aliso*

849 *An. Dep. Invest. cient. Univ. nac. Cuyo*   *Anales Dep. Invest. Ci. Univ. Nac. Cuyo*

850 *An. Inst. bot. A.J.Cavanillo*   *Anales Inst. Bot. Cavanilles*

851 *An. Jard. bot. Madr.*   *Anales Jard. Bot. Madrid*

852 *Annls hist.-nat. Mus. Natn hung.*   *Ann. Hist.-Nat. Mus. Natl. Hung.*

853 *Archo bot. biogeogr. ital.* preceded by *Archo bot. Forlì* and continued as *Archo bot. ital.*   *Arch. Bot. Biogeogr. Ital.* preceded by *Arch. Bot. (Forlì)* and continued as *Arch. Bot. Ital.*

854 *Aroideana*   *Aroideana*

855 *Austr. J. Bot.*   *Austral. J. Bot.*

856 *Baileya*   *Baileya*

857 *Beih. bot. Zbl.*   *Beih. Bot. Centralbl.*

858 *Belg. Hort.*   *Belgique Hort.*

859 *Ber. Arbeitsgem. sächs. Bot.*   *Ber. Arbeitsgem. Sächs. Bot.*

860 *Bot. Arch., Berlin*   *Bot. Arch.*

861 *Bot. Jb.*   *Bot. Jahrb. Syst.*

862 *Bot. Tidsskr.*   *Bot. Tidsskr.*

863 *Bothalia*   *Bothalia*

864 *Brittonia*   *Brittonia*

World list	B-P-H
865 *Bull. Herb. Boissier*	*Bull. Herb. Boissier*
866 *Bull. Soc. bot. Genève*	*Bull. Soc. Bot. Genève*
867 *Can. Fld Nat.*	*Canad. Field-Naturalist*
868 *Can. J. Pl. Sci.*	*Canad. J. Pl. Sci.*
869 *Contr. Qd Herb.*	*Contr. Queensland Herb.*
870 *Denkschr. Akad. Wiss. Wein*	*Akad. Wiss. Wein, Math.-Naturwiss. Kl., Denkschr.*
871 *Deserta*	*Deserta*
872 *Dt. Baumsch.*	*Deutsche Baumschule*
873 *Dt. bot. Mschr.*	*Deutsche Bot. Monatsschr.*
874 *Dumortiera*	*Dumortiera*
875 *E. Cape Nat.*	*E. Cape Naturalist*
- - - *(Engl. Jahrb.)* See **861**	
876 *Fld Mus. Publs Bot.*	*Field Mus. Nat. Hist., Bot. Ser.*
877 *Flora Medit.*	*Fl. Medit.*
878 *Fragm. flor. geobot.*	*Fragm. Florist. Geobot.*
879 *Gartenschönheit.*	*Gart.-Schönheit*
880 *Gartenwelt, Berl.*	*Gartenwelt*
881 *Genetica*	*Genetica*
882 *Gentes Herb.*	*Gentes Herb.*
883 *Gleditschia*	*Gleditschia*
884 *Holmbergia*	*Holmbergia*
885 *Israel J. Bot.*	*Israel J. Bot.*
886 *J. Acad. nat. Sci. Philad.*	*J. Acad. Nat. Sci. Philadelphia*
887 *J. Adelaide bot. Gdns*	*J. Adelaide Bot. Gard.*
888 *Jardin*	*Jardin*
889 *J. Dep. Agric. Vict.*	*J. Dept. Agric. Victoria*
890 *J. Fac. Sci. Tokyo Univ. (Bot.)*	*J. Fac. Sci. Univ. Tokyo, Sect. 3, Bot.*
891 *J. Jap. Bot.*	*J. Jap. Bot.*
892 *Kirkia*	*Kirkia*
893 *Kurtziana*	*Kurtziana*
894 *Lagascalia*	*Lagascalia*
895 *Lazaroa*	*Lazaroa*
896 *Levende Nat.*	*Levende Natuur*
897 *Mem. Boston Soc. nat. Hist.*	*Mem. Boston Soc. Nat. Hist.*
898 *Mem. N.Y. bot. Gdn*	*Mem. New York Bot. Gard.*
899 *Mitt. naturw. Inst. Sofia*	*Mitt. Naturwiss. Inst. Sofia*
900 *Möller's dt. Gärtn.-Ztg*	*Möller's Deutsche Gärtn.-Zeitung*
901 *N.Z. Jl Agric.*	*New Zealand J. Agric.*
902 *Natn. hort. Mag.*	*Natl. Hort. Mag.*
903 *Natura mosana*	*Nat. Mosana*
904 *Ned. kruidk. Archf*	*Ned. Kruidk. Arch.*
905 *Notas Mus. La Plata*	*Notas Mus. La Plata, Bot.* continued as *Notas Mus. La Plata*
906 *Nuytsia*	*Nuytsia*

World list	*B-P-H*
907 *Physis, B. Aires*	*Physis (Buenos Aires)*
908 *Preslia*	*Preslia*
909 *Proc. R. Soc. Vict.*	*Proc. Roy. Soc. Victoria*
910 *Publs Mich. St. Univ. Mus.,*	*Publ. Mus. Michigan State Univ.,*
Biol. ser.	*Biol. Ser.*
911 *Revta agron. NE Argent.*	*Revista Agron. Noroeste Argent.*
912 *Revta argent. Agron.*	*Revista Argent. Agron.*
913 *Rozpr. čsl. Akad. Věd*	*Rozpr. Českoslov. Akad. Věd*
914 *Ruizia*	*Ruizia*
915 *Svensk bot. Tidskr.*	*Svensk Bot. Tidskr.*
916 *Symb. bot. upsal.*	*Symb. Bot. Upsal.*
917 *Syst. Bot.*	*Syst. Bot.*
918 *Syst. bot. Monogr.*	*Syst. Bot. Monogr.*
919 *Telopea*	*Telopea*
920 *Trans. Proc. R. Soc.*	*Trans. & Proc. Roy. Soc. South*
S. Australia	*Australia*
921 *Wentia*	*Wentia*
922 *Wien. ill. Gartenztg*	*Wiener Ill. Gart.-Zeitung*
923 *Willdenowia*	*Willdenowia*

SUPPLEMENTARY LIST

1000 Andrews, H.C. (1797-1815). *Botanists repository*, **1-10**. London.
1001 Andrews, S.B. (1990). *Ferns of Queensland*. Brisbane.
1002 Anon. (1980). *Wild flowers of South Africa*. Cape Town.
1003 Anon. (1991). *The Hillier manual of trees and shrubs*, 6th ed. Newton Abbot.
1004 Armitage, A.M. (1989). *Herbaceous perennial plants*. Athens, Georgia.
1005 Ascherson, P.F.A. & Schweinfurth, G.A. (1887). *Illustration de la flore d'Egypte*. Vienna.
1006 Aubréville, A. *et al.*, eds. (1967-). *Flore de la Nouvelle-Calédonie et dépendances*, **1-**. Paris.
1007 Babcock, E.B. (1947). *The genus Crepis*, parts 1-2. University of California publications in botany, **21**. Berkeley.
1008 Babington, C.C. (1860). *Flora of Cambridgeshire*. London.
1009 Backer, C.A. & Bakhuizen van den Brink, R.C., Jr. (1963-1968). *Flora of Java*, **1-3**. Groningen.
1010 Bailey, F.M. (1906). *The weeds and suspected poisonous plants of Queensland*. Brisbane.
1011 Bailey, L.H. (1949). *Manual of cultivated plants*, rev. ed. New York.
1012 Baillon, H.E. (1858). *Étude générale du groupe des Euphorbiacées*. Paris.
1013 Baillon, H.E. (1866-1895). *Histoire des plantes*, **1-13** (incomplete). Paris.
1013a Baillon, H.E. (1871-1888). *The natural history of plants*, **1-8** (incomplete). London.
1014 Baker, J.G. (1892). *Handbook of the Irideae*. London.
1015 Ballantyne, G.H. (1990). Flowers of west Fife: a select annotated list. *Forth Naturalist & Historian*, **12**: 67-98.
1016 Barneby, T.P. (1967). *European alpine flowers in colour*. London.
1017 Bassett, I.J., *et al.* (1983). *The genus Atriplex (Chenopodiaceae) in Canada*. Agriculture Canada monograph No.31. Ottawa.
1018 Battandier, J.A. & Trabut, L.C. (1888-1897). *Flore de l'Algérie*, **1-3**. Alger & Paris.
1019 Batten, A. & Bokelmann, H. (1966). *Wild flowers of the Eastern Cape Province*. Cape Town.
1020 Beaugé, A. (1974). *Chenopodium album et espèces affines. Étude historique et statistique*. Paris.
1021 Beckett, E. (1988). *Wild flowers of Majorca, Minorca and Ibiza*. Rotterdam.
1022 Berger, A. (1908). *Mesembrianthemen und Portulacaceen*. Stuttgart.
1023 Blanchard, J.W. (1990). *Narcissus. A guide to wild daffodils*. Woking.
1024 Blatter, E. (1927-1928). *Beautiful flowers of Kashmir*, **1-2**. London.
1025 Blombery, A.M. (1972). *What wildflower is that?* New York.
1026 Böcher, T.W., Holmen, K., & Jakobsen, K. (1968). *The flora of Greenland*, trans. Elkington, T.T. & Lewis, M.C. Copenhagen.
1027 Bolòs, O.de, *et al.* (1990). *Flora manual dels països Catalans*. Barcelona.
1028 Bolòs, O.de & Vigo, J. (1984-). *Flora dels països Catalans*, **1-**. Barcelona.

1029 Bolus, H.M.L. (1928-1958). *Notes on Mesembryanthemum and some allied genera*, **1-3**. Cape Town.

1030 Bonafè Barceló, F. (1977-1980). *Flora de Mallorca*, **1-4** (incomplete). Palma de Mallorca.

1031 Bonnard, B. (1988). *A new check list of the flowering plants, trees, and ferns wild on Alderney, and its off-islets*. Privately published. Le Petit Val, Alderney.

1032 Boom, B.K. (1968). *Flora van kamer - en kasplanten. Flora der cultuurgewassen van Nederland, deel 3*. Wageningen.

1033 Boom, B.K. (1975). *Flora der gekweekte, kruidachtige gewassen*, 3rd ed. *Flora der cultuurgewassen van Nederland, deel 2*. Wageningen.

1034 Boom, B.K. (1975). *Nederlandse dendrologie*, 9th ed. *Flora der cultuurgewassen van Nederland, deel 1*. Wageningen.

1035 Bosser, J., *et al.*, eds. (1976-). *Flore des Mascareignes*, fasc.1-. Mauritius, Paris & Kew.

1036 Bouchard, J. (1977). *Flore pratique de la Corse*, 3rd ed. Bastia.

1037 Boullemier, L.B. (1985). *Checklist of species, hybrids and cultivars of the genus Fuchsia*. Poole.

1038 Bowles, E.A. (1934). *A handbook of Narcissus*. London.

1039 Bown, D. (1988). *Aroids. Plants of the Arum family*. London.

1040 Boyce, P. (1993). *The genus Arum*. London.

1041 Braithwaite, M.E. & Long, D.G. (1990). *The botanist in Berwickshire*. Berwickshire Naturalists Club. Hawick.

1042 Britton, N.L., *et al.*, eds. (1905-1957). *North American flora*, **1-34** (incomplete). New York.

    1042a Rogerson, C.T., ed. (1954-). *North American flora, series II*, parts 1-. New York.

1043 Brotero, F.A. (1816-1827). *Phytographia lusitaniae selectior*, **1-2**, 3rd ed. Lisbon.

1044 Brown, P. & Stratton, G.B. (1963-1965). *World list of scientific periodicals published in the years 1900-1960*, **1-3**, 4th ed. London.

1045 Brownsey, P.J. & Smith-Dodsworth, J.C. (1989). *New Zealand ferns and allied plants*. Glenfield, Auckland.

1046 Bryan, J.E. (1989). *Bulbs*, **1-2**. Portland, Oregon.

1047 Burtt Davy, J. (1926-1932). *A manual of the flowering plants and ferns of the Transvaal with Swaziland, South Africa*, parts 1-2 (incomplete). London.

1048 Busch, N. (1913-1931). *Flora sibiriae et orientis extremi*, fasc.1-3. St Petersburg.

1049 Buttler, K.P. (Davies, P., ed.) (1991). *Field guide to orchids of Britain and Europe*. Swindon.

1050 Cabrera, A.L., ed. (1963-1970). *Flora de la provincia de Buenos Aires*, **1-6**. Buenos Aires.

1051 Cabrera, A.L. (1977-). *Flora de la provincia de Jujuy, república Argentina*, **1-**. Buenos Aires.

1052 Cadevall i Diars, J. & Sallent i Gotés. (1913-1937). *Flora de Catalunya*, **1-6**. Barcelona.

1053 Callen, G. (1976-1977). *Les conifères cultivés en Europe*, **1-2**. Paris.
- - - (Candolle) See de Candolle.
1054 Chaudhary, S.A. & Akram, M. (1987). *Weeds of Saudi Arabia and the Arabian peninsula*. Riyadh.
1055 Chippendale, T.M. (1968). *Wildflowers of central Australia*. Jacaranda wildflower guides No.4. Brisbane.
1056 Church, A., comp. (1988). *Arran's flora*. Arran Natural History Society. Brodick.
1057 Clark, L.J. (1976). *Wild flowers of the Pacific northwest*. Sidney, British Columbia.
1058 Clarke, P.M. & Clarke, J. (1991). *The flowering plants of Colonsay and Oransay*. World Wide Fund for Nature.
1059 Clarke, R. (1982). *Seventy years of nature notes in Warlingham and Chelsham: Arthur Beadell's notes on flowers*. Privately published. Warlingham, Surrey.
1060 Clausen, R.T. (1959). *Sedum of the trans-Mexican volcanic belt*. Ithaca, New York.
1061 Clement, E.J. & Foster, M.C. (1983). *Alphabetical tally-list of alien and adventive plants*, 3rd ed. Privately published. Kingston upon Thames.
1062 Clements, F.E. & Clements, E.S. (1914). *Rocky mountain flowers*. New York.
1063 Collett, H. (1902). *Flora Simlensis*. Calcutta.
1064 Compton, R.H. (c.1941). *Our South African flora*. Kirstenbosch.
1065 Cook, C.D.K. (1990). *Aquatic plant book*. The Hague.
1066 Correa, M.N. (1969-). *Flora Patagónica*, **1-**. Buenos Aires.
1067 Correll, D.S. (1962). *The potato and its wild relatives*. Renner, Texas.
1068 Correll, D.S. & Correll, H.B. (1982). *Flora of the Bahama archipelago*. Vaduz.
1069 Costa, A.da & Franquinho, L.de (1992). *Madeira. Plantas e flores*, 12th ed. Funchal.
1070 Courtenay, B. & Zimmerman, J.H. (1972). *Wildflowers and weeds. A guide in full colour*. New York.
1071 Crackles, F.E. (1990). *Flora of the East Riding of Yorkshire*. Hull.
1072 Crockett, J.U., ed. (1971). *Annuals*. New York.
1073 Crookes, M. & Dobbie, H.B. (1963). *New Zealand ferns*, 6th ed. Christchurch.
1074 Cunningham, G.M. *et al.* (1981). *Plants of western New South Wales*. Wagga Wagga.
1075 Dandy, J.E. (1958). *List of British vascular plants*. London.
1076 Dandy, J.E. (1969). *Watsonian vice-counties of Great Britain*. Ray Society publication No.146. London.
1077 Darnell, A.W. (1929-1932). *Hardy and half-hardy plants*, **1-2**. Privately published. Hampton Wick, Middlesex.
1078 Dashorst, G.R.M. & Jessop, J.P. (1990). *Plants of the Adelaide plains and hills*. Kenthurst, New South Wales.
1079 Davies, P. & Gibbons, B. (1993). *Field guide to wild flowers of southern Europe*. Marlborough.

1080 de Candolle, A.P. (1802). *Astragalogiae*. Paris.

1081 Descole, H. *et al.* (1943-1956). *Genera et species plantarum argentinarum*, 1-5(incomplete). Buenos Aires.

1082 De Vos, M.P. (1972). The genus Romulea in South Africa. *J. S. African Bot.*, supplementary volume No.9. Kirstenbosch.

1083 Dickson, J.H. (1991). *Wild plants of Glasgow*. Aberdeen.

1084 Dimitri, M.J. (1974). Pequeña flora ilustrada de los Parques Nacionales Andino-Patagónicos. *Anales Parques Nac.*, **13**:1-122.

1085 Dippel, L. (1889-1893). *Handbuch der Laubholzkunde*, 1-3. Berlin.

1086 Dony, J.G. *et al.* (1986). *English names of wild flowers*, 2nd ed. Botanical Society of the British Isles. London.

1087 Dony, J.G. & Dony, C.M. (1991). *The wild flowers of Luton*. Luton.

1088 Dormon, C. (1934). *Wild flowers of Louisiana*. New York.

1089 Dostál, J. (1989). *Nová květena ČSSR [A new flora of Czechoslovak SSR]*, 1-2. Praha.

1090 Dupias, G. (c.1990). *Fleurs du Parc National des Pyrenees*. Tarbes.

1091 Du Plessis, N. & Duncan, G. (1989). *Bulbous plants of southern Africa*. Cape Town.

1092 Dyer, R.A. *et al.*, eds. (1963-). *Flora of southern Africa*, 1-. Pretoria.

1093 Edees, E.S. (1972). *Flora of Staffordshire*. Newton Abbot.

1094 Edees, E.S. & Newton, A. (1988). *Brambles of the British Isles*. Ray Society. London.

1095 Edgecombe, W.S. (1970). *Weeds of Lebanon*, 3rd ed. Beirut.

1096 Eliovson, S. (1965). *South African wild flowers for the garden*, 4th ed. Cape Town.

1097 Ellis, R.G. (1993). *Aliens in the British flora*. British plant life No.2. Cardiff.

1098 Engler, A. (1908-1925). *Die Pflanzenwelt Afrikas*, 1-5. In Engler, A. & Drude, O. *Die Vegetation der Erde*, 9. Leipzig.

1099 Evans, A.H. (1939). *A flora of Cambridgeshire*. London & Edinburgh.

1100 Everett, T.H. (1980-1982). *The New York Botanical Garden illustrated encyclopedia of horticulture*, 1-10. New York.

1101 Ewart, R. (1982). *Fuchsia lexicon*. Poole.

1102 Exell, A.W. *et al.*, eds. (1960-). *Flora Zambesiaca*, 1-. London.

1103 Featon, E.H. & Featon, S. (1889). *Art album of New Zealand flora*. Wellington.

1104 Fedtschenko, B.A. & Flerow, A.F. (1910). *Flora evropejskoj rossii*. St Petersburg.

1105 Fedtschenko, B.A. & Shishkin, B.K. (1927-1938). *Flora jugo-vostoka evropejskoj časti SSSR [Flora rossiae austro-orientalis]*, 1-3 (in 6 parts). Leningrad.

1106 Fisher, J. (1991). *A colour guide to rare wild flowers*. London.

1107 Fitch, W.H. & Smith, W.G. (1919). *Illustrations of the British flora*, 4th ed. London.

1108 Font Quer, P. (1961). *Plantas medicinales. El Dioscórides renovado*. Barcelona.

1109 Fournet, J. & Hammerton, J.L. (1991). *Weeds of the Lesser Antilles*. Paris.

1110 Fournier, P. (1951-1952). *Flore illustrée des jardins et des parcs. Arbres, arbustes et fleurs de pleine terre*, 1-3 (*Texte*) & 4 (*Atlas*). Paris.

1111 Fryxell, P.A. (1993). *Flora del Valle de Tehuacán-Cuicatlán. Fasc. 1. Malvaceae A.L.Jussieu*. Mexico D.F.

1112 Gachathi, F.N. (1989). *Kikuyu botanical dictionary*. Nairobi.

1113 Gandhi, K.N. & Thomas, R.D. (1989). Asteraceae of Louisiana. *Sida, Bot. Misc.*, 4.

1114 García Rollán, M. (1981). *Claves de la flora de España (Península y Baleares)*, 1-2. Madrid.

1115 Garven, H.S.D. (1937). *Wild flowers of North China and South Manchuria*. Peking Natural History bulletin handbook No.5. Peking.

1116 Gault, S.M. (1976). *The dictionary of shrubs in colour*. London.

1117 Gibson, F. & Hunt, D. [1984]. *A colour guide to the wild flowers of Scilly*. ?St Ives, Cornwall.

1118 Gillam, B., ed. (1993). *The Wiltshire flora*. Newbury.

1119 Graf, A.B. (1978). *Exotica*, series 3, 9th ed. East Rutherford, New Jersey.

1120 Graf, A.B. (1992). *Hortica*. East Rutherford, New Jersey.

1121 Graham, G.G. & Primavesi, A.L. (1993). *Roses of Great Britain and Ireland*. B.S.B.I. handbook No.7. London.

1122 Green, P.S., ed. (1973). *Plants: wild and cultivated*. B.S.B.I. conference report No.13. Hampton, Middlesex.

1123 Greuter, W., Burdet, H.M. & Long, G., eds. (1984-). *Med-checklist*, 1-. Geneva.

1124 Grey-Wilson, C. (1993). *Poppies. A guide to the poppy family in the wild and in cultivation*. London.

1125 Grossheim, A.A. *et al.*, eds. (1939-). *Flora Kavkaza*, 2nd ed., 1-. Baku.

1126 Guerra, A.S. (1983). *Vegetacion y flora de la Palma*. Santa Cruz de Tenerife.

1127 Guinea Lopez, E. (1974). *Flora Española iconografia selecta. V. Brassiceae*. Madrid.

1128 Gupton, O.S. & Swope, F.C. (1987). *Fall wildflowers of the Blue Ridge and Great Smoky Mountains*. Charlottesville.

1129 Haager, J. (1980). *The Hamlyn book of house plants*. London.

1130 Haines, R.W. & Lye, K.A. (1983). *The sedges and rushes of East Africa*. East African Natural History Society. Nairobi.

1131 Harrison, R.E. (1965). *Trees and shrubs*. Rutland, Victoria.

1132 Harron, J. (1986). *Flora of Lough Neagh*. Belfast & Coleraine.

1133 Hay, R. (1969). *The colour dictionary of flowers and plants for home and garden*. New York.

1134 Henderson, D.M. (1991). *Annotated checklist of the flora of West Ross*. Privately published. Wester Ross.

1135 He Shiyuan, ed. (1986-). *Flora Hebeiensis*, 1-. Hebei.

1136 Heimans, E., Heinsius, H.W. & Thijsse, J.P. (1983). *Geïllustreerde flora van Nederland*, 22nd ed. Amsterdam.

1137 Heiser Jr., C.B. (1976). *The sunflower*. Norman, Oklahoma.
1138 Hejný, S. *et al.* (1973). *Quarantine weeds of Czechoslovakia*. Studie ČSAV No.8. Praha.
1139 Hessayon, D.G. (1983). *The tree and shrub expert*. Waltham Cross.
1140 Heukels, H. (1909-1911). *De flora van Nederland*, 1-3. Leiden.
1141 Heukels, H. (1942). *Geïllustreerde schoolflora voor Nederland*, 12th ed., rev. Wachter, W.H. Rotterdam.
1142 Heyn, C.C. (1963). *The annual species of Medicago*. Scripta Hierosolymitana, publications of the Hebrew University, Jerusalem, 12.
1143 Hickman, J.C., ed. (1993). *The Jepson manual. Higher plants of California*. Berkeley.
1144 Hilliard, O.M. (1977). *Compositae in Natal*. Pietermaritzburg.
1145 Hobson, N.K. *et al.* (1975). *Veld plants of southern Africa*. Johannesburg.
1146 Hoffmann, G.F. (1816). *Genera plantarum umbelliferarum*, 2nd ed. Moscow.
1147 Hoffmann, J.A. (1982). *Flora silvestre de Chile zona austral*. ?Santiago, Chile.
1148 Hooker, J.D. (1844-1860). *The botany of the Antarctic voyage*, 1-6. London.
1149 Hooker, J.D. (1884). *The student's flora of the British Islands*, 3rd ed. London.
1150 Hösel, J. (1969). *Wildflowers of south-east Australia*. Melbourne.
1151 Humboldt, F.H.A.von & Bonpland, A.J.A. (1805-1818). *Plantae Aequinoctiales*, 1-2. Paris.
1152 Hunt, P.[F.], ed. (1968-1970). *The Marshall Cavendish encyclopedia of gardening*, 1-22. London.
1153 Hutchinson, J. (1955). *British wild flowers*, 1-2, 2nd ed. Harmondsworth.
1154 Hutchinson, J. (1964-1967). *The genera of flowering plants*, 1-2 (incomplete). Oxford.
1155 Huxley, A. (1971). *Garden annuals and bulbs*. New York.
1156 Hvass, E. (1973). *Plants that feed and serve us*, new ed. London.
1157 Hylander, N. (1953-1966). *Nordisk Kärlväxtflora*, 1-2 (incomplete). Stockholm.
1158 Isley, P.T. (1987). *Tillandsia, the world's most unusual air plants*. Gardena, California.
1159 Jackson, W.P.U. (1977). *Wild flowers of Table Mountain*. Cape Town.
1160 Jaques, H.E. (1975). *Plants we eat and wear*, rev. ed. New York.
1161 Jacquin, J.F.Baron von (1811-1844). *Eclogae plantarum rariorum*, 1-2. Vienna.
1162 Jacquin, N.J.Baron von (1797-1804). *Plantarum rariorum horti caesarei schoenbrunensis*, 1-4. Vienna.
1163 Jelitto, L. & Schacht, W. (1990). *Hardy herbaceous perennials*, 1-2, 3rd ed., rev. Schacht, W. & Fessler, A., trans. Epp, M.E. Portland, Oregon.
1164 Jeppe, B. (1989). *Spring and winter flowering bulbs of the Cape*. Cape Town.
1165 Jepson, W.L. (1909-). *A flora of California*, 1-4 (incomplete). Berkeley.

1166 Jepson, W.L. (1925). *A manual of the flowering plants of California.* Berkeley.

1167 Jermy, A.C. *et al.*, eds. (1978). *Atlas of ferns of the British Isles.* Botanical Society of the British Isles & British Pteridological Society. London.

1168 Jermy, [A.]C. & Camus, J. (1991). *The illustrated field guide to ferns and allied plants of the British Isles.* London.

1169 Jessop, J.P., ed. (1981). *Flora of central Australia.* Sydney.

1170 Johnson, P.N. (1989). *Wetland plants in New Zealand.* Wellington.

1171 Jones, D.L. & Goudey, C.J. (1981). *Exotic ferns in Australia.* Sydney.

1172 Jordanov, D. *et al.*, eds. (1963-). *Flora reipublicae popularis bulgaricae,* 1-. Sofia.

1173 Karjagin, I.I. *et al.*, eds. (1950-1961). *Flora Azerbajdžana,* **1-8**. Baku.

1174 Kasasian, L. (1964). *Common weeds of Trinidad.* St Augustine.

1175 Kelsey, H.P. & Dayton, W.A. (1942). *Standardized plant names,* 2nd ed. Harrisburg, Pennsylvania.

1176 Kent, D.H., comp. (1967). *Index to botanical monographs.* London.

1177 Kent, D.H. (1992). *List of vascular plants of the British Isles.* Botanical Society of the British Isles. London.

1178 Kent, D.H. & Allen, D.E. (1984). *British and Irish herbaria.* Botanical Society of the British Isles. London.

1179 Kiaer, E. (1955). *Indoor plants in colour.* London.

1180 King, R. (1985). *Tresco. England's island of flowers.* London.

1181 Kirk, J.W.C. (1927). *A British garden flora.* London.

1182 Knowles, M.C. (1906). A contribution towards the alien flora of Ireland. *Irish Naturalist,* **15**:143-150.

1183 Köhlein, F. (1984). *Saxifrages and related genera,* trans. Winstanley, D. London.

1184 Koorders, S.H. (1911-1937). *Exkursionsflora von Java,* **1-4** (incomplete). Jena.

1185 Kubitzki, K., ed. (1990-). *The families and genera of vascular plants,* 1-. Berlin.

1186 Labillardière, J.H.H.de (1791-1812). *Icones plantarum syriae rariorum,* **1-5**. Paris.

1187 Lamarck, J.B.A.P.de Monnet de (1791-1823). *Tableau encyclopédique et méthodique des trois règnes de la nature. Botanique,* **1-3**. Paris.

1188 Lang, D.C. (1980). *Orchids of Britain.* Oxford.

1189 Lange, J.M.C. (1864-1866). *Descriptio iconibus illustrata plantarum novarum,* fasc.1-3. Copenhagen.

1190 Lauber, K. & Wagner, G. (1992). *Flora des kantons Bern,* 2nd ed. Bern.

1191 Lawalrée, A. (1952-1956). *Flore générale de Belgique. Spermatophytes,* **1-5** (incomplete). Brussels.

1191a Lawalrée, A. (1950). *Flore générale de Belgique. Ptéridophytes.* Brussels.

1192   Lawrence, G.H.M. *et al.*, eds. (1968). *B-P-H Botanico-Periodicum-Huntianum*. Pittsburgh.

    1192a   Bridson, G.D.R. & Smith, E.R., eds. (1991). *B-P-H/S Botanico-Periodicum-Huntianum / Supplementum*. Pittsburgh.

1193   Leigh, J.H. & Mulham, W.E. (1965). *Pastoral plants of the riverine plain*. Brisbane.

1194   Levyns, M.R. (1929). *A guide to the flora of the Cape peninsula*. Cape Town.

    1194a   Levyns, M.R. (1966). *A guide to the flora of the Cape peninsula*, 2nd ed. Cape Town.

1195   Li, Hui-Lin *et al.*, eds. (1975-1979). *Flora of Taiwan*, **1-6**. Taipei.

1196   Lindman, C.A.M. (1900). *Vegetationen i Rio Grande do Sul (Sydbrasilien)*. Stockholm.

1197   Lindman, C.A.M. (1926). *Svensk fanerogamflora*, 2nd ed. Stockholm.

1198   Liu Shen-O (Liou Tchen-Ngo), ed. (1931-1936). *Flore illustrée du nord de la Chine*, fasc.1-4. Peking.

1199   Lorenzi, H. (1991). *Plantas daninhas do Brasil*, 2nd ed. Nova Odessa.

1200   Loudon, J.W. (1840). *The ladies flower garden. Ornamental annuals*. London.

1201   Loudon, Mrs [J.W.], ed. (1872). *Loudon's encyclopaedia of plants*, new impression. London.

1202   Lousley, J.E., ed. (1953). *The changing flora of Britain*. B.S.B.I. conference report No.3. Oxford.

1203   Löve, Á. (1983). *Flora of Iceland*. Reykjavík.

1204   Lucas, M.J. & Middleton, J. (1985). *Flowers and ferns around Huddersfield*. Huddersfield.

1205   Lundell, C.L. *et al.* (1942-1961). *Flora of Texas*, **1-3** (incomplete). Renner.

1206   Mabey, R. (1976). *Street flowers*. Harmondsworth.

1207   Macoboy, S. (1971). *What flower is that?* New York.

1208   Maheshwari, J.K. (1963). *The flora of Delhi*. New Delhi.

    1208a   Maheshwari, J.K. (1966). *Illustrations to The flora of Delhi*. New Delhi.

1209   Makins, F.K. (1952). *The identification of trees and shrubs*, 2nd ed. London.

1210   Marie-Victorin, F. (1964). *Flore Laurentienne*, 2nd ed., rev. Rouleau,E. Montreal.

1211   Marshall, E.S. (1914). *A supplement to the Flora of Somerset*. Taunton.

1212   Marticorena, C. & Quezada, M. (1985). Catalogo de la flora vascular de Chile. *Gayana, Bot.*, **42**(1-2):1-157.

1213   Martin, W.Keble (1982). *The new concise British flora*, 3rd ed. London.

1214   Matthei, O.R. (1963). *Manual ilustrado de las malezas de la provincia de Ñuble*. Chillán, Chile.

1215   Matthew, K.M. (1981-1983). *The flora of the Tamilnadu Carnatic*, **1-4**. Tiruchirapalli.

1216   Maund, B. (1825-1851). *The botanic garden*, **1-13**. London.

1217   Maximowicz, C.J. (1859). *Primitiae florae amurensis*. St Petersburg.

1218 McClintock, D. (1987). *Supplement to The wild flowers of Guernsey* (*Collins, 1975*). St Peter Port.

1219 McClintock, D. (1988). The wild and naturalised plants of the island of Brecqhou. *Rep. Trans. Soc. Guernésiaise*, **22**:435-452.

1220 McClintock, D. & Fitter, R.S.R. (1956). *The pocket guide to wild flowers*. London.

1221 McDermott, L.F. (1910). *An illustrated key to the North American species of Trifolium*. San Francisco.

1222 Meadly, G.R.W. (1965). *Weeds of Western Australia*. Perth.

1223 [Meikle, R.D.] (1980). *Draft index of author abbreviations compiled at The Herbarium, Royal Botanic Gardens, Kew*. Basildon.

1224 Merino, B. (1897-1904). *Contribución á la flora de Galicia*, with suppls 2-4. Tuy (suppls Madrid).

1225 Merino, B. (1905-1909). *Flora descriptiva é ilustrada de Galicia*, 1-3. Santiago.

1226 Mohlenbrock, R.H. (1967-). *The illustrated flora of Illinois*, 1-. Carbondale & Edwardsville.

1227 Molinari, E.P. & Petetin, C.A. (1977). *Clave illustrada para el reconocimiento de malezas en el campo al estado vegetativo*. Buenos Aires.

1228 Moore, L.B. & Irwin, J.B. (1978). *The Oxford book of New Zealand plants*. Wellington.

- - - (Morley, B.D. & Everard, B.) See 566.

1229 Mueller, F.J.H. (1889-1891). *Iconography of Australian salsolaceous plants*. Melbourne.

1230 Muenscher, W.C. (1944). *Aquatic plants of the United States*. New York.

1231 Muñoz Pizarro, C. (1959). *Sinopsis de la flora Chilena*. Santiago.

1232 Muñoz Pizarro, C. (1966). *Flores silvestres de Chile*. Santiago.

1233 Nasir, E. & Ali, S.I., eds. (1970-). *Flora of West Pakistan*, 1-131. (Retitled as *Flora of Pakistan*, 132-). Karachi.

1234 Neher, R.T. (1966). *Monograph of the genus Tagetes (Compositae)*. Indiana Univ. Ph.D. Botany. Pp.306. (Available as a xerox from University Microfilms Inc., Ann Arbor, Michigan, USA.)

1235 Nehmeh, M. (1978). *Wild flowers of Lebanon*, trans. Dagher, C.N. & Bitar-Ghanem, G. Beirut.

1236 Newton, A. (1990). *Supplement to Flora of Cheshire*. Privately published. Leamington Spa.

1237 Nicholson, B.E., *et al.* (1963). *The Oxford book of garden flowers*. Oxford.

1238 Nicholson, B.E., *et al.* (1969). *The Oxford book of food plants*. Oxford.

1239 Niehaus, T.F. & Ripper, C.L. (1976). *A field guide to Pacific states wildflowers*. Peterson field guide series No.22. Boston.

1240 Niehaus, T.F., Ripper, C.L. & Savage, V. (1984). *A field guide to southwestern and Texas wildflowers*. Peterson field guide series No.31. Boston.

1241 Nikitin, V.V. (1983). *Sornye rastenija flory SSR* [*Weed plants of the USSR flora*]. Leningrad.

440 REFERENCES

1242 Nordhagen, R. (1944-1979). *Norsk flora. Illustrasjonsbind*, parts 1-4 (incomplete). Oslo.
1243 Organization for Flora Neotropica. (1968-). *Flora neotropica: a series of monographs*, 1-. New York.
1244 Pallas, P.S. (1771-1776). *Reise durch verschiedene Provinzen des russischen Reichs*, 1-3. St Petersburg.
1245 Pallas, P.S. (1803-1806). *Illustrationes plantarum imperfecte vel nondum cognitarum*, parts 1-4. Leipzig.
1246 Pankhurst, R.J. & Mullin, J.M. (1991). *Flora of the Outer Hebrides*. London.
1247 Parker, C. (1992). *Weeds of Bhutan*. Simtokha.
1248 Parker, K.F. (1972). *An illustrated guide to Arizona weeds*. Tucson.
1249 Parsa, A. (1943-1974). *Flore de l'Iran*, 1-10. Teheran.
1250 Patrick, S.J. & Hollick, K.M. (1975). *Supplement to Flora of Derbyshire 1969*. Derby.
1251 Perring, F.H., ed. (1968). *Critical supplement to the Atlas of the British flora*. London.
1252 Perring, F.[H.], ed. (1974). *The flora of a changing Britain*. B.S.B.I. conference report No.11. London.
1253 Perring, F.H. & Farrell, L. (1977). *Vascular plants*. British red data books No.1. Nettleham.
1254 Perring, F.H. & Walters, S.M. (1962). *Atlas of the British flora*. London.
1255 Perry, F. (1972). *Flowers of the world*. London.
1256 Peterson, R.T. & McKenny, M. (1968). *A field guide to wildflowers of northeastern and north-central North America*. Peterson field guide series No.17. Boston.
1257 Phillips, R. & Rix, M. (1991). *Perennials, 1(Early perennials)-2(Late perennials)*. London.
1258 Pizzetti, I. & Cocker, H. (1975). *Flowers. A guide for your garden*, 1-2. New York.
1259 Plitmann, U. *et al.* (1983). *Pictorial flora of Israel*. Givatayim.
1260 Pope, W.T. (1968). *Manual of wayside plants of Hawaii*. Vermont & Tokyo.
1261 Porterfield, W.M. (1933). *Wayside plants and weeds of Shanghai*. Shanghai.
1262 Post, G.E. (1932). *Flora of Syria, Palestine and Sinai*, 1-2, 2nd ed., rev. Dinsmore, J.E. Beirut.
1263 Prescott, A. (1988). *It's blue with five petals. Wildflowers of the Adelaide region*. Prospect.
1264 Probst, R. (1949). *Wolladventivflora Mitteleuropas*. Solothurn.
1265 Prodan, I. & Buia, A. (1966). *Flora mica ilustrata a României*, 5th ed. Bucharest.
1266 Pugsley, H.W. (1948). A prodromus of the British Hieracia. *J. Linn. Soc., Bot.*, 54:1-356.
1267 Raciborski, M. *et al.*, eds. (1919-1980). *Flora Polska*, 1-14. Cracow.
1268 Radford, A.E., Ahles, H.E. & Bell, C.R. (1968). *Manual of the vascular flora of the Carolinas*. Chapel Hill.

1269 Rauh, W. (Temple, P., ed.) (1979). *Bromeliads*, trans. Temple, P. & Kendall, H.L. Poole, Dorset.

1270 Rayner, J.F. (1925). A list of the alien plants of Hampshire and the Isle of Wight. *Proc. Isle of Wight Nat. Hist. Soc.*, **1**:229-274.

1271 Rechinger, K.H., ed. (1963-). *Flora Iranica*, fasc.1-. Graz, Austria.

1272 Reed, C.F. (1977). *Economically important foreign weeds*. Agriculture handbook No.498. United States Department of Agriculture. Washington DC.

1273 Reichenbach, H.G.L. (1824-1830). *Iconographia botanica exotica sive hortus botanicus*, **1-3**. Leipzig.

1274 Rice, E.G. & Compton, R.H. (1951). *Wild flowers of the Cape of Good Hope*. Kirstenbosch.

1275 Rich, T.C.G. *et al.* (1988). *Plant Crib*. Botanical Society of the British Isles. London.

1276 Ridley, H.N. (1922-1925). *Flora of the Malay peninsula*, **1-5**. London.

1277 Robyns, W. *et al.*, eds. (1948-1970). *Flore du Congo Belge et du Ruanda-Urundi*, **1-10**. (From 1960, retitled as *Flore du Congo, du Rwanda et du Burundi*.) Brussels.

1278 Rodway, L. (1907). *Trees and shrubs of Tasmanian forests*. Hobart.

1279 Roles, S.J. (1957-1965). *Illustrations to Flora of the British Isles*, **1-4**. Cambridge.

1280 Rougemont, G.M.de (1989). *A field guide to the crops of Britain and Europe*. London.

1281 Rouy, G. & Foucaud, J. (1893-1913). *Flore de France*, **1-14**. Paris.

1282 Rubtsov, N.I., ed. (1972). *Opredelitel' vysšikh rastenij Kryma* [*Key to the higher plants of Crimea*]. Leningrad.

1283 Rycroft, H.B. (1963). *Our flower paradise*. Kirstenbosch.

1284 Rydberg, P.A. (1971). *Flora of the prairies and plains of central North America*, rev. ed. New York.

1285 Salisbury, E.J. (1964). *Weeds and aliens*, 2nd ed. New Naturalist No.43. London.

1286 Salmon, J.T. (1986). *The native trees of New Zealand*, rev. ed. Auckland.

1287 Salm-Reifferscheid-Dyck, J. (1836-1863). *Monographia generum Aloes et Mesembrianthemi*. Bonn.

1288 Sampaio, G. (1949). *Iconografia selecta de flora Portuguesa*. Lisbon.

1289 Sargent, C.S. (1933). *Manual of the trees of North America (exclusive of Mexico)*. Boston.

1290 Saule, M. (1991). *La grande flore illustrée des Pyrénées*. Milan.

1291 Saunders, W.W., ed. (1868-1873). *Refugium Botanicum*, **1-5**. London.

1292 Scannell, M.J.P. & Synnott, D.M. (1987). *Census catalogue of the flora of Ireland*, 2nd ed. Dublin.

1293 Schkuhr, C. (1806-1814). *Botanisches Handbuch*, **1-4**, 2nd ed. Leipzig.

1294 Schmitz, M. (1982). *Wild flowers of Lesotho*. Roma, Lesotho.

1295 Schneider, C.K. (1906-1912). *Illustriertes Handbuch der Laubholzkunde*, **1-2** (and *Register*). Jena.

1296 Schnizlein, A. (1843-1870). *Iconographia familiarum naturalium regni vegetabilis*, **1-4**. Bonn.

1297   Semple, J.C. & Heard, S.B. (1987). *The Asters of Ontario: Aster L. and Virgulus Raf.* Waterloo.

1298   Seymour, F.C. (1969). *The flora of New England.* Rutland, Vermont.

1299   Sfikas, G. (1987). *Wild flowers of Crete.* Athens.

1300   Sharma, B.M. & Jamwal, P.S. (1988-). *Flora of upper Liddar valleys of Kashmir Himalaya,* 1-. Jodhpur.

1301   Shaw, M.[R.], ed. (1988). A flora of the Sheffield area. *Sorby Rec. Special Ser.,* **8.**

1302   Sherff, E.E. (1936). Revision of the genus Coreopsis. *Field Mus. Nat. Hist., Bot. Ser.,* **11**(6). Publication No.366. Chicago.

1303   Shreve, F. & Wiggins, I.L. (1964). *Vegetation and flora of the Sonoran Desert,* 1-2. Stanford, California.

1304   Silverside, A.J. & Jackson, E.H., eds. (1988). *A check-list of the flowering plants and ferns of East Lothian.* Edinburgh.

1305   Širjaev, G. (1928-1933). Generis Trigonella L. revisio critica, 1-6. *Spisy Přír. Fak. Masarykovy Univ.,* **1928**:1-57; **1929**:1-37; **1930**:1-31; **1931**:1-33; **1932**:1-48; **1933**:1-37.

1306   Smith, J.E. (1804-1805). *Exotic botany,* 1-2. London.

1307   Smith, R.A.H. *et al.* (1992). *Checklist of the plants of Perthshire.* Perthshire Society of Natural Science. Perth.

1308   Sowerby, J. [& Smith, J.E.] (1790-1814). *English botany,* 1-36. London.

1309   Sowerby, J. [& Smith, J.E.] (1832-1846). *English botany,* 1-12, 2nd ed. London.

1310   Sowerby, J. [& Smith, J.E.] (1863-1872). *English botany,* 1-11, 3rd ed., rev. Syme, J.T.B. London.

1311   Stace, C.A., ed. (1975). *Hybridization and the flora of the British Isles.* London.

1312   Stapf, O. (1929-1931). *Index Londinensis,* 1-6. Oxford. See also **1365.**

1313   Stechmann, J.P. (1775). *Dissertatio inauguralis botanico-medica de Artemisiis.* Göttingen.

1314   Stevens, G.T. (1930). *Illustrations of flowering plants of the middle Atlantic and New England states.* New York.

1315   Stewart, O. (1990). *Flowering plants and ferns of Kirkcudbrightshire.* Dumfries. (Reprinted from *Trans. Dumfriesshire Galloway Nat. Hist. Antiq. Soc.,* 3rd ser., **65.**)

1316   St John, H. & Hosaka, E.Y. (1932). *Weeds of the pineapple fields of the Hawaiian islands.* University of Hawaii research publication No.6. Honolulu.

1317   Stojanov, N. & Stefanov, B. (1925). *Flore de la Bulgarie.* Sofia.

1318   Straka, H. *et al.* (1987). *Führer zur Flora von Mallorca. Guide to the flora of Majorca.* Stuttgart.

1319   Strausbaugh, P.D. & Core, E.L. (1952-1964). Flora of West Virginia, parts 1-4. *West Virginia Univ. Bull.,* **ser. 52,** No.12-2; **53** No.12-1; **58** No.12-3; **65** No.3-2. Morgantown.

1320   Sturm, J. (1797-1862). *Deutschlands Flora. Phanerogamen,* parts 1-96. Nürnberg.

1321   Sudre, H. (1902). *Hieracium du centre de la France.* Albi.

1322 Sudre, H. (1908-1913). *Rubi Europae*. Paris.
1323 Sutton, D.A. (1988). *A revision of the tribe Antirrhineae*. London.
1324 Suzuki, M. (1982). *Alpine plants in eastern Japan*. Tokyo.
1325 Sweet, R. (1825-1830). *Cistineae, the natural order of Cistus or Rock Rose*. London.
1326 Swindells, P. (1983). *Waterlilies*. London.
1327 Synge, P.M. & Platt, J.W.O. (1964). *Some good garden plants*, new ed. London.
1328 Synnott, D. (1984). An outline of the flora of Mayo. *Glasra*, **9**:13-117.
1329 Szafer, W., Kulczynski, S. & Pawlowski, B. (1953). *Rósliny Polskie*. Warsaw.
1330 Tarpey, T. & Heath, J. (1990). *Wild flowers of north east Essex*. Colchester.
1331 Taylor, W.K. (1992). *The guide to Florida wildflowers*. Dallas.
- - - (Tchang Bok Lee) See **638**
1332 Thellung, A. (1912). *La flore adventice de Montpellier*. Cherbourg.
1333 Townsend, F. (1904). *Flora of Hampshire including the Isle of Wight*, 2nd ed. London.
1334 Trehane, P. (1989-). *Index hortensis*, 1-. Wimborne, Dorset.
1335 Troupin, G. (1971). *Syllabus de la flore du Rwanda. Spermatophytes*. Tervuren, Belgium.
1336 Troupin, G. *et al.* (1978-). *Flore du Rwanda. Spermatophytes*, 1-. Tervuren, Belgium.
1337 Tsunoda, S., Hinata, K. & Gómes-Campo, C., eds. (1980). *Brassica crops and wild allies*. Tokyo.
1338 Turland, N.J., Chilton, L. & Press, J.R. (1993). *Flora of the Cretan area*. London.
1339 Turner, W. (1548). *Names of herbes*. London.
1340 Urban, A. (1990). *Wildflowers and plants of central Australia*. Port Melbourne.
1341 Urquhart, B.L., ed. (1958-1962). *The Rhododendron*, 1-2. Sharpthorne, Sussex.
1342 Vaga, A. *et al.* (1966). *Eesti taimede määraja* [*Key to Estonian plants*]. Tallinn.
1343 van der Spuy, U. (1971). *Wild flowers of South Africa for the garden*. Johannesburg.
1344 Vedel, H. (1978). *Trees and shrubs of the Mediterranean*. Harmondsworth.
1345 Vines, R.A. (1960). *Trees, shrubs, and woody vines of the southwest*. Austin, Texas.
1346 Wade, G. (1794). *Catalogus systematicus plantarum indigenarum in comitatu dublinensi*. Dublin.
1347 Wagner, W.L., Herbst, D.R. & Sohmer, S.H. (1990). *Manual of the flowering plants of Hawai'i*, 1-2. Honolulu.
1348 Walcott, M.V. (1925-1929). *North American wild flowers*, 1-5. Washington.

1349   Watson, H.C. (1868-1870). *A compendium of the Cybele Britannica*, parts 1-3. London.

1350   Watson, P.W. (1825). *Dendrologia Britannica*, 1-2. London.

1351   Webster, M.McCallum. [Card index of alien plant herbarium specimens, coll. MMcCW, 1954-1966, in possession of E.J.Clement.]

1352   Weevers, T. *et al.*, eds. (1948-). *Flora Neerlandica: Flora van Nederland*, 1-. Amsterdam.

1353   Welch, D. (1993). *Flora of North Aberdeenshire*. Privately published. Banchory.

1354   Wendelbo, P. (1977). *Tulips and irises of Iran*. Tehran.

1355   Wharton, M.E. & Barbour, R.W. (1971). *A guide to the wildflowers and ferns of Kentucky*. Lexington.

1356   Wharton, M.E. & Barbour, R.W. (1973). *Trees and shrubs of Kentucky*. Lexington.

1357   Wiggins, I.L. (1980). *Flora of Baja California*. Stanford.

1358   Wiggins, I.L. & Porter, D.M. (1971). *Flora of the Galápagos Islands*. Stanford, California.

1359   Wigginton, M.J. & Graham, G.G. (1981). *Guide to the identification of some of the more difficult vascular plant species*. England field unit occasional paper No.1. Nature Conservancy Council. Banbury.

1360   Wight, R. (1840-1856). *Icones plantarum indiae orientalis*, 1-6. Madras.

1361   Willkomm, H.M. (1852-1856[1862]). *Icones et descriptiones plantarum novarum*, 1-2. Leipzig.

1362   Wilson, H.D. (1978). *Wild plants of Mount Cook National Park*. Christchurch.

1363   Wilson, H.D. (1982). *Stewart Island plants*. Christchurch.

1364   Wit, H.C.D.de (1963-1965). *Plants of the world*, 1-3, trans. Pomerans, A.J. London.

1365   Worsdell, W.C. (1941). *Index Londinensis. Supplement for the years 1921-35*, 1-2. Oxford.

1366   Wyse Jackson, P.[S.] & Skeffington, M.S. (1984). *Flora of inner Dublin*. Dublin.

1367   Yuncker, T.G. (1921). North American and West Indian species of Cuscuta. *Illinois Biol. Monogr.*, 6(2 & 3).

1368   Zohary, M. (1982). *Plants of the Bible*. Cambridge.

1369   Swan, G.A. (1993). *Flora of Northumberland*. Newcastle upon Tyne.

1370   D'arcy, W.G., ed. (1986). *Solanaceae: biology and systematics*. London.

1371   Gmelin, S.G. ([1770]1774-1784). *Reise durch Russland*, 1-4. St Petersburg.

1372   Mabberly, D.J. (1987). *The plant-book*. Cambridge.

1373   Greuter, W. *et al.*, eds. (1988). International code of botanical nomenclature. Adopted by the Fourteenth International Botanical Congress, Berlin, July-Aug. 1987. *Regnum Veg.*, 118.

INDEX

Compiled by R. Gwynn Ellis
Department of Botany, National Museum of Wales, Cardiff

Bonaveria securidaca (L.) Desv., 167
Bonjeanea hirsuta (L.) Reichb., 165
 recta (L.) Reichb., 165
Borage, 249
 Slender, 249
BORAGINACEAE, 244
BORAGO L., 249
 laxiflora (DC.) Fischer, non
  Poiret, 249
 officinalis L., 249
 orientalis L., 250
 pygmaea (DC.) Chater & Greuter,
  249
BOREAVA Jaub. & Spach, 92
 orientalis Jaub. & Spach, 92
BORRERIA G.Meyer, 289
 verticillata (L.) G.Meyer, 289
Boston-ivy, 202
BOTRYCHIUM Sw., 1
 lanceolatum (S.Gmelin)
  Angström, 1
 lunaria (L.) Sw., 1
 matricariae (Schrank) Sprengel, 1
 matricariifolium A.Braun ex
  Koch, 1
 multifidum (S.Gmelin) Rupr., 1
 rutaceum Sw., 1
Boussingaultia baselloides auct., non
  Kunth, 55
 cordifolia Ten., non (Moq.)
  Volkens, 55
 gracilis Miers var.
  pseudobaselloides
  (Hauman) L.Bailey, 55
Bow-flower, Common, 321
 Tiny, 321
BOWLESIA Ruíz Lopez & Pavón,
  218
 incana Ruíz Lopez & Pavón, 218
 tenera Sprengel, 218
Box, Balearic, 198
Boxthorn, Chinese, 230
Boxwood, African, 116
BOYKINIA Nutt., 129
 major A.Gray, 129
Brachycome Cass. (1825), 326
BRACHYGLOTTIS Forster &
  Forster f., 348
 compacta (Kirk) R.Nordenstam ×
  B. *laxifolia* (J.Buch.)
  R.Nordenstam, 348

Dunedin Hybrid group, 348
 monroi (Hook.f.) R.Nordenstam,
  348
 repanda Forster & Forster f., 348
 'Sunshine', 348
BRACHYSCOME Cass., 326
 ciliaris (Labill.) Less., 326
 collina (Sonder) Benth., 326
 graminea (Labill.) F.Muell., 326
 iberidifolia Benth., 326
 perpusilla (Steetz) J.Black, 326
Brachyscome Cass. (1816), 326
Brake, Ladder, 2
 Spider, 2
 Tender, 2
Bramble, Chinese, 135
 Korean, 133
 Slender-spined, 134
 White-stemmed, 133
BRASSICA L., 107
 adpressa Boiss., 109
 alba (L.) Rabenh., 108
 arvensis L., 106
 balearica Pers., 107
 barrelieri (L.) Janka, 107
 brevipes Syme, nom. illeg., pro
  parte, 107
 carinata A.Braun, 107
 cheiranthos Villars, 109
 chinensis L., 108
 dissecta (Lagasca) Boiss., 108
 elongata Ehrh., 107
  subsp. *integrifolia* (Boiss.)
   Breistr., 107
 eruca L., 108
 erucastrum Villars, vix L., 109
 fruticulosa Cirillo, 107
 gallica (Willd.) Druce, 109
 geniculata (Desf.) Ball, 109
 griquana N.E.Br., 109
 × *harmsiana* O.Schulz, 108
 hirta Moench, 108
 hispida (Schousboe) Boiss., 108
 incana (L.) Meigen, non Ten., 109
 integrifolia (West) O.Schulz, 107
  var. *carinata* (A.Braun)
   O.Schulz, 107
 juncea (L.) Czernj., 107
 kaber (DC.) Wheeler var.
  schkuhriana (Reichb.)
  Wheeler, 108

Ragwort, Broad-leaved, 345
    Chamois, 345
    Chinese, 348
    Golden, 345
    Hedge, 348
    Magellan, 346
    Monro's, 348
    Narrow-leaved, 345
    Oxford, 347
    Purple, 345
    Shoddy, 346
    Shrub, 348
    Silver, 344
    Wood, 346
Raimannia humifusa (Nutt.) Rose, 192
Rain-daisy, Cape, 351
RAMONDA Rich., 284
    myconi (L.) Reichb., 284
Rampion, Black, 287
    Oxford, 287
RANUNCULACEAE, 16
RANUNCULUS L., 21
    aconitifolius L., 21
    aconitifolius 'Flore Pleno', 21
    alpestris L., 21
    cordiger Viv., 22
    falcatus L., 22
    gramineus L., 22
    marginatus Urv., 22
    monspeliacus L., 22
    muricatus L., 22
    penicillatus (Dumort.) Bab. subsp.
        pseudofluitans (Syme)
        S.Webster, 22
    pensylvanicus L.f., 22
    rupestris Guss., 22
    sardous Crantz, 22
        subsp. *trilobus* (Desf.) Rouy,
        22
    sessiliflorus R.Br. ex DC., 22
        var. *pilulifer* (Hook.) Melville,
        22
    sphaerospermus Boiss. & Blanche,
        22
    spicatus Desf., 22
    trachycarpus Fischer & C.Meyer,
        22
    trilobus Desf., 22
Rape, 107
    Ethiopian, 107
    Long-stalked, 107
    Oil-seed, 107

RAPHANUS L., 111
    caudatus L., 111
    landra Moretti ex DC., 111
    × *micranthus* (Uechtr.) O.Schulz,
        111
    raphanistrum L. subsp. **landra**
        (Moretti ex DC.) Bonnier
        & Layens, 111
        subsp. *raphanistrum*, 111
        subsp. **rostratus** (DC.) Thell.,
        111
        × *R. sativus*, 111
    rostratus DC., 111
    sativus L., 111
RAPISTRUM Crantz, 110
    glabrum Host, 110
    hispanicum (L.) Crantz, 110
    linnaeanum Boiss. & Reuter, nom.
        illeg., 110
    orientale (L.) Crantz, 110
    perenne (L.) All., 110
    rugosum (L.) Bergeret, *108*,110
        subsp. *linnaeanum* (Cosson)
        Rouy & Fouc., 110
    tenuifolium (Sibth. & Smith)
        Benth. & Hook.f., 110
Raspberry, Purple-flowered, 134
    Rocky Mountain, 133
Raspwort, Creeping, 187
RATIBIDA Raf., 354
    columnifera (Nutt.) Wooton &
        Standley, 354
Rattle-snake-weed, 227
Rauli, 34
Red-bartsia, Southern, 282
Red-cedar, Japanese, 12
    Western, 13
Red-hot-pokers, 378
Red-knotgrass, 67
    Lesser, 67
Red-maids, 55
Red-pine, Japanese, 10
Red-ribbons, 194
Red-Rose-of-Lancaster, 141
Redscale, 47
Redwood, Coastal, 11
    Dawn, 12
Redwood-ivy, 24
REICHARDIA Roth, 309
    tingitana (L.) Roth, 309
REINECKEA Kunth, 381
    carnea (Andrews) Kunth, 381

Moss, 141
Noisette, 142
Rambler, 142
ROSMARINUS L., 263
 officinalis L., 263
Roubieva multifida (L.) Moq., 43
Rowan, Hupeh, 147
 Sargent's, 148
 Vilmorin's, 148
RUBIA L., 291
 tinctorum L., 291
RUBIACEAE, 288
RUBUS L., 133
 allegheniensis Porter, 133
 argenteus Weihe & Nees subsp.
   elegantispinosus
   A.Schum., 134
 armeniacus Focke, 133
 canadensis L., 133
 candicans Weihe ex Reichb.
   group, 134
 cockburnianus Hemsley, 133
 commersonii Poiret var.
   illecebrosus (Focke)
   Makino, 134
 aff. **coreanus** Miq., 133
 deliciosus Torrey, 133
 discolor auct., non Weihe & Nees,
   133
 discolor Weihe & Nees, 133
 elegans Utsch, non P.J.Mueller,
   134
 elegantispinosus (A.Schum.)
   H.E.Weber, 134
 × **fraseri** Rehder, 134
 fruticosus L. 'Flore Pleno', 134
   var. *laciniatus* Weston, 134
 giraldianus Focke, 133
 hedycarpus Focke forma *linkianus*
   Zabel, 134
 'Himalaya Giant', 133
 idaeus L. group, 134
   subsp. *strigosus* (Michaux)
    Focke × *R. ursinus* Cham.
    & Schldl. subsp. *vitifolius*
    Cham. & Schldl., 134
   × *R. phoenicolasius* , 135
 illecebrosus Focke, 134
 laciniatus Willd., 134
   × *R. ulmifolius*, 134
 linkianus Ser., 134

loganobaccus L.Bailey, 134
nutkanus Moçiño ex Ser., 134
odoratus L., 134
 × *R. parviflorus*, 134
oplothyrsus Sudre, 133
parviflorus Nutt., 134
× *paxii* Focke, 135
pergratus Blanchard, 135
phoenicolasius Maxim., 135
polytrichus Franchet, non Progel,
  135
procerus Muller, 133
robustus G.Fraser, non Presl, 134
spectabilis Pursh, 135
thibetanus Franchet, 133
thyrsigeriformis (Sudre) D.Allen,
  135
tokkura Siebold, 133
tricolor Focke, 135
ulmifolius Schott, 133
villosus Aiton, non Thunb., 133
RUDBECKIA L., 354
 amplexicaulis Vahl, 354
 bicolor Nutt., 354
 columnaris Sims, 354
 fulgida Aiton, 354
 hirta L., 354
 laciniata L., 354
 serotina Nutt., non Sweet, 354
 speciosa Wender., 354
 triloba L., 354
Rue, 206
RUMEX L., 69
 acetosa hort., non L., 69
 acetosa L. subsp. **ambiguus**
   (Gren.) Á.Löve, 69
 alpinus L. (1759), non L. (1753),
   71
 altissimus Alph.Wood, 69
 altissimus auct., non Alph.Wood,
   69
 ambiguus Gren., 69
 arifolius All., 69
 bequaertii De Wild., 69
 × *bontei* Danser, 70
 ×*borbasii* Blocki, 69
 brownianus Schultes, 69
 brownii Campderá, 69
 bucephalophorus L., 69
 callosissimus Meissner, 70
 camptodon Rech.f., 69

BSBI PUBLICATIONS

HANDBOOKS

Each handbook deals in depth with one or more difficult groups of British and Irish plants.

No. 1 *SEDGES OF THE BRITISH ISLES*
A.C. Jermy, A.O. Chater and R.W. David. 1982. 268 pages, with a line drawing and distribution map for every species. Paperback. ISBN 0 901158 05 4.

No. 2 *UMBELLIFERS OF THE BRITISH ISLES*
T.G. Tutin. 1980. 197 pages, with a line drawing for each species. Paperback. ISBN 0 901158 02 X. Out of print: new edition in preparation.

No. 3 *DOCKS AND KNOTWEEDS OF THE BRITISH ISLES*
J.E. Lousley and D.H. Kent. 1981. 205 pages, with many line drawings of native and alien taxa. Paperback. ISBN 0 901158 04 6. Out of print: new edition in preparation.

No. 4 *WILLOWS AND POPLARS OF GREAT BRITAIN AND IRELAND*
R.D. Meikle. 1984. 198 pages, with 63 line drawings of all species, subspecies, varieties and hybrids. Paperback. ISBN 0 901158 07 0.

No. 5 *CHAROPHYTES OF GREAT BRITAIN AND IRELAND*
J.A. Moore. 1986. 144 pages with line drawings of 39 species and 17 distribution maps. Paperback. ISBN 0 901158 16 X.

No. 6 *CRUCIFERS OF GREAT BRITAIN AND IRELAND*
T.C.G. Rich. 1991. 336 pages with descriptions of 140 taxa, most illustrated with line drawings and 60 with distribution maps. Paperback. ISBN 0 901158 20 8.

No. 7 *ROSES OF GREAT BRITAIN AND IRELAND*
G.G. Graham and A.L. Primavesi. 1993. 208 pages with descriptions and illustrations of 12 native and eight introduced species, and descriptions of 83 hybrids. Distribution maps are included of 31 selected species and hybrids. Paperback. ISBN 0 901158 22 4.

No. 8 *PONDWEEDS OF GREAT BRITAIN AND IRELAND*
C.D. Preston. Due 1995. c.350 pages. 50 full page illustrations and distribution maps. Paperback. ISBN 0 901158 24 0.

OTHER PUBLICATIONS

LIST OF VASCULAR PLANTS OF THE BRITISH ISLES
D.H. Kent, 1992. 400 pages. Nomenclature and sequence followed by Stace in *New Flora of the British Isles*. Paperback. ISBN 0 901158 21 6.

ATLAS OF THE BRITISH FLORA
F.H. Perring and S.M. Walters, 1990. Reprint of 3rd Edn, 1982. 468 pages. Distribution maps of over 1700 species including updated maps for 321 Red Data Book species. New bibliography of updated distribution maps published elsewhere since 1st Edn, 1962. Paperback. ISBN 0 901158 19 4.

Available from the official agents for BSBI publications:

F.& M. Perring, Green Acre, Wood Lane, Oundle, Peterborough, England PE8 4JQ
Tel.: 01832 273388   Fax: 01832 274568